physica status solidi (b)
basic research

Volumes 157 to 162

Subject and Author Index 1990

Volumes 117 to 122

physica status solidi (a)
applied research

AKADEMIE VERLAG

ISSN 0370-1972/0031-8965 phys. stat. sol. (b)/(a), Berlin **R23** (1990) 1 – 564

Substance Classification*)

S1. Transition metals and their alloys and compounds (for oxides, see S10)
- S1.1. Fe
- S1.2. Other 3d metals
- S1.3. Noble metals
- S1.4. Lanthanides
- S1.5. Actinides
- S1.61. Carbides, silicides, germanides
- S1.62. Nitrides, phosphides, arsenides
- S1.63. Sulfides, selenides, tellurides

S2. Alkali metals and their alloys and compounds (for alkali halides, see S9.11)

S3. Alkaline earth metals and their alloys and compounds (for alkaline earth halides, see S9.12)

S4. Other metals and their alloys

S5. Nonmetallic elements
- S5.11. Silicon
- S5.12. Germanium
- S5.2 Solidified gases

S6. Group IV compounds (for transition metal compounds, see S1.61)

S7. Group V compounds (for transition metal compounds, see S1.62)
- S7.1. III–V compounds
- S7.11. Phosphides
- S7.12. Arsenides
- S7.13. Antimonides
- S7.15. Mixed crystals
- S7.16. Higher compounds

S8. Chalcogenides (for transition metal compounds, see S1.63)
- S8.1. II–VI compounds
- S8.11. Sulfides
- S8.12. Selenides
- S8.13. Tellurides
- S8.15. Mixed crystals
- S8.16. Higher compounds

S9. Halides
- S9.11. Alkali halides
- S9.12. Alkaline earth halides
- S9.15. Mixed crystals
- S9.16. Higher compounds

S10. Oxides
- S10.1. Simple oxides
- S10.15. Mixed crystals

S11. Other higher compounds
- S11.1. Salts of oxy-acids
- S11.2. Spinels, garnets, ferrites

S12. Organic compounds
- S12.1. Quasi-one-dimensional systems

For comments on the use of the classification scheme see p. 13.

*) Outside the ICSTI Classification for Physics.

physica status solidi (b)

basic research

physica status solidi (a)

applied research

SUBJECT AND AUTHOR INDEX

Volumes 157 to 162, 1990 — Volumes 117 to 122, 1990

AKADEMIE VERLAG

How to order

Please place your order with

Kunst und Wissen, Erich Bieber GmbH, a Faxon Company, Wilhelmstr. 4, Postfach 10 28 44, W-7000 Stuttgart 10, Federal Republic of Germany; telephone: 21 07 70, telex: 721 929, telefax: 24 74 39

or with Akademie Verlag GmbH, Leipziger Str. 3–4; Postfach 12 33, O-1086 Berlin, Federal Republic of Germany; telephone: 2 23 63 50 or 2 23 63 51, telex: 114 420

or with your book-shop.

Schriftleiter und verantwortlich für den Inhalt der Zeitschriften:

Professor Dr. E. Gutsche, Leipziger Str. 3–4, Postfach 1233, O-1086 Berlin.

Verantwortlich für den Inhalt der Register:

Dipl.-Phys. K. Müller, Leipziger Str. 3–4, Postfach 1233, O-1086 Berlin.

Verlag:

Akademie Verlag GmbH, Leipziger Str. 3–4, O-1086 Berlin; Fernruf: 2 23 63 50 und 2 23 63 51; Telex-Nr.: 114 420; Telefax: 2 23 63 57; Bank: Dresdner Bank Weinheim, Kto.-Nr.: 754 624 500, BLZ 670 800 50.

Geschäftsführer:

Dr. Gerd Giesler, Dr. Bernhard Tesche.

Gesamtherstellung:

Druckhaus „Thomas Müntzer" GmbH, O-5820 Bad Langensalza.

Bestellnummer des Registers: 1068/85 R 23.

AN (EDV) 20 831/20 735

Preface

The twenty third index volume contains volumes 157 to 162 of "physica status solidi (b) – basic research" as well as volumes 117 to 122 of "physica status solidi (a) – applied research", i.e. all volumes having appeared in 1990 containing 1332 papers.

Beginning from January 1986 the "International Classification System for Physics" (Condensed Matter, items 60 and 70) of the International Council for Scientific and Technical Information, Abstracting Board completed by item 85 and a Substance Classification (both outside the ICSTI Classification for Physics) will be used for quoting papers in physica status solidi. Nevertheless, before taking over these subject classifications into the index volume they were once more carefully revised and in some cases improved and completed in order to get a most uniform and clear classification for all papers. The average number of quotations per paper in this index volume is about 2.7.

I have to express my thanks to the readers for their interest, and I will always readily take up any critical remarks and proposals for improvements.

Berlin, August 1991

Karin Müller

Contents

International Classification System for Physics 6
Substance Classification (according to physica status solidi) 11
Some Comments on the Use of the Subject and Author Index 13
60. Condensed Matter: Structure, Mechanical, and Thermal Properties 17
61. Structure of Liquids and Solids; Crystallography 17
62. Mechanical and Acoustical Properties of Condensed Matter 63
63. Lattice Dynamics and Crystal Statistics 73
64. Equations of State, Phase Equilibria, and Phase Transitions 80
65. Thermal Properties of Condensed Matter 92
66. Transport Properties of Condensed Matter (Nonelectronic) 97
67. Quantum Fluids and Solids; Liquid and Solid Helium 103
68. Surfaces and Interfaces; Thin Films and Whiskers 104
70. Condensed Matter; Electronic Structure; Electrical, Magnetic, and Optical Properties 121
71. Electronic States 121
72. Electronic Transport in Condensed Matter 165
73. Electronic Structure and Electrical Properties of Surfaces, Interfaces, and Thin Films 186
74. Superconductivity 208
75. Magnetic Properties and Materials 222
76. Magnetic Resonances and Relaxation in Condensed Matter; Mössbauer Effect 246
77. Dielectric Properties and Materials 259
78. Optical Properties and Condensed-Matter Spectroscopy and Other Interactions of Matter with Particles and Radiation 266
79. Electron and Ion Emission by Liquids and Solids; Impact Phenomena 310
85. Devices 314
S. Substances 319
Author Index 475

International Classification System for Physics*)

60. Condensed matter: structure, mechanical and thermal properties
61. Structure of liquids and solids; crystallography
    (see also 68.20. Solid surface structures, 71. Electron states)
    61.10. X-ray determination of structures
        (for specific determinations, see 61.55 to 61.80)
    61.12. Neutron determination of structures
        (for specific determinations, see 61.55 to 61.80)
    61.14. Electron determination of structures
        (for specific determinations, see 61.55 to 61.80)
    61.16. Other determination of structures
        (for specific determination, see 61.55 to 61.80)
    61.20. Classical, semiclassical, and quantum theories of liquid structure
        (for electronic states, see 71; for liquid helium, see 67)
    61.25. Studies of specific liquid structures
    61.30. Liquid crystals
    61.40. Amorphous and polymeric materials
    61.50. Crystalline state (including molecular motions in solids)
        (for magnetic structure and spin systems, see 75.25)
    61.55. Specific structures of elements and alloys
    61.60. Specific structures of inorganic compounds
    61.65. Specific structures of organic compounds
    61.70. Defects in crystals
        (see also 61.80. Radiation damage, 62. Mechanical and acoustical properties,
        71.55. Impurities and defect levels, 76.30. EPR of color centers and other defects,
        78.50. Impurity and defect absorption in solids)
    61.80. Radiation damage and other irradition effects
        (for techniques of structure determination, see 61.10 to 61.16; for electron and ion impact phenomena, see 79.20)
    61.90. Other topics in structure of liquids and solids

62. Mechanical and acoustical properties of condensed matter
    (see also 61.70. Defects in crystals, 68.30. Surfaces and interfaces)
    62.10. Mechanical properties of liquids
        (for viscosity of liquids, see also 66.20)
    62.20. Mechanical properties of solids (related to microscopic structure)
    62.30. Mechanical and elastic waves
    62.40. Anelasticity, internal friction, and mechanical resonances
        For thermomechanical effects, see 65.70
        For magnetomechanical effects, see 75.80
        For piezoelectric effects, see 77.60
        For elasto-optical effects, see 78.20
    62.50. High-pressure and shock-wave effects in solids
    62.60. Acoustical properties of liquids
        For lattice dynamics, phonons, see 63
        For second sound in quantum fluids, see 67.40
    62.65. Acoustical properties of solids
        For magnetoacoustic effects in metals, see 75.80
        For acoustoelectric effects, see 72.50
        For acousto-optical effects, see 78.20
    62.80. Ultrasonic relaxation
        (see also 74.30. Ultrasonic attenuation in superconductors)
    62.90. Other topics in mechanical and acoustical properties of condensed matter

*) Excerpt, reproduced with permission of International Council for Scientific and Technical Information (ICSTI).

63. Lattice dynamics and crystal statistics
(see also 65. Thermal properties, 66.70. Thermal conduction, 68.30. Dynamics of surface and interface vibrations, 78.30. Infrared and Raman spectra)
63.10. General theory
63.20. Phonons and vibrations in crystal lattices
63.50. Vibrational states in disordered systems
63.70. Statistical mechanics of lattice vibrations
(see also 65. Thermal properties of condensed matter, and 66.70. Thermal conduction)
63.75. Statistical mechanics of displacive phase transitions
For order–disorder and statistical mechanics of model systems, see 64.60
For crystallographic aspects of polymorphic and order–disorder transformations, see 61.50
63.90. Other topics in lattice dynamics and crystal statistics

64. Equations of state, phase equilibria, and phase transitions
64.10. General theory of equations of state and phase equilibria
64.30. Equations of state of specific substances
(see also 65.70. Thermal expansion)
64.60. General studies of phase transitions
(for magnetic, superconducting, and quantum fluid critical phenomena, see 75, 74, and 67, respectively)
64.70. Phase equilibria, phase transitions, and critical points of specific substances
64.75. Solubility, segregation, and mixing
64.80. Other phase properties of systems
64.90. Other topics in equations of state, phase equilibria, and phase transitions

65. Thermal properties of condensed matter
(see also 63. Lattice dynamics; for thermodynamic properties of quantum fluids, see 67.40; for thermal properties of solid helium, see 67.80)
65.20. Heat capacities of liquids
65.40. Heat capacities of solids
(for specific heat of superconductors, see 74.30; for specific heat of magnetic systems, see 75.40)
65.50. Thermodynamic properties and entropy
65.70. Thermal expansion and thermomechanical effects
(see also 64.30. Equations of state)
For thermal conduction in nonmetallic liquids, see 66.60; for nonmetallic solids, see 66.70
For electronic thermal conduction, see 72.10 and 72.20
For thermal conductivity of superconductors, see 74.30
For pyroelectric and electrocaloric effects, see 77.70
65.90. Other topics in thermal properties of condensed matter

66. Transport properties of condensed matter (nonelectronic)
66.10. Diffusion and ionic conduction in liquids
66.20. Diffusive momentum transport
(for viscosity of liquids, see also 62.10)
66.30. Diffusion in solids
66.60. Thermal conduction in nonmetallic liquids
(for thermal conduction in liquid metals, see 72.15)
66.70. Nonelectronic thermal conduction and heat-pulse propagation in nonmetallic solids
(for thermal conduction in solid metals, see 72.15; for statistical mechanics of lattice vibrations, see 63.70)
66.90. Other topics in nonelectronic transport properties

67. Quantum fluids and solids; liquid and solid helium
    - 67.20. Quantum effects on the structure and dynamics of nondegenerate fluids
    - 67.40. Boson degeneracy and superfluidity of helium-4
    - 67.50. Fermi fluids; liquid helium-3
    - 67.60. Mixed systems; liquid helium-3, -4 mixtures
    - 67.70. Films (including physical adsorption)
    - 67.80. Solid helium and related quantum crystals
    - 67.90. Other topics in quantum fluids (e.g., neutron-star matter)

68. Surfaces and interfaces; thin films and whiskers (for impact phenomena, see 79; for crystals growth, see 61.50)
    - 68.10. Fluid surfaces and fluid–fluid interfaces
    - 68.15. Liquid thin films
    - 68.20. Solid surface structures
    - 68.25. Mechanical and acoustical properties of solid surfaces and interfaces
    - 68.30. Dynamics of solid surfaces and interface vibrations
    - 68.40. Surface energy of solids; thermodynamic properties
    - 68.45. Solid–fluid interface processes
    - 68.48. Solid–solid interfaces (including bicrystals) (for grain boundaries, see 61.70)
    - 68.55. Thin film growth, structure, and epitaxy
    - 68.60. Physical properties of thin films: nonelectronic
    - 68.70. Whiskers and dendrites: growth, structure, and nonelectronic properties
    - 68.90. Other topics in the structure and nonelectronic properties of surfaces and thin films

70. Condensed matter: electronic structure; electrical, magnetic, and optical properties

71. Electronic states (see also 63. Lattice dynamics, 73. Electronic structure and electrical properties of surfaces, interfaces, and thin films)
    - 71.10. General theories and computational techniques
    - 71.20. Electronic density of states determinations (including energy states of liquid semiconductors) (see also 65.40. Electronic heat capacity)
    - 71.25. Nonlocalized single-particle electronic states
    - 71.30. Metal–insulator transitions
    - 71.35. Excitons and related phenomena (including electron–hole drops)
    - 71.36. Polaritons (including photon–phonon and photon–magnon interactions)
    - 71.38. Polarons and electron–phonon interactions (see also 63.20, phonon–electron interactions in lattices)
    - 71.45. Collective effects
    - 71.50. Localized single-particle electronic states (excluding impurities)
    - 71.55. Impurity and defect levels
    - 71.60. Positron states (see also 78.70. Positron annihilation)
    - 71.70. Level splitting and interactions (see also 75.10. — in magnetic phenomena, 75.30. Exchange and superexchange interactions, 73.20. Electronic surface states)
    - 71.90. Other topics in electron states

72. Electronic transport in condensed matter
(for surfaces, interfaces, and thin films, see 73)
- 72.10. Theory of electronic transport; scattering mechanism
- 72.15. Electronic conduction in metals and alloys
- 72.20. Conductivity phenomena in semiconductors and insulators
(for nonelectronic conduction, see 66.70)
- 72.30. High-frequency effects; plasma effects
- 72.40. Photoconduction and photovoltaic effects; photodielectric effects
- 72.50. Acoustoelectric effects
- 72.55. Magnetoacoustic effects
- 72.60. Mixed conductivity and conductivity transitions
- 72.70. Noise processes and phenoma
- 72.80. Conductivity of specific semiconductors and insulators
- 72.90. Other topics in electronic transport in condensed matter

73. Electronic structure and electrical properties of surfaces, interfaces, and thin films
- 73.20. Electronic surface states
(for emission and impact phenomena, see 79)
- 73.25. Surface conductivity
- 73.30. Surface double layers, Schottky barriers, and work functions
- 73.40. Interfaces
- 73.60. Electronic properties of thin films
- 73.90. Other topics in electrical properties of surfaces, interfaces, and thin films

74. Superconductivity
- 74.10. Occurrence, critical temperature
- 74.20. Theory
- 74.30. General properties
- 74.40. Fluctuations and critical effects
- 74.50. Tunneling phenomena, Josephson effect, and proximity effects
- 74.55. Type-I superconductivity
- 74.60. Type-II superconductivity
- 74.70. Superconducting materials
- 74.90. Other topics in superconductivity

75. Magnetic properties and materials
- 75.10. General theory and models of magnetic ordering
(see also 71.25. Nonlocalized single-particle electronic states, 71.70. Level splitting and interactions)
- 75.20. Diamagnetism and paramagnetism
- 75.25. Spin arrangements in magnetically ordered materials
(neutron studies, etc.)
- 75.30. Magnetically ordered materials, other intrinsic properties
(for critical-point effects, see 75.40)
- 75.40. Critical-point effects, specific heats, short-range order
(see also 65.40. Heat capacities)
- 75.50. Studies of specific magnetic materials
- 75.60. Domain effects, magnetization curves, and hysteresis
- 75.70. Magnetic films and plates
- 75.80. Magnetomechanical and magnetoelectric effects, magnetostriction
For galvanomagnetic effects, see 72.15 and 72.20
For magneto-optical effects, see 78.20
- 75.90. Other topics in magnetic properties and materials

76. Magnetic resonances and relaxation in condensed matter; Mössbauer effect
   - 76.20. General theory of resonances and relaxation
   - 76.30. Electron paramagnetic resonance and relaxation
   - 76.40. Diamagnetic and cyclotron resonances
   - 76.50. Ferromagnetic, antiferromagnetic, and ferrimagnetic resonances; spin-wave resonance (see also 75.30. Spin waves)
   - 76.60. Nuclear magnetic resonance and relaxation
   - 76.70. Magnetic double resonances and cross effects
   - 76.80. Mössbauer effect; other γ-ray spectroscopy
   - 76.90. Other topics in magnetic resonances and relaxation

77. Dielectric properties and materials
   (for conductivity phenomena, see 72.20 and 72.80)
   - 77.20. Permittivity
   - 77.30. Polarization and depolarization effects
   - 77.40. Dielectric loss and relaxation
   - 77.50. Dielectric breakdown and space-charge effects
   - 77.55. Dielectric thin films
   - 77.60. Piezoelectricity and electrostriction
     (for piezo-optical effects, see 78.20)
   - 77.70. Pyroelectric and electrocaloric effects
   - 77.80. Ferroelectricity and antiferroelectricity
   - 77.85. Electrical resonances
   - 77.90. Other topics in dielectric properties and materials

78. Optical properties and condensed-matter spectroscopy and other interactions of matter with particles and radiation
   (for phonon spectra, see 63)
   - 78.20. Optical properties and materials
   - 78.30. Infrared and Raman spectra and scattering
   - 78.35. Brillouin and Rayleigh scattering
   - 78.40. Visible and ultraviolet spectra
   - 78.45. Stimulated emission
   - 78.50. Impurity and defect absorption in solids
   - 78.55. Photoluminescence
   - 78.60. Other luminescence and radiative recombination
     For photoconduction and photovoltaic effects, see 72.40
   - 78.65. Optical properties of thin films
   - 78.70. Other interactions of matter with particles and radiation
   - 78.90. Other topics in optical properties of condensed matter and other interactions of matter with particles and radiation

79. Electron and ion emission by liquids and solids; impact phenomena
   - 79.20. Impact phenomena, including electron spectra and sputtering
   - 79.40. Thermoionic emission
   - 79.60. Photoemission and photoelectron spectra
   - 79.70. Field emission and field ionization
   - 79.75. Exoelectron emission
   - 79.80. Resonance tunneling
   - 79.90. Other topics in emission and impact phenomena in condensed matter

85. Devices*)

*) Outside the ICSTI Classification for Physics.

Substance Classification (according to physica status solidi)*)

S1. Transition metals and their alloys and compounds
(for oxides, see S10)
- S1.1. Fe
- S1.2. Other 3d metals
- S1.3. Noble metals
- S1.4. Lanthanides
- S1.5. Actinides
- S1.61. Carbides, silicides, germanides
- S1.62. Nitrides, phosphides, arsenides
- S1.63. Sulfides, selenides, tellurides

S2. Alkali metals and their alloys and compounds
(for alkali halides, see S9.11)

S3. Alkaline earth metals and their alloys and compounds
(for alkaline earth halides, see S9.12)

S4. Other metals and their alloys

S5. Nonmetallic elements
- S5.11. Silicon
- S5.12. Germanium
- S5.2. Solidified Gases

S6. Group IV compounds
(for transition metal compounds, see S1.61)

S7. Group V compounds
(for transition metal compounds, see S1.62)
- S7.1. III–V compounds
- S7.11. Phosphides
- S7.12. Arsenides
- S7.13. Antimonides
- S7.15. Mixed crystals
- S7.16. Higher compounds

S8. Chalcogenides
(for transition metal compounds, see S1.63)
- S8.1. II–VI compounds
- S8.11. Sulfides
- S8.12. Selenides
- S8.13. Tellurides
- S8.15. Mixed crystals
- S8.16. Higher compounds

S9. Halides
- S9.11. Alkali halides
- S9.12. Alkaline earth halides
- S9.15. Mixed crystals
- S9.16. Higher compounds

S10. Oxides
- S10.1. Simple oxides
- S10.15. Mixed crystals

S11. Other higher compounds
- S11.1. Salts of oxy-acids
- S11.2. Spinels, garnets, ferrites

S12. Organic compounds
- S12.1 Quasi-one-dimensional systems

*) Outside the ICSTI Classification for Physics.

Some Comments on the Use of the Subject and Author Index

The "International Classification System for Physics" (the condensed matter items 60 and 70) completed by item 85 and a Substance Classification (according to physica status solidi) is used. On cover two the second-level headings of the condensed matter items and on pages 6 to 10 the third-level headings are reproduced. As a help for using the subject index on pages 14 and 15 in addition also a series of important fourth-level headings is given. The substance classification is published on cover three and on page 11.

All papers will be quoted under each relevant section. There is made a distinction between one or two principal entries (refering to the main subject matter of the paper) and secondary entries (refering to subsidiary subject matter and substance, respectively).

On the line following the title the subject classification is given in the same manner as in the volumes. Each entry gives the volume, the number of the issue, the initial and final page numbers. A paper of physica status solidi (b) or physica status solidi (a) is marked by (b) or (a) in front of the line, respectively. The letter "K" before the page numbers indicates a short note published in the offset part of the journal.

The entries for review articles, original papers, short notes, and errata are made together in a single list for each subject. These are ordered according to data of publication, irrespective of being a paper of physica status solidi (b) or physica status solidi (a). Original papers are sometimes published in several parts under the same principal title (in general with special subtitles). The various parts of papers of this type are treated separately and each part is classified according to its particular contents.

The author index gives the names of the authors in alphabetical order. Each name is followed by the bibliographic details of the paper. A paper of physica status solidi (b) or physica status solidi (a) is again marked by (b) or (a) in front of the number of the volume, respectively. In the last column the subject classification of the paper is presented in the same manner as in the volumes. When several papers have been published by the same author they are arranged according to data of publication, irrespective of being a paper of physica status solidi (b) or physica status solidi (a). To find a paper of your interest please choose first the special entry of the subject index and then use the volume number and page.

In using the subject index attention should be given to the following details:

61.10. to 61.16. Theory and experimental techniques of determination of structures
- for specific determinations see 61.55 to 61.80

61.16. Other determination of structures
- electron microscopy and field-ion microscopy determinations
- EPR and NMR determinations

61.25. Studies of specific liquid structures
- macromolecular and polymer solutions
- molten salts
- liquid metals

61.40. Amorphous and polymeric materials
- glasses, polymers, elastomers, plastics

61.50. Crystalline state (including molecular motions in solids)
- physics of crystal growth
- crystal symmetry: models, space groups, crystalline systems and classes
- crystal morphology and orientation
- crystallographic aspects of polymorphic and order–disorder transformations

61.70. Defects in crystals
- grain and twin boundaries
- stacking faults and other planar or extended defects
- doping and implantation of impurities
- impurity concentration, distribution, and gradients

61.80. Radiation damage and other irradiation effects
- channeling, blocking, energy loss of particles
- X-rays, γ-rays, electrons and positrons, neutrons, atoms and molecules, ions
- for ion implantation see 61.70

63.20. Phonons and vibrations in crystal lattices
- phonon–phonon, phonon–electron, and phonon–defect interactions

64.60. General studies of phase transitions
- order–disorder and statistical mechanics of model systems
- equilibrium properties near single critical points, critical exponents
- dynamic critical phenomena
- multicritical points
- metastable phases

64.80. Other phase properties of systems
- stoichiometry and homogeneity
- microstructure

66.30. Diffusion in solids
- theory of diffusion and ionic conduction in solids
- self-diffusion and ionic conduction in non-metals
- diffusion, migration, and displacement of impurities and other defects

71.25. Nonlocalized single-particle electronic states
- techniques of band-structure calculation
- measurements of Fermi surface parameters (including dHvA, magnetoacoustic, positron annihilation, and cyclotron resonance studies, etc.)
- effective mass and g-factors
- band structure of crystalline metals, semiconductors, and insulators

71.45. Collective effects
- exchange, correlation, dielectric and magnetic functions, plasmons
- Fermi-Thomas models
- calculations of total electronic binding energy

71.70. Level splitting and interactions
- crystal and ligand fields
- spin–orbit coupling, Zeeman, Stark, and strain splitting
- exchange interactions
- nuclear states and interactions

72.10. Theory of electronic transport; scattering mechanism
- general formulation of transport theory
- scattering by phonons, magnons, and other nonlocalized excitations
- scattering by point defects, dislocations, surfaces, and other imperfections (including Kondo effect)

72.15. Electronic conduction in metals and alloys
- galvanomagnetic and other magnetotransport effects
- thermomagnetic and thermoelectric effects
- collective modes
- scattering mechanism and Kondo effect

72.20. Conductivity phenomena in semiconductors and insulators
- scattering mechanism
- piezoresistance
- galvanomagnetic and other magnetotransport effects
- thermomagnetic and thermoelectric effects

73.40. Interfaces
- contact resistance, contact potential, rectification
- tunneling: general
- p–n junctions
- MIS, MIM, MSM, SIS, and SMS structures

75.30. Magnetically ordered materials, other intrinsic properties
- saturation moments and magnetic susceptibility
- spin waves
- exchange and superexchange interactions
- anisotropy
- magnetic impurity interactions
- magnetic phase boundaries
- magnetocaloric effect

75.50. Studies of specific magnetic materials
- ferromagnetism
- antiferromagnetics, ferrimagnetics
- amorphous magnetic materials
- magnetic liquids

75.60. Domain effects, magnetization curves, and hysteresis
- high coercitivity materials
- fine-particle systems
- magnetic aftereffects
- magnetic annealing and temperature-hysteresis effects

78.20. Optical properties and materials
- general theory
- optical constants and parameters
- optical rotatory power
- birefringence
- piezo-, elasto-, acousto-, electro-, magneto-, and thermo-optical effects

78.60. Other luminescence and radiative recombination
- electroluminescence
- cathodo- and iono-luminescence
- thermoluminescence
- sono- and tribo-luminescence
- chemiluminescence

78.70. Other interactions of matter with particles and radiation
- positron annihilation
- X-ray scattering, absorption, and emission
- microwave and radiofrequency (including resonances)

79.20. Impact phenomena, including electron spectra and sputtering
- electron impact: Auger and secondary emission
- atom, molecule, and ion impact
- atomic and molecular beam interactions

Subject Index

60. Condensed Matter: Structure, Mechanical, and Thermal Properties

61. Structure of Liquids and Solids; Crystallography
(see also 68.20. Solid surface structures, 71. Electron states)

61.10. X-ray determination of structures
(for specific determinations, see 61.55 to 61.80)

S. MATTHIES, K. HELMING, and K. KUNZE
On the Representation of Orientation Distributions in Texture Analysis by sigma-Sections. I. General Properties of ...
61.10; 61.12; 61.14
(b) Vol.157,No.1, 71-84

V. GEIST, D. STEPHAN, and D. FINK
The Effect of the Lateral and Longitudinal Spread of Fast Protons on Strain Production in GaP and GaAs
61.10 and 61.80; S7.11; S7.12
(a) Vol.117,No.1, 67-72

S.K. PLACHKOVA, L.E. SHELIMOVA, and O.G. KARPINSKII
X-Ray Study of the Structural Phase Transitions in $Ge_{1-x}Ag_{x/2}Bi_{x/2}Te$ Solid Solutions
64.70; 61.10; S8
(a) Vol.117,No.1, 155-162

S. MATTHIES, K. HELMING, and K. KUNZE
On the Representation of Orientation Distributions in Texture Analysis by sigma-Sections. II. Consideration of Crystal ...
61.10; 61.12; 61.14
(b) Vol.157,No.2, 489-508

A.M. AFANASEV, S.S. FANCHENKO, and A.V. MASLOV
On the Equivalent Strain and Damage Distributions of Thin Subsurface Layers in the Triple-Crystal X-Ray Diffractometry
61.10; 68.20
(a) Vol.117,No.2, 341-350

A.O. ABOYAN, A.A. KHZARDZHYAN, P.A. BEZIRGANYAN, and S.E. BEZIRGANYAN
X-Ray Interferometric Pictures of Dislocations
61.10 and 61.70
(a) Vol.118,No.1, 11-20

A.O. ABOYAN, P.A. BEZIRGANYAN, and A.A. KHZARDZHYAN
Study of X-Ray Diffraction Pictures of Crystal Imperfections by Means of Triple Interferometer
61.10
(a) Vol.118,No.1, 21-28

M. ERMRICH and H. HERMANN
Microabsorption Correction in Quantitative Phase Analysis. Experimental Results
61.10; 61.50
(a) Vol.118,No.1, K1-K4

H. SCHAEBEN, H. SIEMES, and S. AUERBACH
Entropy Optimization in Texture Goniometry. II. Practical Applications
61.50; 61.10; 61.12
(b) Vol.158,No.2, 407-426

N.M. OLEKHNOVICH and A.V. PUSHKAREV
Birefringence and Depolarization Effects of X-Ray Scattering in Laue Geometry for Highly Distorted Crystals
61.10; 61.70; S9.11
(a) Vol.119,No.1, 27-34

A. CATICHA and S. CATICHA-ELLIS
Dynamical Diffraction of X-Rays by Thin Crystals at Bragg Angles near $\pi/2$
61.10; S1.2
(a) Vol.119,No.1, 47-54

A. CATICHA and S. CATICHA-ELLIS
A Fabry-Perot Interferometer for Hard X-Rays
61.10 and 78.70
(a) Vol.119,No.2, 643-654

P. ZAUMSEIL and U. WINTER
Characterization of Boron Implanted Silicon by X-Ray Triple-Crystal Diffractometry
61.10 and 61.70; S5.11
(a) Vol.120,No.1, 67-76

V.V. ARISTOV, V.N. MORDKOVICH, A.YU. NIKULIN, A.A. SNIGIREV, U. WINTER, YU.N. EROKHIN, and P. ZAUMSEIL
Possibilities of X-Ray Interference Diffractometry for the Investigation of Ion-Doped Layers
61.10 and 61.70; S5.11
(a) Vol.120,No.1, K1-K6

N.N. FALEEV, L.I. FLAKS, S.G. KONNIKOV, I.K. SOLOMIN, and S.V. BATASHOVA
The Influence of the Directed Displacement of Atomic Planes on the X-Ray Diffraction Rocking Curves
61.10 and 68.55; S7.12; S7.15
(a) Vol.120,No.2, 327-338

O. PACHEROVÁ
The Calculation of the Coplanar Four-Beam X-Ray Dynamical Diffraction
61.10; S5.12
(a) Vol.120,No.2, 339-350

V.I. KHRUPA, D.O. GRIGOREV, V.V. NIKOLAEV, L.I. DATSENKO, and V.B. MOLODKIN
Influence of Defects on the Thickness Dependences of Integral Reflectivity of Dislocation-Free Silicon Crystals
61.10 and 61.70; S5.11
(a) Vol.121,No.1, 11-20

M. GUYMONT, A. TOMAS, M. JULIEN-POUZOL, S. JAULMES, and M. GUITTARD
Structure and Vacancy Ordering in $Cu_2Er_{2/3}S_2$. Electron and X-Ray Diffraction and High Resolution Electron Microscopy ...
61.60; 61.10; S8
(a) Vol.121,No.1, 21-28

E.L. GARTSTEIN
Multibeam Calculation of the Intensity Distribution for an Imperfect Crystal in Laue Geometry
61.10; 61.70; S7.12
(a) Vol.121,No.1, 37-44

E.A. KONDRASHKINA, D.V. NOVIKOV, and S.A. STEPANOV
Study of X-Ray Surface Back Diffraction. Recording of a Diffracted Beam
61.10; 68.20; S5.12
(a) Vol.121,No.1, K9-K14

M.YA. GAMARNIK
Size Changes of Lattice Parameters in Ultradisperse Diamond and Silicon
61.55; 61.10; S5; S5.11
(b) Vol.161,No.2, 457-462

D. WANG and P.D. BRISTOWE
Calculation of Diffraction Effects Due to Double Positioning in (111) Gold Bicrystals
61.70; 61.10; S1.3
(b) Vol.161,No.2, 501-512

C. LANDRON, D. RUFFIER, B. DUBOIS, P. ODIER, D. BONNIN, and H. DEXPERT
Etude par EXAFS du rôle des precurseurs dans la synthèse de $ZrO_2-Y_2O_3$
61.10 and 61.55; S10.15
(a) Vol.121,No.2, 359-368

K. WOKULSKA, J. HAERTWIG, and S. GROSSWIG
The Bond Method Used to Determine the Value of the Lattice Parameters a for alpha-Fe
68.70; 61.10; S1.1
(a) Vol.121,No.2, 539-546

P. ZAUMSEIL, U. WINTER, I.A. VARTANYANTS, and I.YU. KHARITONOV
Comparison between Two X-Ray Methods for Deformation Profile Determination
61.10; S5.11
(a) Vol.121,No.2, K149-K154

E.Z. ARSHAKYAN and A.O. ABOYAN
Investigation of Internal Stresses in Silicon Semiconductor Crystals in Constant Electric and Magnetic Fields by Means..
61.10
(a) Vol.122,No.1, 43-50

V.G. BARYSHEVSKII and S.A. STEPANOV
Effect of Two Critical Angles of X-Ray Surface Diffraction on Secondary Emission Yield
61.10 and 78.70; 68.20
(a) Vol.122,No.1, 69-72

V.L. INDENBOM and V.M. KAGANER
X-Ray Analysis of Internal Stresses in Crystals. II. Lattice Distortions Due to Residual Strains in Crystals Grown ...
61.70; 61.10
(a) Vol.122,No.1, 97-110

V.I. KUSHNIR and E.V. SUVOROV
X-Ray Backscattering on Perfect Crystals
78.70; 61.10; S5.11; S5.12
(a) Vol.122,No.1, 391-404

V. HOLÝ, J. KUBĚNA, and K. PLOOG
X-Ray Analysis of Structural Defects in a Semiconductor Superlattice
61.10 and 61.70; S7.12
(b) Vol.162,No.2, 347-362

V.I. PUNEGOV, A.P. PETRAKOV, and N.A. TIKHONOV
X-Ray Diffraction on Laser Disturbed Near-Surface Crystal Layers
61.10; S5.11
(a) Vol.122,No.2, 449-458

A.M. AFANASEV and O.G. MELIKYAN
A Modified Dynamical Theory (MDT) of X-Ray Diffraction in Extremely Asymmetric Schemes
61.10
(a) Vol.122,No.2, 459-468

P. YANG, S.-S. JIANG, and D. FENG
The Behaviour of the First Fringes in the X-Ray Diffraction Topographs of Planar Defects. I. General Case
61.10 and 61.70
(a) Vol.122,No.2, 475-480

C. BOULESTEIX, Z.C. KANG, P. AUDIBERT, S. NITSCHE, and O. MONNEREAU
Evolution of BiSrCaCuO (2212) Samples by Air Exposure, Showing a Drastic Change of Modulation Very Likely Due to Water..
61.10; S10.15
(a) Vol.122,No.2, K101-K104

61.12. Neutron determination of structures

(for specific determinations, see 61.55 to 61.80)

S. MATTHIES, K. HELMING, and K. KUNZE
On the Representation of Orientation Distributions in Texture Analysis by sigma-Sections. I. General Properties of ...
61.10; 61.12; 61.14
(b) Vol.157,No.1, 71-84

S. MATTHIES, K. HELMING, and K. KUNZE
On the Representation of Orientation Distributions in Texture Analysis by sigma-Sections. II. Consideration of Crystal ...
61.10; 61.12; 61.14
(b) Vol.157,No.2, 489-508

P.S. GOYAL, R. CHAKRAVARTHY, B.A. DASANNACHARYA, S.K. KULSHRESHTHA, M.S. SASTRY, and J. TOMKINSON
Neutron Incoherent Inelastic Scattering from $(NH_4)_xRb_{1-x}H_2PO_4$
64.60; 61.12; S11
(b) Vol.157,No.2, 547-556

H. SCHAEBEN, H. SIEMES, and S. AUERBACH
Entropy Optimization in Texture Goniometry. II. Practical Applications
61.50; 61.10; 61.12
(b) Vol.158,No.2, 407-426

G.R. AUGST and YU.I. VLADISLAVSKII
A Theoretical Study of Diffuse Scattering by Clusters in Alkali Halides
61.12 and 61.70; S9.11
(b) Vol.160,No.1, 71-82

61.14. Electron determination of structures
(for specific determinations, see 61.55 to 61.80)

S. MATTHIES, K. HELMING, and K. KUNZE
On the Representation of Orientation Distributions in Texture Analysis by sigma-Sections. I. General Properties of ...
61.10; 61.12; 61.14
(b) Vol.157,No.1, 71-84

J.G. WEN and R. WANG
Two-Beam Dynamical Theory of the Displacement and Splitting of Diffraction Fringes Caused by a Stacking Fault
61.14 and 61.70; S5
(a) Vol.117,No.1, 73-84

S. MATTHIES, K. HELMING, and K. KUNZE
On the Representation of Orientation Distributions in Texture Analysis by sigma-Sections. II. Consideration of Crystal ...
61.10; 61.12; 61.14
(b) Vol.157,No.2, 489-508

E. TAMURA, R. FEDER, G. WALLER, and U. GRADMANN
Determination of the Surface Magnetization of Fe(110) by Spin-Polarized Low-Energy Electron Diffraction
68.20 and 75.30; 61.14; S1.1
(b) Vol.157,No.2, 627-634

N.A. SEMIKOLENOVA, V.A. BOGDANOVA, and A.S. SEMIKOLENOV
The Ordering Phenomenon in the Highly Doped III-V Semiconductor Materials
61.14; S7.1
(a) Vol.120,No.2, K121-K124

R.K. KARAKHANYAN, P.L. ALEKSANYAN, and Y.K. MANUCHAROVA
On Electron Double Diffraction in the Formation of Kikuchi Patterns
61.14; S5.11
(a) Vol.121,No.1, K1-K4

N.I. BORGARDT and W. SCHROETER
Effect of Defocusing on Black-and-White Contrast of Small Defects
61.14 and 61.70
(a) Vol.121,No.2, 369-382

R. PEREZ
On the Characterization of Crystalline Defects Using CBED Techniques
61.14 and 61.70; S1.3; S5.11
(a) Vol.122,No.1, 51-68

M. MAEDER, J. RICHTER, and A. ZEHE
Surface Reordering of (111)Si 7x7 to 1x1 Reconstruction during In-Situ Annealing
68.20; 61.14; S5.11
(a) Vol.122,No.2, K133-K134

61.16. Other determination of structures
(for specific determinations, see 61.55 to 61.80)

T. KAMINO, M. TOMITA, I. OHTSUKA, and H. SAKA
High Resolution Transmission Electron Microscopy Observation of the Motion of a Surface Dislocation on a Gold Particle
61.16; 61.70; S1.3
(a) Vol.122,No.2, K105-K108

W. GRUENEWALD and J. ULLMANN
Cross-Sectional Transmission Electron Microscopy Investigations of RF-Sputtered a-C Films
68.55; 61.16; 68.60; S5
(a) Vol.122,No.2, K129-K132

61.20. Classical, semiclassical, and quantum theories of liquid structure
(for electron states, see 71; for liquid helium, see 67)

R.V. GOPALA RAO and R. VENKATESH
Theoretical Method of Evaluation of Heats of Vaporisation of Fused Salts
61.20; 65.50; S9.11
(b) Vol.157,No.1, 65-70

YU.M. MISHIN and I.M. RAZUMOVSKII
A Model of Metal Melting and Premelting
61.20 and 61.70; 61.50
(a) Vol.117,No.1, 91-98

T. DAS, L.B. BHUIYAN, and R.N. JOARDER
Ashcroft-Langreth Prescription for Effective Core Diameter in Liquids: A Variational Analysis
61.20; S2; S5.2
(b) Vol.161,No.2, 473-482

61.25. Studies of specific liquid structures

A. DEROUICHE and J.L. BRETONNET
The Long-Wavelength Limit of the Structure Factor of Liquid Alkali Metals Using a One-Component Plasma System
61.25; 65.50; S2
(b) Vol.157,No.1, 61-64

A.K. MISHRA, B.B. SAHAY, and K.K. MUKHERJEE
Structure and Electrical Resistivity of Alkali-Alkali and Lithium-Based Liquid Binary Alloys
61.25; 72.15; S2
(b) Vol.157,No.1, 85-92

T. DAS, T. NAMMALVAR, L.B. BHUIYAN, and R.N. JOARDER
Computation of Vacancy Formation Energy of Liquid Metals
61.25; 61.70; S2; S4
(b) Vol.157,No.1, 93-100

R.V. GOPALA RAO and U. BANDYOPADHYAY
Application of the One-Component Plasma Approximation to Structural Studies of Like Charges in Molten Salts
61.25; S9.12
(a) Vol.118,No.2, 365-370

R.V. GOPALA RAO and A. SATPATHY
Computation of the Structure Factors of the Peculiar Metals Sb, Bi, Ga, and Sn Using Lebowitzs Solution of Hard-Sphere ...
61.25; S4
(b) Vol.159,No.2, 545-550

O. AKINLADE
A Study of the Structure and Thermodynamics of Liquid Metals Based on Some Reference Systems
61.25 and 65.50; S2; S3; S4
(b) Vol.161,No.1, 75-84

61.40. Amorphous and polymeric materials

Y.F. CHEN and C.K. WONG
Microscopic Model of Metastable Changes in Hydrogenated Amorphous Silicon
61.40; S5.11
(b) Vol.157,No.1, 101-106

A. CZACHOR
Locally Compensated Orderless Structures
61.40 and 61.50
(b) Vol.157,No.1, K1-K6

V. SREERAMALU, V. RAVINDRACHARY, H.R. SREEPAD, A. CHANDRASHEKARA, S. GOPAL, H. SANJEEVAIAH, and B. VISWANATHAN
Structural Relaxation and Crystallization of Metallic Glasses $Fe_{35}Co_{52}V_{13}$ and $Fe_{45}Co_{45}V_6Ni_4$. Positron Annihilation Study
61.40; 78.70; S1.1; S1.2
(a) Vol.117,No.1, 53-60

B.X. GU, B.G. SHEN, and H.R. ZHAI
Magnetic Properties and Crystallization of Amorphous $Nd_{14}Fe_{81}B_5$ Alloys
75.50; 61.40; S1.1
(a) Vol.117,No.1, 283-290

E. MATSUBARA, Y. WASEDA, S. HASHIMOTO, and K. INOMATA
Structure of Ferromagnetic $0.5La_{1-x}Sr_xMnO_3*0.5B_2O_3$ (x=0.3,0.5,and 1.0) Oxide Glasses by X-Ray Diffraction
61.40; S10
(a) Vol.117,No.1, K1-K4

K. KRIŠTIAKOVA, G. TUMBEV, and J. KRIŠTIAK
Annihilation of Positrons in $Fe_{80-x}Ni_xB_{20}$ Metallic Glasses
61.40 and 78.70; S1.1
(a) Vol.117,No.1, K5-K8

J. SCHMELZER, R. PASCOVA, and I. GUTZOW
Cluster Growth and Ostwald Ripening in Viscoelastic Media
61.40; 64.75
(a) Vol.117,No.2, 363-376

M.N. KHAN
A Study of the Effects of CuO on the Structure and Optical Absorption of Tellurite Glasses
78.30; 61.40; S10.15
(a) Vol.117,No.2, 593-600

H. RATAJCZAK, I. GOŚCIAŃSKA, and T. LUCIŃSKI
Influence of Annealing on Saturation Magnetization and Hall Resistivity in Amorphous $Nd_{12}Co_{82}B_6$ Thin Films
61.40 and 68.55; 73.60; 75.70; S1.2; S1.4
(a) Vol.118,No.1, 189-196

M.P. PETROV and P.D. SIMOVA
Textural and Electrooptical Investigations of the HOBA and OOBA Cholesteric Phases Induced by Chiral Dopants
61.40; 78.20; S12
(a) Vol.118,No.2, 357-364

V.G. CHUDINOV and V.V. ANDREEV
Atomic Mechanism of Solid-Liquid Phase Transition in F.C.C. Metals with a Large Stacking-Fault Energy
64.70; 61.40; 61.70; S4
(a) Vol.118,No.2, 415-424

P. ROTH and E. HEGENBARTH
How Universal Are the Glassy States in Solids with a Structural Instability?
61.40; 77.70; 77.80
(b) Vol.159,No.1, K1-K6

V.A. GANSHIN and YU.N. KORKISHKO
Proton Exchange in Lithium Niobate and Lithium Tantalate Single Crystals: Regularities and Specific Features (Review Article)
61.40 and 68.48; 78.20; S11.1
(a) Vol.119,No.1, 11-26

S.U. JEN, M.C. CHIEN, and C.Y. LEE
Viscous Flow in Amorphous Alloys
61.40 and 64.70; 62.20; S1.1; S1.2
(a) Vol.119,No.1, 35-40

Y.Z. ZHANG, L.H. ZHE, X.Z. DU, R.M. YANG, and R. XU
Effect of Yb on Electrodeposited Amorphous Fe-P Alloys
61.40; S1.1
(a) Vol.119,No.1, K1-K4

M.N. GHATIKAR, K.V.S. MAHESH, and S.N. BHATIA
Improved Calculations of Phase Shifts for EXAFS Spectroscopy
61.40; 78.70; S1.1; S1.2
(b) Vol.159,No.2, 551-556

A.M. PRITULOV, A.P. SURZHIKOV, V.A. KOZHEMYAKIN, YU.N. AFANASIEV, A.P. VORONIN, O.S. GRIBKOV, and G.R. KARAGEDOV
Radiation-Thermal Packing of Lithium Ferrite Compacts
61.40 and 61.80; S11.2
(a) Vol.119,No.2, 417-422

P.M. OSSI
Crystal-Glass Phase Transition in Ion Irradiated Binary Systems
61.80 and 64.75; 61.40; S1
(a) Vol.119,No.2, 463-470

V.N. BOGDANOV, I.M. BROVCHENKO, L.V. MAKSIMOV, A.R. SILIN, and O.V. YANUSH
Spectroscopic Optical and Acoustic Investigations of Radiation Modified Silica Glasses
61.40 and 78.20; 61.80; 62.65; S10.1
(a) Vol.119,No.2, 621-630

A. KUMAR, P. AGARWAL, and J.S.P. RAI
Crystallization Kinetics in Glassy $Se_{80}Te_{20-x}Sb_x$
61.40; S8
(a) Vol.119,No.2, K97-K102

R.V. GOPALA RAO and U. BANDYOPADHYAY
An Evaluation of Structural and Transport Properties of Molten Potassium-Rubidium Alloy
61.40; 62.20; 66.30; S2
(b) Vol.160,No.2, 403-414

S.A. SERGEENKOV
On Logarithmic CVC of High-T_c Superconductivity Glass Model
74.20; 61.40; S10.15
(b) Vol.160,No.2, K137-K140

H.R. SREEPAD, V. SREERAMALU, A. CHANDRASHEKARA, V. RAVINDRACHARY, and S. GOPAL
Structural Phase Transition in a Superionic Conductor - $Na_2(La,Al)Zr(PO_4)_3$. Positron Annihilation Study
64.70; 61.40; 78.70; S11
(a) Vol.120,No.2, 411-418

V.T. MASLYUK
Stables States and Ordering in Amorphous Polymer Semiconductors
61.40; 65.50; 71.45; S8
(b) Vol.161,No.1, 43-54

E.M. DIANOV, A.V. EGIBYAN, S.A. AKOPYAN, A.E. RUSTAMYAN, V.O. SOKOLOV, and V.B. SULIMOV
Computer Simulation on the Threefold Coordinated Oxygen Atom in Vitreous Silicon Dioxide
61.40; 61.70; 71.50; S10.1
(b) Vol.161,No.1, 55-64

H. KAHNT and J.M. REAU
Mixed Mobile Ion Effect of Anionic Conduction in Lead Halogenosilicate Glasses
61.40 and 66.30; S9.16
(a) Vol.121,No.1, 73-80

J. PEREZ, S. ETIENNE, and J. TATIBOUËT
Determination of Glass Transition Temperature by Internal Friction Measurements
61.40 and 64.60; 62.40
(a) Vol.121,No.1, 129-138

J. DUBOWIK, YU.V. KUDTYAVTSEV, and E.E. KRASSOVSKII
Spectroscopic Ellipsometric Investigation of Multilayered and Solid-State Amorphized Co/Zr Films
68.55 and 78.65; 61.40; S1.2
(a) Vol.121,No.1, 173-180

A. VOHRA and K.K. SRIVASTAVA
Microwave Switching and AC Conductivity of Memory Glass $As_{10}Ge_{15}Te_{75}$
73.40 and 85; 61.40; S8
(a) Vol.121,No.1, 341-348

B.X. GU, B.G. SHEN, and H.R. ZHAI
Crystallization and Magnetic Properties of Amorphous $Nd_{0.19}Fe_{0.81}$ Alloy
75.50; 61.40; S1.1; S1.4
(a) Vol.121,No.1, K99-K104

BAO-GEN SHEN, HUI-QUN GUO, LIN-YUAN YANG, JUN-XIAN ZHANG, and JIAN-GAO ZHAO
Crystallization of Amorphous $Co_{84-x}B_xZr_{16}$ Alloys and Its Influence on Hard Magnetic Properties
75.30; 61.40; 75.60; S1.2
(a) Vol.121,No.1, K105-K110

V.S. STEPANYUK, A.A. KATSNELSON, S. SZASZ, and O.S. TRUSHIN
B-B Direct Contact in Amorphous $Ni_{76}B_{24}$ System Based on Molecular Dynamics
61.40; S1.2
(b) Vol.161,No.2, K77-K80

GUANGHUA CHEN, JINZHANG XU, and FANGQING ZHANG
The ESR Studies of B-Doped a-Si:H Thin Films Prepared by Reactive-Sputtering
61.40 and 76.30; 68.55; S5.11
(a) Vol.121,No.2, 501-506

L.I. VERSHININA, S.Z. SKLUEV, V.S. ZHIGALOV, A.G. STEPANOV, G.I. FROLOV, A.S. AVILOV, and S.V. OREKHOV
The Structure of Amorphous Dy-Co Films with Magnetic Anisotropy
61.40 and 75.50; S1.2
(a) Vol.121,No.2, K145-K148

DIANFEN WANG and R.M.J. COTTERILL
A Four-Dimensional Lennard-Jones Glass
61.40
(b) Vol.162,No.1, K1-K4

M. BAUER, J. MAYER, M. HOHENSTEIN, and M. WILKENS
Quantitative Studies on the Amorphization of Quasicrystals by Means of High Voltage Electron Microscopy
61.40 and 61.80; 64.70; S1.2; S4
(a) Vol.122,No.1, 79-86

V.N. NOVIKOV
Microcreep of Metallic Glasses at Room Temperature
62.20; 61.40; S1.1; S1.2
(a) Vol.122,No.1, 215-220

H. MIZUBAYASHI, H. AGARI, and S. OKUDA
Hydrogen Internal Friction Peak in Amorphous $Cu_{50}Ti_{50}$
61.40 and 62.40; S1.2
(a) Vol.122,No.1, 221-234

V.S. STEPANYUK, A.A. KATSNELSON, O.S. TRUSHIN, and A. SZÁSZ
Structure of the Amorphous Fe-B Alloy Modeled by Molecular Dynamics
61.40; S1.1
(a) Vol.122,No.1, K7-K10

I.A. GOHAR, W. SCHMITZ, P. PAUFLER, and R.M. ELSHAZLY
Comments on the Structure of Mixed Alkali-Borosilicate Glasses Containing Chromium and/or Manganese
61.40 and 61.60; S10.15
(a) Vol.122,No.2, 469-474

J. SERNA
Positronium Diffusion in Crystalline Polyethylene
61.40 and 61.70; 78.70; S12
(a) Vol.122,No.2, 489-494

I. GOŚCIAŃSKA, H. RATAJCZAK, P. SOVÁK, P. SPIŠÁK, and M. KONČ
The Structure of Fe-Rich Vapor Quenched Fe-Ti Thin Films
68.55; 61.40; 73.60; S1.1; S1.2
(a) Vol.122,No.2, 575-582

61.50. Crystalline state (including molecular motion in solids)
(for magnetic structure and spin systems, see 75.25)

A. CZACHOR
Locally Compensated Orderless Structures
61.40 and 61.50
(b) Vol.157,No.1, K1-K6

YU.N. DEVYATKO, V.N. TRONIN, and V.I. TROYAN
Kinetics of the New Phase Nucleus Growth on a Magnet near the Curie Point
61.50; 75.30; S1.2
(a) Vol.117,No.1, 61-66

YU.M. MISHIN and I.M. RAZUMOVSKII
A Model of Metal Melting and Premelting
61.20 and 61.70; 61.50
(a) Vol.117,No.1, 91-98

JIAN-JUN XU
Global Wave Mode Theory for the Formation of Dendritic Structure on a Growing Needle Crystal
68.70; 61.50
(b) Vol.157,No.2, 577-592

U. PIETSCH
Lattice Compression from Conduction Band Electrons in As-Implanted Silicon
61.50; 62.30; S5.11
(b) Vol.157,No.2, K73-K76

T. FUKAMI
Refinement of the Crystal Structure of KH_2PO_4 in the Ferroelectric Phase
61.50 and 77.80; S11.1
(a) Vol.117,No.2, K93-K96

E.M. GOLOLOBOV, N.A. PRYTKOVA, ZH.M. TOMILO, D.M. TURTSEVICH, M.S. TSELUEVSKII, and N.M. SHIMANSKAYA
Properties of High T_c Superconducting Metal Oxide $Bi_2Sr_2CaCu_2O_y$ with Boron Nitride Additions
61.50 and 74.70; S10.15
(a) Vol.117,No.2, K121-K126

W. SALEJDA and N.A. DZHAVADOV
Phase Transition in $Rb_3H(SeO_4)_2$-Type Crystals. I. The Symmetry Analysis of Proton Ordering
64.60; 61.50; 65.50; S11
(b) Vol.158,No.1, 119-126

M. ERMRICH and H. HERMANN
Microabsorption Correction in Quantitative Phase Analysis. Experimental Results
61.10; 61.50
(a) Vol.118,No.1, K1-K4

H. SCHAEBEN, H. SIEMES, and S. AUERBACH
Entropy Optimization in Texture Goniometry. II. Practical Applications
61.50; 61.10; 61.12
(b) Vol.158,No.2, 407-426

G. BEHR, G. KRABBES, J. WERNER, P. DORDOR, and J.-P. DOUMERC
Crystal Growth and Electrical Properties of Sb-Doped SnO_2 Single Crystals
72.80; 61.50; S10.1
(a) Vol.118,No.2, K91-K94

JIZHONG ZHANG
Fractal Growth of MoO_3 Crystals during Vapor-Phase Deposition
61.50 and 68.45; S10.1
(a) Vol.119,No.1, 41-46

R. KRAIENHEMKE, P. SEMIG, and F. FISCHER
Travelling Solvent Zone Melting Growth and Optical Properties of Lead Activated Calcite Crystals
78.50 and 78.55; 61.50; S11.1
(a) Vol.119,No.1, 327-336

XICHUNNG JIN, LINJIANG SHEN, ZUGANG LI, HUIMIN SHAO, HUAQIN WANG, YITONG ZHANG, LONG HAI, and MEISHENG HU
Microdomain Structure and Temperature Behaviour of $YBa_2(Cu_{1-x}Fe_x)_3O_{7-y}$
61.50; 74.50; S10.15
(a) Vol.119,No.1, K5-K8

S.K. GHOSHAL
Photoinduced Crystallization of Organic Molecular Solids from Vapour and Solution Phases
61.50; 64.60; S12
(a) Vol.120,No.1, 61-66

A.A. REMPEL and A.I. GUSEV
Short-Range Order in Superstructures
61.50; 61.55; S1.2; S1.3
(b) Vol.160,No.2, 389-402

O. RADULESCU and R. MANAILA
On the Bond Orientational Order in Icosahedral Phases
61.50; 61.55; S1.1; S1.2; S4
(b) Vol.161,No.1, 65-74

I. RUPSKY, I. IORDANOVA, and A. APOSTOLOV
Monte Carlo Simulation of Normal Grain Growth by Consideration of the Diffusion Character of the Processes
61.50; 61.70; 66.30
(a) Vol.121,No.1, 45-56

S.D. ALTEKAR and A.K. RAY
A Correlation Study of Small Potassium Cationic Clusters
61.50; 71.45; S2
(b) Vol.161,No.2, 463-472

V.M. TALANOV
Structural Modelling of Low-Symmetry Phases of Spinels. I. Phases with 1:1 Octahedral Order
61.50; 64.60; S11.2
(b) Vol.162,No.1, 61-74

U. WALZER
Systematization of Binary Intermetallic Compounds
61.50; 71.10; 71.45; S2
(b) Vol.162,No.1, 75-88

Y. TAHTAMONI and Ş. ERKOÇ
Structural Stability and Energetics of X_n(X=V,Cr,Nb,n=3 to 7) Microclusters: Empirical Many-Body Potential Energy ...
61.50; S1; S1.2
(b) Vol.162,No.1, K5-K8

H. ATMANI and C. VAUTIER
Crystallization of Amorphous BiSe Alloys: Advantages and Limits of a Proposed Theoretical Model
61.50 and 68.55; S8
(a) Vol.122,No.1, 73-78

B.P. SINGH
Analysis of Ionic Radii, Second Neighbour Interactions, and Electronegativity Difference for Ionic Crystals
61.50; 71.45; S7.1; S8.1; S9.11
(b) Vol.162,No.2, 329-338

H. LU, C.L. BAO, X.N. YU, Q.K. XUE, D.S. TANG, P.J. WU, and Y.Y. XIONG
Crystallization on CuZr Amorphous Alloy Surface. An X-Ray Photoelectron Spectroscopy Study
61.50; 79.60; S1.2
(a) Vol.122,No.2, 481-488

61.55. Specific structures of elements and alloys

E.P. ELSUKOV, YU.N. VOROBEV, A.V. TRUBACHEV, and V.A. BARINOV
Structure and Magnetic Properties of Fe-P Electrodeposited Alloys
61.55 and 75.50; 76.80; S1.1
(a) Vol.117,No.1, 291-298

A. BUDKOWSKI, V. MARINKOVIC, A. PRODAN, and F.W. BOSWELL
The Two-Dimensional Modulation in Bulk and Thin-Film $Au_{2+x}Cd_{1-x}$
61.55 and 68.55; S1.3
(a) Vol.117,No.2, 351-362

A.Z. MENSHIKOV, A.P. VOKHMYANIN, and YU.A. DOROFEEV
Magnetic Structure and Phase Transformations in Mn_5Si_3
75.30; 61.55; S1.61
(b) Vol.158,No.1, 319-328

M.YA. GAMARNIK
On the Physical Nature of the Decrease in Lattice Parameters of Small Metal Particles
61.55; S1.3
(b) Vol.160,No.1, K1-K4

A.A. REMPEL and A.I. GUSEV
Short-Range Order in Superstructures
61.50; 61.55; S1.2; S1.3
(b) Vol.160,No.2, 389-402

Z.T. SERIKBAEVA, T.M. ZHANTIKIN, A.K. ZHETBAEV, and K.G. BINNATOV
Moessbauer Investigations of Ordering of the $Ni_3(Fe_{1-x}Mn_x)$ Alloy System
61.55; 76.80; S1.1; S1.2
(a) Vol.120,No.2, 305-314

N.P. KULISH and P.V. PETRENKO
Short-Range Order in the Binary Solid Solutions. Ordering and Its Change on Heating in Fe-Al, Cu-Al, and Ag-Al Alloys
61.55; S1.1; S1.2; S1.3; S4
(a) Vol.120,No.2, 315-326

V.G. SHEPELEVICH
Structure and Electrical Properties of Quickly-Quenched Foils of Antimony and Its Alloys
61.55 and 72.15; S4
(a) Vol.120,No.2, K173-K176

O. RADULESCU and R. MANAILA
On the Bond Orientational Order in Icosahedral Phases
61.50; 61.55; S1.1; S1.2; S4
(b) Vol.161,No.1, 65-74

Ş. ERKOÇ
Structural Stability and Energetics of F.C.C. Metal Microclusters: Empirical Many-Body Potential Energy ...
63.10; 61.55; 62.20; S1.2; S1.3; S4
(b) Vol.161,No.1, 211-216

A.M. SHAMAH, N.K. MINA, and M.K. EL-MOUSLY
Structural Investigation of the High T_c Superconducting Oxide System $La_{1-x}Pr_xCaBaCu_3O_7$
61.55 and 74.70; S10.15
(a) Vol.121,No.1, 235-240

M.YA. GAMARNIK
Size Changes of Lattice Parameters in Ultradisperse Diamond and Silicon
61.55; 61.10; S5; S5.11
(b) Vol.161,No.2, 457-462

C. LANDRON, D. RUFFIER, B. DUBOIS, P. ODIER, D. BONNIN, and H. DEXPERT
Etude par EXAFS du rôle des precurseurs dans la synthèse de ZrO_2-Y_2O_3
61.10 and 61.55; S10.15
(a) Vol.121,No.2, 359-368

R. BRUN DEL RE, J.C. WOOLLEY, M. QUINTERO, and R. TOVAR
T(z) Diagram of $CuIn_{1-z}Fe_zS_2$ Alloys
64.75; 61.55; S1.1
(a) Vol.121,No.2, 483-488

Y. KHAN
Phases Obtained by Continuous Heating of the Amorphous Permanent Magnet Alloy $Nd_{4.4}Fe_{77.8}B_{17.8}$
61.55 and 75.50; S1.1
(a) Vol.122,No.1, K1-K6

I. MAHMOUD, A.M. HAFEZ, H.A. MOTAWEH, and N.S. ALY
Effect of Nickel on the C_{14} Structure Stability in the $Er(Mn_{1-x}Ni_x)_2$ Quasibinary System
61.55; S1.4
(a) Vol.122,No.2, K113-K116

61.60. Specific structures of inorganic compounds

XI CHEN, CHAO DING, YU WANG, XIAN-TING ZEN, and BAO-SHENG LUO
The Structures of $YBa_2Cu_3O_{6+\varepsilon}$ Superconducting and Nonsuperconducting Phases
61.60 and 74.70; S10.1
(a) Vol.117,No.2, 335-340

J.M. RÉAU, J. SÉNÉGAS, M. ELOMARI, J.P. LAVAL, and B. FRIT
Ordre à courte distance et conductivité ionique au sein de la solution $Pb_{1-x}In_xF_{2+x}$
61.60 and 66.30; 64.75; S9.15
(a) Vol.117,No.2, 409-416

M. GUYMONT, A. TOMAS, and M. PALAZZI
Short-Range Order in $EuLiS_2$ by Electron Microscopy
61.60; S8.16
(a) Vol.118,No.1, 29-40

J.J. BARA, B.F. BOGACZ, A. POLACZEK, M. PEKAŁA, and A. GRODZIŃSKI
The Temperature Induced Transition in the theta-Type Vanadium Oxide Bronzes
61.60; 64.70; 76.80; S10
(a) Vol.118,No.1, 41-52

B. GILLOT and A. ROUSSET
On the Limit of Aluminium Substitution in Fe_3O_4 and gamma-Fe_2O_3
61.60; 78.30; S10.15
(a) Vol.118,No.1, K5-K8

J. ECHIGOYA, S. HIRATSUKA, and H. SUTO
Structure and Low-Temperature Phase of $LaCoO_3$
61.60; S11.1
(a) Vol.118,No.2, 371-378

D. WINDISCH and P. BECKER
Silicon Lattice Parameters as an Absolute Scale of Length for High Precision Measurements of Fundamental Constants
61.60 and 61.70; S5.11
(a) Vol.118,No.2, 379-388

Y. NAKAMURA, H. KOIKE, and O. NITTONO
Structure of Long Period Superstructure in Cu-Rich gamma-Brass Investigated by Electron Diffraction
61.60 and 61.70; 68.48; S1.2
(a) Vol.118,No.2, 389-400

U.-C. BOEHNKE, H. BOERNER, J. HERRMANN, B. LIPPOLD, CH. SEMMELHACK, L. WALDOESTL, and M. WURLITZER
Some Investigations on Praseodymium-Substituted Yttrium-Barium-Copper-Oxide
74.70; 61.60; 75.30; S10.15
(a) Vol.120,No.2, 557-566

T. KANOMATA, T. KAWASHIMA, T. KANEKO, H. TAKAHASHI, and T. MŌRI
Pressure Effect on the Lattice Parameters of RMn_2X_2 (R=Y,La,Nd and X=Ge,Si)
61.60 and 62.20; S8.16
(a) Vol.120,No.2, K117-K120

M. GUYMONT, A. TOMAS, M. JULIEN-POUZOL, S. JAULMES, and M. GUITTARD
Structure and Vacancy Ordering in $Cu_2Er_{2/3}S_2$. Electron and X-Ray Diffraction and High Resolution Electron Microscopy ...
61.60; 61.10; S8
(a) Vol.121,No.1, 21-28

S.K. SHARMA, N. SHUKLA, and G.C. TRIGUNAYAT
TEM and X-Ray Studies on Melt-Grown CdI_2 and $CdBr_2$ Crystals
61.60; 61.70; S9
(a) Vol.121,No.1, 29-36

R.A. YAKSHIBAEV, N.N. MUKHAMADEEVA, and R.F. KADRGULOV
Phase Relations, Ionic Transport, and Diffusion in the Alloys of Ag_2Te-Cu_2Te Mixed Conductors
61.60 and 64.80; 66.30; S8.16
(a) Vol.121,No.1, 111-118

R.A. YAKSHIBAEV, A.F. NADEZHDINA, and R.F. ALMUKHAMETOV
Structural Features of Niobium Diselenide Intercalated by Copper at High Temperatures
61.60; 65.70; S1.63
(a) Vol.121,No.1, K5-K8

QIANG WANG, BINGLIN GU, and XIAOWEN ZHANG
Preliminary Studies on the Stabilities of Three Types of Ordered Structures in the $A(B^{1}_{0.5}B^{2}_{0.5})O_3$ System
64.70; 61.60; S10
(b) Vol.161,No.2, 537-542

I. ŚLEDZIŃSKA, A. MURASIK, and A. WIŚNIEWSKI
Linear-Chain Antiferromagnetism in alpha-NiC_2O_4*$2D_2O$
61.60 and 75.40; S10
(b) Vol.161,No.2, K99-K102

T. FUKAMI
X-Ray Study of Crystal Structure of $NH_4H_2AsO_4$ in the Paraelectric and Antiferroelectric Phases
61.60; S11
(a) Vol.121,No.2, 383-390

J. ECHIGOYA and K. NARITA
A Study of $La(Mn_{1-x}Ni_x)O_3$ and Its Bonding with Cubic ZrO_2
61.60; 64.75; S10.15
(a) Vol.122,No.1, 87-96

F. SAJOVEC, R. WOLF, A. FATTAH, K. BICKMANN, H. WENZL, G. NAGEL, H. RUEFER, E. TOMZIG, and P. DE BIÈVRE
Defect Analysis on GaAs Crystals by Precision Measurements of Density and Lattice Parameter
61.70; 61.60; S7.12
(a) Vol.122,No.1, 139-152

V.M. TALANOV
Structural Modelling of Low-Symmetry Phases of Spinels.
II. Phases with 1:3 Octahedral Order
61.60; S11.2
(b) Vol.162,No.2, 339-346

I.A. GOHAR, W. SCHMITZ, P. PAUFLER, and R.M. ELSHAZLY
Comments on the Structure of Mixed Alkali-Borosilicate Glasses Containing Chromium and/or Manganese
61.40 and 61.60; S10.15
(a) Vol.122,No.2, 469-474

M. POPESCU, G. ALDICA, C. SÂRBU, P. PĂUŞESCU, E. CRUCEANU, F. VASILIU, and M. APOSTOL
Crystalline Phases in Superconducting $Bi:Sr:Ca:Cu_2:O_x$ Ceramic
61.60 and 74.70; S10.15
(a) Vol.122,No.2, 637-644

V.V. IVANOV and V.M. TALANOV
Modeling of the Structure of the Ordered Spinel-Like Phases (of Type 2:1)
61.60; S11.2
(a) Vol.122,No.2, K109-K112

T. FUKAMI
Refinement of the Crystal Structure of RbH_2PO_4 in the Ferroelectric Phase
61.60 and 77.80; S11.1
(a) Vol.122,No.2, K117-K120

61.65. Specific structures of organic compounds

S. SOEDERHOLM, J. HELLBERG, J. KRZYSTEK, G. OLOVSSON, I. OLOVSSON, and J.U. VON SCHUETZ
Electronic Properties and Crystal Structure of a Methylthio Substituted Naphthalene Cation Radical Salt: $(DMbTN)_3(ClO_4)_2$
61.65 and 72.20; 71.55; 75.30; 76.30; S12.1
(b) Vol.161,No.1, 245-256

61.70. Defects in crystals
(see also 61.80. Radiation damage,
62. Mechanical and acoustical properties,
71.55. Impurity and defect levels,
76.30. EPR of color centres and other defects,
78.50. Impurity and defect absorption in solids)

T. DAS, T. NAMMALVAR, L.B. BHUIYAN, and R.N. JOARDER
Computation of Vacancy Formation Energy of Liquid Metals
61.25; 61.70; S2; S4
(b) Vol.157,No.1, 93-100

Y. ESTRIN, C. ABROMEIT, and E.C. AIFANTIS
Possible Role of Deformation-Induced Point Defects in Dislocation Patterning
61.70
(b) Vol.157,No.1, 117-128

R. CYWIŃSKI, E.MUGEŃSKI, and I. SOKÓLSKA
Eu^{2+}-Mn^{2+} Energy Transfer in NaCl and KCl Crystals
61.70; 78.50; S9.11
(b) Vol.157,No.1, 137-144

K.D. TOVSTYUK, I.O. POLYAKOV, E.I. SLINKO, N.K. TOVSTYUK, and A.G. KHANDOZHKO
Lithium Impurity States in PbTe
64.60; 61.70; 76.60; S8
(b) Vol.157,No.1, 151-158

A.K. ZHETBAEV, K.M. DONBAEV, and M.B. ABLANOV
Spin Reorientation in Irradiated Hematite at 7 to 60 K
61.70 and 76.80; S10.1
(b) Vol.157,No.1, K55-K60

J.G. WEN and R. WANG
Two-Beam Dynamical Theory of the Displacement and Splitting of Diffraction Fringes Caused by a Stacking Fault
61.14 and 61.70; S5
(a) Vol.117,No.1, 73-84

YU.M. MISHIN and I.M. RAZUMOVSKII
A Model of Metal Melting and Premelting
61.20 and 61.70; 61.50
(a) Vol.117,No.1, 91-98

J. SAID, H. JAOUEN, G. GHIBAUDO, and I. STOEMENOS
Electrical and Physical Investigation of Defect Annihilation in Arsenic Implanted Silicon
61.70; 61.80; S5.11
(a) Vol.117,No.1, 99-104

XIAN-FANG ZHU, LI-DE ZHANG, and ZI-YING JIANG
Effect of Yttrium on Grain Boundary Internal Friction in Commericial Aluminium
62.40; 61.70; S4
(a) Vol.117,No.1, 105-112

CH. GENZEL and A. GENZEL
Shear Stress Distribution in Crystals Induced by Mechanical Surface Load
62.20; 61.70
(a) Vol.117,No.1, 141-154

A. BEL HADJ, B. EL JANI, M. GUITTARD, R. BENNACEUR, and P. GIBART
Photoluminescence of Heavily Doped Sn-GaAs Grown by Metal Organics Vapor Phase Epitaxy
61.70 and 78.65; 73.60; S7.12
(a) Vol.117,No.1, 169-176

R.K. BHAN, S.K. LOMASH, and K.C. CHHABRA
Determination and Study of Field Dependent Defect Density in Thermal SiO_2 Using 1,1,1 Trichloroethane (TCA)
61.70 and 68.48; S5.11; S12
(a) Vol.117,No.1, 203-212

T. KURIHARA and S. TANIGAWA
The Anomalous Temperature Dependence of Positron Annihilation in Dilute Al-Zn Alloy near the Melting Point
61.70 and 78.70; S4
(a) Vol.117,No.1, K9-K12

I.S. BAKSHEE, L.A. KARACHEVTSEVA, A.V. LYUBCHENKO, V.A. PETRYAKOV, E.A. SALKOV, and B.I. KHIZHNYAK
Superposition of Burst Noises as the Origin of 1/f Noise Due to Grain Boundaries in $Cd_xHg_{1-x}Te$
72.70; 61.70; S8.15
(a) Vol.117,No.1, K37-K42

T. SOMA, S. TAHATA, and H.-MATSUO KAGAYA
Interstitial Atom in the Zincblende-Type Lattice
61.70; S7.12
(b) Vol.157,No.2, 509-518

A.K. BANDYOPADHYAY and S.K. SEN
Calculation of Self-Interstitial Formation Energy (Both Split and Non-Split) in Noble Metals
61.70; S1.2; S1.3
(b) Vol.157,No.2, 519-530

K.R. MURALI and Y.V.G.S. MURTI
Color Centers in Irradiated $NaF:Mn^{++}$ Crystals
61.70; 78.55; 78.60; S9.11
(a) Vol.117,No.2, 377-386

M.D. DE COTEAU, P.R. WILSHAW, and R. FALSTER
Gettering of Copper to Oxidation Induced Stacking Faults in Silicon
61.70; S5.11
(a) Vol.117,No.2, 403-408

H. ALEXANDER, S. DIETRICH, M. HUEHNE, M. KOLBE, and G. WEBER
EBIC Microscopy Applied to Glide Dislocations
61.70; 71.55; 72.20; S7.12
(a) Vol.117,No.2, 417-428

A.D. SHEIKH-ALI and R.Z. VALIEV
Dislocation Analysis of the Coupling of Grain Boundary Sliding and Migration during the Deformation of Zn Bicrystals
61.70; S1.2
(a) Vol.117,No.2, 429-436

T. KUROBORI, E. YOSHIDA, S. NAKAMURA, N. TAKEUCHI, K. KAWABATA, and Y. FUKUDA
Optical Properties of a Modified F^+ Center in CaO Crystal Doped with Impurities
78.55; 61.70; S10.1
(a) Vol.117,No.2, K177-K180

E. PIGORSCH and W.E. STEGER
Identification of LiO Bands in the Infrared Spectra of the Insertion Compound delta-LiV_2O_5
61.70 and 78.30; S10.1
(a) Vol.117,No.2, K189-K192

T. TAKEUCHI
Electronic Structure of F Centers in LiH by SCF X_α Method
71.55; 61.70; S9
(b) Vol.158,No.1, 277-282

P.S. GUIMARÃES and M.L. DE SIQUEIRA
The Crystalline Cluster Method for Ionic Solids: The F Center in NaCl
61.70 and 71.55; S9.11
(b) Vol.158,No.1, 283-292

A. TOMITA, Y. FUKUDA, Y. KUTOMI, and N. TAKEUCHI
Emission Processes in Thermally Stimulated Exoelectron Emission and Thermoluminescence of KCl:Ag Single Crystals
61.70 and 78.60; 79.75; S9.11
(b) Vol.158,No.1, 383-390

A.O. ABOYAN, A.A. KHZARDZHYAN, P.A. BEZIRGANYAN, and S.E. BEZIRGANYAN
X-Ray Interferometric Pictures of Dislocations
61.10 and 61.70
(a) Vol.118,No.1, 11-20

I.M. FILIPPOV, L.A. KAZAKEVICH, and P.F. LUGAKOV
The Effect of Irradiation and Heat Treatment on the Dislocation Recombination Activity in n-Type Silicon
61.70 and 71.55; S5.11
(a) Vol.118,No.1, 53-60

J. DRYZEK and E. ROGOWSKA
Positron Annihilation in Sodium and Copper beta-Vanadium Oxide Bronzes
61.70; 71.25; 78.70; S10.15
(a) Vol.118,No.1, 61-64

I.I. KOLKOVSKII, P.F. LUGAKOV, V.V. LUKYANITSA, and V.V. SHUSHA
Formation of Thermal Donors during Annealing of Radiation Defects in Float-Zone Dislocations-Free n-Si
61.70 and 71.55; S5.11
(a) Vol.118,No.1, 65-70

V.L. INDENBOM and V.M. KAGANER
X-Ray Analysis of Internal Stresses in Crystals. I. General Equations
61.70; 62.20
(a) Vol.118,No.1, 71-84

J. REICHEL, M. REICHE, and E. ZUBER
Formation of New Donors in Cz-Grown Silicon at Temperatures below 500 °C
61.70; S5.11
(a) Vol.118,No.1, 85-98

G. SCHOECK and W. PICHL
Bond Trapping of Cracks
61.70
(a) Vol.118,No.1, 109-116

Y.Y. LOGINOV, V.M. LENCHENKO, P.D. BROWN, and J. WOODS
TEM Study of the Formation of Structural Defects in Ion Implanted Silicon at Depths of $h \lessapprox R_p$ and $h \gtrapprox 2R_p$
61.70; 61.80; S5.11
(a) Vol.118,No.1, 117-130

T. OHGAKU and N. TAKEUCHI
Study on Dislocation-Impurity Interaction by the Blaha Effect
62.20; 61.70; S9.11
(a) Vol.118,No.1, 153-160

H. GERSTENBERG and W. GLAESER
Transmutation Doping and Lattice Defects in Degenerate InSb
61.70 and 72.20; 71.55; S7.13
(a) Vol.118,No.1, 241-252

S.P. KUILA
Photostimulation Studies in Pure and Li-Doped Alkali Halides
78.60; 61.70; 71.55; S9.11
(a) Vol.118,No.1, 311-316

DEYAN HE, FANGQING ZHANG, and GUANGHUA CHEN
A Direct Measurement of Thermal Equilibrium Defects in Hydrogenated Amorphous Silicon-Nitrogen Alloy Films
61.70 and 76.30; S5.11
(a) Vol.118,No.1, K39-K42

HONG WANG, JINKE WEN, BIN LI, and HUAFU WANG
Infrared Absorption Study of OH^- in $LiNbO_3$:Mg and $LiNbO_3$:Mg:Fe Crystals
61.70 and 78.30; S11.1
(a) Vol.118,No.1, K47-K50

B.A. GREENBERG, M.I. KATSNELSON, V.G. KORESHKOV, YU.N. OSETSKII, G.V. PESCHANSKIKH, A.V. TREFILOV, YU.F. SHAMANAEV et al.
On the Possibility of Describing Lattice Properties of Iridium in Terms of Pseudopotential Theory
63.20 and 71.10; 61.70; 62.20; S1
(b) Vol.158,No.2, 441-456

D. WINDISCH and P. BECKER
Silicon Lattice Parameters as an Absolute Scale of Length for High Precision Measurements of Fundamental Constants
61.60 and 61.70; S5.11
(a) Vol.118,No.2, 379-388

Y. NAKAMURA, H. KOIKE, and O. NITTONO
Structure of Long Period Superstructure in Cu-Rich gamma-Brass Investigated by Electron Diffraction
61.60 and 61.70; 68.48; S1.2
(a) Vol.118,No.2, 389-400

N.N. KHUSNATDINOV, V.F. PETRENKO, and A.N. TURANOV
Intrinsic Photoconductivity of Hexagonal Ice
61.70 and 72.40; S10
(a) Vol.118,No.2, 401-408

V.G. CHUDINOV and V.V. ANDREEV
Atomic Mechanism of Solid-Liquid Phase Transition in F.C.C. Metals with a Large Stacking-Fault Energy
64.70; 61.40; 61.70; S4
(a) Vol.118,No.2, 415-424

N.M. OLEKHNOVICH and A.V. PUSHKAREV
Birefringence and Depolarization Effects of X-Ray Scattering in Laue Geometry for Highly Distorted Crystals
61.10; 61.70; S9.11
(a) Vol.119,No.1, 27-34

D. DRAHEIM, A. TEMPEL, A. ZEHE, and D. BAITHER
TEM Investigations of MBE Grown CaF_2 Strained Layers on (111) Silicon
68.55; 61.70; S9.12
(a) Vol.119,No.1, 209-214

M.R. MURTI and K.V. REDDY
Grain Boundary Effects on the Carrier Mobility of Polysilicon
72.20; 61.70; S5.11
(a) Vol.119,No.1, 237-240

R. HERGT, W. ANDRAE, K. FISCHER, N.M. TCHEBOTAEV, and S.L. TOWN
Pinning in Twinned Single Crystals of Superconducting $REBa_2Cu_3O_{7-x}$ Investigated by Torque Magnetometry
74.60 and 74.70; 61.70; S10.15
(a) Vol.119,No.1, 241-250

XIAO-GUANG WANG and HONG ZHANG
A Vacancy Complex in the Dilute Fe-V-N Alloy Identified by Positron Annihilation Spectroscopy
61.70 and 78.70; S1.1
(a) Vol.119,No.1, K15-K20

A. SARCE and E.J. SAVINO
Grain Boundary Sink Strength in Anisotropic Materials
61.70 and 66.30
(b) Vol.159,No.2, 557-570

T. KAMIKAWA
A New Hole Centre in Neutron Irradiated LiF Crystals
61.70; 76.30; S9.11
(b) Vol.159,No.2, 571-576

F. PIO, M. GUZZI, G. SPINOLO, and M. MARTINI
Intrinsinc and Impurity-Related Point Defects in Amorphous Silica: A Spectroscopic Study
61.70; 78.50; 78.55; S10.1
(b) Vol.159,No.2, 577-588

C.H. WOO and B.N. SINGH
The Concept of Production Bias and Its Possible Role in Defect Accumulation under Cascade Damage Conditions
61.80; 61.70
(b) Vol.159,No.2, 609-616

Y.M. GU, L. FRITSCHE, and N.E. CHRISTENSEN
Chemical Trends in the Formation and Binding Energy of Native Defects in III-V Semiconductors
61.70 and 71.55; S7.1
(b) Vol.159,No.2, 617-628

P.G. BARANOV, V.V. DYAKONOV, N.G. ROMANOV, and V.A. VETROV
Frenkel Defect Formation by Intense UV Light Irradiation in the Absorption Bands of Tl^+-Like Ions in Alkali Halides
61.70; S9.11
(b) Vol.159,No.2, K33-K36

G. LIFANTE, P. SILFSTEN, F. CUSSÓ, and F. JAQUE
$(F_2^+)_H$-Centre Production in Molecule-Doped KCl
61.70; S9.11
(b) Vol.159,No.2, K37-K42

G.A. MALYGIN
Dislocation Density Evolution Equation and Strain Hardening of F.C.C. Crystals
61.70; 62.20; S4
(a) Vol.119,No.2, 423-436

A.S. TAHA and F.H. HAMMAD
The Applications of the Hall-Petch Relation to Microhardness Measurements on Al, Cu, Al-MD 105, and Al-Cu Alloys
62.20; 61.70; 64.80; S1.2; S4
(a) Vol.119,No.2, 455-462

A. FEDOTOV, B. EVTODYI, L. FIONOVA, YU. ILYASHUK, E. KATZ, and L. POLYAK
Electrical Activity of Grain Boundaries in Shaped Grown Silicon
61.70 and 68.55; 73.20; 73.40; S5.11
(a) Vol.119,No.2, 523-534

J.E. BISBEE and N.C. HALDER
Activation Energy and Distribution Function of the EL2 Defect Level in Si-Implanted GaAs
61.70 and 71.55; S7.12
(a) Vol.119,No.2, 545-554

J. MOLENDA, A. STOKŁOSA, S. MROWEC, and DO THAN
On the Defect Structure and Electronic Properties of Titanium Disulfide
72.15; 61.70; S1.63
(a) Vol.119,No.2, 571-582

J. REICHEL and M. REICHE
The Importance of Self-Interstitials to the Defect Formation in Silicon during Quenching
61.70; S5.11
(a) Vol.119,No.2, K103-K106

M. JAIN
A Theoretical Consideration of Dislocation Dynamics during Steady State Cyclic Plastic Straining of Cell Forming Metals
61.70
(a) Vol.119,No.2, K107-K112

I.M. MIKHAILOVSKII and V.B. RABUKHIN
Energy of Boundaries in the Vicinity of a Triple Junction
61.70; S1
(a) Vol.119,No.2, K113-K116

G.R. AUGST and YU.I. VLADISLAVSKII
A Theoretical Study of Diffuse Scattering by Clusters in Alkali Halides
61.12 and 61.70; S9.11
(b) Vol.160,No.1, 71-82

E.M. DIANOV, V.O. SOKOLOV, and V.B. SULIMOV
Theory of Silicon Vacancy in Vitreous Silica
61.70 and 71.55; S10.1
(b) Vol.160,No.1, 263-274

A.V. EGRANOV, V.V. OTROSHOK, and B.P. CHERNYAGO
Infrared Absorption Due to U-Centres in Highly Doped $LiF:H^-$ Crystals
61.70 and 78.30; S9.11
(b) Vol.160,No.1, K75-K78

U. WOGGON, M. MUELLER, I. RUECKMANN, J. KOLENDA, and M. PETRAUSKAS
Influence of Hydrogenation of Defects at the Microcrystallite-Glass Interface on the Nonlinear Optics ...
61.70 and 78.20; S8.12
(b) Vol.160,No.1, K79-K84

P. ZAUMSEIL and U. WINTER
Characterization of Boron Implanted Silicon by X-Ray Triple-Crystal Diffractometry
61.10 and 61.70; S5.11
(a) Vol.120,No.1, 67-76

S.V. DUBONOS and S.V. KOVESHNIKOV
Formation of High Mobile Defects in GaAs under Ar-Plasma Etching
61.70; 71.55; S7.12
(a) Vol.120,No.1, 77-82

M.I. MOLOTSKII and S.Z. SHMURAK
Electron Emission at Plastic Deformation of Colored Crystals
61.70; 79.75; S9
(a) Vol.120,No.1, 83-96

F.G. REQUEJO, A.G. BIBILONI, H. SAITOVITCH, and P.R.J. SILVA
Study of Point Defects in Irradiated and Reduced MoO_3 through Hyperfine Interactions on ^{99}Mo
61.70; 61.80; S10.1
(a) Vol.120,No.1, 105-118

Y.B. XIA, Z.G. WANG, and R.H. WANG
Secondary Hardening and Dislocation Evolution in Low Cycle Fatigue of Polycrystalline Aluminium
62.20; 61.70; S4
(a) Vol.120,No.1, 125-132

S. KAHL, R. NEUHAUS, H. PINKVOS, and CH. SCHWINK
The Influence of Tensile Deformation and Atomic Short-Range Order on the Susceptibility of CuMn Spin Glasses
62.20 and 75.40; 61.70; S1.2
(a) Vol.120,No.1, 221-232

V.V. ARISTOV, V.N. MORDKOVICH, A.YU. NIKULIN, A.A. SNIGIREV, U. WINTER, YU.N. EROKHIN, and P. ZAUMSEIL
Possibilities of X-Ray Interference Diffractometry for the Investigation of Ion-Doped Layers
61.10 and 61.70; S5.11
(a) Vol.120,No.1, K1-K6

E.A. BRENER and D.E. TEMKIN
Grain Boundary Migration Induced by Diffusion Flow across the Boundary
61.70 and 66.30
(a) Vol.120,No.1, K7-K10

Y. KAMIURA, M. YONETA, and F. HASHIMOTO
Hole Traps Produced by High-Temperature Heat Treatments of Silicon above 1300 °C
61.70; 71.55; 66.30; S5.11
(a) Vol.120,No.1, K11-K14

V.V. EMTSEV, V.A. KHRAMTSOV, and K. SCHMALZ
On the A-Centre Formation in Heat-Treated Cz-Silicon
61.70; S5.11
(a) Vol.120,No.1, K15-K18

S.V. NISTOR, GH. STOICESCU, and C.D. MATEESCU
Incorporation and Aggregation Properties of Bismuth Ions in KCl Crystals
61.70; 66.30; 78.50; S9.11
(b) Vol.160,No.2, 423-432

W. WALZ and H. KRONMUELLER
Point Defects in Electron-Irradiated Stoichiometric Magnetite
61.70 and 75.50; S10
(b) Vol.160,No.2, 661-672

V.I. GOSTISHEV, S.E. DEMYANOV, V.R. SOBOL, and D.V. PASHIK
On the Anisotropy of Electron-Dislocation Scattering in Aluminum in Charge Transport
61.70 and 72.15; S4
(b) Vol.160,No.2, K133-K136

J. ČERMÁK
Diffusion along Moving Grain Boundaries
66.30; 61.70; S1.1; S1.2
(a) Vol.120,No.2, 351-362

I.T. SUPRUN
Determination of the Dislocation Structure Parameters from the Data on Amplitude Dependence of Internal Friction
61.70; 62.40
(a) Vol.120,No.2, 363-370

A.M. MAZZONE
Monte-Carlo Simulation of the Backscattering Effects of Loop Dislocations in a Silicon Crystal
61.70; 61.80; S5.11
(a) Vol.120,No.2, 379-386

S.V. KOVESHNIKOV, S.V. NOSENKO, and E.B. YAKIMOV
The Peculiarities of Deep Level Defect Passivation in Si by Atomic Hydrogen
61.70 and 71.55; 66.30; S5.11
(a) Vol.120,No.2, 391-396

B.M. LOGINOV and V.T. DEGTYAREV
Regularities Governing the Dislocation Forest Effects on Crystal Resistance to Deformation under Complex Loading ...
61.70; S1.2; S9.11
(a) Vol.120,No.2, 397-410

L.F. CHEN, S. PATU, and C.X. SHI
Effect of Phosphorus Ion Implantation on Fatigue Behaviour of Polycrystalline Nickel
62.20; 61.70; S1.2
(a) Vol.120,No.2, 429-440

K.M. LIN and S. LEE
Dislocations near a Bimetallic Interface
61.70 and 68.25
(a) Vol.120,No.2, 497-506

O.O. AWADELKARIM, A. HENRY, B. MONEMAR, and J.L. LINDSTROEM
The Role of Group-V Impurity in Defect Formation in Irradiated Silicon
71.55; 61.70; S5.11
(a) Vol.120,No.2, 539-546

S. CHABIK
Investigation of Interactions of Electrically Charged Hydrogen Atoms with Defects in Nickel by the Positron ...
61.70 and 78.70; S1.2
(a) Vol.120,No.2, K125-K130

Y.V.G.S. MURTI, G.A. POPSON, and A.L. LASKAR
Correlation of Melting Temperature and Formation Enthalpy of Schottky Defects in Ionic Solids
61.70; S9.11; S10.1
(a) Vol.120,No.2, K133-K138

ZHIFANG PENG
On the Type of Point Defects Generated by Motion of the Jogged Srew Dislocations
61.70
(a) Vol.120,No.2, K139-K144

XIAOTONG LIN, ZHIPU YOU, and HUIFEN GUO
An Enhancement Phenomenon of the Minority Carrier Lifetime in Annealed Silicon
72.40; 61.70; S5.11
(a) Vol.120,No.2, K177-K180

C. KISIELOWSKI-KEMMERICH
Vacancies and Their Complexes in the Core of Screw Dislocations. Models which Account for ESR-Investigation ... (Review Article)
61.70; 76.30; S5.11
(b) Vol.161,No.1, 11-42

E.M. DIANOV, A.V. EGIBYAN, S.A. AKOPYAN, A.E. RUSTAMYAN, V.O. SOKOLOV, and V.B. SULIMOV
Computer Simulation on the Threefold Coordinated Oxygen Atom in Vitreous Silicon Dioxide
61.40; 61.70; 71.50; S10.1
(b) Vol.161,No.1, 55-64

A. KALNIŅŠ, I. PLAVIŅA, A.I. POPOV, and A. TALE
Determination of the Effective Absorption Cross-Section of F-Centres in KBr-In by Photostimulated Luminescence
61.70; 78.55; S9.11
(b) Vol.161,No.1, 85-90

G.S. MYAKENKAYA, G.L. GUTSEV, N.P. AFANASEVA, V.A. EVSEEV, and R.F. KONOPLEVA
Study of Lithium Interaction with Lattice Defects in Silicon
61.70 and 61.80; 72.20; S5.11
(b) Vol.161,No.1, 91-104

O. SILD
The Model of Rotation of NO_2^- Impurity Molecule in KCl Crystal
61.70; 78.30; 78.55; S9.11
(b) Vol.161,No.1, 105-112

H. GUENTHER
Tachyons in a Crystalline Structure
62.20; 61.70
(b) Vol.161,No.1, 179-190

H.-B. ABEL
Low Frequency SERS from Ag Colloids in Alkali Halide Crystals
78.30; 61.70; S9.11
(b) Vol.161,No.1, 435-446

F. MAES, F. CALLENS, P. MATTHYS, and E. BOESMAN
EPR of S_2^- in NaBr
61.70 and 76.30; S9.11
(b) Vol.161,No.1, K1-K4

V.I. KHRUPA, D.O. GRIGOREV, V.V. NIKOLAEV, L.I. DATSENKO, and V.B. MOLODKIN
Influence of Defects on the Thickness Dependences of Integral Reflectivity of Dislocation-Free Silicon Crystals
61.10 and 61.70; S5.11
(a) Vol.121,No.1, 11-20

S.K. SHARMA, N. SHUKLA, and G.C. TRIGUNAYAT
TEM and X-Ray Studies on Melt-Grown CdI_2 and $CdBr_2$ Crystals
61.60; 61.70; S9
(a) Vol.121,No.1, 29-36

E.L. GARTSTEIN
Multibeam Calculation of the Intensity Distribution for an Imperfect Crystal in Laue Geometry
61.10; 61.70; S7.12
(a) Vol.121,No.1, 37-44

I. RUPSKY, I. IORDANOVA, and A. APOSTOLOV
Monte Carlo Simulation of Normal Grain Growth by Consideration of the Diffusion Character of the Processes
61.50; 61.70; 66.30
(a) Vol.121,No.1, 45-56

D.P. ERCHAK, R.B. GELFAND, N.M. PENINA, V.F. STELMAKH, V.P. TOLSTYKH, A.G. ULYASHIN, V.S. VARICHENKO, and A.M. ZAITSEV
Point Paramagnetic Defects in Diamond Irradiated by High-Energy Ions
61.70; 76.30; 78.60; S5
(a) Vol.121,No.1, 63-72

A. FOITZIK, W. SKROTZKI, and P. HAASEN
Dislocation Microstructure on {100} and {110} Slip Planes in Lead Sulphide
61.70; S8
(a) Vol.121,No.1, 81-94

H. MENNIGER, M. BEER, R. GLEICHMANN, H. RAIDT, B. ULRICI, and G. VOIGT
Electrical Homogeneity of Semi-Insulating LEC GaAs by Post-Growth Annealing
61.70 and 78.60; 85; S7.12
(a) Vol.121,No.1, 95-104

SIE CHIN TJONG
Dislocation Microstructures in Fatigued Fe-30Mn-10Al-0.4C Alloy
62.20; 61.70; S1.1
(a) Vol.121,No.1, 119-128

CHANGHE HUANG, ZHENZHONG YU, and DINGYUAN TANG
Defect Related Absorption in p-Type $Hg_{1-x}Cd_xTe$ Alloys
78.50; 61.70; S8.15
(a) Vol.121,No.1, 273-280

C.Q. TANG and Y.F. ZHANG
Positron Lifetime Study on the Precipitation Phenomenon in 4Cr3Mo2NiVNb Alloy Steel
61.70 and 78.70; S1.2
(a) Vol.121,No.1, K21-K26

V.I. BARBASHOV and T.V. RODZINA
The Reverse Motion of Screw Dislocations in Indium Antimonide Crystals
61.70; S7.13
(a) Vol.121,No.1, K27-K30

ZHIFANG PENG and YUEHUANG XU
On the Reaction between Perfect Dislocations on {110}<111> Slip Systems in the B.C.C. Lattice
61.70
(a) Vol.121,No.1, K37-K42

D. WANG and P.D. BRISTOWE
Calculation of Diffraction Effects Due to Double Positioning in (111) Gold Bicrystals
61.70; 61.10; S1.3
(b) Vol.161,No.2, 501-512

A.CH. LUSHCHIK and A.G. FRORIP
Thermalized and Hot Interstitial Halogen Ions in Alkali Halides
61.70 and 78.55; 61.80; 76.30; S9.11
(b) Vol.161,No.2, 525-536

A.P. PECHONII, T.V. ANTIMIROVA, M.D. GLINCHUK, and I.M. SMOLYANINOV
An ESR Study of Li^+ Impurity Ion Deformational Fields in $KTaO_3$
76.30; 61.70; S11
(b) Vol.161,No.2, 843-852

N.I. BORGARDT and W. SCHROETER
Effect of Defocusing on Black-and-White Contrast of Small Defects
61.14 and 61.70
(a) Vol.121,No.2, 369-382

E.S. TSIKHOTSKII, I.N. ZAKHARCHENKO, and P.V. DUDKEVICH
Linear Defects and Ferroelectric Phase Transition in $BaTiO_3$ Crystals
61.70; 77.80; S10.15
(a) Vol.121,No.2, 391-398

H. OPYRCHAŁ and M. MANFREDI
Spectroscopic Evidences of Eu-V Dipole Precipitation in KCl:Eu Crystals
61.70; 71.55; 78.55; S9.11
(a) Vol.121,No.2, 407-414

QIREN ZHANG and XIQI FENG
Defect Structures and Densities of Mg-Doped Lithium Niobate
61.70; S11.1
(a) Vol.121,No.2, 429-436

L.B. ZUEV, V.E. GROMOV, and L.I. GUREVICH
The Effect of Electric Current Pulses on the Dislocation Mobility in Zinc Single Crystals
61.70; S1.2
(a) Vol.121,No.2, 437-444

E.V. TATYANIN and V.G. KURDYUMOV
The Nucleation of the Deformation Induced Amorphous Phase at Twin Boundaries in TiNi Alloy
61.70 and 62.20; 64.70; S1.2
(a) Vol.121,No.2, 455-460

H. MURAKAMI and M. SANO
Positron-Trapping Centers in Potassium-Intercalated Graphite, KC_8 and $KC_8H_{2/3}$
61.70; 78.70; S2; S6
(a) Vol.121,No.2, 461-466

F.S. NASREDINOV, E.YU. TURAEV, P.P. SEREGIN, H.B. RAKHMATULLAEV, and M.K. BAKHADIRKHANOV
Mechanism of the Two-Electron Exchange between Neutral and Ionized Centres of Tin in the $PbS_{1-x}Se_x$ Solid Solutions
61.70 and 71.55; 76.80; S8.15
(a) Vol.121,No.2, 571-578

J. DRYZEK and M. ROMEK
Positron Annihilation Study of Oxygen Vacancies in the High Temperature Superconductor Y-Ba-Cu-O
61.70 and 78.70; 74.50; S10.15
(a) Vol.121,No.2, K159-K162

N.YU. ARUTYUNOV, V.V. EMTSEV, K. SCHMALZ, and V.YU. TRASHCHAKOV
A Study of Defects in Heat Treated Cz-Silicon by Positron Annihilation
61.70 and 78.70; S5.11
(a) Vol.121,No.2, K163-K166

S. DATTA ROY (PAUL) and S. SENGUPTA
A Greens Function Estimation of Lattice Relaxation for Isovalent Substitutional Defects in Alkali Halide Crystals ...
61.70; S9.11
(b) Vol.162,No.1, 89-98

N. SMETNIANSKY-DEGRANDE, C.N. TOMÉ, and R. PASSIANOT
Anisotropic Diffusion of Point Defects to Spheroidal Cavities
61.70 and 66.30
(b) Vol.162,No.1, 99-110

Y. MADHAVAN, K. PONNAMMAL, K. RAMACHANDRAN, and T.M. HARIDASAN
Mean Square Displacement (MSD) of Neighbours around a Vacancy in Some Zinc Based Tellurides
63.20; 61.70; S8.15
(b) Vol.162,No.1, 119-124

A.B. SOBOLEV, A.N. VARAKSIN, O.A. KEDA, and A.P. KHAIMENOV
Electronic Structure and Charge State of Oxygen Vacancies in ZrO_2 Perfect Crystals
71.55; 61.70; S10.1
(b) Vol.162,No.1, 165-172

A.O. DONGHOL DIALLO, E. MROSAN, and P. ZIESCHE
Model Calculation of the Specific Dislocation Resistivity for Copper
61.70 and 72.10; S1.2
(b) Vol.162,No.1, 237-242

R. PEREZ
On the Characterization of Crystalline Defects Using CBED Techniques
61.14 and 61.70; S1.3; S5.11
(a) Vol.122,No.1, 51-68

V.L. INDENBOM and V.M. KAGANER
X-Ray Analysis of Internal Stresses in Crystals. II. Lattice Distortions Due to Residual Strains in Crystals Grown ...
61.70; 61.10
(a) Vol.122,No.1, 97-110

I.E. BONDARENKO and E.B. YAKIMOV
EBIC Investigation of Electrical Activity of Dislocations with Different Impurity Atmospheres in Si
61.70; 78.70; S5.11
(a) Vol.122,No.1, 121-128

F. SAJOVEC, R. WOLF, A. FATTAH, K. BICKMANN, H. WENZL, G. NAGEL, H. RUEFER, E. TOMZIG, and P. DE BIÈVRE
Defect Analysis on GaAs Crystals by Precision Measurements of Density and Lattice Parameter
61.70; 61.60; S7.12
(a) Vol.122,No.1, 139-152

G. D'ANNA and W. BENOIT
Dislocation Internal Friction Investigation in High Purity Cold Worked Tantalum
61.70 and 62.40; S1
(a) Vol.122,No.1, 153-162

Y. NISHINO, T. NISHIKAWA, and S. ASANO
Strain Aging in Czochralski-Grown Silicon Crystals
61.70; 62.20; S5.11
(a) Vol.122,No.1, 163-170

J.D. STEPHENSON
Step-Scanning WSR Section Topography for Indirect (Point Defect) Characterization of Dislocation-Free Si Wafers
61.70; S5.11
(a) Vol.122,No.1, 171-186

V.G. CHUDINOV, R.M.J. COTTERILL, and V.V. ANDREEV
Possible Mechanism of the Solid-Liquid Phase Transition in H.C.P. and B.C.C. Structures
64.70; 61.70
(a) Vol.122,No.1, 187-194

J.Y. KIM and R.E. HUMMEL
Hole and Hillock Formation in Gold Metallizations at Elevated Temperatures Deposited on Titanium, Vanadium and Other ...
61.70 and 68.55; S1.3
(a) Vol.122,No.1, 255-274

C.Q. TANG, G.Y. LI, and J.Y. TANG
Positron Lifetime Study on Thermal Fatigue in 5CrNiMo Steel
61.70 and 78.70; 62.20; S1.1
(a) Vol.122,No.1, K11-K14

M.R. SUROWIEC
Possibility of a New Type Lomer-Cottrell Lock Formation
61.70; S7.13
(a) Vol.122,No.1, K15-K18

R. HERGT, W. ANDRAE, and W. SCHUEPPEL
Pinning of Vortices at Twin Boundaries in $YBa_2Cu_3O_{7-x}$ Single Crystals
74.60 and 74.70; 61.70; S10.15
(a) Vol.122,No.1, K51-K54

T. KUROBORI and N. TAKEUCHI
On the Lifetime in Magnesium-Doped Calcium Oxide Crystals
78.55; 61.70; 71.55; S10.1
(a) Vol.122,No.1, K91-K96

V. HOLÝ, J. KUBĚNA, and K. PLOOG
X-Ray Analysis of Structural Defects in a Semiconductor Superlattice
61.10 and 61.70; S7.12
(b) Vol.162,No.2, 347-362

W. PUESCHL
The Energy Constructions in Extended Dislocations
61.70
(b) Vol.162,No.2, 363-370

P. YANG, S.-S. JIANG, and D. FENG
The Behaviour of the First Fringes in the X-Ray Diffraction Topographs of Planar Defects. I. General Case
61.10 and 61.70
(a) Vol.122,No.2, 475-480

J. SERNA
Positronium Diffusion in Crystalline Polyethylene
61.40 and 61.70; 78.70; S12
(a) Vol.122,No.2, 489-494

A.A. NAZAROV, A.E. ROMANOV, and R.Z. VALIEV
On the Hierarchy of Dislocation Descriptions of Grain Boundary Structures
61.70
(a) Vol.122,No.2, 495-502

Y. LI and Q.P. KONG
The Concentration of Extended Jogs and Its Influence on the Creep Rate
62.20; 61.70
(a) Vol.122,No.2, 511-516

J. BARTOŠ and L. TESAŘ
Some Electrical and Optical Properties of Nickel-Related Deep Levels in Silicon
61.70 and 71.55; 73.40; S5.11
(a) Vol.122,No.2, 607-616

T. KAMINO, M. TOMITA, I. OHTSUKA, and H. SAKA
High Resolution Transmission Electron Microscopy Observation of the Motion of a Surface Dislocation on a Gold Particle
61.16; 61.70; S1.3
(a) Vol.122,No.2, K105-K108

J.D. STEPHENSON
Quasi-Realtime Characterization of Microchip Wafers Using Scanning WSR-Section Topography
61.70; S5.11
(a) Vol.122,No.2, K121-K122

61.80. Radiation damage and other irradiation effects
(for techniques of structure determination, see 61.10 to 61.16; for electron and ion impact phenomena, see 79.20)

V.G. BARYSHEVSKII and A.M. ZAITSEVA
Orientational Effects Due to Traveling of Fast Neutrons and Hard gamma-Quanta through Crystals
61.80; 78.70; S5.12
(b) Vol.157,No.1, 129-136

V. GEIST, D. STEPHAN, and D. FINK
The Effect of the Lateral and Longitudinal Spread of Fast Protons on Strain Production in GaP and GaAs
61.10 and 61.80; S7.11; S7.12
(a) Vol.117,No.1, 67-72

L.F. ZAKHARENKOV, V.V. KOZLOVSKII, and B.A. SHUSTROV
Transmutation Doping of Indium Phosphide and Gallium Arsenide Due to Protons and alpha-Particles
61.80; 71.55; S7.11; S7.12
(a) Vol.117,No.1, 85-90

J. SAID, H. JAOUEN, G. GHIBAUDO, and I. STOEMENOS
Electrical and Physical Investigation of Defect Annihilation in Arsenic Implanted Silicon
61.70; 61.80; S5.11
(a) Vol.117,No.1, 99-104

M.E. KASSEM, N.G. GOMAA, A.M. EL-KHATIB, and S.H. KANDIL
Fast Neutron Irradiation Effect and Thermal Properties of Doped Nonstoichiometric Lithium Potassium Sulphate Crystals
65.40; 61.80; S11
(a) Vol.117,No.1, 119-124

V. GRAEFE, D. SCHALCH, and A. SCHARMANN
Hydrogenation of Reactively Sputtered AlN_xO_y Films
61.80 and 78.65; S6
(a) Vol.117,No.1, K23-K28

KH.B. GEZALOV, A.M. GASANOV, A.A. GARIBOV, M.M. ALIEV, and N.I. MUSAEV
The Nature of Paramagnetic Centers in gamma-Irradiated Boron Oxides
61.80 and 76.30; S10.1
(a) Vol.117,No.1, K57-K60

P. MORETTI, P. THEVENARD, G. GODEFROY, R. SOMMERFELDT, P. HERTEL, and E. KRAETZIG
Waveguides in Barium Titanate by Helium Implantation
78.20; 61.80; S11.1
(a) Vol.117,No.1, K85-K88

N.A. KUDRYASHOV, S.V. PETROVSKII, V.M. SAMSONOV, and M.N. STRIKHANOV
Energy Loss Distribution of Relativistic Charged Particles in Crystals
61.80
(b) Vol.157,No.2, 531-546

V.YU. PETUKHOV, I.B. KHAIBULLIN, M.M. ZARIPOV, E. WIESER, R. GROETZSCHEL, and H. BARTSCH
Formation of Epitaxial $NiSi_2$ by High Dose Implantation and Rapid Thermal Annealing
61.80 and 68.55; S1.61
(a) Vol.117,No.2, 477-484

S.I. KHLUPIN and A.F. AKKERMAN
Radial Energy Transfer Density Distribution around the Fast Ion Tracks in Silicon and Germanium
61.80; S5.11; S5.12
(b) Vol.158,No.1, 63-66

A.F. AKKERMAN and V.E. FAIN
Calculation of Energy Loss Characteristics of Accelerated Ions Passing through Thin Targets
61.80; S5
(b) Vol.158,No.1, 67-72

L.A. GEVORGYAN and K.Z. ATSAGORTSYAN
Suppression of Transparency Effect and Enhancement of Particle Reflection in an Asymmetric Potential Well
61.80 and 71.10
(b) Vol.158,No.1, 213-222

A. BUKALUK
AES Depth Profile Studies of Interdiffusion in the Ag-Cu Bilayer and Multilayer Thin Films
66.30 and 68.48; 61.80; S1.2; S1.3
(a) Vol.118,No.1, 99-108

Y.Y. LOGINOV, V.M. LENCHENKO, P.D. BROWN, and J. WOODS
TEM Study of the Formation of Structural Defects in Ion Implanted Silicon at Depths of $h \lesssim R_p$ and $h \gtrsim 2R_p$
61.70; 61.80; S5.11
(a) Vol.118,No.1, 117-130

TENG-GE MA, XIAN-JUN YANG, JI-SHENG ZHANG, and PEI-HSIN TSIEN
Localization of the Boundaries in Laser Recrystallization of SOI with Antireflective Stripe
61.80 and 68.55; 73.60; S5.11
(a) Vol.118,No.1, 179-184

S. BAUNACK, S. LANDGRAF, and A. ZEHE
Electron Beam Effects Due to Bombardment of CaF_2-Layers Investigated by Auger Electron Spectroscopy
61.80 and 68.60; 79.20; S9.12
(a) Vol.118,No.1, 203-208

V.N. BRUDNYI and V.V. PESHEV
Electric Field Effect on the Temperature Dependence of the (E_C-0.33eV) Center Introduction Rate under gamma-Irradiation ...
71.55; 61.80; S7.12
(a) Vol.118,No.1, 219-224

A.A. KOLB, I.G. MEGELA, A.P. BUTURLAKIN, and D.B. GOYER
Low-Temperature Annealing of Radiation Defects in Electron-Irradiated Gallium Phosphide
61.80 and 78.30; S7.11
(a) Vol.118,No.1, K9-K12

N.A. DOUMANIS and CH.G. TRIKALINOS
On the Transverse Kinetic Energy of Relativistic Channeled Particles
61.80; 79.20
(b) Vol.158,No.2, K101-K104

V.V. BOGDANOV, V.V. ZAKURKIN, D.V. PETROV, and YU.P. PENKOV
Investigation of Proton-Irradiated Platinum by Moessbauer Spectroscopy
61.80 and 76.80; S1.3
(b) Vol.158,No.2, K105-K110

H. REUTHER, E. RICHTER, and I.S. SZUECS
CEMS Measurements on Ion Implanted Tool Steels
61.80 and 76.80; S1.1
(a) Vol.118,No.2, 409-414

K.A. AL-SALEH, I.J. JABR, and N.S. SALEH
A Comparison of Ion Beam Induced Atomic Mixing Kinetics of Ti/Si, Fe/Si, and Ni/Si Systems. In Situ Electrical ...
61.80 and 73.40; S1.1; S1.2; S5.11
(a) Vol.118,No.2, 467-472

P.H. DIONISIO, B.A.S. DE BARROS, JR., and I.J.R. BAUMVOL
A Moessbauer Spectroscopic Study of the High-Temperature Oxidation Behaviour of an Austenitic Stainless Steel ...
61.80 and 68.20; 76.80; S1.1
(a) Vol.118,No.2, 473-478

T. TSVETKOVA
Modification Effect of Ni Ion Implantation on the Electrical Properties of Vitreous As_2Se_3
61.80 and 72.80; S8
(a) Vol.118,No.2, K83-K86

Z.T. SERIKBAEVA, T.M. ZHANTIKIN, and A.K. ZHETBAEV
Moessbauer Study of 30 MeV Proton Dose Dependence of Radiation Damage in Disordered Ni_3Fe
61.80 and 76.80; S1.1
(a) Vol.118,No.2, K99-K102

T. TSVETKOVA, B. AMOV, E. VATEVA, and V. AVERYANOV
Ion Implantation Induced Modification of Vitreous Chalcogenides
73.60; 61.80; 78.65; S8.12
(a) Vol.119,No.1, 107-112

M.KH. KHOKONOV and H. NITTA
Channeling of Antiprotons and Electrons
61.80; S5.11
(b) Vol.159,No.2, 589-596

M. ABBAS, H. KERKOW, and R. WEDELL
Heavy Ion Channeling Implantation Processes in Single Crystals
61.80; S5.11
(b) Vol.159,No.2, 597-608

C.H. WOO and B.N. SINGH
The Concept of Production Bias and Its Possible Role in Defect Accumulation under Cascade Damage Conditions
61.80; 61.70
(b) Vol.159,No.2, 609-616

A.M. PRITULOV, A.P. SURZHIKOV, V.A. KOZHEMYAKIN, YU.N. AFANASIEV, A.P. VORONIN, O.S. GRIBKOV, and G.R. KARAGEDOV
Radiation-Thermal Packing of Lithium Ferrite Compacts
61.40 and 61.80; S11.2
(a) Vol.119,No.2, 417-422

V.G. CHUDINOV, B.N. GOSHCHITSKII, N.V. MOSEEV, and V.V. ANDREEV
Kinetics of Radiation-Induced Segregation of Impurities in F.C.C. Metals at the Thermal Stage of the Collision Cascade
61.80 and 64.75; S1.1; S4
(a) Vol.119,No.2, 437-442

P.M. OSSI
Crystal-Glass Phase Transition in Ion Irradiated Binary Systems
61.80 and 64.75; 61.40; S1
(a) Vol.119,No.2, 463-470

V.N. BOGDANOV, I.M. BROVCHENKO, L.V. MAKSIMOV, A.R. SILIN, and O.V. YANUSH
Spectroscopic Optical and Acoustic Investigations of Radiation Modified Silica Glasses
61.40 and 78.20; 61.80; 62.65; S10.1
(a) Vol.119,No.2, 621-630

S.N. BARILO, V.I. GATALSKAYA, G.V. GATALSKII, A.P. GES,
S.A. GURETSKII, F.P. KORSHUNOV, L.A. KUROCHKIN et al.
Electron Irradiation of Single-Crystalline $Bi_2Sr_2CaCu_2O_{8+\delta}$
61.80 and 74.70; S10.15
(a) Vol.119,No.2, K151-K154

I.CH. AKOPYAN, B.V. NOVIKOV, R.V. GRIGOREV, S.R. GRIGOREV,
H. ROEPPISCHER, and N. STEIN
Photoluminescence Study of Na and Li Acceptors in ZnSe
78.55; 61.80; S8.12
(a) Vol.119,No.2, K167-K172

F.G. REQUEJO, A.G. BIBILONI, H. SAITOVITCH, and P.R.J. SILVA
Study of Point Defects in Irradiated and Reduced MoO_3 through Hyperfine Interactions on ^{99}Mo
61.70; 61.80; S10.1
(a) Vol.120,No.1, 105-118

S. TAKAMURA and M. KOBIYAMA
New Peak in Internal Friction after Deformation of Irradiated Al Alloys
61.80 and 62.40; S4
(a) Vol.120,No.1, 119-124

N.S. SALEH, K.A. AL-SALEH, and A.M. AL-SAIE
Kinetics of Ion Beam Induced Mixing in Au/Si System
61.80 and 68.55; 73.60; S1.3; S5.11
(a) Vol.120,No.1, 169-174

A.M. MAZZONE
Monte-Carlo Simulation of the Backscattering Effects of Loop Dislocations in a Silicon Crystal
61.70; 61.80; S5.11
(a) Vol.120,No.2, 379-386

P.V. PAVLOV, A.P. PAVLOV, and D.I. TETELBAUM
The Effect of Ion Implantation on the Substructure of Metal Films
68.55; 61.80; S1.2; S4
(a) Vol.120,No.2, 441-446

D.B. TRAN THOAI and E. ZEITLER
Inelastic Scattering of Fast Electrons by Thin Metal Slabs
71.45 and 73.20; 61.80; S4
(a) Vol.120,No.2, 467-474

V.V. ARTAMONOV, V.S. LYSENKO, A.N. NAZAROV, B.D. NECHIPORUK,
V.V. STRELCHUK, and M.YA. VALAKH
Radio-Frequency Plasma Treatment and Thermal Annealing in Implanted Si. Raman Study
61.80 and 68.55; 78.30; S5.11
(a) Vol.120,No.2, 475-484

T.S. SHI, A. DE VEIRMAN, J. VAN LANDUYT, G.Q. YANG, and C.L. LIN
TEM Study of the Molecular Effect in Phosphorus Implanted Silicon
61.80; S5.11
(a) Vol.120,No.2, K131-K132

G.S. MYAKENKAYA, G.L. GUTSEV, N.P. AFANASEVA, V.A. EVSEEV, and R.F. KONOPLEVA
Study of Lithium Interaction with Lattice Defects in Silicon
61.70 and 61.80; 72.20; S5.11
(b) Vol.161,No.1, 91-104

L.L. BALASHOVA, N.M. KABACHNIK, and V.N. KONDRATEV
Charge Asymmetry in the Dependence of Stopping on Impact Parameter
61.80; 78.70; 79.90; S5.11
(b) Vol.161,No.1, 113-122

M. DANZIGER and G. ANDRASSY
Experimental and Theoretical Study of Heavy Ion Irradiation Processes in Polymer Materials
61.80 and 68.60; 73.60; S12
(b) Vol.161,No.1, 223-232

GANGANG CAI, SHOUSHAN WANG, and JIAN ZHANG
Radiation Damage and Recovery of LiF:Mg,Ti-TLD at the Dose of 10^6 Gy Gamma Rays
61.80; 78.60; S9.11
(a) Vol.121,No.1, 57-62

E.V. KABANOVA, S.N. KULKOV, B.P. GRITZENKO, B.Y. YAKOVLEV, and P.P. PERSTNEV
The Change of Optical and Structural Properties of MgO by the Exposition of Laser Pulse Irradiation
61.80; 78.60; S10.1
(a) Vol.121,No.1, 105-110

A.P. TYUTNEV, S.G. BOEV, V.N. ABRAMOV, E.A. KUZMENKOV, and A.V. VANNIKOV
Track Effects in Radiation Induced Conductivity of Polymers
61.80 and 72.20; S12
(a) Vol.121,No.1, 205-212

TH. DITTRICH
Laser Induced Generation of a Micro-Crack Structure on (111) GaP
61.80; S7.11
(a) Vol.121,No.1, K15-K16

A. POGGI, M. MERLI, E. SUSI, R. ANGELUCCI, and A. SARDO
Effects of Silicon Ion-Beam Mixing on p^+/n Diodes: DLTS Analyses of the Induced Defects
61.80 and 85; S5.11
(a) Vol.121,No.1, K135-K140

V.A. ARUTYUNOV, N.A. KUDRYASHOV, O.A. MISHIN, V.M. SAMSONOV, M.N. STRIKHANOV, and YU.A. SYTSKO
Synchrotron Radiation, Radiative Losses, and Polarization Effects for Relativistic Beams in Bent Crystals
61.80 and 78.70; S5.11
(b) Vol.161,No.2, 483-500

K. LENKEIT, CH. TRIKALINOS, L.L. BALASHOVA, N.M. KABACHNIK, and V.I. SHULGA
Monte Carlo Simulation Analysis of Proton-Energy Spectra for Axial Channeling in Silicon in the Intermediate Energy Region
61.80; S5.11
(b) Vol.161,No.2, 513-524

A.CH. LUSHCHIK and A.G. FRORIP
Thermalized and Hot Interstitial Halogen Ions in Alkali Halides
61.70 and 78.55; 61.80; 76.30; S9.11
(b) Vol.161,No.2, 525-536

YA.M. STRELNIKER and O.S. ZINETS
Inelastic Scattering of Low-Energy Electrons by Electrons Localized in Surface Potential Wells
73.20; 61.80
(b) Vol.161,No.2, 597-602

K.K. BOURDELLE, N.G. CHECHENIN, A.S. AKHMANOV, A.YU. POROIKOV, and A.V. SUVOROV
Melting and Damage Production in Silicon Carbide under Pulsed Laser Irradiation
61.80; 64.75; S6
(a) Vol.121,No.2, 399-406

R. HEIDINGER, H. APPEL, G. THEN, and W.-G. THIES
Observation of Radiation Damage in HfV_2 and Its Hybrides by TDPAC
61.80; 64.80; 74.70; 76.80; S1.2
(a) Vol.121,No.2, 445-454

K. LENKEIT, A. PIRRWITZ, A.I. NIKIFOROV, B.Z. KANTER, S.I. STENIN, and V.P. POPOV
Investigations of MBE Grown Superlattices with Sb Delta-Layers in Silicon by Rutherford Backscattering Spectrometry
68.55; 61.80; S5.11
(a) Vol.121,No.2, 523-532

TH. DITTRICH and V.I. TIMOSHENKO
Laser Induced Modification of n-GaAs below the Classical Melting Threshold
61.80 and 78.65; S7.12
(a) Vol.121,No.2, 547-554

N.V. BRILLIANTOV, A.I. RUDENKO, YU.V. SHCHERBAKOV, and V.V. ZVEREV
Anomalous Low Temperature Annealing of Electron Irradiated Lithium-Doped Silicon Solar Cells
61.80; 71.55; 85; S5.11
(a) Vol.121,No.2, K155-K158

M. BAUER, J. MAYER, M. HOHENSTEIN, and M. WILKENS
Quantitative Studies on the Amorphization of Quasicrystals by Means of High Voltage Electron Microscopy
61.40 and 61.80; 64.70; S1.2; S4
(a) Vol.122,No.1, 79-86

MOTOKO KWETE, D. SEGERS, M. DORIKENS, L. DORIKENS-VANPRAET, and P. CLAUWS
Positron Annihilation Study of Defects Created in Silicon Irradiated with Electrons of High Energy
61.80; 78.70; S5.11
(a) Vol.122,No.1, 129-138

E. SOBESLAVSKY and W. SKORUPA
Peculiarities in the Modeling of High Dose Implantation of Nitrogen on Silicon Targets
61.80; S5.11
(a) Vol.122,No.1, K19-K24

S.S. SIMEONOV, E.I. KAFEDJIISKA, and A.L. GUERASSIMOV
Deep Levels in n-Type Si Substrates Subjected to Ion Cleaning
71.55; 61.80; S5.11
(a) Vol.122,No.2, K147-K150

62. Mechanical and Acoustical Properties of Condensed Matter
(see also 61.70. Defects in crystals, 68.30. Surfaces and interfaces)

62.10. Mechanical properties of liquids
(for viscosity of liquids, see also 66.20)

I. AVRAMOV
Pressure Dependence of Shear Viscosity
62.10; 64.60
(a) Vol.120,No.1, 133-138

62.20. Mechanical properties of solids
(related to microscopic structure)

A.K. SHARMA and S.C. GOYAL
Second-Order Pressure Derivatives of SOE Constants of Fluorite Structure Solids
62.20; S9.12
(b) Vol.157,No.1, 145-150

A.D. POGREBNYAK
Metastable States and Structural Phase Changes in Metals and Alloys Exposed to High Power Pulsed Ion Beams (Review Article)
64.70 and 79.20; 62.20; 68.45; 68.55; S1.1; S1.2
(a) Vol.117,No.1, 17-52

P. FELTHAM and N. KAUSER
Solid-Solution Hardening
62.20; S1.2
(a) Vol.117,No.1, 135-140

CH. GENZEL and A. GENZEL
Shear Stress Distribution in Crystals Induced by Mechanical Surface Load
62.20; 61.70
(a) Vol.117,No.1, 141-154

A. SAKAKIBARA, T. KANADANI, M. YAMADA, and M. OHTA
Effect of Surface on the Reversion of Age-Hardened Al-Zn Alloy and Role of Grain Boundaries
62.20; S4
(a) Vol.117,No.1, K19-K22

Z.M. FARID, S.A. MAHMOUD, A.R. ALI, and N. EL-NAQUIB
On the Transition in Creep Behaviour during Primary Creep of Cu-Zn Alloys
62.20; S1.2
(a) Vol.117,No.2, 437-446

T. KANADANI and A. SAKAKIBARA
Effect of Soft Surface Layer on the Fatigue Strength of Age-Hardened Al-2mass%Cu Alloy
62.20; S4
(a) Vol.117,No.2, K97-K98

S.K. MISHRA and T.N. SINGH
Lattice Dynamics of Alkali Metals
63.20; 62.20; S2
(b) Vol.158,No.1, 153-164

V.L. INDENBOM and V.M. KAGANER
X-Ray Analysis of Internal Stresses in Crystals.
I. General Equations
61.70; 62.20
(a) Vol.118,No.1, 71-84

T. OHGAKU and N. TAKEUCHI
Study on Dislocation-Impurity Interaction by the Blaha Effect
62.20; 61.70; S9.11
(a) Vol.118,No.1, 153-160

Y. DAI, S.M. LIU, and Q.P. KONG
The Changes of Relaxation Parameters of Grain Boundaries in the Course of Constant Rate Deformation
62.20; 62.40; S1.2
(a) Vol.118,No.1, K21-K26

B.A. GREENBERG, M.I. KATSNELSON, V.G. KORESHKOV, YU.N. OSETSKII, G.V. PESCHANSKIKH, A.V. TREFILOV, YU.F. SHAMANAEV et al.
On the Possibility of Describing Lattice Properties of Iridium in Terms of Pseudopotential Theory
63.20 and 71.10; 61.70; 62.20; S1
(b) Vol.158,No.2, 441-456

Q.P. KONG and Y. DAI
The Changes of Relaxation Parameters of Grain Boundaries during Creep in Pure Copper
62.20 and 62.40; S1.2
(a) Vol.118,No.2, 431-440

M. HAMERSKÝ, P. LUKÁČ, and Z. TROJANOVÁ
Stress Relaxation in Cd-Zn Polycrystals
62.20; S1.2
(a) Vol.118,No.2, 455-460

S.U. JEN, M.C. CHIEN, and C.Y. LEE
Viscous Flow in Amorphous Alloys
61.40 and 64.70; 62.20; S1.1; S1.2
(a) Vol.119,No.1, 35-40

S.N. BISWAS, M.J.P. MURINGER, and C.A. TEN SELDAM
Pressure Dependence of the Shear Modulus of Polycrystalline Silver from 77 to 298 K and up to 253 MPa
62.20; S1.3
(a) Vol.119,No.1, 63-70

H.-A. KUHN and H.G. SOCKEL
Contributions of the Different Phases of Two Nickel-Base Superalloys to the Elastic Behaviour in a Wide Temperature Range
62.20; S1.2; S1.61
(a) Vol.119,No.1, 93-106

J.C. UPADHYAYA, S. LEHRI, D.K. SHARMA, and S.C. UPADHYAYA
Angular Force Models in the Lattice Dynamics of Cubic Systems
63.20; 62.20; S1
(b) Vol.159,No.2, 659-674

W. WASSERBAECH
Third-Order Elastic Constants of a Cubic Quasi-Isotropic Solid
62.20; S1
(b) Vol.159,No.2, 689-698

H.C. GUPTA
Shear Elastic Constant c_{44} of Hexagonal Crystals
62.20
(b) Vol.159,No.2, K47-K50

G.A. MALYGIN
Dislocation Density Evolution Equation and Strain Hardening of F.C.C. Crystals
61.70; 62.20; S4
(a) Vol.119,No.2, 423-436

A.S. TAHA and F.H. HAMMAD
The Applications of the Hall-Petch Relation to Microhardness Measurements on Al, Cu, Al-MD 105, and Al-Cu Alloys
62.20; 61.70; 64.80; S1.2; S4
(a) Vol.119,No.2, 455-462

G.A. HASSAN and H.H. HASSAN
Strain Hardening Characteristics of 99.70 % Pure Silver
62.20; S1.3
(a) Vol.119,No.2, 471-478

J. POKLUDA and P. ŠANDERA
On the Relation of Ideal Strength to Other Physical Properties
62.20; S1.1; S1.2
(b) Vol.160,No.1, 89-96

Y.B. XIA, Z.G. WANG, and R.H. WANG
Secondary Hardening and Dislocation Evolution in Low Cycle Fatigue of Polycrystalline Aluminium
62.20; 61.70; S4
(a) Vol.120,No.1, 125-132

Z. WOKULSKI
Mechanical Properties of TiN Whiskers
62.20 and 68.70; S1.62
(a) Vol.120,No.1, 175-184

S. KAHL, R. NEUHAUS, H. PINKVOS, and CH. SCHWINK
The Influence of Tensile Deformation and Atomic Short-Range Order on the Susceptibility of CuMn Spin Glasses
62.20 and 75.40; 61.70; S1.2
(a) Vol.120,No.1, 221-232

M. TANIBAYASHI
A Method to Determine the Internal Stress and the Stress Exponent m of the Dislocation Velocity
62.20; S9.11
(a) Vol.120,No.1, K19-K24

M.A. KENAWY, M.S. SAKR, E.M. SAKR, H.A. ZAYED, and N.O. MOURAD
Effect of Dissolution of the beta-Phase on the Work Hardening Characteristics of Al-4.5wt%Mg Alloy
62.20 and 64.75; S4
(a) Vol.120,No.1, K31-K36

R.V. GOPALA RAO and U. BANDYOPADHYAY
An Evaluation of Structural and Transport Properties of Molten Potassium-Rubidium Alloy
61.40; 62.20; 66.30; S2
(b) Vol.160,No.2, 403-414

G.J. VÁZQUEZ
Specific Heat and Elastic Constants for Martensitic Forms of Lithium
62.20 and 65.40; S2
(b) Vol.160,No.2, K89-K92

L.F. CHEN, S. PATU, and C.X. SHI
Effect of Phosphorus Ion Implantation on Fatigue Behaviour of Polycrystalline Nickel
62.20; 61.70; S1.2
(a) Vol.120,No.2, 429-440

T. KANOMATA, T. KAWASHIMA, T. KANEKO, H. TAKAHASHI, and T. MŌRI
Pressure Effect on the Lattice Parameters of RMn_2X_2 (R=Y,La,Nd and X=Ge,Si)
61.60 and 62.20; S8.16
(a) Vol.120,No.2, K117-K120

M.Z. BUTT, Z. RAFI, and M.A. KHAN
Low-Temperature Anomaly in the Strength of Copper-Manganese Alloy Single Crystals
62.20; S1.2
(a) Vol.120,No.2, K149-K154

V.P. GUPTA and S.K. SIPANI
The Analysis of Rydberg Potential and van der Waals Interactions in Alkali Halides
63.90; 62.20; 65.70; S9.11
(b) Vol.161,No.1, 131-144

G. SINGH
Strength and Stability of Silver
62.20; S1.3
(b) Vol.161,No.1, 145-152

S.V. FOMICHEV, M.I. KATSNELSON, V.G. KORESHKOV, and A.V. TREFILOV
On the Effect of Three-Body Interactions and Proximity of the Fermi Level to the Brillouin Zone Faces on Elastic ...
62.20 and 71.25; S2; S3; S4
(b) Vol.161,No.1, 153-164

H. GUENTHER
Tachyons in a Crystalline Structure
62.20; 61.70
(b) Vol.161,No.1, 179-190

Ş. ERKOÇ
Structural Stability and Energetics of F.C.C. Metal Microclusters: Empirical Many-Body Potential Energy ...
63.10; 61.55; 62.20; S1.2; S1.3; S4
(b) Vol.161,No.1, 211-216

SIE CHIN TJONG
Dislocation Microstructures in Fatigued Fe-30Mn-10Al-0.4C Alloy
62.20; 61.70; S1.1
(a) Vol.121,No.1, 119-128

S. DIXIT and R. NIWAS
Cohesive Energies, Equation of State, and Thermoelastic Properties of Mixed Alkali Halide Crystals
62.20 and 64.30; 65.70; S9.11
(b) Vol.161,No.2, 583-596

E.V. TATYANIN and V.G. KURDYUMOV
The Nucleation of the Deformation Induced Amorphous Phase at Twin Boundaries in TiNi Alloy
61.70 and 62.20; 64.70; S1.2
(a) Vol.121,No.2, 455-460

M.A. KENAWY, M.S. SAKR, E.M. SAKR, H.A. ZAYED, and N.O. MOURAD
Transient and Steady-State Creep of Al-4.5wt%Mg during Phase Transformation
62.20; S4
(a) Vol.121,No.2, 467-474

Y. NISHINO, T. NISHIKAWA, and S. ASANO
Strain Aging in Czochralski-Grown Silicon Crystals
61.70; 62.20; S5.11
(a) Vol.122,No.1, 163-170

J. YAN, Z.G. WANG, T.Y. ZHANG, Y.Y. GUO, and C.H. SHIH
Latent Hardening in Aluminum Single Crystals under Cyclic Loading
62.20; S4
(a) Vol.122,No.1, 195-206

P. SISODIA and M.P. VERMA
A Graphical Interpretation of the Bounds of the Cubic Polycrystalline Shear Moduli. Application to Hexahalometallates
62.20
(a) Vol.122,No.1, 207-214

V.N. NOVIKOV
Microcreep of Metallic Glasses at Room Temperature
62.20; 61.40; S1.1; S1.2
(a) Vol.122,No.1, 215-220

B. FERNÁNDEZ and S.M. WASIM
Sound Velocities and Elastic Moduli in $CuInTe_2$ and $CuInSe_2$
62.20 and 62.65; S8.16
(a) Vol.122,No.1, 235-242

C.Q. TANG, G.Y. LI, and J.Y. TANG
Positron Lifetime Study on Thermal Fatigue in 5CrNiMo Steel
61.70 and 78.70; 62.20; S1.1
(a) Vol.122,No.1, K11-K14

Y. LI and Q.P. KONG
The Concentration of Extended Jogs and Its Influence on the Creep Rate
62.20; 61.70
(a) Vol.122,No.2, 511-516

P. SISODIA and M.P. VERMA
Polycrystalline Elastic Moduli of Some Hexagonal and Tetragonal Materials
62.20; S1; S8; S10
(a) Vol.122,No.2, 525-534

T. ARAI and S. OKUDA
Relaxation Spectrum Analysis of the Bordoni Peak in Cu by Nonlinear Least-Squares Fitting
62.40; 62.20; S1.2
(a) Vol.122,No.2, 535-544

A. KARA, J.J. COUDERC, C. LEVADE, and G. VANDERSCHAEVE
Memory Photoelastic Effects in ZnS Sphalerite
62.20 and 78.20; S8.11
(a) Vol.122,No.2, 545-554

62.30. Mechanical and elastic waves

U. PIETSCH
Lattice Compression from Conduction Band Electrons in As-Implanted Silicon
61.50; 62.30; S5.11
(b) Vol.157,No.2, K73-K76

V.E. GUSEV
Generation of Picosecond Acoustic Pulses by Laser Radiation in Piezosemiconductors
62.30 and 78.20
(b) Vol.158,No.1, 367-382

TS.S. VELINOV and K.P. BRANSALOV
Interference Analysis of Photothermal Generation of Waves in Layered Structures
62.30 and 68.20; 78.20
(a) Vol.121,No.2, 555-562

62.40. Anelasticity, internal friction, and mechanical resonances

For thermomechanical effects, see 65.70
For magnetomechanical effects, see 75.80
For piezoelectric effects, see 77.60
For elasto-optical effects, see 78.20

XIAN-FANG ZHU, LI-DE ZHANG, and ZI-YING JIANG
Effect of Yttrium on Grain Boundary Internal Friction in Commericial Aluminium
62.40; 61.70; S4
(a) Vol.117,No.1, 105-112

H. MIZUBAYASHI, H. AKINAGA, and K. TAKITA
Low-Temperature Elastic Properties of $Nd_{1+x}Ba_{2-x}Cu_3O_{7-\delta}$
62.40; 74.70; S10.15
(a) Vol.117,No.1, 125-134

Y. DAI, S.M. LIU, and Q.P. KONG
The Changes of Relaxation Parameters of Grain Boundaries in the Course of Constant Rate Deformation
62.20; 62.40; S1.2
(a) Vol.118,No.1, K21-K26

Q.P. KONG and Y. DAI
The Changes of Relaxation Parameters of Grain Boundaries during Creep in Pure Copper
62.20 and 62.40; S1.2
(a) Vol.118,No.2, 431-440

S. PILECKI, V.I. IVANOV, and A.B. LEBEDEV
Amplitude-Dependent Internal Friction in Silicon Bronze at Low Temperatures
62.40; S1.2
(a) Vol.119,No.1, 87-92

V.I. MITROKHIN, S.I. REMBEZA, V.V. SVIRIDOV, and N.P. YAROSLAVTSEV
Acousting Probing of Deep Centres in III-V Semiconductors
71.55; 62.40; S7.1
(a) Vol.119,No.2, 535-544

S. TAKAMURA and M. KOBIYAMA
New Peak in Internal Friction after Deformation of Irradiated Al Alloys
61.80 and 62.40; S4
(a) Vol.120,No.1, 119-124

I.T. SUPRUN
Determination of the Dislocation Structure Parameters from the Data on Amplitude Dependence of Internal Friction
61.70; 62.40
(a) Vol.120,No.2, 363-370

M.L. NÓ, A. OLEAGA, C. ESNOUF, and J. SAN JUAN
Internal Friction at Medium Temperatures in High Purity Magnesium
62.40; S3
(a) Vol.120,No.2, 419-428

J. PEREZ, S. ETIENNE, and J. TATIBOUËT
Determination of Glass Transition Temperature by Internal Friction Measurements
61.40 and 64.60; 62.40
(a) Vol.121,No.1, 129-138

T.S. KÊ and Q.F. FANG
Evolution of the Anomalously Amplitude Dependent Internal Friction Peaks in Al-Mg Solid Solutions and the Discovery ...
62.40; S4
(a) Vol.121,No.1, 139-148

S.P. WU, L.D. ZHANG, G.Y. TIAN, and J.Z. LU
High-Temperature Internal Friction Peaks in Dilute Al-Sn Polycrystals
62.40; S4
(a) Vol.121,No.1, 149-156

G.Y. TIAN, J.Z. LU, S.P. WU, and L.D. ZHANG
High-Temperature Internal Friction Peak in Al-0.008wt%Sn Single Crystals
62.40; S4
(a) Vol.121,No.2, 475-482

SIMING HE, MINGZHI LIU, and LINCHANG TANG
Internal Friction Peak Associated with Hydrogen Diffusion in Amorphous $Pd_{77.5}Ag_6Si_{16.5}$ Alloy
62.40; 66.30; S1.3
(a) Vol.121,No.2, K167-K170

G. D ANNA and W. BENOIT
Dislocation Internal Friction Investigation in High Purity Cold Worked Tantalum
61.70 and 62.40; S1
(a) Vol.122,No.1, 153-162

H. MIZUBAYASHI, H. AGARI, and S. OKUDA
Hydrogen Internal Friction Peak in Amorphous $Cu_{50}Ti_{50}$
61.40 and 62.40; S1.2
(a) Vol.122,No.1, 221-234

Q. TAN and T.S. KÊ (GE TINGSUI)
Activation Energy Associated with the Nonlinear Internal Friction Peaks (P1′ and P1′′ Peaks) in Cold-Worked and ...
62.40; S4
(a) Vol.122,No.1, K25-K30

FANG YUAN, YENING WANG, and HIUMIN SHEN
Study of the Incommensurate Phase in NiTi(Fe) Alloys
64.70; 62.40; 64.80; S1.2
(a) Vol.122,No.2, 517-524

T. ARAI and S. OKUDA
Relaxation Spectrum Analysis of the Bordoni Peak in Cu by Nonlinear Least-Squares Fitting
62.40; 62.20; S1.2
(a) Vol.122,No.2, 535-544

62.65. Acoustical properties of solids

For magnetoacoustic effects in metals, see 75.80
For acoustoelectric effects, see 72.50
For acoust-optical effects, see 78.20

M. DEY (BOSE) and S. GHOSH
Amplification of Acoustic Waves in Magnetised High Resistivity Piezoelectric Semiconductors. Effect of Nonuniform ...
62.65 and 72.50; S8.11
(b) Vol.157,No.1, 159-166

R. SANGHVI and S. GHOSH
Amplitude Modulation and Demodulation of Electromagnetic Waves in Magnetised Material with Strain-Dependent Dielectric..
72.30; 62.65; S11.1
(a) Vol.118,No.1, 235-240

F. MICHARD and A. ZAREMBOWITCH
Changes in the Propagation of Ultrasonic Waves in Silica Aerogels
62.65; 78.35; S10
(a) Vol.118,No.2, 461-466

V.N. BOGDANOV, I.M. BROVCHENKO, L.V. MAKSIMOV, A.R. SILIN, and O.V. YANUSH
Spectroscopic Optical and Acoustic Investigations of Radiation Modified Silica Glasses
61.40 and 78.20; 61.80; 62.65; S10.1
(a) Vol.119,No.2, 621-630

B. FERNÁNDEZ and S.M. WASIM
Sound Velocities and Elastic Moduli in $CuInTe_2$ and $CuInSe_2$
62.20 and 62.65; S8.16
(a) Vol.122,No.1, 235-242

E.M. GOLOLOBOV, E.L. MAGER, and Z.V. MESHEVICH
Ultrasonic Investigations of Structural Instabilities of ZrV_2 under High Hydrostatic Pressures
62.65 and 64.70; S1.2
(a) Vol.122,No.2, K123-K128

62.80. Ultrasonic relaxation

(see also 74.30. Ultrasonic attenuation in superconductors)

A.D. BELYAEV, V.E. RODIONOV, and N.S. CHERNAYA
Effect of Ultrasound Oscillations on Electroluminescent Properties of MISIM Structures
78.60; 62.80; 73.40; S8.11
(a) Vol.118,No.1, 327-332

63. Lattice Dynamics and Crystal Statistics

(see also 65. Thermal properties,
66.70. Thermal conduction,
68.30. Dynamics of surface and interface vibrations,
78.30. Infrared and Raman spectra)

63.10. General theory

T. TAKAI, D.K. CHOI, Y.T. THATHACHARI, T. HALICIOGLU, and W.A. TILLER
A Potential Energy Function for GaAs Systems
63.10; 68.20; S7.12
(b) Vol.157,No.1, K13-K18

T. TAKAI, T. HALICIOGLU, and W.A. TILLER
A Model Potential Energy Function for Systems Containing Al, Ga, and As Atoms
63.10; 68.20; S7.15
(b) Vol.157,No.2, K77-K82

Ş. ERKOÇ
Structural Stability and Energetics of F.C.C. Metal Microclusters: Empirical Many-Body Potential Energy ...
63.10; 61.55; 62.20; S1.2; S1.3; S4
(b) Vol.161,No.1, 211-216

63.20. Phonons and vibrations in crystal lattices

M.H. BROOKER
A Search for a Phase Transition in $Sr(NO_3)_2$ by Raman Spectroscopy. Evidence for Partial Disorder
64.60 and 78.30; 63.20; 63.50; S11.1
(b) Vol.157,No.1, 167-182

I.N. YURIICHUK, S.V. MELNICHUK, and V.A. SHENDEROVSKII
Local Modes in CdTe Doped with Transition Element Impurities
63.20; S8.13
(b) Vol.157,No.1, K19-K22

J.P. SRIVASTAVA, ASITA KULSHRESHTHA, and H. RAUH
FIR Reflectivity and Vibrational Spectra of $(CH_3NH_3)_2ZnCl_4$ Crystals
78.30; 63.20; S12
(a) Vol.117,No.1, K77-K80

Z.A. GURSKII and YA.G. CHUSHAK
Lattice Dynamics of Binary Alloys
63.20; S2
(b) Vol.157,No.2, 557-566

K. KRÁL, B. HEJDA, and Z. KHÁS
Hot LO-Phonon Quasi-Steady State Distribution Function in Polar Semiconductors
63.20 and 72.10; S7.1; S8.1
(b) Vol.157,No.2, 667-676

R.M. SARDARLY, R.A. ALIEV, E.M. KERIMOVA, N.A. BAKHYSHOV, and I.SH. SADYKHOV
Rearrangement Features of the Vibrational Spectra of $TlInS_{2x}Se_{2(1-x)}$ and $TlGa_xIn_{1-x}Se_2$ Solid Solutions
63.20 and 78.30; 64.75; S8.16
(b) Vol.158,No.1, 73-80

V.V ZUBRITSKII and R.P. PURLYS
The Relation between the Orientation of Streamer Discharges and Anisotropy of Thermal Vibrations of Atoms in II-VI ...
63.20; 71.55; 72.20; 77.50; S8.11
(b) Vol.158,No.1, 91-102

M. SIESKIND, D. AYACHOUR, J.-C. MERLE, and J.-C. BOULOU
Crystal Growth, Raman and Infrared Spectra of BaFI Single Crystals
63.20 and 78.30; S9.16
(b) Vol.158,No.1, 103-112

L.P. WANG, L.Z. Zhang, W.X. ZHU, X.T. LU, and G.G. QIN
The Behaviour of the Hydrogen-Localized Vibrational Mode in Proton-Implanted GaAs
63.20 and 78.50; S7.12
(b) Vol.158,No.1, 113-118

S.K. MISHRA and T.N. SINGH
Lattice Dynamics of Alkali Metals
63.20; 62.20; S2
(b) Vol.158,No.1, 153-164

I. YANETKA
The One-Dimensional Plasma Model
63.20 and 71.45
(b) Vol.158,No.1, 181-192

N.M. GASANLY, S.A. EL-HAKIM, L.G. GASANOVA, and A.Z. MAGOMEDOV
Lattice Vibrations in $CuIn_5S_8$ Crystals
63.20 and 78.30; S8.16
(b) Vol.158,No.1, K1-K4

N.M. GASANLY, A.G. GUSEINOV, E.A. ASLANOV, and S.A. EL-HAMID
Infrared Reflection Spectra of $Cu_3B_5^{III}C_9^{VI}$ Single Crystals
78.30; 63.20; S8.16
(b) Vol.158,No.1, K85-K88

N.S. ORLOVA, G.A. TURTSEVICH, O.E. KOCHKARIK, and I.V. BODNAR
X-Ray Study of Debye-Waller Factors and Debye Characteristic Temperatures in $AgInC_2^{IV}$ Compounds
63.20; S8.16
(a) Vol.118,No.1, 141-146

N. SINGH and S.P. SINGH
Lattice Dynamics of Ca and Sr in B.C.C. and F.C.C. Phases. A Dynamical Treatment
63.20; S3
(b) Vol.158,No.2, 433-440

B.A. GREENBERG, M.I. KATSNELSON, V.G. KORESHKOV, YU.N. OSETSKII, G.V. PESCHANSKIKH, A.V. TREFILOV, YU.F. SHAMANAEV et al.
On the Possibility of Describing Lattice Properties of Iridium in Terms of Pseudopotential Theory
63.20 and 71.10; 61.70; 62.20; S1
(b) Vol.158,No.2, 441-456

E. HAQUE and A.K.M.A. ISLAM
Hydrides and Deuterides of Lithium and Sodium. I. Model Potentials and Their Use in Perfect Lattice
63.20; S2
(b) Vol.158,No.2, 457-468

I.V. BODNAR, G.F. SMIRNOVA, A.G. KAROZA, and A.P. CHERNYAKOVA
Vibrational Spectra of $CuGaS_2$ and $CuGaSe_2$ Compounds and $CuGaS_{2x}Se_{2(1-x)}$ Solid Solutions
63.20; 64.75; 78.30; S8.16
(b) Vol.158,No.2, 469-474

M. MAĆKOWIAK, N. WEIDEN, and A. WEISS
Phase Transitions and Molecular Dynamics in Tris(Methylammonium) Nonabromodiantimonate(III) and Its ...
63.20 and 64.70; 76.60; S12
(a) Vol.119,No.1, 77-86

K. KUMAZAKI, K. IMAI, and A. ODAJIMA
Estimation of Strains in MBE Grown ZnSe Films by Raman Scattering
63.20 and 68.30; 68.60; S8.12
(a) Vol.119,No.1, 177-182

V.F. LEMBERG and S.G. PSAKHIE
Analytical Representation of the Thermodynamic Functions of Simple Metals in the Model of Self-Consistent Einstein...
65.50; 63.20; 64.10; S4
(b) Vol.159,No.2, 645-658

J.C. UPADHYAYA, S. LEHRI, D.K. SHARMA, and S.C. UPADHYAYA
Angular Force Models in the Lattice Dynamics of Cubic Systems
63.20; 62.20; S1
(b) Vol.159,No.2, 659-674

S. VES
High Pressure Raman Study of the Chain Chalcogenide $TlInTe_2$
63.20 and 78.30; 64.60; S8.16
(b) Vol.159,No.2, 699-706

S.E. SARKISOV, A.A. KAMINSKII, A.V. BUTASHIN, YU.E. PERLIN, V.N. ENAKII, and M.G. BLAZHA
Selective Spectroscopy, Luminescence Kinetics of Pr^{3+} Ions, and Phonon Density Calculation in Multicentred and ...
78.55; 63.20; 71.55; 78.30; S11.2
(a) Vol.119,No.2, 607-620

H. RUECKER, M. MORA-RAMOS, and F. COMAS
Magnetopolaron in a Quantum Well. LO-Phonon Confinement Effects
71.38 and 73.40; 63.20; S7.12
(b) Vol.160,No.1, 117-126

H.C. GUPTA, A. PARASHAR, V.B. GUPTA, and B.B. TRIPATHI
Lattice Vibrations in Spinel-Type Compounds $CdCr_2S_4$ and $CdCr_2Se_4$
63.20; S8.16
(b) Vol.160,No.1, K19-K24

N. KRISTOFFEL
On the Vibronic Nature of the Tetragonal-to-Orthorhombic Structural Transition in $La_{2-x}M_xCuO_4$
64.70; 63.20; 74.70; S10.15
(b) Vol.160,No.2, K93-K96

R. ZIMMERMANN and K. KRÁL
LO-Phonon Generation and Electron Energy Collision Broadening in Polar Semiconductors
63.20 and 72.10
(b) Vol.160,No.2, K109-K112

R. CHATTERJEE
Phonon-Phonon Interaction in a One-Dimensional Anharmonic Lattice
63.20
(b) Vol.161,No.1, 191-200

W. JUNKER and M. WAGNER
Mori Approach to the Dynamics and Thermodynamics of Tunneling Centers
63.20 and 65.50; 64.70; 65.40; 66.30
(b) Vol.161,No.2, 543-564

R. SHYAM, S.C. UPADHYAYA, and J.C. UPADHYAYA
First-Principles Calculation of the Lattice Dynamics of the $Co_{0.92}Fe_{0.08}$ Alloy
63.20; S1.1; S1.2
(b) Vol.161,No.2, 565-574

E. ANACHKOVA, M. GOSPODINOV, A. NIKOLOV, P. SVESTAROV, N. PETKOV, YU. MARKOV, M. LIMONOV, and G. BRUCHMAN
Fundamental Vibrations in $Pb_5(GeO_4)(VO_4)_2$ Crystals
63.20; 78.30; S10.15
(b) Vol.161,No.2, 575-582

W.Y. ZENG and A. ANDERSON
Lattice Dynamics of Ammonia
63.20; S5.2
(b) Vol.162,No.1, 111-118

Y. MADHAVAN, K. PONNAMMAL, K. RAMACHANDRAN, and T.M. HARIDASAN
Mean Square Displacement (MSD) of Neighbours around a Vacancy in Some Zinc Based Tellurides
63.20; 61.70; S8.15
(b) Vol.162,No.1, 119-124

W. KULLMANN, G. EICHHORN, H. RAUH, R. GEICK, G. ECKOLD, and U. STEIGENBERGER
Lattice Dynamics and Phonon Dispersion in the Narrow Gap Semiconductor Bi_2Te_3 with Sandwich Structure
63.20; S8
(b) Vol.162,No.1, 125-140

A. LEHMANN, E. BROESE, and B. SCHROETER
Vibrational Modes of Ge_xSe_{1-x} Films
63.20 and 78.30; S5
(b) Vol.162,No.1, K53-K58

L.N. GUMEN and O.V. USATENKO
Phonon Spectrum of the One-Dimensional Two-Component Quasiperiodic Crystal
63.20; S12.1
(b) Vol.162,No.2, 387-394

O.S. BINBREK and A. ANDERSON
Lattice Dynamics of Sulphur Dioxide
63.20; S5.2
(b) Vol.162,No.2, 421-428

V.I. TYUTYUNNIK and O.I. TYUTYUNNIK
Phonon Structure of Raman Scattering Spectra of LiH Crystals
63.20 and 78.30; S9.11
(b) Vol.162,No.2, 597-604

M.K. MISHRA
New Approach to the Lattice Dynamical Study of Cubic Metals
63.20; S1.2
(b) Vol.162,No.2, K73-K76

63.50. Vibrational states in disordered systems

M.H. BROOKER
A Search for a Phase Transition in $Sr(NO_3)_2$ by Raman Spectroscopy. Evidence for Partial Disorder
64.60 and 78.30; 63.20; 63.50; S11.1
(b) Vol.157,No.1, 167-182

H. BOETTGER and G. KASNER
Self-Similarity of the Density of States of Electron and Phonon Models in Fibonacci Chains. An Analytic Study
63.50 and 71.20
(b) Vol.158,No.1, 143-152

D. BHANDARI, A. PRATAP, N.S. SAXENA, and M.P. SAKSENA
Collective Excitations in a Liquid Alloy
63.50; S2
(b) Vol.160,No.1, 83-88

K.C. JAIN, N. GUPTA, A. PRATAP, and N.S. SAXENA
Dynamical Correlations in Liquid Sodium
63.50; S2
(b) Vol.160,No.2, 433-442

K.C. JAIN, NEELAM GUPTA, and N.S. SAXENA
Dynamical Correlations in Liquid Metals: K and Cs
63.50; S2
(b) Vol.161,No.1, 201-210

K.C. JAIN, N. GUPTA, and N.S. SAXENA
Dynamics of Liquid $Na_{0.5}K_{0.5}$ Alloy
63.50; S2
(b) Vol.162,No.2, 395-406

63.75. Statistical mechanics of displacive phase-transitions

For order-disorder and statistsical mechanics of model systems, see 64.60
For crystallographic aspects of polymorphic and order-disorder transformations, see 61.50

J. SHANKER and M. KUMAR
Studies on Melting of Alkali Metals (Review Article)
64.70; 63.75; S9.11
(b) Vol.158,No.1, 11-50

63.90. Other topics in lattice dynamics and crystal statistics

P.S. GOYAL, R. CHAKRAVARTHY, B.A. DASANNACHARYA, and C.J. CARLILE
Reorientations of NH^{4+} Ions in the Ferroelectric Phase of $(NH_4)_2SO_4$
63.90; 64.70; 77.80; S11.1
(a) Vol.118,No.2, 425-430

V.P. GUPTA and S.K. SIPANI
The Analysis of Rydberg Potential and van der Waals Interactions in Alkali Halides
63.90; 62.20; 65.70; S9.11
(b) Vol.161,No.1, 131-144

NGUYEN TANG and VU VAN HUNG
Investigation of the Thermodynamic Properties of Anharmonic Crystals by the Momentum Method. IV. The Limiting of ...
65.50; 63.90; 64.60; S5.2
(b) Vol.162,No.2, 379-386

64. Equations of State, Phase Equilibria, and Phase Transitions

64.10. General theory of equations of state and phase equilibria

V.F. LEMBERG and S.G. PSAKHIE
Analytical Representation of the Thermodynamic Functions of Simple Metals in the Model of Self-Consistent Einstein...
65.50; 63.20; 64.10; S4
(b) Vol.159,No.2, 645-658

64.30. Equations of state of specific substances
(see also 65.70. Thermal expansion)

T. SOMA, N. MORI, and H.-MATSUO KAGAYA
Pressure-Volume Relation and Pressure-Induced Phase Transition of the Ge-Sn System
64.30; 64.70; S6
(b) Vol.159,No.2, K51-K54

S. DIXIT and R. NIWAS
Cohesive Energies, Equation of State, and Thermoelastic Properties of Mixed Alkali Halide Crystals
62.20 and 64.30; 65.70; S9.11
(b) Vol.161,No.2, 583-596

64.60. General studies of phase transitions
(for magnetic, superconducting, and quantum fluid critical phenomena, see 75, 74, and 67, respectively)

K.D. TOVSTYUK, I.O. POLYAKOV, E.I. SLINKO, N.K. TOVSTYUK, and A.G. KHANDOZHKO
Lithium Impurity States in PbTe
64.60; 61.70; 76.60; S8
(b) Vol.157,No.1, 151-158

M.H. BROOKER
A Search for a Phase Transition in $Sr(NO_3)_2$ by Raman Spectroscopy. Evidence for Partial Disorder
64.60 and 78.30; 63.20; 63.50; S11.1
(b) Vol.157,No.1, 167-182

B.E. VUGMEISTER and T.V. ANTIMIROVA
Simple Model for Ferroelectric Phase Transition in $KTa_{1-f}Nb_fO_3$
64.60 and 77.80; S10.15
(b) Vol.157,No.1, 183-188

P.S. GOYAL, R. CHAKRAVARTHY, B.A. DASANNACHARYA, S.K. KULSHRESHTHA, M.S. SASTRY, and J. TOMKINSON
Neutron Incoherent Inelastic Scattering from $(NH_4)_xRb_{1-x}H_2PO_4$
64.60; 61.12; S11
(b) Vol.157,No.2, 547-556

A.S. BAKAI and M.P. FATEEV
Equations of Order-Disorder Kinetics in Alloys with B.C.C. and F.C.C. Lattices. Tetrahedron Approximation
64.60
(b) Vol.158,No.1, 81-90

W. SALEJDA and N.A. DZHAVADOV
Phase Transition in $Rb_3H(SeO_4)_2$-Type Crystals.
I. The Symmetry Analysis of Proton Ordering
64.60; 61.50; 65.50; S11
(b) Vol.158,No.1, 119-126

ZHENGJU YANG and GUANGQUEN MA
The Kinetic Theory of the Orientational Ordering Phase Transition: The Model and the Kinetic Behaviour
64.60
(b) Vol.158,No.1, 127-142

G. REITER and P. KINDL
Positron Lifetime Investigations on Linear Polyethylene Compared to Branched Polyethylene
64.60; 78.70; S12
(a) Vol.118,No.1, 161-168

B. SALEJDA and N.A. DZHAVADOV
Phase Transition in $Rb_3H(SeO_4)_2$-Type Crystals.
II. The Molecular Field Approximation
64.60; 65.50; S11
(b) Vol.158,No.2, 475-480

V.K. JAIN and G. LEHMANN
Electron Paramagnetic Resonance of Mn^{2+} in Orthorhombic and Higher Symmetry Crystals (Review Article)
76.30; 64.60; S11
(b) Vol.159,No.2, 495-544

ZHENGJU YANG, J. ORTIZ-LOPEZ, and F. LUTY
Theoretical Investigation of Ferroelastic Phase Transition of Pure and Mixed Alkali Cyanide Systems by ... III. The ...
64.60; S9
(b) Vol.159,No.2, 629-644

YU.M. MISHIN and D.YA. KATS
On the Thermodynamics of Surface Transitions Close to the Bulk Critical Point
64.60 and 68.40
(b) Vol.159,No.2, 675-688

S. VES
High Pressure Raman Study of the Chain Chalcogenide $TlInTe_2$
63.20 and 78.30; 64.60; S8.16
(b) Vol.159,No.2, 699-706

S.K. GHOSHAL
Photoinduced Crystallization of Organic Molecular Solids from Vapour and Solution Phases
61.50; 64.60; S12
(a) Vol.120,No.1, 61-66

I. AVRAMOV
Pressure Dependence of Shear Viscosity
62.10; 64.60
(a) Vol.120,No.1, 133-138

S.I. KLOKISHNER, A.V. KORYACHENKO, and B.S. TSUKERBLAT
Charge Ordering of Multielectronic Clusters in Mixed-Valence Molecular Crystals
71.70 and 75.30; 64.60; S10
(b) Vol.160,No.2, 641-648

S.M. AKHMIN, V.P. MEIKLYAR, A.E. USACHEV, and YU.V. YABLOKOV
$^{59}Co^{2+}$ ENDOR in the Tetragonal Phase of $CsCaCl_3$
76.30 and 76.70; 64.60; S9.16
(b) Vol.160,No.2, K149-K152

J. PEREZ, S. ETIENNE, and J. TATIBOUËT
Determination of Glass Transition Temperature by Internal Friction Measurements
61.40 and 64.60; 62.40
(a) Vol.121,No.1, 129-138

A.A. KISELEV and A.V̌. LIAPTSEV
A Model Theory of Structural and Superconducting Phase Transitions in Chain-Like Systems
74.20; 64.60; S12.1
(b) Vol.161,No.2, 717-730

V.M. TALANOV
Structural Modelling of Low-Symmetry Phases of Spinels. I. Phases with 1:1 Octahedral Order
61.50; 64.60; S11.2
(b) Vol.162,No.1, 61-74

S. DEFREGGER, G.F. ENGEL, and P.W. KREMPL
The Temperature Dependence of the Optical Second Harmonic Generation Coefficient d_{11} of Quartz Homeotypic $AlPO_4$
78.20; 64.60; S11.1
(b) Vol.162,No.1, 311-320

NGUYEN TANG and VU VAN HUNG
Investigation of the Thermodynamic Properties of Anharmonic Crystals by the Momentum Method. IV. The Limiting of ...
65.50; 63.90; 64.60; S5.2
(b) Vol.162,No.2, 379-386

L. SKALA, V.M. KENKRE, M.W. WEISER, and J.D. KATZ
Remarks on the Thermodynamics of Sintered Ceramics
65.50; 64.60; S6; S10.1
(b) Vol.162,No.2, 429-434

G.V. IONOVA and A.V. NIKOLAEV
Isostructural gamma-alpha Transition in Metallic Cerium as a Mott Transition
64.60 and 71.20; S1.4
(b) Vol.162,No.2, 451-460

64.70. Phase equilibria, phase transitions, and critical points of specific substances

A.D. POGREBNYAK
Metastable States and Structural Phase Changes in Metals and Alloys Exposed to High Power Pulsed Ion Beams (Review Article)
64.70 and 79.20; 62.20; 68.45; 68.55; S1.1; S1.2
(a) Vol.117,No.1, 17-52

J.L. MACQUERON, LL. MAÑOSA, and G. GUÉNIN
High Temperature Martensite Induced by Quenching
64.70; S1.2
(a) Vol.117,No.1, 113-118

S.K. PLACHKOVA, L.E. SHELIMOVA, and O.G. KARPINSKII
X-Ray Study of the Structural Phase Transitions in $Ge_{1-x}Ag_{x/2}Bi_{x/2}Te$ Solid Solutions
64.70; 61.10; S8
(a) Vol.117,No.1, 155-162

S. AHMED, A.M. SHAMAH, H.A. KHALIK, A. IBRAHIM, and F. HANNA
Electrical and Thermal Studies of $[(NH_4)_{1-x}K_x]_2SO_4$ Crystals at High Temperature
64.70; 72.20; S11
(a) Vol.117,No.1, 163-168

V.D. KUSHKOV, A.V. ZVERLIN, A.M. ZASLAVSKII, and A.V. MELNIKOV
Cubic-Tetragonal Phase Transformation in Films in the Zirconia-Rare Earth Oxide Binary Systems
64.70; S10.15
(a) Vol.117,No.1, K13-K14

N.T. MAMEDOV and A.M. PANICH
Phase Transition in $TlGaTe_2$
64.70 and 76.60; S8.16
(a) Vol.117,No.1, K15-K18

S. WAPLAK and N.N. KOLPAKOVA
Phase Transitions in $Cd_2Nb_2O_7$ Studied by EPR of Cr^{3+} Ion
64.70; 76.30; S10.15
(a) Vol.117,No.2, 461-466

L.M. DI, P.I. LOEFF, and H. BAKKER
Mechanical Amorphisation of Ni_3Sn
64.70; S1.2
(a) Vol.117,No.2, K99-K102

J. SHANKER and M. KUMAR
Studies on Melting of Alkali Metals (Review Article)
64.70; 63.75; S9.11
(b) Vol.158,No.1, 11-50

J.J. BARA, B.F. BOGACZ, A. POLACZEK, M. PĘKAŁA, and A. GRODZIŃSKI
The Temperature Induced Transition in the theta-Type Vanadium Oxide Bronzes
61.60; 64.70; 76.80; S10
(a) Vol.118,No.1, 41-52

I. LICEA, A. IOANID, and A. DAFINEI
High-Temperature Phase Transition in KH_2PO_4
64.70; 65.40; S11.1
(a) Vol.118,No.1, 131-140

E.I. SHREDER, I.A. POPOV, E.E. KRASOVSKII, and I.I. SASOVSKAYA
The Optical Properties and Electronic Structure of AgZn Alloy Undergoing $\beta \rightleftharpoons \zeta$ Structural Transformation
64.70 and 78.20; 71.25; S1.2; S1.3
(a) Vol.118,No.1, 147-152

V.G. CHUDINOV and V.V. ANDREEV
Atomic Mechanism of Solid-Liquid Phase Transition in F.C.C. Metals with a Large Stacking-Fault Energy
64.70; 61.40; 61.70; S4
(a) Vol.118,No.2, 415-424

P.S. GOYAL, R. CHAKRAVARTHY, B.A. DASANNACHARYA, and C.J. CARLILE
Reorientations of NH^{4+} Ions in the Ferroelectric Phase of $(NH_4)_2SO_4$
63.90; 64.70; 77.80; S11.1
(a) Vol.118,No.2, 425-430

Z. TYLCZYŃSKI and P. PISKUNOWICZ
Dilatometric Studies of $(NH_4)_2ZnCl_4$ in the Temperature Range from 100 to 450 K
64.70 and 65.70; S9.16
(a) Vol.118,No.2, 441-446

W. ZAPART
Possibility of Simultaneously Incommensurate and Ferroelastic Phase in $RbIn(MoO_4)_2$ by EPR of Cr^{3+} Ion
64.70 and 76.30; S11.1
(a) Vol.118,No.2, 447-454

H. KRESSE and F. KREMER
Dielectric Relaxation of a Swallow-Tailed Compound at the Phase Transition from the Nematic into the Isotropic Phase
77.40; 64.70; S12
(a) Vol.118,No.2, K119-K122

S.U. JEN, M.C. CHIEN, and C.Y. LEE
Viscous Flow in Amorphous Alloys
61.40 and 64.70; 62.20; S1.1; S1.2
(a) Vol.119,No.1, 35-40

YIFENG WANG, WENHUI SU, HONGJIAN LIU, and LIPING LI
Orthorhombic Phase and Oxygen Deficiencies in the Electron Type Superconductor $Ln_{2-x}Ce_xCuO_{4-y}$
64.70 and 74.70; S10.15
(a) Vol.119,No.1, 55-62

D. BALTRŪNAS, R. MIKAITIS, V.YU. SLIVKA, and YU.M. VYSOCHANSKII
Manifestation of the Ferroelectric Phase Transition in the Moessbauer Parameters of the System $(Sn_xPb_{1-x})_2P_2S_6$
64.70 and 76.80; S8.16
(a) Vol.119,No.1, 71-76

M. MAĆKOWIAK, N. WEIDEN, and A. WEISS
Phase Transitions and Molecular Dynamics in Tris(Methylammonium) Nonabromodiantimonate(III) and Its ...
63.20 and 64.70; 76.60; S12
(a) Vol.119,No.1, 77-86

M. WAUTELET
On the Transitions between the Crystalline, Amorphous, and Liquid Phases of Silicon and Germanium, when Their Size ...
64.70; 68.60; S5.11; S5.12
(b) Vol.159,No.2, K43-K46

T. SOMA, N. MORI, and H.-MATSUO KAGAYA
Pressure-Volume Relation and Pressure-Induced Phase Transition of the Ge-Sn System
64.30; 64.70; S6
(b) Vol.159,No.2, K51-K54

I.KH. AKOPYAN, A.A. KLOCHIKHIN, B.V. NOVIKOV, M.YA. VALAKH, A.P. LITVINCHUK, and I. KOSAZKII
Optical Spectroscopy of the Superionic Crystals (Review Article)
78.30 and 78.40; 64.70; 71.25; 71.35; 78.55; S9
(a) Vol.119,No.2, 363-416

P. SAINT-GRÉGOIRE, R. ALMAIRAC, J. LAPASSET, and J. MORET
Structural Phase Transitions in $KFeF_4$ and Related Materials
64.70; S9
(a) Vol.120,No.1, K25-K30

N. KRISTOFFEL
On the Vibronic Nature of the Tetragonal-to-Orthorhombic Structural Transition in $La_{2-x}M_xCuO_4$
64.70; 63.20; 74.70; S10.15
(b) Vol.160,No.2, K93-K96

H.R. SREEPAD, V. SREERAMALU, A. CHANDRASHEKARA, V. RAVINDRACHARY, and S. GOPAL
Structural Phase Transition in a Superionic Conductor - $Na_2(La,Al)Zr(PO_4)_3$. Positron Annihilation Study
64.70; 61.40; 78.70; S11
(a) Vol.120,No.2, 411-418

W. RUDZIŃSKA, ST. CHABIK, and CZ. SZYMAŃSKI
Investigation of the Solid-Liquid Phase Transition for Sn-5wt%In Alloy by the Positron Annihilation Method
64.70; 78.70; S4
(a) Vol.120,No.2, K145-K148

W. ZAPART and M.B. ZAPART
Incommensurate Phase in $KSc(MoO_4)_2$ by EPR of Cr^{3+}
64.70 and 76.30; S11.1
(a) Vol.121,No.1, K43-K46

S.K. GUPTA, V.V. GLADKII, and K. LAL
Relaxation Effects in the Vicinity of the Incommensurate-Commensurate Phase Transition in Rb_2ZnCl_4 ...
64.70; S9.15
(a) Vol.121,No.1, K47-K50

QIANG WANG, BINGLIN GU, and XIAOWEN ZHANG
Preliminary Studies on the Stabilities of Three Types of Ordered Structures in the $A(B^1_{0.5}B^2_{0.5})O_3$ System
64.70; 61.60; S10
(b) Vol.161,No.2, 537-542

W. JUNKER and M. WAGNER
Mori Approach to the Dynamics and Thermodynamics of Tunneling Centers
63.20 and 65.50; 64.70; 65.40; 66.30
(b) Vol.161,No.2, 543-564

E.V. TATYANIN and V.G. KURDYUMOV
The Nucleation of the Deformation Induced Amorphous Phase at Twin Boundaries in TiNi Alloy
61.70 and 62.20; 64.70; S1.2
(a) Vol.121,No.2, 455-460

T. TSUBOI and P. SILFSTEN
Detection of Structural Phase Transition in $BaMnF_4$ by Luminiscence Lifetime
64.70 and 78.55; S9.16
(b) Vol.162,No.1, K15-K20

M. BAUER, J. MAYER, M. HOHENSTEIN, and M. WILKENS
Quantitative Studies on the Amorphization of Quasicrystals by Means of High Voltage Electron Microscopy
61.40 and 61.80; 64.70; S1.2; S4
(a) Vol.122,No.1, 79-86

V.G. CHUDINOV, R.M.J. COTTERILL, and V.V. ANDREEV
Possible Mechanism of the Solid-Liquid Phase Transition in H.C.P. and B.C.C. Structures
64.70; 61.70
(a) Vol.122,No.1, 187-194

FANG YUAN, YENING WANG, and HIUMIN SHEN
Study of the Incommensurate Phase in NiTi(Fe) Alloys
64.70; 62.40; 64.80; S1.2
(a) Vol.122,No.2, 517-524

E.I. TOCHITSKII, O.G. SVIRIDOVICH, E.M. GOLOLOBOV, and D.M. TURTSEVICH
Phase Transformations in $Y_1Ba_2Cu_3O_{7-x}$ Films Prepared by Pulsed Laser Deposition
64.70 and 68.55; 74.70; S10.15
(a) Vol.122,No.2, 555-562

E.M. GOLOLOBOV, E.L. MAGER, and Z.V. MESHEVICH
Ultrasonic Investigations of Structural Instabilities of ZrV_2 under High Hydrostatic Pressures
62.65 and 64.70; S1.2
(a) Vol.122,No.2, K123-K128

64.75. Solubility, segregation, and mixing

J. SCHMELZER, R. PASCOVA, and I. GUTZOW
Cluster Growth and Ostwald Ripening in Viscoelastic Media
61.40; 64.75
(a) Vol.117,No.2, 363-376

J.M. RÉAU, J. SÉNÉGAS, M. ELOMARI, J.P. LAVAL, and B. FRIT
Ordre à courte distance et conductivité ionique au sein de la solution $Pb_{1-x}In_xF_{2+x}$
61.60 and 66.30; 64.75; S9.15
(a) Vol.117,No.2, 409-416

R.M. SARDARLY, R.A. ALIEV, E.M. KERIMOVA, N.A. BAKHYSHOV, and I.SH. SADYKHOV
Rearrangement Features of the Vibrational Spectra of $TlInS_{2x}Se_{2(1-x)}$ and $TlGa_xIn_{1-x}Se_2$ Solid Solutions
63.20 and 78.30; 64.75; S8.16
(b) Vol.158,No.1, 73-80

I.V. BODNAR, G.F. SMIRNOVA, A.G. KAROZA, and A.P. CHERNYAKOVA
Vibrational Spectra of $CuGaS_2$ and $CuGaSe_2$ Compounds and $CuGaS_{2x}Se_{2(1-x)}$ Solid Solutions
63.20; 64.75; 78.30; S8.16
(b) Vol.158,No.2, 469-474

V.G. CHUDINOV, B.N. GOSHCHITSKII, N.V. MOSEEV, and V.V. ANDREEV
Kinetics of Radiation-Induced Segregation of Impurities in F.C.C. Metals at the Thermal Stage of the Collision Cascade
61.80 and 64.75; S1.1; S4
(a) Vol.119,No.2, 437-442

P.M. OSSI
Crystal-Glass Phase Transition in Ion Irradiated Binary Systems
61.80 and 64.75; 61.40; S1
(a) Vol.119,No.2, 463-470

M.A. KENAWY, M.S. SAKR, E.M. SAKR, H.A. ZAYED, and N.O. MOURAD
Effect of Dissolution of the beta-Phase on the Work Hardening Characteristics of Al-4.5wt%Mg Alloy
62.20 and 64.75; S4
(a) Vol.120,No.1, K31-K36

U. BECKER, P. RUDOLPH, R. BOYN, M. WIENECKE, and I. UTKE
Characterization of p-Type CdTe Bridgman Crystals by Infrared Extinction Spectra
64.75 and 78.30; S8.13
(a) Vol.120,No.2, 653-660

J. SCHMELZER
On the Kinetics of Decomposition of Dilute Al-(Zn,Mg)Alloys
64.75; S1.2; S4
(b) Vol.161,No.1, 173-178

D. VLADIKOVA, L. ILKOV, and S. KARBANOV
Optimal Prefiring Conditions of Substituted Nickel Ferrites for Microwave Applications
64.75 and 75.50; S11.2
(a) Vol.121,No.1, 249-256

G. KISSINGER, K. TITTELBACH-HELMRICH, J. KNOPKE, and U. RETZLAFF
On the Gettering Effiency of Buried Layers in Dielectrically Insulated Structures
64.75 and 85; 71.55; S5.11
(a) Vol.121,No.1, K141-K144

E.M. LEVIN, B.S. KUZHEL, O.I. BODAK, B.D. BELAN, and I.N. STETS
Concentrated Kondo Systems in Solid Solutions on the Base of Europium Ternary Compounds
72.10; 64.75; 71.25; S1.4
(b) Vol.161,No.2, 783-796

K.K. BOURDELLE, N.G. CHECHENIN, A.S. AKHMANOV, A.YU. POROIKOV, and A.V. SUVOROV
Melting and Damage Production in Silicon Carbide under Pulsed Laser Irradiation
61.80; 64.75; S6
(a) Vol.121,No.2, 399-406

M. EL OMARI, J.M. RÉAU, and J. SÉNÉGAS
Corrélations entre propriétés electriques et ordre à courte distance au sein de quelques solutions solides basées sur...
66.30; 64.75; S9.15
(a) Vol.121,No.2, 415-428

R. BRUN DEL RE, J.C. WOOLLEY, M. QUINTERO, and R. TOVAR
T(z) Diagram of $CuIn_{1-z}Fe_zS_2$ Alloys
64.75; 61.55; S1.1
(a) Vol.121,No.2, 483-488

H.L. PARK, S.Y. GEE, S.K. CHANG, S. PARK, S.K. LEE, J.Y. LEE, C.S. YOON, J.E. KIM, H.Y. PARK, W.T. KIM, and H. LIM
Solid Solubility of Selenium in $AgGaSe_2$
64.75; 78.30; S8.16
(b) Vol.162,No.1, K9-K14

J. ECHIGOYA and K. NARITA
A Study of $La(Mn_{1-x}Ni_x)O_3$ and Its Bonding with Cubic ZrO_2
61.60; 64.75; S10.15
(a) Vol.122,No.1, 87-96

64.80. Other phase properties of systems

A.S. TAHA and F.H. HAMMAD
The Applications of the Hall-Petch Relation to Microhardness Measurements on Al, Cu, Al-MD 105, and Al-Cu Alloys
62.20; 61.70; 64.80; S1.2; S4
(a) Vol.119,No.2, 455-462

YUPING SUN, JIANYI JIANG, FANCHUN ZHENG, HUAQING YIN, and JIAJU DU
Synthetic Condition and Properties of the Bi-Pb-Sr-Ca-Cu-O Superconductor with Single High-T_c Phase
74.70; 64.80; S10.15
(a) Vol.119,No.2, 555-560

A.H. WAFIK and A.Z. MOHAMED
Barkhausen Jumps and Microstructure in Molybdenum Eutectoid Steel
75.60; 64.80; S1.1
(a) Vol.120,No.2, 601-608

G.F. ZHOU, X.K. SUN, S.Y. FU, Y.C. CHUANG, R. GROESSINGER, and H.R. KIRCHMAYR
Relationship between Microstructure and Coercivity in (Nd,Dy)-(Fe,Co)-B Based Sintered Magnets with Minor Nb and Al ...
75.50; 64.80; S1.1
(a) Vol.120,No.2, 627-634

R.A. YAKSHIBAEV, N.N. MUKHAMADEEVA, and R.F. KADRGULOV
Phase Relations, Ionic Transport, and Diffusion in the Alloys of Ag_2Te-Cu_2Te Mixed Conductors
61.60 and 64.80; 66.30; S8.16
(a) Vol.121,No.1, 111-118

R. HEIDINGER, H. APPEL, G. THEN, and W.-G. THIES
Observation of Radiation Damage in HfV_2 and Its Hybrides by TDPAC
61.80; 64.80; 74.70; 76.80; S1.2
(a) Vol.121,No.2, 445-454

A. LARGETEAU, J.M. REAU, and J. RAVEZ
Dispersions diélectrique et d'impédance et énergies d'activation de ceramiques à couches d'arret de type ...
77.20; 64.80; 77.40; S11.2
(a) Vol.121,No.2, 627-634

FANG YUAN, YENING WANG, and HIUMIN SHEN
Study of the Incommensurate Phase in NiTi(Fe) Alloys
64.70; 62.40; 64.80; S1.2
(a) Vol.122,No.2, 517-524

O. EIBL, J.M. TRISCONE, and O. FISCHER
Imaging of $YBa_2Cu_3O_7/PrBa_2Cu_3O_7$ Superlattices by TEM
68.55 and 73.40; 64.80; S10.15
(a) Vol.122,No.2, 589-598

65. Thermal Properties of Condensed Matter

(see also 63. Lattice dynamics;
for thermodynamic properties of quantum fluids, see 67.40;
for thermal properties of solid helium, see 67.80)

65.40. Heat capacities of solids

(for specific heat of superconductors, see 74.30;
for specific heat of magnetic systems, see 75.40)

M.E. KASSEM, N.G. GOMAA, A.M. EL-KHATIB, and S.H. KANDIL
Fast Neutron Irradiation Effect and Thermal Properties of Doped Nonstoichiometric Lithium Potassium Sulphate Crystals
65.40; 61.80; S11
(a) Vol.117,No.1, 119-124

I. LICEA, A. IOANID, and A. DAFINEI
High-Temperature Phase Transition in KH_2PO_4
64.70; 65.40; S11.1
(a) Vol.118,No.1, 131-140

U. WENSCHUH and E. HEINER
Optical Properties of Gold in the Nonequilibrium Thermalized Stage
71.10; 65.40; 71.38; 78.20; S1.3
(b) Vol.160,No.1, 225-236

G.J. VÁZQUEZ
Specific Heat and Elastic Constants for Martensitic Forms of Lithium
62.20 and 65.40; S2
(b) Vol.160,No.2, K89-K92

V.M. BURLAKOV, S.A. KISELEV, V.N. PETUKHOV, and A.A. ANIKIEV
The Dynamics of the Domain Boundaries and Soft Lattice Mode in SbSI Crystals
77.80; 65.40; S9.16
(b) Vol.161,No.1, 387-394

D.A. GUSEINOV and M.A. ALYANOV
Heat Capacity and Thermal Expansion of $TiFe_2Se_4$
65.40; 75.40; S11.2
(a) Vol.121,No.1, K51-K54

W. JUNKER and M. WAGNER
Mori Approach to the Dynamics and Thermodynamics of Tunneling Centers
63.20 and 65.50; 64.70; 65.40; 66.30
(b) Vol.161,No.2, 543-564

U. WENSCHUH and E. HEINER
Nonequilibrium Thermoreflectance Evolution of Laser Irradiated Noble Metals
78.20; 65.40; 71.10; S1.3
(b) Vol.162,No.1, 303-310

NGUYEN TANG and VU VAN HUNG
Investigation of the Thermodynamic Properties of Anharmonic Crystals by the Momentum Method. III. Thermodynamic ...
65.50; 65.40; 65.70; S5.2
(b) Vol.162,No.2, 371-378

H. WADA, T. INOUE, M. HADA, M. SHIGA, and Y. NAKAMURA
Effect of Magnetic Phase Transition on the Electronic Specific Heat Coefficient of $Lu(Co_{1-x}Al_x)_2$
65.40; 75.30; S1.2; S1.4
(b) Vol.162,No.2, 407-412

65.50. Thermodynamic properties and entropy

A. DEROUICHE and J.L. BRETONNET
The Long-Wavelength Limit of the Structure Factor of Liquid Alkali Metals Using a One-Component Plasma System
61.25; 65.50; S2
(b) Vol.157,No.1, 61-64

R.V. GOPALA RAO and R. VENKATESH
Theoretical Method of Evaluation of Heats of Vaporisation of Fused Salts
61.20; 65.50; S9.11
(b) Vol.157,No.1, 65-70

W. SALEJDA and N.A. DZHAVADOV
Phase Transition in $Rb_3H(SeO_4)_2$-Type Crystals.
I The Symmetry Analysis of Proton Ordering
64.60; 61.50; 65.50; S11
(b) Vol.158,No.1, 119-126

B. SALEJDA and N.A. DZHAVADOV
Phase Transition in $Rb_3H(SeO_4)_2$-Type Crystals.
II. The Molecular Field Approximation
64.60; 65.50; S11
(b) Vol.158,No.2, 475-480

V.F. LEMBERG and S.G. PSAKHIE
Analytical Representation of the Thermodynamic Functions of Simple Metals in the Model of Self-Consistent Einstein...
65.50; 63.20; 64.10; S4
(b) Vol.159,No.2, 645-658

V.T. MASLYUK
Stables States and Ordering in Amorphous Polymer Semiconductors
61.40; 65.50; 71.45; S8
(b) Vol.161,No.1, 43-54

O. AKINLADE
A Study of the Structure and Thermodynamics of Liquid Metals Based on Some Reference Systems
61.25 and 65.50; S2; S3; S4
(b) Vol.161,No.1, 75-84

NGUYEN TANG and VU VAN HUNG
Investigation of the Thermodynamic Properties of Anharmonic Crystals by the Momentum Method. II. Comparison of ...
65.50; S5.2
(b) Vol.161,No.1, 165-172

W. JUNKER and M. WAGNER
Mori Approach to the Dynamics and Thermodynamics of Tunneling Centers
63.20 and 65.50; 64.70; 65.40; 66.30
(b) Vol.161,No.2, 543-564

T. SOMA, M. HASEBE, and H.-MATSUO KAGAYA
Lattice Dynamical Contribution to the Bulk and Thermal Properties of Alkali Metals
65.50; 65.70; S2
(b) Vol.162,No.1, K21-K24

NGUYEN TANG and VU VAN HUNG
Investigation of the Thermodynamic Properties of Anharmonic Crystals by the Momentum Method. III. Thermodynamic ...
65.50; 65.40; 65.70; S5.2
(b) Vol.162,No.2, 371-378

NGUYEN TANG and VU VAN HUNG
Investigation of the Thermodynamic Properties of Anharmonic Crystals by the Momentum Method. IV. The Limiting of ...
65.50; 63.90; 64.60; S5.2
(b) Vol.162,No.2, 379-386

P. ANAGNOSTARAS, I. SIKAKANA, and J.E. LOWTHER
Electronic Structure and Thermodynamic Properties of PdH_x
65.50; 71.20; S1
(b) Vol.162,No.2, 413-420

L. SKALA, V.M. KENKRE, M.W. WEISER, and J.D. KATZ
Remarks on the Thermodynamics of Sintered Ceramics
65.50; 64.60; S6; S10.1
(b) Vol.162,No.2, 429-434

65.70. Thermal expansion and thermomechanical effects

(see also 64.30. Equations of state)
For thermal conduction in nonmetallic liquids, see 66.60; for nonmetallic solids, see 66.70
For electronic thermal conduction, see 72.10 and 72.20
For thermal conductivity in superconductors, see 74.30
For pyroelectric and electrocaloric effects, see 77.70

J. BĄK-MISIUK, J. WOLF, and U. PIETSCH
The Thermal Expansion Coefficient of $A^{III}B^{V}$ Multilayer Structures
65.70 and 68.60; S7.15
(a) Vol.118,No.1, 209-218

J. BĄK-MISIUK, G. KUEHNEL, W. SIEGEL, and U. PIETSCH
The Possible Influence of the Free Carrier Redistribution within the Conduction Band on the Thermal Expansion ...
65.70; 72.20; S7.12
(b) Vol.158,No.2, K111-K114

Z. TYLCZYŃSKI and P. PISKUNOWICZ
Dilatometric Studies of $(NH_4)_2ZnCl_4$ in the Temperature Range from 100 to 450 K
64.70 and 65.70; S9.16
(a) Vol.118,No.2, 441-446

V.P. GUPTA and S.K. SIPANI
The Analysis of Rydberg Potential and van der Waals Interactions in Alkali Halides
63.90; 62.20; 65.70; S9.11
(b) Vol.161,No.1, 131-144

R.A. YAKSHIBAEV, A.F. NADEZHDINA, and R.F. ALMUKHAMETOV
Structural Features of Niobium Diselenide Intercalated by Copper at High Temperatures
61.60; 65.70; S1.63
(a) Vol.121,No.1, K5-K8

S. DIXIT and R. NIWAS
Cohesive Energies, Equation of State, and Thermoelastic Properties of Mixed Alkali Halide Crystals
62.20 and 64.30; 65.70; S9.11
(b) Vol.161,No.2, 583-596

T. SOMA, M. HASEBE, and H.-MATSUO KAGAYA
Lattice Dynamical Contribution to the Bulk and Thermal Properties of Alkali Metals
65.50; 65.70; S2
(b) Vol.162,No.1, K21-K24

NGUYEN TANG and VU VAN HUNG
Investigation of the Thermodynamic Properties of Anharmonic Crystals by the Momentum Method. III. Thermodynamic ...
65.50; 65.40; 65.70; S5.2
(b) Vol.162,No.2, 371-378

65.90. Other topics in thermal properties of condensed matter

P. WUENSCHE and A. MUELLER
Interpretation of Thermally Stimulated Processes by an Experimentally Determined Distribution of the Activation Energy
65.90; S12
(a) Vol.117,No.2, 447-460

P. WUENSCHE
Influence of Kinetics of Thermally Stimulated Processes on Experimentally Determined Distribution of the Activation ...
65.90; S12
(a) Vol.120,No.1, 139-150

D. DĂDÂRLAT, M. CHIRTOC, C. NEAMŢU, R.M. CÂNDEA, and D. BICANIC
Inverse Photopyroelectric Detection Method
65.90 and 77.70; S1.1; S1.2
(a) Vol.121,No.2, K231-K234

66. Transport Properties of Condensed Matter (Nonelectronic)

66.30. Diffusion in solids

N.A. STOLWIJK
On Diffusion by the Dissociative Mechanism in the Case of a Finite Foreign Atom Source
66.30
(b) Vol.157,No.1, 107-116

HUI-FANG DENG
Basic Equation of Evolution of the Effective Probability Density of a Particle or Quasi-Particle Migrating in a ...
66.30
(b) Vol.157,No.1, K7-K12

J. ČERMÁK
Grain Boundary Diffusion of Co-60 in Nickel
66.30; S1.2
(a) Vol.117,No.2, 387-394

H. GDANITZ and K. SCHMALZ
On the Backside Gettering of Au in Silicon
66.30; S5.11
(a) Vol.117,No.2, 395-402

J.M. RÉAU, J. SÉNÉGAS, M. ELOMARI, J.P. LAVAL, and B. FRIT
Ordre à courte distance et conductivité ionique au sein de la solution $Pb_{1-x}In_xF_{2+x}$
61.60 and 66.30; 64.75; S9.15
(a) Vol.117,No.2, 409-416

R. MITDANK, A. STURM, and B. JACOBS
Investigation of Excess Carrier Diffusion in $Ga_{0.35}Al_{0.65}As$ by Cathodoluminescence
73.60; 66.30; S7.15
(a) Vol.117,No.2, 485-494

V. ZUBKUS, P. KUNDROTAS, S. LAPINSKAS, and V. KASHCHEEV
Diffusion in Crystals with Competing Interactions of Migrating Ions
66.30; S11.1
(b) Vol.158,No.1, 51-62

A. BUKALUK
AES Depth Profile Studies of Interdiffusion in the Ag-Cu Bilayer and Multilayer Thin Films
66.30 and 68.48; 61.80; S1.2; S1.3
(a) Vol.118,No.1, 99-108

C.A. LONDOS, K. EFTAXIAS, and V. HADJICONTIS
Correlation of Solubilities of Various Elements in Silicon
66.30; S5.11
(a) Vol.118,No.1, K13-K16

FANGQING ZHANG, DEYAN HE, ZHIZHONG SONG, and GUANGHUA CHEN
Boron Diffusion in Device-Like B-Doped a-SiC:H/a-Si:H Heterojunction
66.30; S5.11
(a) Vol.118,No.1, K17-K20

Y. MADHAVAN, K. RAMACHANDRAN, and T.M. HARIDASAN
Tellurium Diffusion in the Mixed System $Cd_xHg_{1-x}Te$
66.30; S8.15
(b) Vol.158,No.2, 427-432

I. MANNA, S. BADER, W. GUST, and B. PREDEL
Interdiffusion between In Layer and Bulk Cu or Cu-In Alloy
66.30 and 68.48; S1.2
(a) Vol.119,No.1, K9-K14

A. SARCE and E.J. SAVINO
Grain Boundary Sink Strength in Anisotropic Materials
61.70 and 66.30
(b) Vol.159,No.2, 557-570

J. ČERMÁK
Diffusion of ^{51}Cr along High-Diffusivity Paths in Ni-Fe Alloys
66.30; S1.1; S1.2
(a) Vol.119,No.2, 443-454

S.O. GLADKOV and N.YA. KUZNETSOVA
On Diffusion Saturation of Porous Structures by Liquids
66.30
(b) Vol.160,No.1, K5-K8

K. EFTAXIAS and V. HADJICONTIS
Comments on the Diffusion of Ga in Ge
66.30; S5.12
(b) Vol.160,No.1, K9-K12

W. KORNETA and Z. PYTEL
Random Walk near the Surface of Dynamically Disordered Systems
66.30
(b) Vol.160,No.1, K13-K18

L. KOVÁCS, K. POLGÁR, R. CAPELLETTI, and C. MORA
Diffusion of Hydrogen Isotopes in Pure and Mg-Doped $LiNbO_3$ Crystals
66.30; 78.30; S10.15
(a) Vol.120,No.1, 97-104

E.A. BRENER and D.E. TEMKIN
Grain Boundary Migration Induced by Diffusion Flow across the Boundary
61.70 and 66.30
(a) Vol.120,No.1, K7-K10

Y. KAMIURA, M. YONETA, and F. HASHIMOTO
Hole Traps Produced by High-Temperature Heat Treatments of Silicon above 1300 °C
61.70; 71.55; 66.30; S5.11
(a) Vol.120,No.1, K11-K14

R.V. GOPALA RAO and U. BANDYOPADHYAY
An Evaluation of Structural and Transport Properties of Molten Potassium-Rubidium Alloy
61.40; 62.20; 66.30; S2
(b) Vol.160,No.2, 403-414

V.K. DUGAEV and K.D. TOVSTYUK
Particularities of Diffusion in Two-Dimensional Systems with Random Forces
66.30
(b) Vol.160,No.2, 415-422

S.V. NISTOR, GH. STOICESCU, and C.D. MATEESCU
Incorporation and Aggregation Properties of Bismuth Ions in KCl Crystals
61.70; 66.30; 78.50; S9.11
(b) Vol.160,No.2, 423-432

K. SADAIYANDI and K. RAMACHANDRAN
Chlorine Diffusion in CdTe
66.30; S8.13
(b) Vol.160,No.2, K85-K88

J. ČERMÁK
Diffusion along Moving Grain Boundaries
66.30; 61.70; S1.1; S1.2
(a) Vol.120,No.2, 351-362

E. MAJKOVA, V. BOHAC, S. LUBY, R. SANDRIK, and M. VESELY
Diffusion of Au Atoms in Ni-Si-B and Fe-Co-B Metallic Glasses
66.30; S1.1; S1.2
(a) Vol.120,No.2, 371-378

J.E. GARBARCZYK
Study of Protonic Diffusion in $H_{4.5}UO_2(IO_6)_{1.3}*mH_2O$ Using PFG NMR Method
66.30; 76.60; S11
(a) Vol.120,No.2, 387-390

S.V. KOVESHNIKOV, S.V. NOSENKO, and E.B. YAKIMOV
The Peculiarities of Deep Level Defect Passivation in Si by Atomic Hydrogen
61.70 and 71.55; 66.30; S5.11
(a) Vol.120,No.2, 391-396

V.L. TAVGIN and A.V. STEPANOV
Activation Process Model
66.30 and 78.70
(b) Vol.161,No.1, 123-130

G. NEUMANN and D.L. BEKE
Is There a Connection between the Temperature Function of BΩ and the Curvature of LnD versus 1/T for Self-Diffusion in ...
66.30; S1.2; S2
(b) Vol.161,No.1, K5-K10

I. RUPSKY, I. IORDANOVA, and A. APOSTOLOV
Monte Carlo Simulation of Normal Grain Growth by Consideration of the Diffusion Character of the Processes
61.50; 61.70; 66.30
(a) Vol.121,No.1, 45-56

H. KAHNT and J.M. REAU
Mixed Mobile Ion Effect of Anionic Conduction in Lead Halogenosilicate Glasses
61.40 and 66.30; S9.16
(a) Vol.121,No.1, 73-80

R.A. YAKSHIBAEV, N.N. MUKHAMADEEVA, and R.F. KADRGULOV
Phase Relations, Ionic Transport, and Diffusion in the Alloys of Ag_2Te-Cu_2Te Mixed Conductors
61.60 and 64.80; 66.30; S8.16
(a) Vol.121,No.1, 111-118

J. ČERMÁK
Grain Boundary Diffusion of ^{64}Cu in Cu-Al Alloys
66.30; S1.2
(a) Vol.121,No.1, K31-K36

W. JUNKER and M. WAGNER
Mori Approach to the Dynamics and Thermodynamics of Tunneling Centers
63.20 and 65.50; 64.70; 65.40; 66.30
(b) Vol.161,No.2, 543-564

A.O. ALEKSANYAN, V.A. GANSHIN, YU.N. KORKISHKO, and V.A. FEDOROV
Equilibrium and Kinetics of Ion Exchange in Zinc Selenide Single Crystals
66.30 and 68.45; 68.40; S8.12
(b) Vol.161,No.2, 629-636

M. EL OMARI, J.M. RÉAU, and J. SÉNÉGAS
Corrélations entre propriétés électriques et ordre à courte distance au sein de quelques solutions solides basées sur...
66.30; 64.75; S9.15
(a) Vol.121,No.2, 415-428

SIMING HE, MINGZHI LIU, and LINCHANG TANG
Internal Friction Peak Associated with Hydrogen Diffusion in Amorphous $Pd_{77.5}Ag_6Si_{16.5}$ Alloy
62.40; 66.30; S1.3
(a) Vol.121,No.2, K167-K170

N. SMETNIANSKY-DEGRANDE, C.N. TOMÉ, and R. PASSIANOT
Anisotropic Diffusion of Point Defects to Spheroidal Cavities
61.70 and 66.30
(b) Vol.162,No.1, 99-110

V.G. CHUDINOV, R.M.J. COTTERILL, and V.V. ANDREEV
Kinetics of the Diffuse Processes within a Cascade Region in the Sub-Threshold Stages of F.C.C. and H.C.P. Metals
66.30; S1; S4
(a) Vol.122,No.1, 111-120

N. MATHUR and A.N. NIGAM
Temperature Dependence of Electrolytic Hydrogen Diffusion in Nickel: A Thermoelectric Study
66.30 and 72.15; S1.2
(a) Vol.122,No.1, K59-K62

B. FRITZSCH and A. ZEHE
Analysis of Migration Energy Dependence on Atomic Number for Diffusion in Metals
66.30
(b) Vol.162,No.2, K69-K72

K. IWAUCHI
Ionic Conductivity and Dielectric Relaxation of beta-Al_2O_3 Fiber Superionic Conductor
66.30; 77.40; S10
(a) Vol.122,No.2, 503-510

66.70. Nonelectronic thermal conduction and heat-pulse propagation in nonmetallic solids

(for thermal conduction in solid metals, see 72.15; for statistical mechanics of lattice vibrations, see 63.70)

V. BAIER and F. VOELKLEIN
Thermal Conductivity of Thin Films
66.70
(a) Vol.118,No.2, K69-K72

T. KOPTE and E. HEGENBARTH
Thermal Conductivity of High-T_c YBCO-Ceramics in Magnetic Fields up to 10 T
66.70 and 74.70; S10.15
(a) Vol.120,No.2, 547-556

T. KOPTE and E. HEGENBARTH
Long-Time Effects in the Thermal Conductivity of YBCO
66.70 and 74.70; S10.15
(a) Vol.121,No.1, K17-K20

67. Quantum Fluids and Solids; Liquid and Solid Helium

67.50. Fermi fluids; liquid helium-3

M. AHMAD and S.J. GLADYSZ
The Dynamic Spin Susceptibility and l=0.1 Spin Waves in a Charged Fermi Liquid of Alkali Metals in a DC Magnetic Field
75.30; 67.50; S2
(b) Vol.159,No.2, K119-K124

M. AHMAD
Spin Waves with Arbitary l in Two-Dimensional Charged Fermi Liquid Immersed in a DC Magnetic Field
67.50 and 75.30
(b) Vol.160,No.1, K65-K70

68. Surfaces and Interfaces; Thin Films and Whiskers
(for impact phenomena, see 79;
for crystal growth, see 61.50)

68.15. Liquid thin films

P. KLEINERT
A Field-Theoretical Treatment of Third-Sound Localization
68.15; 68.30; S5
(b) Vol.158,No.2, 539-550

68.20. Solid surface structures

T. TAKAI, D.K. CHOI, Y.T. THATHACHARI, T. HALICIOGLU, and W.A. TILLER
A Potential Energy Function for GaAs Systems
63.10; 68.20; S7.12
(b) Vol.157,No.1, K13-K18

S. ERKOC, T. HALICIOGLU, and W.A. TILLER
Simulation Calculations on Energetics and Reconstruction Patterns for the GaAs(100) Surface
68.20; S7.12
(b) Vol.157,No.1, K23-K28

F. BECHSTEDT and D. REICHARDT
Calculation of Surface-Induced Core-Level Shifts for Covalent Semiconductors C, Si, Ge, and alpha-Sn. II. (100)2x1 Surfaces
68.20 and 73.20; S5; S5.11; S5.12
(b) Vol.157,No.2, 567-576

N.L. DMITRUK, N.A. FIDRYA, O.V. SNITKO, and E.F. VENGER
Surface Relief and Dissipation Influence on Plasmon-Phonon Polariton Dispersion of Polar Semiconductors
68.20 and 71.36; 71.45; S7.12
(b) Vol.157,No.2, 601-608

E. TAMURA, R. FEDER, G. WALLER, and U. GRADMANN
Determination of the Surface Magnetization of Fe(110) by Spin-Polarized Low-Energy Electron Diffraction
68.20 and 75.30; 61.14; S1.1
(b) Vol.157,No.2, 627-634

T. TAKAI, T. HALICIOGLU, and W.A. TILLER
A Model Potential Energy Function for Systems Containing Al, Ga, and As Atoms
63.10; 68.20; S7.15
(b) Vol.157,No.2, K77-K82

A.M. AFANASEV, S.S. FANCHENKO, and A.V. MASLOV
On the Equivalent Strain and Damage Distributions of Thin Subsurface Layers in the Triple-Crystal X-Ray Diffractometry
61.10; 68.20
(a) Vol.117,No.2, 341-350

J. PASTRŇÁK, J. OSWALD, O. PETŘIČEK, and V.N. OVSYUK
The Degradation of Band Edge Luminescence of GaAs under Laser Light Excitation
78.55; 68.20; S7.12
(a) Vol.118,No.1, 317-326

P.H. DIONISIO, B.A.S. DE BARROS, JR., and I.J.R. BAUMVOL
A Moessbauer Spectroscopic Study of the High-Temperature Oxidation Behaviour of an Austenitic Stainless Steel ...
61.80 and 68.20; 76.80; S1.1
(a) Vol.118,No.2, 473-478

Z.A. MATYSINA, M.I. MILYAN, D.R. RIZDVYANETSKII, and S.YU. ZAGINAICHENKO
The Surface Atomic and Magnetic Order in Crystals
68.20 and 75.30
(b) Vol.159,No.1, K7-K12

YU.B. BOLKHOVITYANOV and A.E. DOLBAK
Surface Composition of Epitaxial Films of Gallium Arsenide and Related Compounds after Termination of Their Growth ...
68.20 and 68.55; S7.12
(a) Vol.119,No.1, 139-144

C. ENGLER
Adsorption of H_2O (Molecular and Dissociative) on Si(100) and Si(111) Surfaces Considered by an Empirical Pair Potential Method
68.20; S5.11
(b) Vol.159,No.2, 721-730

Z.-J. DING, H. YOSHIKAWA, and R. SHIMIZU
Angular Distribution of Elastically Reflected Electrons from Au
68.20 and 79.20; S1.3
(b) Vol.161,No.1, 257-270

E.A. KONDRASHKINA, D.V. NOVIKOV, and S.A. STEPANOV
Study of X-Ray Surface Back Diffraction. Recording of a Diffracted Beam
61.10; 68.20; S5.12
(a) Vol.121,No.1, K9-K14

TS.S. VELINOV and K.P. BRANSALOV
Interference Analysis of Photothermal Generation of Waves in Layered Structures
62.30 and 68.20; 78.20
(a) Vol.121,No.2, 555-562

V.G. BARYSHEVSKII and S.A. STEPANOV
Effect of Two Critical Angles of X-Ray Surface Diffraction on Secondary Emission Yield
61.10 and 78.70; 68.20
(a) Vol.122,No.1, 69-72

M. MAEDER, J. RICHTER, and A. ZEHE
Surface Reordering of (111)Si 7x7 to 1x1 Reconstruction during In-Situ Annealing
68.20; 61.14; S5.11
(a) Vol.122,No.2, K133-K134

A.H. KADIKOVA, N.V. ERSHOV, A.L. AGEEV, YU.V. RUTS, and S.P. SENTEMOV
Investigation of the Local Atomic Structure of the Copper Surface by the EELFS Method
68.20; S1.2
(a) Vol.122,No.2, K143-K146

68.25. Mechanical and acoustical properties of solid surface and interfaces

WANLU WANG and KEJUN LIAO
Effect of the Surface Condition on the Internal Stress of a-Si:H, a-SiN_x:H, and a-Si:H/a-SiN_x:H Heterojunction Films
68.25; S5.11
(a) Vol.119,No.1, K31-K34

Š. PICK, M. TOMÁŠEK, G.V. GADIYAK, and YU.N. MOROKOV
Interactions on Silicon Surfaces and Theory of Moments
68.25 and 68.40; 73.20; S5.11
(b) Vol.160,No.2, 483-490

K.M. LIN and S. LEE
Dislocations near a Bimetallic Interface
61.70 and 68.25
(a) Vol.120,No.2, 497-506

WANLU WANG
Experimental Studies of Light-Induced Stress in a-Si:H/a-SiN_x:H Multilayer Films
68.25; S5.11; S6
(a) Vol.121,No.2, K189-K192

68.30. Dynamics of solid surfaces and interface vibrations

S. PATRA and A. KUMAR
Effect of Lattice Vibrations on the Change in Work Function
68.30 and 73.30; S2
(b) Vol.157,No.1, 209-220

S.A. LEE and S.M. LINDSAY
Surface Brillouin Scattering from Graphite
68.30 and 78.35; S5
(b) Vol.157,No.2, K83-K86

NINGSHENG LUO, WENLAN XU, and SHUECHU SHEN
Embedded-Atom Method for Surface Lattice Dynamics in Copper
68.30; S1.2
(b) Vol.158,No.2, 493-502

P. KLEINERT
A Field-Theoretical Treatment of Third-Sound Localization
68.15; 68.30; S5
(b) Vol.158,No.2, 539-550

K. KUMAZAKI, K. IMAI, and A. ODAJIMA
Estimation of Strains in MBE Grown ZnSe Films by Raman Scattering
63.20 and 68.30; 68.60; S8.12
(a) Vol.119,No.1, 177-182

V.M. FOMIN and E.P. POKATILOV
Optical Properties of Multi-Layer Structures. III. Raman Scattering in Multi-Layer Structures and Superlattices
68.30 and 78.30
(b) Vol.159,No.2, 707-720

SHAOPING LI and L.E. CROSS
Analysis of Magnetostatic Modes Modified by Exchange and Anisotropy Energies
68.30 and 75.10; 75.70
(b) Vol.159,No.2, 861-872

T. YANAGIHARA and K. YOMOGITA
Potential Energy of an Argon Atom Adsorbed on the Stepped and Kinked Surfaces of a Sodium Chloride Crystal
68.30; 68.45; S9.11
(b) Vol.162,No.2, 443-450

68.40. Surface energy of solids; thermodynamic properties

YU.M. MISHIN and D.YA. KATS
On the Thermodynamics of Surface Transitions Close to the Bulk Critical Point
64.60 and 68.40
(b) Vol.159,No.2, 675-688

Š. PICK, M. TOMÁŠEK, G.V. GADIYAK, and YU.N. MOROKOV
Interactions on Silicon Surfaces and Theory of Moments
68.25 and 68.40; 73.20; S5.11
(b) Vol.160,No.2, 483-490

A.O. ALEKSANYAN, V.A. GANSHIN, YU.N. KORKISHKO, and V.A. FEDOROV
Equilibrium and Kinetics of Ion Exchange in Zinc Selenide Single Crystals
66.30 and 68.45; 68.40; S8.12
(b) Vol.161,No.2, 629-636

68.45. Solid-fluid interface processes

A.D. POGREBNYAK
Metastable States and Structural Phase Changes in Metals and Alloys Exposed to High Power Pulsed Ion Beams (Review Article)
64.70 and 79.20; 62.20; 68.45; 68.55; S1.1; S1.2
(a) Vol.117,No.1, 17-52

V.Z. LOZOVSKII and B.I. KHUDIK
The New Mechanism of Physical Adsorption on Solid Surface. I. Adsorption of Nonpolar Molecules
68.45
(b) Vol.158,No.2, 511-520

JIZHONG ZHANG
Fractal Growth of MoO_3 Crystals during Vapor-Phase Deposition
61.50 and 68.45; S10.1
(a) Vol.119,No.1, 41-46

V.Z. LOZOVSKII and B.I. KHUDIK
The New Mechanism of Physical Adsorption on Solid Surface. II. Adsorption of Polar Molecules
68.45
(b) Vol.160,No.1, 137-142

T. ZHANG, M.C. KANG, and W.C. LU
Impurity Effect on the Chemisorption of Hydrogen on Supported Catalysts
68.45; S1.2; S10.1
(a) Vol.120,No.1, K41-K44

R.F. SABIRYANOV, D.L. NOVIKOV, A.L. IVANOVSKII, and V.A. GUBANOV
Quantum-Chemical Calculations of Hydrogen Adsorption on the Polar Surface of Titanium Carbide
68.45 and 73.20; 79.20; S1.61
(b) Vol.160,No.2, 459-472

A.O. ALEKSANYAN, V.A. GANSHIN, YU.N. KORKISHKO, and V.A. FEDOROV
Equilibrium and Kinetics of Ion Exchange in Zinc Selenide Single Crystals
66.30 and 68.45; 68.40; S8.12
(b) Vol.161,No.2, 629-636

YIFAN HU and RONGCHANG YIN
The Gas-Sensing Characteristics of Bismuth Chromate Material
68.45 and 73.60; S11.1
(a) Vol.122,No.1, 249-254

T. YANAGIHARA and K. YOMOGITA
Potential Energy of an Argon Atom Adsorbed on the Stepped and Kinked Surfaces of a Sodium Chloride Crystal
68.30; 68.45; S9.11
(b) Vol.162,No.2, 443-450

I. JAEGER
The Ising Antiferromagnet with Competing Interactions : Locating the Tricritical Point
75.10; 68.45
(b) Vol.162,No.2, 575-584

68.48. Solid-solid interfaces (including bicrystals)

(for grain boundaries, see 61.70)

R.K. BHAN, S.K. LOMASH, and K.C. CHHABRA
Determination and Study of Field Dependent Defect Density in Thermal SiO_2 Using 1,1,1 Trichloroethane (TCA)
61.70 and 68.48; S5.11; S12
(a) Vol.117,No.1, 203-212

A.P. DOSTANKO, V.V. SHATALOV, and G.M. SHAKHLEVICH
Investigation of the Phase Formation Process in Ni-Based Binary Alloy-Silicon Systems
68.48 and 68.55; S1.61; S5.11
(a) Vol.117,No.1, 213-220

E. NEBAUER
Some Remarks on the Interface Displacement in the AuGeNi-GaAs System
68.48; 85; S7.12
(a) Vol.117,No.2, K111-K114

A. BUKALUK
AES Depth Profile Studies of Interdiffusion in the Ag-Cu Bilayer and Multilayer Thin Films
66.30 and 68.48; 61.80; S1.2; S1.3
(a) Vol.118,No.1, 99-108

Y. NAKAMURA, H. KOIKE, and O. NITTONO
Structure of Long Period Superstructure in Cu-Rich gamma-Brass Investigated by Electron Diffraction
61.60 and 61.70; 68.48; S1.2
(a) Vol.118,No.2, 389-400

V.A. GANSHIN and YU.N. KORKISHKO
Proton Exchange in Lithium Niobate and Lithium Tantalate Single Crystals: Regularities and Specific Features (Review Article)
61.40 and 68.48; 78.20; S11.1
(a) Vol.119,No.1, 11-26

A.E. TSURKAN, T.D. SHEMYAKOVA, S.P. MEDVETSKII, and L.A. NAZARENKO
Spectral Response and Detectivity of InAs Hybrid Structures
68.48; 73.60; S7.12; S10.1
(a) Vol.119,No.1, 191-200

I. MANNA, S. BADER, W. GUST, and B. PREDEL
Interdiffusion between In Layer and Bulk Cu or Cu-In Alloy
66.30 and 68.48; S1.2
(a) Vol.119,No.1, K9-K14

M. LESZCZYNSKI, A. KULIK, and P. CIEPIELEWSKI
Lattice Parameters of GaInP/GaAs Epi-Layers at Temperatures 77 to 752 K
68.48 and 68.55; S7.12; S7.15
(a) Vol.119,No.2, 495-502

X.J. WU, F.H. LI, H.Q. HOU, J.M. ZHOU, and H. HASHIMOTO
Direct Observation of Interfaces and Microstructures in GaAs/Si and GaAs/$Ga_{1-x}Al_xAs$/Si by Transmission Electron Microscopy
68.55; 68.48; S5.11; S7.13; S7.15
(a) Vol.119,No.2, 511-522

T. BERGUNDE, M. WIENECKE, and B. THOMAS
High Mobility HgTe and High Resistivity CdTe Grown by MOVPE on GaAs(001)
68.48 and 73.60; S8.13
(a) Vol.121,No.1, K55-K60

X.D. LI and L.Z. ZENG
Structure and Reactivity of the System Si/SiO_2/Fe
68.48 and 79.60; S1.1; S10.1
(a) Vol.121,No.2, K185-K188

R. KASCHNER and TH. VOJTA
Accurate Numerical Calculation of the Electron Density at a Bijellic Interface within the Gradient Method
68.48 and 73.40; S2
(b) Vol.162,No.1, 141-146

A. REIF, P. STREUBEL, A. MEISEL, and D. ZEISSIG
Characterization of the Diffusion and Reaction Behaviour of Ti/Pt/Au Layer Contacts on GaAs by Means of Auger Electron..
68.48 and 73.40; 79.20; S7.12
(a) Vol.122,No.1, 331-340

D. HOEHL, V. HEERA, H. BARTSCH, K. WOLLSCHLAEGER, W. SKORUPA, and V. VOELSKOW
Temperature Dependence of Electron Beam Induced Epitaxial Crystallization of Silicon
68.48 and 68.55; S5.11
(a) Vol.122,No.1, K35-K38

68.55. Thin film growth, structure, and epitaxy

A.D. POGREBNYAK
Metastable States and Structural Phase Changes in Metals and Alloys Exposed to High Power Pulsed Ion Beams (Review Article)
64.70 and 79.20; 62.20; 68.45; 68.55; S1.1; S1.2
(a) Vol.117,No.1, 17-52

M. BOBETH and W. POMPE
Effect of High-Energetic Particles on Thin Film Growth Due to Thermal Spikes
68.55
(a) Vol.117,No.1, 191-202

A.P. DOSTANKO, V.V. SHATALOV, and G.M. SHAKHLEVICH
Investigation of the Phase Formation Process in Ni-Based Binary Alloy-Silicon Systems
68.48 and 68.55; S1.61; S5.11
(a) Vol.117,No.1, 213-220

A. HU, S.S. JIANG, H. CHEN, W. LIU, Y. QIU, and D. FENG
Anomalous Annealing Behaviour of Ta/Al Periodic Multilayer Films
68.55; S4
(a) Vol.117,No.1, 221-226

YU.A. MANZHOSOV and G.A. KACHURIN
Large Area SOI Films Crystallized by Nanosecond Laser Pulses
68.55; S5.11
(a) Vol.117,No.1, K29-K32

M.M. EL-OCKER, F. SHARAF, S.M. YOSSEF, F. METAWE, and M. EL-DESOUKI
Optical Behaviour of Lead Selenide Films
68.55 and 78.65; S8
(b) Vol.157,No.2, 609-614

A. BUDKOWSKI, V. MARINKOVIC, A. PRODAN, and F.W. BOSWELL
The Two-Dimensional Modulation in Bulk and Thin-Film $Au_{2+x}Cd_{1-x}$
61.55 and 68.55; S1.3
(a) Vol.117,No.2, 351-362

M.D. MIKHAILOV, E.A. KARPOVA, Z. CIMPL, and F. KOSEK
Photo- and Thermally Induced Change in the Volume of Thin Chalcogenide Films
68.55; 68.60
(a) Vol.117,No.2, 467-476

V.YU. PETUKHOV, I.B. KHAIBULLIN, M.M. ZARIPOV, E. WIESER, R. GROETZSCHEL, and H. BARTSCH
Formation of Epitaxial $NiSi_2$ by High Dose Implantation and Rapid Thermal Annealing
61.80 and 68.55; S1.61
(a) Vol.117,No.2, 477-484

R. SATHYAMOORTHY, SA.K. NARAYANDASS, C. BALASUBRAMANIAN, and D. MANGALARAJ
Structure, Dielectric, and AC Conduction Studies on Yttrium Fluoride Thin Films
68.55 and 73.60; 77.55; S9
(a) Vol.117,No.2, 495-500

TENG-GE MA, XIAN-JUN YANG, JI-SHENG ZHANG, and PEI-HSIN TSIEN
Localization of the Boundaries in Laser Recrystallization of SOI with Antireflective Stripe
61.80 and 68.55; 73.60; S5.11
(a) Vol.118,No.1, 179-184

H. RATAJCZAK, I. GOŚCIAŃSKA, and T. LUCIŃSKI
Influence of Annealing on Saturation Magnetization and Hall Resistivity in Amorphous $Nd_{12}Co_{82}B_6$ Thin Films
61.40 and 68.55; 73.60; 75.70; S1.2; S1.4
(a) Vol.118,No.1, 189-196

W.A. BADAWY, R.S. MOMTAZ, and E.M. ELGIAR
Solid State Characteristics of Indium-Incorporated TiO_2 Thin Films
68.55 and 78.65; 73.60; S10.1
(a) Vol.118,No.1, 197-202

J.L. OUDAR, B. SFEZ, R. KUSZELEWICZ, J.C. MICHEL, and R. AZOULAY
Optical Nonlinearities of GaAs-Based Epitaxial Structures for All-Optical Switching
78.20; 68.55; S7.12; S7.15
(b) Vol.159,No.1, 181-190

E.A. REPNIKOVA, V.A. GURTOV, and Z.V. PANOVA
The Effect of Synthesis Conditions on the Short-Range Order Characteristics in Silicon Nitride Layers
68.55; S6
(a) Vol.119,No.1, 113-126

B. THOMAS and T.R.N. KUTTY
Formation of Single-Phase Indium Selenide Thin Films by Elemental Evaporation
68.55; 73.60; 78.65; S8.12
(a) Vol.119,No.1, 127-138

YU.B. BOLKHOVITYANOV and A.E. DOLBAK
Surface Composition of Epitaxial Films of Gallium Arsenide and Related Compounds after Termination of Their Growth ...
68.20 and 68.55; S7.12
(a) Vol.119,No.1, 139-144

D. DRAHEIM, A. TEMPEL, A. ZEHE, and D. BAITHER
TEM Investigations of MBE Grown CaF_2 Strained Layers on (111) Silicon
68.55; 61.70; S9.12
(a) Vol.119,No.1, 209-214

A.G. BUDREVICH, A.S. LYUTOVICH, and B.L. OKSENGENDLER
Microselectivity Energy Dependence of Growing Film Orientation during Ion-Molecular Flow Condensation
68.55; S4
(a) Vol.119,No.1, K27-K30

M. LESZCZYNSKI, A. KULIK, and P. CIEPIELEWSKI
Lattice Parameters of GaInP/GaAs Epi-Layers at Temperatures 77 to 752 K
68.48 and 68.55; S7.12; S7.15
(a) Vol.119,No.2, 495-502

K.-H. BAETHER and G. ZIES
Film Thickness Effects and Their Influence on the Stability of the Cr-Si-O/V-O/Al Film System
68.55 and 73.40; S1.61; S4
(a) Vol.119,No.2, 503-510

X.J. WU, F.H. LI, H.Q. HOU, J.M. ZHOU, and H. HASHIMOTO
Direct Observation of Interfaces and Microstructures in GaAs/Si and GaAs/$Ga_{1-x}Al_xAs$/Si by Transmission Electron Microscopy
68.55; 68.48; S5.11; S7.13; S7.15
(a) Vol.119,No.2, 511-522

A. FEDOTOV, B. EVTODYI, L. FIONOVA, YU. ILYASHUK, E. KATZ, and L. POLYAK
Electrical Activity of Grain Boundaries in Shaped Grown Silicon
61.70 and 68.55; 73.20; 73.40; S5.11
(a) Vol.119,No.2, 523-534

P. MIKULSKI, K. FABISIAK, S. ORZESZKO, and F. ROZPŁOCH
The Effect of Starting Material on the Properties of Soft a-C:H Films Obtained by the Glow Discharge Technique
68.55; 78.30; S5
(a) Vol.119,No.2, K127-K130

Š. CHROMIK, V. ŠTRBIK, S. BEŇAČKA, R. ADAM, and M. JERGEL
Thin $YBa_2Cu_3O_{7-x}$ Films Prepared by Low Temperature Coevaporation on Si Substrates without Buffer Layer
74.70; 68.55; S10.15
(a) Vol.119,No.2, K155-K158

J. GUTOWSKI, N. PRESSER, and G. KUDLEK
Optical Properties of ZnSe Epilayers and Films (Review Article)
78.65; 68.55; 71.35; 73.60; 78.55; S8.12
(a) Vol.120,No.1, 11-60

B.X. GU, H. HOMBURG, S. METHFESSEL, and H.R. ZHAI
Crystallization Behavior and Magnetic Properties of Amorphous Nd-Fe-B Thin Films
68.55 and 75.70; 76.80; S1.1
(a) Vol.120,No.1, 159-168

N.S. SALEH, K.A. AL-SALEH, and A.M. AL-SAIE
Kinetics of Ion Beam Induced Mixing in Au/Si System
61.80 and 68.55; 73.60; S1.3; S5.11
(a) Vol.120,No.1, 169-174

A.M. CHAPLANOV and A.N. SHIBKO
Modification of Metal Films by Pulsed Laser Radiation in Organic Liquids
68.55; S1; S1.61
(a) Vol.120,No.1, K37-K40

S. MUELLER-PFEIFFER, H.J. ANKLAM, and W. HAUBENREISSER
A Generalized Ballistic Aggregation Model for the Simulation of Thin Film Growth with Special Consideration of Nodular ...
68.55
(b) Vol.160,No.2, 491-504

N.N. FALEEV, L.I. FLAKS, S.G. KONNIKOV, I.K. SOLOMIN, and S.V. BATASHOVA
The Influence of the Directed Displacement of Atomic Planes on the X-Ray Diffraction Rocking Curves
61.10 and 68.55; S7.12; S7.15
(a) Vol.120,No.2, 327-338

P.V. PAVLOV, A.P. PAVLOV, and D.I. TETELBAUM
The Effect of Ion Implantation on the Substructure of Metal Films
68.55; 61.80; S1.2; S4
(a) Vol.120,No.2, 441-446

N.V. EDNERAL, V.G. KUTILIN, YU.A. LISOVSKII, M.L. SOKOLOVA, A.M. UMPELEV, A.O. TITOV, and L.K. FIONOVA
Thin Film Structures for Magneto-Optical Information Recording
68.55; 75.70; 78.20; S1.1
(a) Vol.120,No.2, 457-466

V.V. ARTAMONOV, V.S. LYSENKO, A.N. NAZAROV, B.D. NECHIPORUK, V.V. STRELCHUK, and M.YA. VALAKH
Radio-Frequency Plasma Treatment and Thermal Annealing in Implanted Si. Raman Study
61.80 and 68.55; 78.30; S5.11
(a) Vol.120,No.2, 475-484

J.C. BUREAU, A. GLACHANT, and B. BALLAND
Thermal Nitridation of SiO_2 Thin Films in NH_3 + H_2O (or O_2) Mixtures
68.55 and 73.60; 78.65; S10.1
(a) Vol.120,No.2, 507-518

J. ATAEV, O.I. KONKOV, E.I. TERUKOV, V.A. VASSILEV, and A.S. VOLKOV
Influence of Preparation Conditions on the Radiative Recombination in a-Si:H Films
68.55 and 78.55; S5.11
(a) Vol.120,No.2, 519-524

E.V. SVIRIDOV, V.A. ALYOSHIN, YU.I. GOLOVKO, I.N. ZAKHARCHENKO, V.M. MUKHORTOV, and V.P. DUDKEVICH
Roentgenographic and Electron-Microscopic Data on Heteroepitaxial $PbTiO_3$/(001)MgO Films
68.55; 77.80; S10
(a) Vol.121,No.1, 157-162

J. DUBOWIK, YU.V. KUDTYAVTSEV, and E.E. KRASSOVSKII
Spectroscopic Ellipsometric Investigation of Multilayered and Solid-State Amorphized Co/Zr Films
68.55 and 78.65; 61.40; S1.2
(a) Vol.121,No.1, 173-180

Y.S. TANG, J.E. CASTLE, H. LIU, J.F. WATTS, and N.K. HUANG
Rutherford Backscattering and X-Ray Photoelectron Spectroscopy Studies of Dual Ion Beam Deposited Zirconium Oxide Films
68.55; S10.1
(a) Vol.121,No.1, K61-K64

A.I. KURBATOV, M. MAEDER, N.N. SOKOLOV, A. TEMPEL, J. TOBISCH, and A. ZEHE
gamma-Ray Diffraction Study of Thin CaF_2/Si(111) Heterostructures Grown by MBE
68.55; S5.11; S9.12
(a) Vol.121,No.1, K65-K68

GUANGHUA CHEN, JINZHANG XU, and FANGQING ZHANG
The ESR Studies of B-Doped a-Si:H Thin Films Prepared by Reactive-Sputtering
61.40 and 76.30; 68.55; S5.11
(a) Vol.121,No.2, 501-506

K. LENKEIT, A. PIRRWITZ, A.I. NIKIFOROV, B.Z. KANTER, S.I. STENIN, and V.P. POPOV
Investigations of MBE Grown Superlattices with Sb Delta-Layers in Silicon by Rutherford Backscattering Spectrometry
68.55; 61.80; S5.11
(a) Vol.121,No.2, 523-532

R. ASHOKAN, V. GOPAL, and K.C. CHHABRA
IR Absorption Spectra of SiO_2 Films Grown by Photo-CVD
68.55 and 78.30; S10.1
(a) Vol.121,No.2, 533-538

YINYUE WANG, FANGQING ZHANG, and GUANGHUA CHEN
Behavior of the Thermal Equilibrium in Reactively Sputtered a-$Si_{1-x}Ge_x$:H Alloy Films Studied by Conductivity and...
73.60; 68.55; S5.11; S5.12
(a) Vol.121,No.2, K177-K180

V.P. PARKHUTIK, YU.E. MAKUSHOK, S.YU. BORISOV, D.V. YAKOVLEV, J.M. MARTINEZ-DUART, J.M. ALBELLA, and A. CLIMENT
Plasma Anodization of Silicon Nitride
68.55; S6
(a) Vol.121,No.2, K181-K184

L.I. BEREZHINSKII, S.I. VLASKINA, G.I. LYASHENKO, and V.E. RODIONOV
On the Possibility of Obtaining the Crystalline Phase in Amorphous Films of Silicon Carbide on Silicon at Low Temperature
68.55; 73.60; 78.55; S6
(a) Vol.121,No.2, K193-K198

H. ATMANI and C. VAUTIER
Crystallization of Amorphous BiSe Alloys: Advantages and Limits of a Proposed Theoretical Model
61.50 and 68.55; S8
(a) Vol.122,No.1, 73-78

J.Y. KIM and R.E. HUMMEL
Hole and Hillock Formation in Gold Metallizations at Elevated Temperatures Deposited on Titanium, Vanadium and Other ...
61.70 and 68.55; S1.3
(a) Vol.122,No.1, 255-274

K. CHERIFI, P. DONOVAN, C. DUFOUR, PH. MANGIN, and G. MARCHAL
Microstructural Observations of Rare-Earth/Iron Multilayers
68.55; S1.1; S1.4
(a) Vol.122,No.1, 311-322

E. GRANTSCHAROVA and D. DOBREV
Recrystallization Texture and Microstructure Development in Silver Films
68.55; S1.3
(a) Vol.122,No.1, K31-K34

D. HOEHL, V. HEERA, H. BARTSCH, K. WOLLSCHLAEGER, W. SKORUPA, and V. VOELSKOW
Temperature Dependence of Electron Beam Induced Epitaxial Crystallization of Silicon
68.48 and 68.55; S5.11
(a) Vol.122,No.1, K35-K38

E.I. TOCHITSKII, O.G. SVIRIDOVICH, E.M. GOLOLOBOV, and D.M. TURTSEVICH
Phase Transformations in $Y_1Ba_2Cu_3O_{7-x}$ Films Prepared by Pulsed Laser Deposition
64.70 and 68.55; 74.70; S10.15
(a) Vol.122,No.2, 555-562

I. GOŚCIAŃSKA, H. RATAJCZAK, P. SOVÁK, P. SPIŠÁK, and M. KONČ
The Structure of Fe-Rich Vapor Quenched Fe-Ti Thin Films
68.55; 61.40; 73.60; S1.1; S1.2
(a) Vol.122,No.2, 575-582

O. EIBL, J.M. TRISCONE, and O. FISCHER
Imaging of $YBa_2Cu_3O_7$/$PrBa_2Cu_3O_7$ Superlattices by TEM
68.55 and 73.40; 64.80; S10.15
(a) Vol.122,No.2, 589-598

P. DANESH, A. TONEVA, P. PANTCHEV, and V. KUDEYAROVA
Microstructure and Optical Properties in CVD a-Si:H Films
68.55 and 78.65; S5.11
(a) Vol.122,No.2, 599-606

W. GRUENEWALD and J. ULLMANN
Cross-Sectional Transmission Electron Microscopy Investigations of RF-Sputtered a-C Films
68.55; 61.16; 68.60; S5
(a) Vol.122,No.2, K129-K132

K.S. CHOI, B.J. KOO, M. JUNG, H.J. SHIN, J.H. LEE, S.B. KIM, S.K. CHANG, and H.L. PARK
(100) CdTe Thin Film Growth by Temperature Gradient Vapor Transport (TGVT)
68.55; 78.40; 78.55; S8.13
(a) Vol.122,No.2, K135-K138

A.M. BARANOV, S.S. GERASIMOVICH, V.M. ELINSON, A.A. KUZIN, V.V. SLEPTSOV, and P.E. KONDRASHOV
Phase Composition Modeling of alpha-C:H Films on the Basis of Their Optical Properties
68.55 and 78.65; S5
(a) Vol.122,No.2, K139-K142

68.60. Physical properties of thin films: nonelectronic

V.M. KOLESHKO and I.V. KIRYUSHIN
Electromigration, Stress, and Pulse Effects in Thin Film Conductors. Model
68.60; S4
(a) Vol.117,No.1, 177-190

M.D. MIKHAILOV, E.A. KARPOVA, Z. CIMPL, and F. KOSEK
Photo- and Thermally Induced Change in the Volume of Thin Chalcogenide Films
68.55; 68.60
(a) Vol.117,No.2, 467-476

S. BAUNACK, S. LANDGRAF, and A. ZEHE
Electron Beam Effects Due to Bombardment of CaF_2-Layers Investigated by Auger Electron Spectroscopy
61.80 and 68.60; 79.20; S9.12
(a) Vol.118,No.1, 203-208

J. BAK-MISIUK, J. WOLF, and U. PIETSCH
The Thermal Expansion Coefficient of $A^{III}B^{V}$ Multilayer Structures
65.70 and 68.60; S7.15
(a) Vol.118,No.1, 209-218

GUANGHUA CHEN, WEI JIA, and FANGQING ZHANG
ESR Study on Annealing Behavior of Reactively Sputtered a-GeN_x:H Films
68.60 and 76.30; S6
(a) Vol.118,No.2, K115-K118

K. KUMAZAKI, K. IMAI, and A. ODAJIMA
Estimation of Strains in MBE Grown ZnSe Films by Raman Scattering
63.20 and 68.30; 68.60; S8.12
(a) Vol.119,No.1, 177-182

M. WAUTELET
On the Transitions between the Crystalline, Amorphous, and Liquid Phases of Silicon and Germanium, when Their Size ...
64.70; 68.60; S5.11; S5.12
(b) Vol.159,No.2, K43-K46

DEYAN HE, FANGQING ZHANG, and GUANGHUA CHEN
Thermal Equilibration in Boron-Doped Hydrogenated Amorphous Silicon-Carbon Alloy Films
68.60; S5.11
(a) Vol.120,No.1, K49-K52

M. DANZIGER and G. ANDRASSY
Experimental and Theoretical Study of Heavy Ion Irradiation Processes in Polymer Materials
61.80 and 68.60; 73.60; S12
(b) Vol.161,No.1, 223-232

W. GRUENEWALD and J. ULLMANN
Cross-Sectional Transmission Electron Microscopy Investigations of RF-Sputtered a-C Films
68.55; 61.16; 68.60; S5
(a) Vol.122,No.2, K129-K132

68.70. Whiskers and dendrites: growth, structure, and nonelectronic properties

JIAN-JUN XU
Global Wave Mode Theory for the Formation of Dendritic Structure on a Growing Needle Crystal
68.70; 61.50
(b) Vol.157,No.2, 577-592

Z. WOKULSKI
Mechanical Properties of TiN Whiskers
62.20 and 68.70; S1.62
(a) Vol.120,No.1, 175-184

K. WOKULSKA, J. HAERTWIG, and S. GROSSWIG
The Bond Method Used to Determine the Value of the Lattice Parameters a for alpha-Fe
68.70; 61.10; S1.1
(a) Vol.121,No.2, 539-546

68.90. Other topics in the structure and nonelectronic properties of surfaces and thin films

YU.E. LOZOVIK and L.M. POMIRCHY
Shell Structure of Two-Dimensional Electron Clusters
68.90
(b) Vol.161,No.1, K11-K14

70. Condensed Matter; Electronic Structure; Electrical, Magnetic, and Optical Properties

71. Electronic States

(see also 63. Lattice dynamics,
73. Electronic structure and electrical properties of surfaces, interfaces, and thin films)

71.10. General theories and computational techniques

B. ESSER
Excitation Propagation in the Presence of a Driven and Kicked Vibration
71.10 and 71.35
(b) Vol.157,No.1, 253-260

A.A. KORNIENKO, A.A. KAMINSKII, and E.B. DUNINA
Dependence of the Line Strength of f-f Transitions on the Manifold Energy. I. Projector on the Basis Nonorthogonal ...
71.10 and 78.20
(b) Vol.157,No.1, 261-266

A.A. KORNIENKO, A.A. KAMINSKII, and E.B. DUNINA
Dependence of the Line Strength of f-f Transitions on the Manifold Energy. II. Analysis of Pr^{3+} in $KPrP_4O_{12}$
71.10 and 78.20; 71.70; S10
(b) Vol.157,No.1, 267-274

R.A. EVARESTOV and V.A. VERYAZOV
The Electronic Structure of Copper Oxide Crystalline Compounds. I. LUC-CNDO Approach to the Electronic Structure of Cu_2O and ...
71.10 and 71.25; 74.10; S1.2; S10.1
(b) Vol.157,No.1, 281-292

N.C. BACALIS and N.I. PAPANICOLAOU
Electron Momentum Distribution and Compton Profiles of Crystalline Rare Gases by the Augmented-Plane-Wave Method
71.25; 71.10; S5.2
(b) Vol.157,No.1, 293-304

M. KOTRLA and V. DRCHAL
Mean-Field Solution Strongly Correlated Systems Using Hubbard Atomic Operators. Hubbard Model with Infinite U
71.10; 75.10
(b) Vol.157,No.2, 635-640

L.A. GEVORGYAN and K.Z. ATSAGORTSYAN
Suppression of Transparency Effect and Enhancement of Particle Reflection in an Asymmetric Potential Well
61.80 and 71.10
(b) Vol.158,No.1, 213-222

B.A. GREENBERG, M.I. KATSNELSON, V.G. KORESHKOV, YU.N. OSETSKII, G.V. PESCHANSKIKH, A.V. TREFILOV, YU.F. SHAMANAEV et al.
On the Possibility of Describing Lattice Properties of Iridium in Terms of Pseudopotential Theory
63.20 and 71.10; 61.70; 62.20; S1
(b) Vol.158,No.2, 441-456

G. BAUMANN and R. DUSCHER
The Functional Equations for the Kinetic and Exchange Energy Functionals
71.10
(b) Vol.158,No.2, 573-588

H.-J. UNGER
Covalent Bonds in Amorphous Semiconductors and Their Influence on the Structure of the Energy Spectrum
71.50; 71.10; 71.25; S5.11; S5.12
(b) Vol.158,No.2, 609-628

J. GRAEFENSTEIN
An Alternative Approach to Calculate the Hydrostatic Pressure in Solids
71.10
(b) Vol.158,No.2, K133-K136

S.A. OSTANIN, N.A. SHILKOVA, and V.P. SHIROKOVSKII
Optical Absorption in Heavy Transition B.C.C. Metals
78.20; 71.10; S4
(b) Vol.159,No.2, 893-902

U. WENSCHUH and E. HEINER
Optical Properties of Gold in the Nonequilibrium Thermalized Stage
71.10; 65.40; 71.38; 78.20; S1.3
(b) Vol.160,No.1, 225-236

B. TANATAR
Calculation of Local-Field Correction in the Two-Dimensional Electron Gas
71.10; 71.45
(b) Vol.160,No.1, 237-244

E. STEFANOVICH, E. SHIDLOVSKAYA, A. SHLUGER, and M. ZAKHAROV
The Modification of the INDO Calculation Scheme and Parametrization for Ionic Crystals
71.10; 71.25; S9.11; S10; S11.1
(b) Vol.160,No.2, 529-540

M. TROTT, S. TROTT, G. TEICHMANN, and CH. SCHNITTLER
Modified Thomas-Fermi Treatment of delta-Doping Layer-Like Structures
71.10 and 73.90; 71.20; 71.50
(b) Vol.161,No.1, K15-K18

W. HERGERT and P. RENNERT
LCAO Recursion Approach in k-Space Representations for the Treatment of Bulk and Surface Problems
71.10; 71.20; 71.25; 73.20
(b) Vol.161,No.2, 667-676

O.I. DUBROVSKII, S.I. KURGANSKII, L.A. RUBTSOVA, and G.W. WOLF
Non-Relativistic and Scalar-Relativistic LAPW Electronic Energy Bands of (001)Cu Films
71.25 and 73.60; 71.10; S1.2
(b) Vol.161,No.2, 697-704

U. WALZER
Systematization of Binary Intermetallic Compounds
61.50; 71.10; 71.45; S2
(b) Vol.162,No.1, 75-88

U. WENSCHUH and E. HEINER
Nonequilibrium Thermoreflectance Evolution of Laser Irradiated Noble Metals
78.20; 65.40; 71.10; S1.3
(b) Vol.162,No.1, 303-310

P.T. SALO and K. KOKKO
Extraction of Valence Kinetic Energy of Metals from Electron Density
71.10; S1.3; S2
(b) Vol.162,No.1, K29-K32

L. REINING and R. DEL SOLE
A Note on the Calculation of Lattice Sums
71.10
(b) Vol.162,No.1, K37-K42

H. BOETTGER and G. KASNER
Multifractal, Localized, and Extended Wave Functions in Generalized Fibonacci Chains
71.10; 71.20; S12.1
(b) Vol.162,No.2, 489-496

D. ŠOKČEVIĆ and V. ZLATIĆ
Pertubation Approach to the Periodic Anderson Hamiltonian with k-Dependent Hybridization
71.10; 71.20; 75.10
(b) Vol.162,No.2, K83-K86

71.20. Electron density of states determinations (including energy states of liquid semiconductors)
(see also 65.40. Electron heat capacity)

J.W. STASIAK, J. KIERUL, and L. KUENNE
Influence of Phonons on the Calculations of the Density of States in Thin Films
71.20 and 73.20; S1.2
(b) Vol.157,No.1, 229-234

K.L. SOROKINA
The Meyer-Neldel Rule in a-Si:H and the Correlation Energy
71.20 and 72.20; S5.11
(b) Vol.157,No.1, K43-K46

G. KONTRYM-SZNAJD
Three-Dimensional Image Reconstruction with Application in Positron Annihilation
71.20 and 78.70
(a) Vol.117,No.1, 227-240

P. NOZAR, V. SECHOVSKY, M. SOLANKI-MOSER, E. BAUER, and E. GRATZ
Point-Contact Spectroscopy on the $Ce(Cu_xAl_{1-x})_5$ System
71.20; 72.20; 75.30; S1.2; S1.4
(a) Vol.117,No.2, 501-508

H. BOETTGER and G. KASNER
Self-Similarity of the Density of States of Electron and Phonon Models in Fibonacci Chains. An Analytic Study
63.50 and 71.20
(b) Vol.158,No.1, 143-152

R.A. EVARESTOV and V.A. VERYAZOV
The Electronic Structure of Copper Oxide Crystalline Compounds II. Chemical Bonding in Copper-Oxygen Crystals
71.20 and 71.25; 74.10; S1.2; S10
(b) Vol.158,No.1, 201-212

V.A. LOBATCH, B.E. KULYABIN, V.P. ZHUKOV, and N.I. MEDVEDEVA
The Electronic Structure of the Alkaline-Earth Oxides. I.The Differences between MgO and LiF, NaF Electronic Structures
71.20 and 71.25; S9.11; S10.1
(b) Vol.158,No.1, 239-248

H.J. FISCHBECK and R. HAYN
On the Density of States of a Disordered Peierls-Froehlich Chain
71.20
(b) Vol.158,No.2, 565-572

EN-GE WANG, LI-YUAN ZHANG, HUAI-YU WANG, and WEI-MIN JIN
Antisite Defects in Si-AlAs/GaAs Modulation-Doped Superlattices
71.55; 71.20; S5.11; S7.15
(b) Vol.158,No.2, 589-594

C. SCHOBER, V.N. ANTONOV, and P. ZIESCHE
Internal Conversion Coefficient for Valence Electrons of Metals. Theory and Application to ^{110}Ag
71.20 and 71.25; S1.3
(b) Vol.159,No.2, 771-780

J. CHOJCAN
The Moessbauer Effect and Positron Annihilation Studies of Iron-Cobalt Alloys
71.20 and 76.80; S1.1
(b) Vol.159,No.2, K59-K62

J. MUSILOVÁ
Method of the Exact Determination of Transition Energies from Modulated Piezoreflectance Spectra of Cubic Crystals
78.20; 71.20; S5.12
(a) Vol.119,No.2, 669-676

V.A. LOBATCH, A.B. SOBOLEV, and I.R. RUBIN
The Electronic Structure of Alkaline-Earth Oxides. II. Crystalline Cluster Approach Formalism
71.20 and 71.25; S10
(b) Vol.160,No.1, 165-172

V.M. ANISHCHIK and N.N. DOROZHKIN
Electronic Structure of TiB_2 and ZrB_2
71.25; 71.20; S1.2
(b) Vol.160,No.1, 173-178

A.A. KLOCHIKHIN, O. KUEHN, and S.G. OGLOBLIN
The Dependence of the Density of States Tail on the Random Potential Statistics
71.20
(b) Vol.160,No.1, 211-218

V.A. LOBATCH and I.R. RUBIN
Electronic Structure of Alkaline-Earth Oxides. III. Resonant Behaviour of MgO Valence Band Top
71.20 and 71.25; S10
(b) Vol.160,No.2, 505-516

V.A. LOBATCH, I.R. RUBIN, and A.B. SOBOLEV
Electronic Structure of Alkaline-Earth Oxides. IV. CaO and SrO Ground State in the Crystalline Cluster Approach
71.20 and 71.25; S10.1
(b) Vol.161,No.1, 271-280

M. TROTT, S. TROTT, G. TEICHMANN, and CH. SCHNITTLER
Modified Thomas-Fermi Treatment of delta-Doping Layer-Like Structures
71.10 and 73.90; 71.20; 71.50
(b) Vol.161,No.1, K15-K18

V.A. LOBATCH, I.R. RUBIN, and P.V. LUSHNIKOV
Electronic Structure of Alkaline-Earth Oxides.
V. Peculiarities of the Hole-Trapping in NaF and MgO
71.20 and 71.25; S9.11; S10.1
(b) Vol.161,No.2, 647-660

W. HERGERT and P. RENNERT
LCAO Recursion Approach in k-Space Representations for the Treatment of Bulk and Surface Problems
71.10; 71.20; 71.25; 73.20
(b) Vol.161,No.2, 667-676

H. AOURAG, B. SOUDINI, B. KHELIFA, and A. BELAIDI
Electron and Positron Distributions in Grey Tin
71.20 and 71.60; S6
(b) Vol.161,No.2, 685-696

P. ANAGNOSTARAS, I. SIKAKANA, and J.E. LOWTHER
Electronic Structure and Thermodynamic Properties of PdH_x
65.50; 71.20; S1
(b) Vol.162,No.2, 413-420

G.V. IONOVA and A.V. NIKOLAEV
Isostructural gamma-alpha Transition in Metallic Cerium as a Mott Transition
64.60 and 71.20; S1.4
(b) Vol.162,No.2, 451-460

H. BOETTGER and G. KASNER
Multifractal, Localized, and Extended Wave Functions in Generalized Fibonacci Chains
71.10; 71.20; S12.1
(b) Vol.162,No.2, 489-496

R. GRILL
Self-Consistent Calculation of Density of States in Impure Semiconductors
71.20; S8.16
(b) Vol.162,No.2, 509-516

Š. VARGA and M. KRAJČÍ
Calculation of the Electronic Structure of the (Al,Zn)-Mg Quasicrystal System
71.20; S4
(b) Vol.162,No.2, K77-K82

D. ŠOKČEVIĆ and V. ZLATIĆ
Pertubation Approach to the Periodic Anderson Hamiltonian with k-Dependent Hybridization
71.10; 71.20; 75.10
(b) Vol.162,No.2, K83-K86

H. SOMMER, A. WEISS, and H. NEUMANN
Valence Band Density of States of $ZnSiP_2$
71.20 and 79.60; S7.16
(b) Vol.162,No.2, K87-K90

71.25. Nonlocalized single-particle electronic states

DOAN NHAT QUANG
Auger Recombination in Disordered Semiconductors. Effect of the Band-Gap Fluctuation
71.25 and 72.20; S7.1
(b) Vol.157,No.1, 235-246

L. WANG
Two-Dimensional Electron Gas in a Magnetic Field. The Non-Degenerate Limit
71.25; 75.10
(b) Vol.157,No.1, 247-252

R.A. EVARESTOV and V.A. VERYAZOV
The Electronic Structure of Copper Oxide Crystalline Compounds. I. LUC-CNDO Approach to the Electronic Structure of Cu_2O and ...
71.10 and 71.25; 74.10; S1.2; S10.1
(b) Vol.157,No.1, 281-292

N.C. BACALIS and N.I. PAPANICOLAOU
Electron Momentum Distribution and Compton Profiles of Crystalline Rare Gases by the Augmented-Plane-Wave Method
71.25; 71.10; S5.2
(b) Vol.157,No.1, 293-304

L.G. GRECHKO, V.M. OGENKO, and A.YU. SEMCHUK
Field Influence on the Parameters of Ferromagnetic Semiconductor Superlattices Formed by Coherent Light Beams
71.25 and 75.10
(b) Vol.157,No.1, 451-458

I.R. SHEIN and A.L. IVANOVSKII
Electronic Structure of Fluorite-Like TiF_2
71.25; S9
(b) Vol.157,No.1, K29-K32

L. PAVESI and F.-K. REINHART
Transfer Matrix Method to Compute Energy Levels of Superlattices
71.25 and 73.40; S7.1
(b) Vol.157,No.2, 615-626

N.A. GUSHCHINA and V.K. NIKULIN
A Tight-Binding Calculation of Band Discontinuities in the $Al_xGa_{1-x}As/GaAs(AlAs)$ Heterostructures
71.25 and 73.40; S7.15
(b) Vol.158,No.1, 175-180

R.A. EVARESTOV and V.A. VERYAZOV
The Electronic Structure of Copper Oxide Crystalline Compounds II. Chemical Bonding in Copper-Oxygen Crystals
71.20 and 71.25; 74.10; S1.2; S10
(b) Vol.158,No.1, 201-212

A.P. MIKHAILOVSKII, A.M. POLUBOTKO, V.D. PROCHUKHAN, and YU.V. RUD
Gapless State in $CuFeS_2$
71.25 and 75.10; S8
(b) Vol.158,No.1, 229-238

V.A. LOBATCH, B.E. KULYABIN, V.P. ZHUKOV, and N.I. MEDVEDEVA
The Electronic Structure of the Alkaline-Earth Oxides. I.The Differences between MgO and LiF, NaF Electronic Structures
71.20 and 71.25; S9.11; S10.1
(b) Vol.158,No.1, 239-248

G. VIELSACK and R. VON BALTZ
Dielectric Function of La_2CuO_4
71.25; 78.20; S11.1
(b) Vol.158,No.1, 249-260

B.K. PANDA, D.P. MAHAPATRA, and H.C. PADHI
Compton Profile Anisotropies of GaAs
71.25 and 78.70; S7.12
(b) Vol.158,No.1, 261-266

T. PISARKIEWICZ and A. KOLODZIEJ
Nonparabolicity of the Conduction Band Structure in Degenerate Tin Dioxide
71.25; S10.1
(b) Vol.158,No.1, K5-K8

S. FUJITA and S. WATANABE
On the Orientation Dependence of the Cyclotron Resonance Peaks for Holes in Ge and Si
71.25 and 76.30; S5.11; S5.12
(b) Vol.158,No.1, K69-K74

J. DRYZEK and E. ROGOWSKA
Positron Annihilation in Sodium and Copper beta-Vanadium Oxide Bronzes
61.70; 71.25; 78.70; S10.15
(a) Vol.118,No.1, 61-64

E.I. SHREDER, I.A. POPOV, E.E. KRASOVSKII, and I.I. SASOVSKAYA
The Optical Properties and Electronic Structure of AgZn Alloy Undergoing Structural Transformation
64.70 and 78.20; 71.25; S1.2; S1.3
(a) Vol.118,No.1, 147-152

H. RODRIGUEZ-COPPOLA, J.L. MARTINEZ-CUELLAR, and R. PEREZ-ALVAREZ
Two-Band Envelope Function Model for Graded Interfaces
71.25 and 73.40; S7.15
(b) Vol.158,No.2, 481-492

H.-J. UNGER
Covalent Bonds in Amorphous Semiconductors and Their Influence on the Structure of the Energy Spectrum
71.50; 71.10; 71.25; S5.11; S5.12
(b) Vol.158,No.2, 609-628

DOAN NHAT QUANG
Auger Recombination in Heavily Doped Semiconductors
71.25 and 72.20; S7.12
(b) Vol.158,No.2, 669-684

J. D'ALBUQUERQUE E CASTRO, R. BECHARA MUNIZ, and L.E. OLIVEIRA
A Model Calculation of the Dielectric Band Structure of Diamond and Silicon
71.25 and 77.90; S5.11
(b) Vol.158,No.2, 743-752

A.L. IVANOVSKII
Band Structure of Ti, Zr, Hf Cubic Monoborides
71.25; S1
(b) Vol.158,No.2, K129-K132

M. HARTMANN, H. STOLZ, and R. ZIMMERMANN
Kinetics of Screening in Optically Excited Semiconductors
71.25 and 72.10; 78.20
(b) Vol.159,No.1, 35-42

C. ELL and H. HAUG
Absorption and Optical Gain Spectra and Band Gap Renormalization of Highly Excited Quantum Well Systems
71.25 and 78.20; 73.40; S7.11; S7.15
(b) Vol.159,No.1, 117-124

A. STAHL
Coupled Two-Level Systems and the Dynamics of Semiconductor Electrons
71.25 and 78.20; 71.35
(b) Vol.159,No.1, 327-338

K.A. KIKOIN, G.P. NIZHNIKOVA, and O.V. FARBEROVICH
Optical Properties of the Mixed-Valent TmSe
71.25; 71.50; 78.20; S1.4; S8
(b) Vol.159,No.2, 743-754

C. SCHOBER, V.N. ANTONOV, and P. ZIESCHE
Internal Conversion Coefficient for Valence Electrons of Metals. Theory and Application to ^{110}Ag
71.20 and 71.25; S1.3
(b) Vol.159,No.2, 771-780

M. DEY (BOSE) and S. GHOSH
Amplification of Acoustohelicon Waves in Piezoelectric Semiconductors. Effect of Nonlinearity in Electron ...
72.10 and 72.50; 71.25; S7.13
(b) Vol.159,No.2, 817-826

S. BISWAS and R. BHATTACHARYA
Magnetic Properties of Single Crystals of Bismuth Telluride Doped with 0.2 at% Lead and Its Thermoelectric Power
75.20; 71.25; 72.20; S8
(b) Vol.159,No.2, 851-860

I.KH. AKOPYAN, A.A. KLOCHIKHIN, B.V. NOVIKOV, M.YA. VALAKH, A.P. LITVINCHUK, and I. KOSAZKII
Optical Spectroscopy of the Superionic Crystals (Review Article)
78.30 and 78.40; 64.70; 71.25; 71.35; 78.55; S9
(a) Vol.119,No.2, 363-416

J.C. PARLEBAS
Periodic Cluster Anderson Model: Kondo, Intermediate Valence, and Heavy Fermion Systems (Review Article)
71.25 and 71.55
(b) Vol.160,No.1, 11-70

F. RODRÍGUEZ, A. CAMACHO, L. QUIROGA, and R. BAQUERO
Electronic Structure of (001) Semiconducting MTe Surfaces (M=Zn,Cd,Hg)
71.25 and 73.20; S8.13
(b) Vol.160,No.1, 127-136

YU.N. ZHURAVLEV and A.S. POPLAVNOI
Band Structure of Sodium and Lithium Azides
71.25; S2
(b) Vol.160,No.1, 153-164

V.A. LOBATCH, A.B. SOBOLEV, and I.R. RUBIN
The Electronic Structure of Alkaline-Earth Oxides.
II. Crystalline Cluster Approach Formalism
71.20 and 71.25; S10
(b) Vol.160,No.1, 165-172

V.M. ANISHCHIK and N.N. DOROZHKIN
Electronic Structure of TiB_2 and ZrB_2
71.25; 71.20; S1.2
(b) Vol.160,No.1, 173-178

A.R. HASSAN
Three-Photon Interband Process in Solid
71.25; 78.20
(b) Vol.160,No.1, 185-192

H. AOURAG, B. KHELIFA, A. BELAIDI, A. TADJER, M. REZKI, and M. GAMOUDI
Positron Band Structures of GaAs and CdTe
71.25 and 71.60; 78.70; S7.12; S8.13
(b) Vol.160,No.1, 193-200

V.S. STEPANYUK, A. SZÁSZ, A.A. KATSNELSON, A.A. GRIGORENKO, and O.V. FARBEROVICH
On the Electronic Structure of Small Ni-P Clusters
71.25; S1.62
(b) Vol.160,No.1, 219-224

E.V. BOGDANOV, D.A. KOMARKOV, and W. KRAAK
Pressure Dependence of the Direct Energy Gap in $Bi_{1-x}Sb_x$ Semiconductor Alloys
71.25 and 72.20; S4
(b) Vol.160,No.1, K33-K36

L.G. GERCHIKOV and A.V. SUBASHIEV
Interface States in Subband Structure of Semiconductor Quantum Wells
71.25 and 73.40; S7.12; S7.15
(b) Vol.160,No.2, 443-458

V.A. LOBATCH and I.R. RUBIN
Electronic Structure of Alkaline-Earth Oxides.
III. Resonant Behaviour of MgO Valence Band Top
71.20 and 71.25; S10
(b) Vol.160,No.2, 505-516

N.I. MEDVEDEVA, V.P. ZHUKOV, M.YA. KHODOS, and V.A. GUBANOV
The Electronic Structure and Cohesive Energy of HfO_2, ZrO_2, TiO_2, and SnO_2 Crystals
71.25; S10.1
(b) Vol.160,No.2, 517-528

E. STEFANOVICH, E. SHIDLOVSKAYA, A. SHLUGER, and M. ZAKHAROV
The Modification of the INDO Calculation Scheme and Parametrization for Ionic Crystals
71.10; 71.25; S9.11; S10; S11.1
(b) Vol.160,No.2, 529-540

M. ZAŁUŻNY
The Influence of the Size Quantization on the Cyclotron Effective Mass and the Density of States in Bi Films
71.25; S4
(b) Vol.160,No.2, 541-548

R. AHUJA, A.K. SOLANKI, and S. AULUCK
Fermi Surface Characteristics of Palladium
71.25; S1.2
(b) Vol.160,No.2, 549-560

H. AOURAG, B. KHELIFA, A. BELAIDI, and T. KOBAYASHI
Interpretation of the Group-Theoretically Unexpected Shape of the Contour Appearing in the Experimental LCW Data for...
71.25; 71.60; 78.70; S5.11; S5.12
(b) Vol.160,No.2, 561-568

K.P. GHATAK and M. MONDAL
On a Simplified Analysis of the Magnetic Susceptibilities of Electrons in Ultrathin Films of Non-Parabolic ...
75.20 and 75.70; 71.25; S7.16
(b) Vol.160,No.2, 673-682

W. LUBCZYŃSKI, J. CISOWSKI, and J.C. PORTAL
Shubnikov-de Haas Effect in $Cd_{3-x}Zn_xAs_2$ Alloys
71.25; 72.20; S1.62
(a) Vol.120,No.2, 525-530

S.V. FOMICHEV, M.I. KATSNELSON, V.G. KORESHKOV, and A.V. TREFILOV
On the Effect of Three-Body Interactions and Proximity of the Fermi Level to the Brillouin Zone Faces on Elastic ...
62.20 and 71.25; S2; S3; S4
(b) Vol.161,No.1, 153-164

V.A. LOBATCH, I.R. RUBIN, and A.B. SOBOLEV
Electronic Structure of Alkaline-Earth Oxides. IV. CaO and SrO Ground State in the Crystalline Cluster Approach
71.20 and 71.25; S10.1
(b) Vol.161,No.1, 271-280

E. DĘBOWSKA and J. RUDZIŃSKA-GIRULSKA
Examination of the Electronic Structure of $Pd_{1-x}Ni_x$ Alloys by Positron Annihilation
71.25; 78.70; S1.2
(b) Vol.161,No.1, 281-286

S. BHOWMIK, C.K. SARKAR, and K. SANTRA
The Effect of the Quantum Screening on the Energy Loss Rates of Hot Electrons in n-$Hg_{0.8}Cd_{0.2}Te$ in the Extreme Quantum .
71.25 and 72.20; S8.15
(b) Vol.161,No.1, 329-334

M. ZAŁUŻNY
Coupling between Intersubband Modes in Wide Quantum Wells
71.25 and 73.90; S7.1
(b) Vol.161,No.1, K19-K24

NGUYEN QUE HUONG, NGUYEN TOAN THANG, and NGUYEN AI VIET
Electron-Hole Exchange Interaction of Excitons in Direct Band Gap Cubic Semiconductors with Degenerated Valence Band
71.25 and 71.35
(b) Vol.161,No.2, 637-646

V.A. LOBATCH, I.R. RUBIN, and P.V. LUSHNIKOV
Electronic Structure of Alkaline-Earth Oxides.
V. Peculiarities of the Hole-Trapping in NaF and MgO
71.20 and 71.25; S9.11; S10.1
(b) Vol.161,No.2, 647-660

W. HERGERT and P. RENNERT
LCAO Recursion Approach in k-Space Representations for the Treatment of Bulk and Surface Problems
71.10; 71.20; 71.25; 73.20
(b) Vol.161,No.2, 667-676

O.I. DUBROVSKII, S.I. KURGANSKII, L.A. RUBTSOVA, and G.W. WOLF
Non-Relativistic and Scalar-Relativistic LAPW Electronic Energy Bands of (001)Cu Films
71.25 and 73.60; 71.10; S1.2
(b) Vol.161,No.2, 697-704

E.M. LEVIN, B.S. KUZHEL, O.I. BODAK, B.D. BELAN, and I.N. STETS
Concentrated Kondo Systems in Solid Solutions on the Base of Europium Ternary Compounds
72.10; 64.75; 71.25; S1.4
(b) Vol.161,No.2, 783-796

G. DLUBEK and W. GERBER
Estimation of Fermi Momentum in Al from Fourier Transform of Positron Annihilation Angular Correlation Curve
71.25 and 78.70; S4
(b) Vol.161,No.2, K81-K84

K.H. HERRMANN, K.-P. MOELLMANN, and J.W. TOMM
On the Broadening Mechanisms near the E_0 Transition in Narrow-Gap (Hg,Cd)Te
78.20; 71.25; 78.55; S8.15
(a) Vol.121,No.2, 635-640

G. PAASCH, P.H. NGUYEN, and G. GOBSCH
Envelope Equation and Wave Function Matching for Narrow-Gap Semiconductors
71.25 and 73.40
(b) Vol.162,No.1, 155-164

A.I. DMITRIEV, Z.D. KOVALYUK, V.I. LAZORENKO, and G.V. LASHKAREV
Two-Dimensional Character of Electron Gas in Layered InSe Crystals
72.20; 71.25; S8
(b) Vol.162,No.1, 213-226

V.I. GAVRILENKO, A.V. POSTNIKOV, N.I. KLYUI, and V.G. LITOVCHENKO
Energy Band Structure and Optical Properties of Wurtzite-Structure of Silicon Carbide Crystals
71.25 and 78.20; S6
(b) Vol.162,No.2, 477-488

A.K. SOLANKI, R. AHUJA, and S. AULUCK
Fermi Surface and Mass Enhancement Factor for Niobium
71.25; S1
(b) Vol.162,No.2, 497-508

M. QUINTERO, R. TOVAR, C. BELLABARBA, and J.C. WOOLLEY
Temperature Variation of Optical Energy Gap and Deformation Potentials in $AgInTe_2$
71.25; 78.20; S8
(b) Vol.162,No.2, 517-522

YU.I. MAZUR, YU.G. RUBO, O.V. SNITKO, and M.V. STRIKHA
Auger Recombination via Defects in Tellurium
72.10; 71.25; 72.40; S5
(b) Vol.162,No.2, 531-538

71.30. Metal-insulator transitions

B.A. ARONZON and I.M. TSIDILKOVSKII
Magnetic-Field-Induced Localization of Electrons in Fluctuation-Potential Wells of Impurities (Review Article)
71.30; 72.20; 75.40; S7.13; S8.15
(b) Vol.157,No.1, 17-60

M.L. MARÍN, M. ORTUÑO, A. HERNÁNDEZ, and J. ABELLÁN
Percolative Treatment of the Verwey Transition in Cobalt-Iron and Nickel-Iron Ferrites
71.30 and 75.40; S11.2
(b) Vol.157,No.1, 275-280

V.A. KASIYAN, D.D. NEDEOGLO, A.V. SIMASHKEVICH, and I.N. TIMCHENKO
Metal-Dielectric Transition in n-ZnSe Obtained by Doping with Shallow Donor Impurity
72.20; 71.30; S8.12
(b) Vol.157,No.1, 341-350

V.A. KASIYAN, D.D. NEDEOGLO, A.V. SIMASHKEVICH, and I.N. TIMCHENKO
Coulomb Gap in n-ZnSe and Its Critical Behaviour in the Region of the Metal-Dielectric Transition
71.30 and 72.60; S8.12
(b) Vol.158,No.1, 293-300

I.M. TSIDILKOVSKII, YU.G. ARAPOV, N.G. GLUZMAN, N.K. LERINMAN, L.D. SABIRSYANOVA, and M.L. ZVEREVA
Magnetic-Field-Induced Metal-Insulator Transition in $Hg_{1-x}Cd_xTe$ and $Hg_{1-x-y}Cd_xMn_yTe$
71.30; S8.16
(b) Vol.160,No.2, K141-K144

G.A. MATVEEV, I.M. TSIDILKOVSKI, and A.T. LONCHAKOV
On the Anisotropy of Scattering in Doped n-Ge at Low-Temperatures
71.30 and 72.20; S5.12
(b) Vol.161,No.1, 303-308

M. SASAKI, G.X. TAI, and M. INOUE
Transient Thermoelectric Effect in Quasi-Two-Dimensional eta-Mo_4O_{11} Crystal
72.20; 71.30; S10
(b) Vol.162,No.2, 553-558

71.35. Excitons and related phenomena (including electron-hole drops)

B. ESSER
Excitation Propagation in the Presence of a Driven and Kicked Vibration
71.10 and 71.35
(b) Vol.157,No.1, 253-260

YA.O. DOVGYI and I.V. KITYK
Antiresonances in the Fundamental Spectra of CdI_2-4H and $CdBr_2$-6R Crystals
71.35 and 78.40; S9
(b) Vol.157,No.1, 473-480

D. FROEHLICH, ST. KIRCHHOFF, P. KOEHLER, and W. NIESWAND
Magnetic Field Splitting and Dispersion of the Lower 1S-Exciton Polariton in KI and RbI
71.35 and 71.70; 71.36; 78.20; S9.11
(b) Vol.158,No.1, 267-276

V.P. GRIBKOVSKII, L.G. ZIMIN, S.V. GAPONENKO, I.E. MALINOVSKII, P.I. KUZNETSOV, and G.G. YAKUSHCHEVA
Optical Absorption near Excitonic Resonance of the MOCVD-Grown ZnSe Single Crystals
71.35 and 78.65; S8.12
(b) Vol.158,No.1, 359-366

R. LÉVY, M.J.M. GOMES, B. KIPPELEN, and B. HOENERLAGE
Coherent Signal Generation in CuCl by Light-Induced Grating and Induced Biexciton Decay
78.40; 71.35; S9
(b) Vol.158,No.1, 391-396

M.A. DE MOURA and J.R. DE OLIVEIRA
Self-Induced Transparency of Excitons
71.35
(b) Vol.158,No.1, K9-K12

K.-H. PANTKE, H. OVER, and I. BROSER
Excitation of the Impurity Emission Due to the Mixed Mode Exciton Absorption in CdS
78.55; 71.35; S8.11
(a) Vol.118,No.1, 343-348

A.D. BROTHERS, F.J. FEUERBORN, and S.E. BAUER
Excitons in Indium Halide Thin Films
78.65; 71.35; S9
(b) Vol.158,No.2, 503-510

W. LAASCH, H. HAGEDORN, T. KLOIBER, and G. ZIMMERER
Fine Structure of the Luminescence of Solid Neon and Its Relation to Exciton Trapping and Desorption Induced by ...
78.55; 71.35; S5.2
(b) Vol.158,No.2, 753-768

D.S. CHEMLA, S. SCHMITT-RINK, W.H. KNOX, D.A.B. MILLER, K.W. GOOSSEN, and G. HASNAIN
Ultrafast Nonlinear Optical Effects in Biased Semiconductor Quantum Wells
71.35 and 78.20; 73.40; S7.12
(b) Vol.159,No.1, 11-28

I. ABRAM
Photon Echoes and the Structure of Wannier Excitons
71.35 and 78.20; S7.12; S7.15
(b) Vol.159,No.1, 29-34

W. RUDOLPH, J. PULS, F. HENNEBERGER, and D. LAP
Femtosecond Studies of Transient Nonlinearities in Wide-Gap II-VI Semiconductor Compounds
71.35 and 78.20; S8.11; S8.12; S8.15
(b) Vol.159,No.1, 49-58

N.A. GIPPIUS, A.L. IVANOV, S.G. TIKHODEEV, and S.G. VYGOVSKII
Phonoriton Transient Phenomena
78.20; 71.35; 71.36; S8.11
(b) Vol.159,No.1, 71-80

F. JAHNKE and K. HENNEBERGER
Nonequilibrium Description of the Intense Resonant Laser Excitation of a Semiconductor
71.35 and 72.30; 78.20; S8.11
(b) Vol.159,No.1, 81-90

M.J.M. GOMES, B. KIPPELEN, R. LEVY, J.B. GRUN, and B. HOENERLAGE
Time-Resolved Four-Wave Mixing Experiments in CuCl
71.35 and 78.20; S9
(b) Vol.159,No.1, 101-106

A.V. NURMIKKO and R.L. GUNSHOR
Excitons and Phonons in CdTe/MnTe and ZnTe/MnTe Single Quantum Wells. Aspects of Confinement and Optical Phonon Coupling
71.35 and 78.20; 73.40; S8.13
(b) Vol.159,No.1, 107-116

E. LACH, V.D. KULAKOVSKII, A. FORCHEL, T.L. REINECKE, J. STRAKA, D. GRUETZMACHER, and G. WEIMANN
Single and Many Particle Effects in the Emission Spectra of Laterally Homogeneous 2D Plasmas
71.35 and 78.55; 73.40; S7.12; S7.15
(b) Vol.159,No.1, 125-132

Y. MASUMOTO, S. IDESHITA, and T. WAMURA
Nonlinear Optical Study of Semiconductor Microstructures, CuCl Quantum Dots, and GaAs Quantum Wells
71.35 and 78.20; 73.40; S7.12; S9
(b) Vol.159,No.1, 133-142

L. WENDLER, F. BECHSTEDT, and M. FIEDLER
Quasi-Two-Dimensional Screening of the Electron-Hole Interaction in Modulation-Doped Quantum Wells
71.35 and 73.40; S7.12; S7.15
(b) Vol.159,No.1, 143-154

A.I. BOBRYSHEVA and S.S. RUSSU
Optical Properties of Biexcitons in ZnSe/$Zn_{1-x}Mn_xSe$ Quantum Wells
71.35 and 78.20; 73.40; S8.12; S8.15
(b) Vol.159,No.1, 155-164

L.V. BUTOV, V.D. KULAKOVSKII, and T.G. ANDERSSON
Photoluminescence of a Dense Quasi-2D Electron-Hole System in Modulation Doped Heterostructure with Strained InGaAs QW
71.35 and 78.55; 73.40; S7.15
(b) Vol.159,No.1, 165-172

K.-H. SCHLAAD, CH. WEBER, D.S. CHEMLA, J. CUNNINGHAM,
D.A.B. MILLER, C.V. HOOF, G. BORGHS, G. WEIMANN, W. SCHLAPP et al.
Nonlinear Optical Properties of GaAs/(AlGa)As Multiple Quantum Well Structures under Quasistationary High Excitation ...
71.35 and 78.20; 73.40; S7.12; S7.15
(b) Vol.159,No.1, 173-180

I.L. ALEINER, L.M. CHISTYAKOVA, S.I. KOKHANOVSKII, R.P. SEISYAN,
A.A. TOROPOV, and T.V. SHUBINA
Passive and Active Q-Switching in an AlGaAs/GaAs SQW Laser. Giant Stark Shift in the Laser p-n SQW Heterostructure
71.35 and 78.20; S7.12; S7.15
(b) Vol.159,No.1, 197-204

M. MAREYEN, F.-J. SCHUETTE, and R. TIEBEL
Dynamics of Carrier-Screened Photon-Exciton Interaction
71.35 and 78.20
(b) Vol.159,No.1, 235-240

V.A. STADNIK
Multistability of Absorption Domains in Bulk CdS Due to Band Edge Nonlinearities
71.35 and 78.20; S8.11
(b) Vol.159,No.1, 241-248

Y.Z. HU, S.W. KOCH, M. LINDBERG, and N. PEYGHAMBARIAN
Theoretical and Experimental Results on Coulomb Effects in Semiconductor Quantum Dots
71.35 and 78.20; S8.12
(b) Vol.159,No.1, 249-258

P. GILLIOT, B. HOENERLAGE, R. LEVY, and J.B. GRUN
Nonlinearities of CuCl Microcrystals
71.35 and 78.20; S9
(b) Vol.159,No.1, 259-266

L.G. ZIMIN, S.V. GAPONENKO, V.YU. LEBED, I.E. MALINOVSKII,
I.N. GERMANENKO, E.E. PODOROVA, and V.A. TSEKHOMSKII
Nonlinear Optical Absorption in I-VII and II-VI Microcrystallites under Quantum Confinement of Elementary ...
71.35 and 78.65; S8.15; S9
(b) Vol.159,No.1, 267-274

H. ROSSMANN, A. SCHUELZGEN, F. HENNEBERGER, and M. MUELLER
Quantum Confined DC Stark Effect in Microcrystallites
71.35 and 78.20; S8.11
(b) Vol.159,No.1, 287-290

TH. PESCHEL
Possibilities for Intrinsic Optical Bistability in Semiconductor-Doped Glasses
71.35 and 78.20
(b) Vol.159,No.1, 291-296

D. FROEHLICH, CH. NEUMANN, B. UEBBING, and R. WILLE
Experimental Investigation of Three-Level Optical Stark Effect in Semiconductors
71.35 and 78.20; 73.40; S7.1; S10.1
(b) Vol.159,No.1, 297-308

H. HAUG, L. BANYAI, J. LIEBLER, and T. WICHT
The Pulsed Nonresonant Optical Stark Effect and the Urbach Tail in Semiconductors
71.35 and 78.20; S7.12; S8.11; S9
(b) Vol.159,No.1, 309-316

R. ZIMMERMANN
Transverse Relaxation and Polarization Specifics in the Dynamical Stark Effect
71.35 and 78.20; S8.11
(b) Vol.159,No.1, 317-326

A. STAHL
Coupled Two-Level Systems and the Dynamics of Semiconductor Electrons
71.25 and 78.20; 71.35
(b) Vol.159,No.1, 327-338

I. BALSLEV
Induced Absorption and the Dynamical Stark Shift Associated with Biexcitons
71.35 and 78.20; S8.12; S9
(b) Vol.159,No.1, 339-346

M. KUWATA-GONOKAMI, R. SHIMANO, J. IWAMATSU, H. AKIYAMA, T. KUGA, and M. MATSUOKA
Optical Nonlinearity and Dynamics of Exciton-Biexciton System in CuCl in the Weak Excitation Regime
71.35 and 78.20; S9
(b) Vol.159,No.1, 347-352

CH. SPIEGELBERG, J. PULS, and F. HENNEBERGER
Nonlinear Optical Properties of CdS at Room Temperature
71.35 and 78.20; S8.11
(b) Vol.159,No.1, 353-362

T. ISHIHARA
The Origin of Third-Order Optical Nonlinearity in Exciton Systems
71.35 and 78.20; 78.65; S9
(b) Vol.159,No.1, 371-378

A. MYSYROWICZ, D.W. SNOKE, and J.P. WOLFE
Progress on Bose-Einstein Condensation in Excitons
71.35 and 78.55; S10.1
(b) Vol.159,No.1, 387-402

V.D. EGOROV, HOANG XUAN NGUYEN, R. ZIMMERMANN, V.S. DNEPROVSKII, M. KASCHKE, and D.S. KHECHINASHVILI
Mott Transition in CdS Revealed by Picosecond Nonlinear Absorption Measurements
71.35 and 78.20; S8.11
(b) Vol.159,No.1, 403-410

H. FIESELER, R. SCHWABE, and J.L. STAEHLI
Electron-Hole Plasma in Cross-Over-Near Ga(As,P)
71.35 and 78.55; S7.15
(b) Vol.159,No.1, 411-416

A.L. IVANOV and V.V. PANASHCHENKO
The Exciton-Biexciton Spectrum Renormalization in the Presence of a Polariton Pumping Wave. The New Mechanism, Exciton ...
71.35 and 78.20; 71.36; S8.11
(b) Vol.159,No.1, 417-424

V.A. KHARCHENKO, G.V. MIKHAILOV, D.K. NELSON, and B.S. RAZBIRIN
Impact Excitation and Auger Decay of Bound Exciton Complexes in a Dense Exciton Gas
71.35 and 78.55; S8.11
(b) Vol.159,No.1, 425-430

N. PRESSER, G. KUDLEK, J. GUTOWSKI, S. DURBIN, D.R. MENKE, M. KOBAYASHI, and R.L. GUNSHOR
High-Density Spectroscopy of ZnSe/GaAs Epilayers in the Near-Band-Edge Region
71.35 and 78.65; 78.55; S7.12; S8.12
(b) Vol.159,No.1, 443-448

L.G. ZIMIN, S.V. GAPONENKO, I.E. MALINOVSKII, V.YU. LEBED, P.I. KUZNETSOV, G.G. YAKUSHCHEVA, and A.V. KUZNETSOV
Excitonic Optical Nonlinearity in II-VI MOCVD-Grown Crystals
71.35 and 78.20; S8.1
(b) Vol.159,No.1, 449-456

A. FORCHEL, B.E. MAILE, H. LEIER, G. MAYER, and R. GERMANN
Optical Spectroscopy on One-Dimensional Compound Semiconductor Structures
71.35 and 78.20; S7.12; S7.15
(b) Vol.159,No.1, 457-470

M.T. PRIMATAROWA and D.I. PUSHKAROV
On the Theory of Solitary Excitons in One-Dimensional Systems
71.35; S12.1
(b) Vol.159,No.1, K17-K22

M.I. SHMIGLYUK, P.I. BARDETSKII, and S.D. TIRON
Polariton-Biexciton Nonlinear Nutation and Stationary Spectrum Renormalization
71.35 and 71.36
(b) Vol.159,No.2, 755-764

A. BECKMANN, K. DETTMER, and F.R. KESSLER
Determination of the Indirect Exciton Energy Gap of GaP Using Thermomodulation in the Temperature Region 80 to 350 K
71.35; 78.20; S7.11
(b) Vol.159,No.2, 765-770

I.KH. AKOPYAN, A.A. KLOCHIKHIN, B.V. NOVIKOV, M.YA. VALAKH, A.P. LITVINCHUK, and I. KOSAZKII
Optical Spectroscopy of the Superionic Crystals (Review Article)
78.30 and 78.40; 64.70; 71.25; 71.35; 78.55; S9
(a) Vol.119,No.2, 363-416

T.A. KUDYKINA
Boundary Conditions in Case of Electromagnetic Wave Absorption
78.20; 71.35
(b) Vol.160,No.1, 365-374

J. GUTOWSKI, N. PRESSER, and G. KUDLEK
Optical Properties of ZnSe Epilayers and Films (Review Article)
78.65; 68.55; 71.35; 73.60; 78.55; S8.12
(a) Vol.120,No.1, 11-60

G. SCHRAMM
Determination of the Free Exciton Capture Cross Section of As Doped Silicon by Photoluminescence
78.55; 71.35; S5.11
(a) Vol.120,No.1, K101-K104

E.L. IVCHENKO, V.P. KOCHERESHKO, I.N. URALTSEV, and D.R. YAKOVLEV
Reflectivity and Photoreflectivity in Superlattices and Quantum Wells
73.40 and 78.20; 71.35; S7.12; S7.15
(b) Vol.161,No.1, 217-222

M.A. JAKOBSON, V.D. KAGAN, R. KATILIUS, and G.O. MUELLER
Impact Ionization of Free and Bound Excitons by Free Carriers in an Electric Field in CdS: Investigation by Optical ...
78.55; 71.35; S8.11
(b) Vol.161,No.1, 395-404

YU.P. GNATENKO, YU.I. ZHIRKO, and Z.D. KOVALYUK
Optical Properties of GaSe Crystals Containing Mn Impurity Atoms. I. Exciton-Phonon Interaction
71.35 and 78.30; S8
(b) Vol.161,No.1, 419-426

YU.P. GNATENKO, YU.I. ZHIRKO, and Z.D. KOVALYUK
Optical Properties of GaSe Crystals Containing Mn Impurity Atoms. II. Excitonic and Impurity Emission
78.55; 71.35; 71.55; S8
(b) Vol.161,No.1, 427-434

E.P. POKATILOV, S.I. BERIL, V.M. FOMIN, and V.V. KALINOVSKII
The Size-Quantized States of the Wannier-Mott Exciton in Structures with Superthin Films of CdTe
71.35 and 78.65; S8.13
(b) Vol.161,No.2, 603-612

NGUYEN QUE HUONG, NGUYEN TOAN THANG, and NGUYEN AI VIET
Electron-Hole Exchange Interaction of Excitons in Direct Band Gap Cubic Semiconductors with Degenerated Valence Band
71.25 and 71.35
(b) Vol.161,No.2, 637-646

A.A. O'CONNEL-BRONIN and A.B. TRESHCHALOV
Exciton Reflection Spectra of NaI Crystals. Effect of Surface Irradiation by Photons and Incidence Angle Dependence
71.35 and 78.40; 73.20; S9.11
(b) Vol.162,No.1, 289-296

71.36. Polaritons (including photon-phonon and photon-magnon interactions)

N.L. DMITRUK, N.A. FIDRYA, O.V. SNITKO, and E.F. VENGER
Surface Relief and Dissipation Influence on Plasmon-Phonon Polariton Dispersion of Polar Semiconductors
68.20 and 71.36; 71.45; S7.12
(b) Vol.157,No.2, 601-608

D. FROEHLICH, ST. KIRCHHOFF, P. KOEHLER, and W. NIESWAND
Magnetic Field Splitting and Dispersion of the Lower 1S-Exciton Polariton in KI and RbI
71.35 and 71.70; 71.36; 78.20; S9.11
(b) Vol.158,No.1, 267-276

N.A. GIPPIUS, A.L. IVANOV, S.G. TIKHODEEV, and S.G. VYGOVSKII
Phonoriton Transient Phenomena
78.20; 71.35; 71.36; S8.11
(b) Vol.159,No.1, 71-80

I.V. BELOUSSOV and YU.M. SHVERA
Kinetics of Partially Coherent Polaritons in Semiconductors. Schwinger-Dyson Equations
71.36 and 78.20
(b) Vol.159,No.1, 91-100

T. HOENIG, J. GUTOWSKI, and I. BROSER
Carrier-Induced Pulse Narrowing in Indium-Doped CdS
71.55 and 78.20; 71.36; S8.11
(b) Vol.159,No.1, 363-370

A.L. IVANOV and V.V. PANASHCHENKO
The Exciton-Biexciton Spectrum Renormalization in the Presence of a Polariton Pumping Wave. The New Mechanism, Exciton ...
71.35 and 78.20; 71.36; S8.11
(b) Vol.159,No.1, 417-424

K.-H. PANTKE, H. OVER, and I. BROSER
Dissipation of Excitonic Polaritons in CdS
71.36 and 78.20; S8.11
(b) Vol.159,No.1, 437-442

V.R. MISKO, S.A. MOSKALENKO, A.H. ROTARU, and YU.M. SHVERA
The Statistical Properties of the Coherently Excited Polariton Wave in a Small Volume Crystal
71.36
(b) Vol.159,No.1, 477-484

M.I. SHMIGLYUK, P.I. BARDETSKII, and S.D. TIRON
Polariton-Biexciton Nonlinear Nutation and Stationary Spectrum Renormalization
71.35 and 71.36
(b) Vol.159,No.2, 755-764

N.F. GASHIMZADE
Guided Polariton Waves in Semiconductor Superlattices
71.36; 78.65
(b) Vol.160,No.2, K113-K116

T. DUMELOW, A.R. EL GOHARY, K.A. MASLIN, T.J. PARKER, D.R. TILLEY, and S.N. ERSHOV
Analysis of Far Infrared Spectra Showing Bulk and Surface Phonon-Polaritons in CdTe Epilayers on GaAs Substrates
78.30 and 78.65; 71.36; S8.13
(b) Vol.161,No.1, 233-244

ZHANG-LONG XU, KA-DI ZHU, and SHI-WEI GU
A Quantum Mechanical Treatment of Surface Polaritons in Semi-Infinite Dielectric Systems
71.36 and 73.90; S7.11
(b) Vol.161,No.1, K25-K28

M.I. KAGANOV and T.I. SHALAEVA
Resonance Polaritons
71.36; 76.20
(b) Vol.162,No.2, 469-476

71.38. Polarons and electron-phonon interactions
(see also 63.20. Phonon-electron interactions in lattices)

L.N. KRISTOFOROV and V.N. KHARKYANEN
Resonance Electron Tunneling under Strong Electron-Vibrational Coupling with a Medium
71.55; 71.38
(b) Vol.157,No.2, K99-K102

N.K. GHOSH, S.C. GHOSH, and R.L. SARKAR
Electron-Phonon Interaction Induced Hybridisation in Mixed-Valence Systems
71.38; S1.4
(b) Vol.158,No.1, 223-228

G.P. TRIBERIS
On the Importance of Correlations in the Study of the Conductivity Due to Small Polaron Hopping Motion in Amorphous ...
72.10; 71.38
(b) Vol.158,No.2, K149-K154

P. KLEINERT
Weak Localization Due to Quasi-Elastic Electron-Phonon Interaction
71.38
(b) Vol.159,No.1, K13-K16

H. RUECKER, M. MORA-RAMOS, and F. COMAS
Magnetopolaron in a Quantum Well. LO-Phonon Confinement Effects
71.38 and 73.40; 63.20; S7.12
(b) Vol.160,No.1, 117-126

U. WENSCHUH and E. HEINER
Optical Properties of Gold in the Nonequilibrium Thermalized Stage
71.10; 65.40; 71.38; 78.20; S1.3
(b) Vol.160,No.1, 225-236

G. BOCHNACKA, V.A. OSIPOV, and V.K. FEDYANIN
Polaron States of the Linear Diatomic Polymer
71.38; S12.1
(b) Vol.160,No.1, K29-K32

A.K. MEIKAP, A.R. JANA, S.K. DE, and S. CHATTERJEE
Weak Localization and Interaction Effects in Thin Disordered Zinc Films
71.38 and 73.60; S1.2
(b) Vol.160,No.2, 473-482

S.C. GHOSH, N.K. GHOSH, and R.L. SARKAR
Role of Electron-Phonon Interaction Induced Hybridisation in Spin Included Mixed-Valence Phenomena
71.38
(b) Vol.161,No.2, 661-666

S. LIU and J. SAK
Microscopic Derivation of a Transport Equation for the Acoustic Polaron
71.38
(b) Vol.161,No.2, 677-684

M. ISTRATE and S. ISTRATE
Drag Effects in Semiconductors with Position-Dependent Band Structure
72.20; 71.38
(b) Vol.161,No.2, 823-830

P.A. APANASEVICH, V.I. KRUGLOV, and I.E. TRALLE
Absorption of Impurity Centres of a Crystal in the Field of a Strong Monochromatic Wave
71.55 and 78.50; 71.38
(b) Vol.161,No.2, 869-882

71.45. Collective effects

L.S. BRIZHIK, A.S. DAVYDOV, and V.N. ERMAKOV
The Magnetic Field Influence on a Superconducting Bisoliton Condensate
74.20; 71.45; S12.1
(b) Vol.157,No.1, 417-424

N.L. DMITRUK, N.A. FIDRYA, O.V. SNITKO, and E.F. VENGER
Surface Relief and Dissipation Influence on Plasmon-Phonon Polariton Dispersion of Polar Semiconductors
68.20 and 71.36; 71.45; S7.12
(b) Vol.157,No.2, 601-608

L.S. BRIZHIK
Bisoliton in Constant Magnetic Field
74.20; 71.45; S12.1
(b) Vol.157,No.2, 649-656

I. YANETKA
The One-Dimensional Plasma Model
63.20 and 71.45
(b) Vol.158,No.1, 181-192

D.B. TRAN THOAI, E. ZEITLER, and DO NAM
Self-Energy of an Electron in the Vacuum Gap between Two Metals
71.45 and 73.20
(b) Vol.158,No.2, 557-564

U. NEUKIRCH, I. BROSER, and R. RASS
Nonlinear Luminescence of Heavily Doped CdS:In Crystals
78.55; 71.45; S8.11
(b) Vol.159,No.1, 431-436

P.F. ROBUSTO and R. BRAUNSTEIN
Optical Measurements of the Highly Damped Surface Plasmon of SnTe
71.45 and 78.65; S8
(a) Vol.119,No.1, 145-154

P.F. ROBUSTO and R. BRAUNSTEIN
Optical Measurements of the Surface Plasmon of Indium-Tin Oxide
71.45 and 78.65; S10
(a) Vol.119,No.1, 155-168

E.M. EPSHTEIN and G.M. SHMELEV
Parametric Interaction of Two-Dimensional Electron Systems in a Strong Electromagnetic Field
71.45
(b) Vol.160,No.1, 179-184

B. TANATAR
Calculation of Local-Field Correction in the Two-Dimensional Electron Gas
71.10; 71.45
(b) Vol.160,No.1, 237-244

A.S. DAVYDOV
Foundations of the Quasi-One-Dimensional Bisoliton Model of Superconductivity of Ceramic Oxides
71.45 and 74.20; S12.1
(b) Vol.160,No.1, 305-312

D.B. TRAN THOAI and E. ZEITLER
Inelastic Scattering of Fast Electrons by Thin Metal Slabs
71.45 and 73.20; 61.80; S4
(a) Vol.120,No.2, 467-474

V.T. MASLYUK
Stables States and Ordering in Amorphous Polymer Semiconductors
61.40; 65.50; 71.45; S8
(b) Vol.161,No.1, 43-54

S.D. ALTEKAR and A.K. RAY
A Correlation Study of Small Potassium Cationic Clusters
61.50; 71.45; S2
(b) Vol.161,No.2, 463-472

V.M. LOKTEV
On the Carrier Spectrum of CuO_2 Planes in High-T_c Superconductors and Its Dependence on Stoichiometry
74.20; 71.45; S12.1
(b) Vol.161,No.2, 731-744

V.N. ERMAKOV, S.P. KRUCHININ, and E.A. PONEZHA
Superconducting Current in a Bisoliton Superconductivity Model
74.20; 71.45
(b) Vol.161,No.2, 745-754

V.N. KADANTSEV and L.N. LUPICHEV
The Capture of an Extra Electron by Supersonic Acoustic Soliton in a Molecular Chain in Presence of Thermal Vibrations
71.45 and 72.10; S12.1
(b) Vol.161,No.2, 769-776

U. WALZER
Systematization of Binary Intermetallic Compounds
61.50; 71.10; 71.45; S2
(b) Vol.162,No.1, 75-88

V.N. KADANTSEV, L.N. LUPICHEV, and A.V. SAVIN
Stationary Autolocalized States in Thermalized Molecular Chains
71.45
(b) Vol.162,No.1, K33-K36

B.P. SINGH
Analysis of Ionic Radii, Second Neighbour Interactions, and Electronegativity Difference for Ionic Crystals
61.50; 71.45; S7.1; S8.1; S9.11
(b) Vol.162,No.2, 329-338

J.X. ZHONG, J.Q. YOU, and J.R. YAN
Collective Charge-Density Excitations in Generalized Fibonacci Semiconductor Superlattices
71.45
(b) Vol.162,No.2, 461-468

71.50. Localized single-particle electronic states (excluding impurities)

E.P. POKATILOV, S.I. BERIL, N.N. SEMENOVSKAYA, and M. FAHOOD
Charge Carrier Energy Spectrum in Multilayer Structures and Superlattices in the Field of the Self-Action Potential
71.50 and 73.40
(b) Vol.158,No.1, 165-174

H.-J. UNGER
Covalent Bonds in Amorphous Semiconductors and Their Influence on the Structure of the Energy Spectrum
71.50; 71.10; 71.25; S5.11; S5.12
(b) Vol.158,No.2, 609-628

B. BOULIL, M. GIRY, and O. HENRI-ROUSSEAU
Non-Resonant Energy Transfer between Donor and Acceptor Systems in Presence of Relaxation
71.50
(b) Vol.158,No.2, 629-642

K.A. KIKOIN, G.P. NIZHNIKOVA, and O.V. FARBEROVICH
Optical Properties of the Mixed-Valent TmSe
71.25; 71.50; 78.20; S1.4; S8
(b) Vol.159,No.2, 743-754

P. LAIS and M. KRAUS
Dynamics of the Pseudo-E $\otimes \varepsilon$ Jahn-Teller Hamiltonian. A Quantum Optical Approach
71.50 and 71.55
(b) Vol.160,No.1, 275-290

L.I. KURKINA and O.V. FARBEROVICH
Electron Structure and Magnetic Properties of Small Iron Particles
71.50; 73.20; 75.25; S1.1
(b) Vol.160,No.1, K37-K42

E.M. DIANOV, A.V. EGIBYAN, S.A. AKOPYAN, A.E. RUSTAMYAN, V.O. SOKOLOV, and V.B. SULIMOV
Computer Simulation on the Threefold Coordinated Oxygen Atom in Vitreous Silicon Dioxide
61.40; 61.70; 71.50; S10.1
(b) Vol.161,No.1, 55-64

M. TROTT, S. TROTT, G. TEICHMANN, and CH. SCHNITTLER
Modified Thomas-Fermi Treatment of delta-Doping Layer-Like Structures
71.10 and 73.90; 71.20; 71.50
(b) Vol.161,No.1, K15-K18

T. IKARI, H. YOKOYAMA, K. MAEDA, and K. FUTAGAMI
Near Band Edge Photoacoustic Spectra of GaAs below Room Temperature
78.20; 71.50; S7.12
(a) Vol.121,No.1, K125-K128

71.55. Impurity and defect levels

A. ZEHE and D. DE LA CRUZ
Effect of Fluctuations in delta-Doping Superlattices on Donor-Acceptor Pair Luminescence
71.55; 78.55; S7.12
(b) Vol.157,No.1, 305-310

M. SAID and M.A. KANEHISA
Excited Acceptor States in II-VI and III-V Semiconductors
71.55; S7.12; S8.13
(b) Vol.157,No.1, 311-322

L.F. ZAKHARENKOV, V.V. KOZLOVSKII, and B.A. SHUSTROV
Transmutation Doping of Indium Phosphide and Gallium Arsenide Due to Protons and alpha-Particles
61.80; 71.55; S7.11; S7.12
(a) Vol.117,No.1, 85-90

V.I. MIKLA, D.G. SEMAK, A.V. MATELESHKO, and A.A. BAGANICH
Photosensitized Deep Gap Center Relaxation in Amorphous Selenium Layers Containing As Additions
71.55; 72.40; 73.60; S5
(a) Vol.117,No.1, 241-250

K.S.R. KOTESWARA RAO and V. KUMAR
Estimation of Trap Concentration in Linearly Graded Junctions Using DLTS
71.55; 73.40
(a) Vol.117,No.1, 251-258

B. SORKIN and H. KÄÄMBRE
Exoemission Mechanisms as Discerned via X-Ray Exposure Variations
71.55 and 79.75; S9.11; S10.1
(a) Vol.117,No.1, K33-K36

R. DEBNATH, J. CHAUDHURY, and S. CHANDRA BERA
Optical Properties and Nature of Coordination of Cu^+ Ions in Calcium-Metaphosphate Glass
78.50 and 78.55; 71.55; S11.1
(b) Vol.157,No.2, 723-734

N. KRISTOFFEL and P. RUBIN
Localized One-Electron States of Impurities on Cu Sites in the CuO_2 Planes of Oxide Superconductors
71.55 and 74.70; S10.15
(b) Vol.157,No.2, K93-K98

L.N. KRISTOFOROV and V.N. KHARKYANEN
Resonance Electron Tunneling under Strong Electron-Vibrational Coupling with a Medium
71.55; 71.38
(b) Vol.157,No.2, K99-K102

H. ALEXANDER, S. DIETRICH, M. HUEHNE, M. KOLBE, and G. WEBER
EBIC Microscopy Applied to Glide Dislocations
61.70; 71.55; 72.20; S7.12
(a) Vol.117,No.2, 417-428

E.G. SALMAN, A.N. KORSHUNOV, and V.N. VERTOPRAKHOV
A DLTS Study of InAs MIS Structures
71.55 and 73.40; S7.12
(a) Vol.117,No.2, 509-514

M. KARAI, K. KIDO, H. NAITO, K. KUROSAWA, M. OKUDA, T. FUJINO, and M. KITAGAWA
Deep Trap Levels in Zn-Annealed ZnSe Single Crystals
71.55; 73.30; S8.12
(a) Vol.117,No.2, 515-526

H. WITTE, W. SIEGEL, G. KUEHNEL, T. FLADE, and H.A. SCHNEIDER
Medium-Deep Acceptors in LEC GaAs
71.55; 72.20; S7.12
(a) Vol.117,No.2, 527-534

R. SALAM
Trapping Parameters of Electronic Defect States in Indium Tin Oxide from AC Conductivity
73.60; 71.55; S10
(a) Vol.117,No.2, 535-540

B.G. DICK and M. GUTOWSKI
Quantum Chemical Calculation of Possible Forms of Electronic-Vibrational Coupling for $F_H(CN^-)$ Defects
71.55; S9.11
(a) Vol.117,No.2, K115-K118

V.V ZUBRITSKII and R.P. PURLYS
The Relation between the Orientation of Streamer Discharges and Anisotropy of Thermal Vibrations of Atoms in II-VI ...
63.20; 71.55; 72.20; 77.50; S8.11
(b) Vol.158,No.1, 91-102

B. SUKUMAR and K. NAVANEETHAKRISHNAN
Diamagnetic Susceptibility of a Donor in a GaAs/$Ga_{1-x}Al_xAs$ Quantum Well Heterostructure and Its Pressure Dependence
73.40 and 75.20; 71.55; S7.12; S7.15
(b) Vol.158,No.1, 193-200

T. TAKEUCHI
Electronic Structure of F Centers in LiH by SCF Xα Method
71.55; 61.70; S9
(b) Vol.158,No.1, 277-282

P.S. GUIMARÃES and M.L. DE SIQUEIRA
The Crystalline Cluster Method for Ionic Solids: The F Center in NaCl
61.70 and 71.55; S9.11
(b) Vol.158,No.1, 283-292

M. WATTENBACH, C. KISIELOWSKI-KEMMERICH, H. ALEXANDER, V.V. KVEDER, T.R. MCHEDLIDZE, and YU.A. OSIPYAN
Electric-Dipole Spin Resonance of Dislocations in Plastically Deformed p-Type Silicon
71.55 and 76.30; S5.11
(b) Vol.158,No.1, K49-K54

I.M. FILIPPOV, L.A. KAZAKEVICH, and P.F. LUGAKOV
The Effect of Irradiation and Heat Treatment on the Dislocation Recombination Activity in n-Type Silicon
61.70 and 71.55; S5.11
(a) Vol.118,No.1, 53-60

I.I. KOLKOVSKII, P.F. LUGAKOV, V.V. LUKYANITSA, and V.V. SHUSHA
Formation of Thermal Donors during Annealing of Radiation Defects in Float-Zone Dislocations-Free n-Si
61.70 and 71.55; S5.11
(a) Vol.118,No.1, 65-70

V.N. BRUDNYI and V.V. PESHEV
Electric Field Effect on the Temperature Dependence of the (E_C-0.33eV) Center Introduction Rate under gamma-Irradiation ...
71.55; 61.80; S7.12
(a) Vol.118,No.1, 219-224

H. ZIMMERMANN, R. BOYN, C. MICHEL, and P. RUDOLPH
Absorption-Calibrated Determination of Impurity Concentrations in CdTe from Excitonic Photoluminescence
71.55; 78.55; S8.13
(a) Vol.118,No.1, 225-234

H. GERSTENBERG and W. GLAESER
Transmutation Doping and Lattice Defects in Degenerate InSb
61.70 and 72.20; 71.55; S7.13
(a) Vol.118,No.1, 241-252

S.P. KUILA
Photostimulation Studies in Pure and Li-Doped Alkali Halides
78.60; 61.70; 71.55; S9.11
(a) Vol.118,No.1, 311-316

EN-GE WANG, LI-YUAN ZHANG, HUAI-YU WANG, and WEI-MIN JIN
Antisite Defects in Si-AlAs/GaAs Modulation-Doped Superlattices
71.55; 71.20; S5.11; S7.15
(b) Vol.158,No.2, 589-594

R. PAESSLER
Novel Method for Detecting Thermal Trap Depths from Phonon-Broadened Photoneutralization Cross Section Bands
71.55; S7.11
(b) Vol.158,No.2, K143-K148

Z. KUČERA, P. HLIDEK, P. HOESCHL, V. KOUBELE, V. PROSSER, and M. ZVÁRA
Magneto-Optical Spectroscopy of p-Type $Hg_{0.8}Cd_{0.2}Te$
72.40; 71.55; S8.15
(b) Vol.158,No.2, K173-K178

M. KANIEWSKA and J. KANIEWSKI
Influence of DX Centers on Photoresponse of GaAs/AlGaAs Double Heterostructures
73.40; 71.55; 72.40; S7.12; S7.15
(a) Vol.118,No.2, 487-490

G.A. ADEGBOYEGA, A. POGGI, and E. SUSI
Some Electrical Characteristics of Fe and Its Annealing Behaviour in FZ Silicon
71.55; 72.20; S5.11
(a) Vol.118,No.2, 491-496

J.A. BAQUEDANO and J.M. CABRERA
Absorption Bands Associated to Photorefractive Erasure in $Bi_{12}SiO_{20}$
71.55; 72.40; 78.50; S10
(a) Vol.118,No.2, 497-504

V.N. BRUDNYI, S.I. BORISENKO, and A.I. POTAPOV
Electrical and Optical Properties and Fermi Level Pinning in Electron Irradiated $ZnSnAs_2$
72.20 and 78.20; 71.55; S7.16
(a) Vol.118,No.2, 505-512

P. SIVAPRASAD, K. RAMESH, and Y.P. REDDY
Optical and EPR Studies of Cu^{2+} in $MgNH_4PO_4 * 6 H_2O$
76.30 and 78.50; 71.55; S11.1
(a) Vol.118,No.2, K103-K106

T. HOENIG, J. GUTOWSKI, and I. BROSER
Carrier-Induced Pulse Narrowing in Indium-Doped CdS
71.55 and 78.20; 71.36; S8.11
(b) Vol.159,No.1, 363-370

D.M. ZAYACHUK, D.D. IVANCHUK, R.D. IVANCHUK, S.S. MASLYANCHUK, and V.I. MIKITYUK
The Effect of Gadolinium Doping on the Physical Properties of Lead Telluride
71.55; 72.80; 75.50; S8
(a) Vol.119,No.1, 215-220

M. STIEBLER, J. REICHARDT, R. HIRRLE, and S. KEMMLER-SACK
Cathodo- and Photoluminescence in the System $Y_{2-x}Tb_xSiO_5:Eu^{3+}$. A Case with an Unusual Effective Energy Transfer in ...
78.55 and 78.60; 71.55; S10.15
(a) Vol.119,No.1, 317-326

J.M. SALLESE, D. LAVIELLE, J. SINGLETON, A. LEVCURAS, J.-C. GRENET, P. GIBART, and J.-C. PORTAL
Non Emergence of DX States in GaAs:Te under Hydrostatic Pressure up to 1.5 GPa
71.55; S7.12
(a) Vol.119,No.1, K41-K46

Y.M. GU, L. FRITSCHE, and N.E. CHRISTENSEN
Chemical Trends in the Formation and Binding Energy of Native Defects in III-V Semiconductors
61.70 and 71.55; S7.1
(b) Vol.159,No.2, 617-628

J. HUGEL and J.C. PARLEBAS
Optical Spectrum Model for a Substitutional Isoelectronic Impurity in Binary Semiconductors
71.55 and 78.50
(b) Vol.159,No.2, 927-936

P. DEMKIN, V. LAZAUSKAS, and Z. JANUŠKEVIČIUS
Investigation of Isovalent Impurity States in GaP Using the Extended Hueckel and the Mulliken-Wolfsberg-Helmholtz Method
71.55; S7.11
(b) Vol.159,No.2, K63-K66

A. MZERD, D. SAYAH, I.J. SAUNDERS, and B.K. JONES
Electrical Properties of $Cd_{1-y}Zn_yS$ and CdS(In) Thin Films Prepared by Pyrolytic Spray Technique
73.60; 71.55; S8.15
(a) Vol.119,No.2, 487-494

V.I. MITROKHIN, S.I. REMBEZA, V.V. SVIRIDOV, and N.P. YAROSLAVTSEV
Acousting Probing of Deep Centres in III-V Semiconductors
71.55; 62.40; S7.1
(a) Vol.119,No.2, 535-544

J.E. BISBEE and N.C. HALDER
Activation Energy and Distribution Function of the EL2 Defect Level in Si-Implanted GaAs
61.70 and 71.55; S7.12
(a) Vol.119,No.2, 545-554

S.E. SARKISOV, A.A. KAMINSKII, A.V. BUTASHIN, YU.E. PERLIN, V.N. ENAKII, and M.G. BLAZHA
Selective Spectroscopy, Luminescence Kinetics of Pr^{3+} Ions, and Phonon Density Calculation in Multicentred and ...
78.55; 63.20; 71.55; 78.30; S11.2
(a) Vol.119,No.2, 607-620

J. REICHARDT, M. STIEBLER, R. HIRRLE, and S. KEMMLER-SACK
Cathodo- and Photoluminescence in Oxyorthosilicates of X_1 and X_2 Type: System $Y_{2-x}Gd_xSiO_5:Tb^{3+}$
78.55 and 78.60; 71.55; S10
(a) Vol.119,No.2, 631-642

B. PÖDÖR
On the Concentration and Composition Dependence of the Acceptor Energy in $Al_xGa_{1-x}As:Ge$ (x<0.40)
71.55; S7.15
(a) Vol.119,No.2, K135-K140

J.C. PARLEBAS
Periodic Cluster Anderson Model: Kondo, Intermediate Valence, and Heavy Fermion Systems (Review Article)
71.25 and 71.55
(b) Vol.160,No.1, 11-70

E.M. DIANOV, V.O. SOKOLOV, and V.B. SULIMOV
Theory of Silicon Vacancy in Vitreous Silica
61.70 and 71.55; S10.1
(b) Vol.160,No.1, 263-274

P. LAIS and M. KRAUS
Dynamics of the Pseudo-$E \otimes \mathcal{E}$ Jahn-Teller Hamiltonian.A Quantum
A Quantum Optical Approach
71.50 and 71.55
(b) Vol.160,No.1, 275-290

E.P. SINYAVSKII and E.YU. SAFRONOV
The Optical Properties of Impurity Center Short-Range
Potentials in the Magnetic Field
78.50 and 78.55; 71.55; S5.12
(b) Vol.160,No.1, 357-364

S.V. DUBONOS and S.V. KOVESHNIKOV
Formation of High Mobile Defects in GaAs under Ar-Plasma
Etching
61.70; 71.55; S7.12
(a) Vol.120,No.1, 77-82

A.A. KAMINSKII, B.V. MILL, A.V. BUTASHIN, K. KURBANOV,
and L.A. POLYAKOVA
Spectroscopy of Disordered $La_3Ga_5SiO_{14}$ Single Crystals Doped
with Pr^{3+}, Ho^{3+}, and Er^{3+} Ions
78.55; 71.55; 78.50; S10
(a) Vol.120,No.1, 253-266

Y. KAMIURA, M. YONETA, and F. HASHIMOTO
Hole Traps Produced by High-Temperature Heat Treatments of
Silicon above 1300 °C
61.70; 71.55; 66.30; S5.11
(a) Vol.120,No.1, K11-K14

T. HATTORI, T. TAKADA, K. KAGAWA, A. MITSUISHI, and Y. KAMIURA
Copper Related Acceptors in Quenched Germanium
71.55 and 78.50; S5.12
(a) Vol.120,No.1, K63-K66

H. KITAGAWA and S. TANAKA
Electrically Active Nickel in Silicon Studied by DLTS in
Several Kinds of Silicon Diodes
71.55; S5.11
(a) Vol.120,No.1, K67-K70

A.I. OLEMSKOI
Self-Consistent Theory of Delocalization
71.55; 71.70
(b) Vol.160,No.2, 569-582

S.V. KOVESHNIKOV, S.V. NOSENKO, and E.B. YAKIMOV
The Peculiarities of Deep Level Defect Passivation in Si by
Atomic Hydrogen
61.70 and 71.55; 66.30; S5.11
(a) Vol.120,No.2, 391-396

H. HEUKENKAMP and W. HOERSTEL
Trap Tunneling in HgCdTe n^{+}-p Junctions Fabricated by Ion Implantation
73.40; 71.55; S8.15
(a) Vol.120,No.2, 485-496

J. KANIEWSKI, M. KANIEWSKA, W. JUNG, and T. PIOTROWSKI
Iron-Related Deep States in Chemical Vapour Deposited Si
71.55 and 73.40; S5.11
(a) Vol.120,No.2, 531-538

O.O. AWADELKARIM, A. HENRY, B. MONEMAR, and J.L. LINDSTROEM
The Role of Group-V Impurity in Defect Formation in Irradiated Silicon
71.55; 61.70; S5.11
(a) Vol.120,No.2, 539-546

R. HIRRLE, J. WIEHL, W. WISCHERT, and S. KEMMLER-SACK
Cathodo- and Photoluminescence in Nd^{3+} and Tb^{3+} Activated Garnets of Type $Cd_{3-x}Ca_xGa_2Ge_3O_{12}$
78.55 and 78.60; 71.55; S11.2
(a) Vol.120,No.2, 643-652

S. SOEDERHOLM, J. HELLBERG, J. KRZYSTEK, G. OLOVSSON, I. OLOVSSON, and J.U. VON SCHUETZ
Electronic Properties and Crystal Structure of a Methylthio Substituted Naphthalene Cation Radical Salt: $(DMbTN)_3(ClO_4)_2$
61.65 and 72.20; 71.55; 75.30; 76.30; S12.1
(b) Vol.161,No.1, 245-256

E.YU. TURAEV, P.P. SEREGIN, and F.S. NASREDINOV
Tin and Iron Impurity Atom States in Modified As_2Se_3 Films
71.55 and 73.60; 78.65; S8
(b) Vol.161,No.1, 287-292

YU.P. GNATENKO, YU.I. ZHIRKO, and Z.D. KOVALYUK
Optical Properties of GaSe Crystals Containing Mn Impurity Atoms. II. Excitonic and Impurity Emission
78.55; 71.35; 71.55; S8
(b) Vol.161,No.1, 427-434

G.A. ADEGBOYEGA and A. POGGI
Intrinsic Gettering of Cr Impurities in p-Type Cz Silicon
71.55; 72.80; 78.30; S5.11
(a) Vol.121,No.1, 181-186

T. FIGIELSKI
Electron Emission from Extended Defects: DLTS Signal in Case of Dislocation Traps
71.55
(a) Vol.121,No.1, 187-194

A. POEPPL and G. VOELKEL
ESR and Photo-ESR Investigations of the V_0^+ Centre in ZnO Raw Material and Li-Doped ZnO Ceramic Powder
71.55; 76.30; S10.1
(a) Vol.121,No.1, 195-204

R. HIRRLE, W. WISCHERT, and S. KEMMLER-SACK
Cathodo- and Photoluminescence in the Mn^{2+} Activated Garnets of Type $Cd_{3-x}Ca_xGa_2Ge_3O_{12}$
78.55 and 78.60; 71.55; S11.2
(a) Vol.121,No.1, 289-298

A. POEPPL and G. VOELKEL
ESR and Photo-ESR Investigations of the Pb^{3+} Centre in Undoped ZnO Ceramics
71.55; 76.30; S10.1
(a) Vol.121,No.1, K69-K72

G. KISSINGER, K. TITTELBACH-HELMRICH, J. KNOPKE, and U. RETZLAFF
On the Gettering Effiency of Buried Layers in Dielectrically Insulated Structures
64.75 and 85; 71.55; S5.11
(a) Vol.121,No.1, K141-K144

A.N. GRUZINTSEV, S.V. DEMISHEV, and T.V. NIKIFOROVA
The Influence of Complex Luminescence Centres on the Luminescent Properties of CaS:Eu
71.55 and 78.55; 78.60; S8
(b) Vol.161,No.2, 705-710

P.A. APANASEVICH, V.I. KRUGLOV, and I.E. TRALLE
Absorption of Impurity Centres of a Crystal in the Field of a Strong Monochromatic Wave
71.55 and 78.50; 71.38
(b) Vol.161,No.2, 869-882

A.D. VISHNOI, V.B. SAPRE, and C. MANDE
A Study of the White Lines Associated with the Rare Earth L_{II} and L_{III} Absorption Discontinouities in Some Oxides ...
78.50; 71.55; S11.1
(b) Vol.161,No.2, 891-896

H. OPYRCHAŁ and M. MANFREDI
Spectroscopic Evidences of Eu-V Dipole Precipitation in KCl:Eu Crystals
61.70; 71.55; 78.55; S9.11
(a) Vol.121,No.2, 407-414

F.S. NASREDINOV, E.YU. TURAEV, P.P. SEREGIN, H.B. RAKHMATULLAEV, and M.K. BAKHADIRKHANOV
Mechanism of the Two-Electron Exchange between Neutral and Ionized Centres of Tin in the $PbS_{1-x}Se_x$ Solid Solutions
61.70 and 71.55; 76.80; S8.15
(a) Vol.121,No.2, 571-578

O.A. GOLIKOVA, N. AMANDZHANOV, M.M. KAZANIN, G.M. KLIMASHIN, and V.V. KUTASOV
Electrical Activity of Impurities in beta-Rhombohedral Boron
71.55; 72.80; 76.80; S5
(a) Vol.121,No.2, 579-586

A. MZERD, D. SAYAH, P.C. RUSSELL, and B.K. JONES
Characterization of Electron Traps in $Au-Cd_{1-y}Zn_yS$ Polycrystalline Schottky Diodes by Capacitance Transient ...
71.55 and 73.40; 85; S8.15
(a) Vol.121,No.2, 587-594

F. TAKEDA, K. MATSUURA, and S. SETO
Admittance Spectroscopy and Photoluminescence Characterization of High-Quality p-Type Cadmium Telluride
71.55 and 78.55; 72.80; S8.13
(a) Vol.121,No.2, 595-606

N.V. BRILLIANTOV, A.I. RUDENKO, YU.V. SHCHERBAKOV, and V.V. ZVEREV
Anomalous Low Temperature Annealing of Electron Irradiated Lithium-Doped Silicon Solar Cells
61.80; 71.55; 85; S5.11
(a) Vol.121,No.2, K155-K158

A.B. SOBOLEV, A.N. VARAKSIN, O.A. KEDA, and A.P. KHAIMENOV
Electronic Structure and Charge State of Oxygen Vacancies in ZrO_2 Perfect Crystals
71.55; 61.70; S10.1
(b) Vol.162,No.1, 165-172

E.M. DIANOV, V.O. SOKOLOV, and V.B. SULIMOV
Theory of Germanium-Related Defects in Vitreous Silicon Dioxide. I. Germanium Atom in Silicon Site
71.55; S10.1
(b) Vol.162,No.1, 173-182

A. DARGYS, N. ŽURAUSKIENĖ, and S. ŽURAUSKAS
Temperature Dependent Field Ionization of Phosphorus Atoms in Silicon
71.55 and 79.70; S5.11
(b) Vol.162,No.1, 183-192

SH.M. EFENDIEV, V. GAVRYUSHIN, G. RACIUKAITIS, V.E. BAGIEV,
V. KAZLAUSKAS, G. PUZONAS, R.A. ALIEV, and E.R. MUSTAFAEV
Induced Absorption Spectroscopy by Lattice Defects in Sillenite-Type Crystals
78.50; 71.55; S11
(b) Vol.162,No.1, 281-288

R. PAESSLER
Zero-Phonon versus Classical Absorption Thresholds Due to Photoexcitation Cross-Section Bands of the A Centre in GaAs
71.55; S7.12
(b) Vol.162,No.1, K47-K52

I. THURZO, V. ŃADAŽDY, and E. PINČÍK
Electric-Field-Enhanced Emission from a Discrete Energy Level at the GaAs-Oxide Interface
73.40; 71.55; S7.12
(a) Vol.122,No.1, 275-284

M.D. GLINCHUK, I.P. BYKOV, V.M. KURLIAND, M. BOUDYŠ,
and K. NEJEZCHLEB
Valency States and Distribution of Manganese Ions in PZT Ceramics Simultaneously Doped with Mn and Nb
71.55; 76.30; S10.15
(a) Vol.122,No.1, 341-346

V.K. ADAMCHUK, V.V. AFANASEV, and A.V. AKULOV
The Electron Trap Activation in Thermal SiO_2
71.55; 73.40; S10.1
(a) Vol.122,No.1, 347-354

T. KUROBORI and N. TAKEUCHI
On the Lifetime in Magnesium-Doped Calcium Oxide Crystals
78.55; 61.70; 71.55; S10.1
(a) Vol.122,No.1, K91-K96

A. KANOUN, S. ALAYA, and H. MAAREF
Spectroscopic Properties of Pr^{3+} and Nd^{3+} Ions in Zinc Tellurite Glasses
71.55; 78.55; S8.13
(b) Vol.162,No.2, 523-530

V. HIZHNYAKOV, M. KARKLINYA, V. NAGIRNYI, L. NAGLI,
and S. ZAZUBOVICH
On the Reasons of the Off-Centre Position of Excited Ga^+ and In^+ Ions in Alkali Halides
71.55 and 78.55; S9.11
(b) Vol.162,No.2, K91-K94

J. BARTOŠ and L. TESAŘ
Some Electrical and Optical Properties of Nickel-Related Deep Levels in Silicon
61.70 and 71.55; 73.40; S5.11
(a) Vol.122,No.2, 607-616

H. LEMKE
Properties of Silicon Crystals Doped with Zirconium or Hafnium
71.55; S5.11
(a) Vol.122,No.2, 617-630

S.S. SIMEONOV, E.I. KAFEDJIISKA, and A.L. GUERASSIMOV
Deep Levels in n-Type Si Substrates Subjected to Ion Cleaning
71.55; 61.80; S5.11
(a) Vol.122,No.2, K147-K150

71.60. Positron states
(see also 78.70. Positron annihilation)

H. AOURAG, B. KHELIFA, A. BELAIDI, A. TADJER, M. REZKI, and M. GAMOUDI
Positron Band Structures of GaAs and CdTe
71.25 and 71.60; 78.70; S7.12; S8.13
(b) Vol.160,No.1, 193-200

H. AOURAG, B. KHELIFA, A. BELAIDI, and Z. BELARBI
Positron Distribution in Semiconductors
71.60; 78.70; S5.11; S5.12; S7.1; S8.1
(b) Vol.160,No.1, 201-210

H. AOURAG, B. KHELIFA, A. BELAIDI, and T. KOBAYASHI
Interpretation of the Group-Theoretically Unexpected Shape of the Contour Appearing in the Experimental LCW Data for...
71.25; 71.60; 78.70; S5.11; S5.12
(b) Vol.160,No.2, 561-568

H. AOURAG, B. SOUDINI, B. KHELIFA, and A. BELAIDI
Electron and Positron Distributions in Grey Tin
71.20 and 71.60; S6
(b) Vol.161,No.2, 685-696

H. AOURAG, B. KHELIFA, and A. BELAIDI
Positron Charge Distribution in Aluminium
71.60; 78.70; S4
(b) Vol.162,No.1, K43-K46

71.70. Level splitting and interactions
(see 75.10. ... in magnetic phenomena,
75.30. Exchange and superexchange interactions,
73.20. Electronic surface states)

A.A. KORNIENKO, A.A. KAMINSKII, and E.B. DUNINA
Dependence of the Line Strength of f-f Transitions on the Manifold Energy. II. Analysis of Pr^{3+} in $KPrP_4O_{12}$
71.10 and 78.20; 71.70; S10
(b) Vol.157,No.1, 267-274

D. BOULANGER and R. PARROT
Molecular Perturbation Model for the Spin-Lattice Coupling Coefficients of $^6A_1$ States to E_θ Strains
71.70; S8.11
(b) Vol.157,No.1, 323-330

WEN-CHEN ZHENG
Spin-Lattice Coupling Coefficients F_{ij} for $SrTiO_3:Cr^{3+}$ Crystal
71.70; 76.30; S11.1
(b) Vol.157,No.1, K33-K38

A.G. MAKHANEK
On the Mechanism of Two-Photon Absorption by Nd^{3+} Ions in Yttrium Aluminium Garnet and Yittrium Lithium Fluoride in ...
71.70; 78.50; S11.2
(b) Vol.157,No.2, 641-648

D. FROEHLICH, ST. KIRCHHOFF, P. KOEHLER, and W. NIESWAND
Magnetic Field Splitting and Dispersion of the Lower 1S-Exciton Polariton in KI and RbI
71.35 and 71.70; 71.36; 78.20; S9.11
(b) Vol.158,No.1, 267-276

S.G. KOZLOVA, YU.G. KRIGER, and S.P. GABUDA
Anisotropic Reorientation of Octahedral Ions According to NMR Relaxation in Solid State
76.60; 71.70; S11
(b) Vol.158,No.1, 329-336

V.K. VORONKOVA, L.V. MOSINA, A.E. USACHEV, and YU.V. YABLOKOV
Weak Pair Exchange Interactions between Cobalt Ions in the Quasi-One-Dimensional Antiferromagnet $CsCoCl_3$
76.30; 71.70; S9.15
(b) Vol.158,No.1, 337-346

WAN-LUN YU
Crystal Field Effect on the g-Factors of ^6S-State Ions
71.70; 76.30; S8.11; S10.1; S11.1
(b) Vol.158,No.1, K13-K16

L.K. AMINOV and YU.YU. KOSTETSKII
Temperature Dependence of EPR Linewidth of Impurities in Van-Vleck Paramagnets
71.70 and 76.30; S9.16
(b) Vol.158,No.2, 595-602

K. ZINK, A. KROST, H. NELKOWSKI, J. SAHM, and H. STUTENBECKER
Energy Transfer Mechanisms in $Zn_{1-x}Mn_xSe$ (0.005<x<0.017) Detected by ODMR
71.70; 76.30; 78.55; S8.15
(b) Vol.158,No.2, 603-608

WEN-CHEN ZHENG
An Estimate of the Spin-Lattice Coupling Coefficients for Trigonal $CdS:Mn^{2+}$ Crystals
71.70; 76.30; S8.11
(b) Vol.158,No.2, K137-K142

S.KH. KARPENKOV
s-d Spin-Orbit Interaction between Conduction Electrons and d-Spin System
72.15; 71.70; S1.1; S1.2
(b) Vol.158,No.2, K169-K172

GUOYIN SHEN, CHANGQING XU, and GUIRU BAI
Cu^{2+}-Cu^{2+} Pair Effect on EPR Parameters in C_{2v} Symmetry Crystals
71.70 and 76.30; S11
(b) Vol.158,No.2, K185-K190

N.P. KOLMAKOVA, I.B. KRYNETSKII, M.M. LUKINA, and A.A. MUKHIN
Crystal Field and Metamagnetic Behavior of Rare Earth Orthoaluminates: $DyAlO_3$
75.30; 71.70; S10
(b) Vol.159,No.2, 845-850

WEN-CHEN ZHENG
Spin-Lattice Coupling Coefficients F_{ij} for d^8 Ions in Cubic Symmetry
71.70; 76.30; S10.1
(b) Vol.159,No.2, K67-K72

M.F. OMELYAN, B.A. LUKIYANETS, and K.D. TOVSTYUK
Change in the Electronic Subsystem Energy by Intercalation of Layered Crystals
71.70
(b) Vol.159,No.2, K73-K76

I.YA. FUGOL, A.M. RATNER, and E.M. YURTAEVA
On Conditions of Spectroscopic Manifestation of Dynamical Jahn-Teller Effect
71.70; 78.50; S5.2
(b) Vol.160,No.1, 245-254

Z.Z. AKSELROD, B.A. KOMISSAROVA, L.N. KRYUKOVA, G.K. RYASNYI, L.G. SHPINKOVA, and A.A. SOROKIN
Electron Quadrupole Interaction of ^{181}Ta in Hf_2Fe Remeasured
71.70; 76.80; S1.1; S4
(b) Vol.160,No.1, 255-262

H.A. BUCKMASTER and K. KUMAR
The C_{3h} Symmetry EPR Spectral Parameters for $(0.005Gd,0.995La)(C_2H_5SO_4)_3 * 9\ H_2O$ and 9 D_2O
76.30; 71.70; S11
(b) Vol.160,No.1, 349-356

WEN-CHEN ZHENG
Spin-Lattice Coupling Coefficients for $CaO:Ni^{2+}$ Crystals
71.70; 76.30; S10.1
(b) Vol.160,No.1, K43-K48

A.I. OLEMSKOI
Self-Consistent Theory of Delocalization
71.55; 71.70
(b) Vol.160,No.2, 569-582

V. GOMATHY, C. BASU, and U.S. GHOSH
Configuration Interaction and Crystal Field Splitting of Cu^{2+} Ion in D_{2d} and D_{3h} Symmetry
71.70; S9.16; S11
(b) Vol.160,No.2, 583-590

S.A. MARSHALL, D.R. YODER-SHORT, C. YU, Y.N. ZHANG, and J.K. FURDYNA
Super Hyperfine Structure in the EPR Spectrum of Divalent Manganese in Single Crystal Zinc Selenide
71.70 and 76.30; S8.12
(b) Vol.160,No.2, 591-600

S.I. KLOKISHNER, A.V. KORYACHENKO, and B.S. TSUKERBLAT
Charge Ordering of Multielectronic Clusters in Mixed-Valence Molecular Crystals
71.70 and 75.30; 64.60; S10
(b) Vol.160,No.2, 641-648

A. GOLTZENÉ and C. SCHWAB
Qualitative and Quantitative Differentiation of Paramagnetic Anion-Antisite-Related Spectra in GaAs
71.70 and 76.30; S7.12
(b) Vol.160,No.2, 649-660

FU-ZHEN LI and ZHAO-MIN LI
The Eight-Order Perturbation Formula of Zero-Field Splitting for $3d^5$ Ions in D_{3d} Site and Its Application to $CaCO_3:Mn^{2+}$
71.70; 76.30; S11.1
(b) Vol.161,No.1, K29-K34

WEN-CHEN ZHENG
Spin-Lattice Coupling Coefficients G_{ij} for Trigonal $Al_2O_3:Fe^{3+}$ and $ZnSiF_6 * 6\ H_2O:Ni^{2+}$ Crystals
71.70; 76.30; S9.16; S10
(b) Vol.161,No.2, 711-716

A.A. SOROKIN, D.D. PERLOV, R.Z. LEVITIN, N.P. KOLMAKOVA, and A.P. DODOKIN
Determination of Ground Quasidoublet Parameters for Ho^{3+} Ion in $(Ho_xY_{1-x})_3Al_5O_{12}$ from Low-Temperature Magnetic ...
71.70 and 75.50; S11.2
(b) Vol.161,No.2, K103-K106

KANG-WEI ZHOU, SANG-BO ZHAO, PING-FENG WU, and JUN-KAI XIE
EPR Parameters and Spectral Fine Structure of Ni^{2+} in $LiNbO_3$
71.70 and 76.30; 78.50; S11.1
(b) Vol.162,No.1, 193-198

72. Electronic Transport in Condensed Matter

(for surfaces, interfaces, and thin films, see 73)

72.10. Theory of electronic transport; scattering mechanism

R. LENK
First Order Quantum Corrections to the Current Driven by a Force and Einstein Equivalence
72.10
(b) Vol.157,No.1, 331-340

K. KRÁL, B. HEJDA, and Z. KHÁS
Hot LO-Phonon Quasi-Steady State Distribution Function in Polar Semiconductors
63.20 and 72.10; S7.1; S8.1
(b) Vol.157,No.2, 667-676

A. URBANIAK-KUCHARCZYK, T. BALCERZAK, J. MIELNICKI, and G. WIATROWSKI
The Magnetic Contribution to Electrical Resistivity in Diluted Ising Ferromagnets
72.10 and 75.10
(b) Vol.157,No.2, K103-K108

R. LENK
Theory of the Residual Resistivity Dipole.Flow of Non-Interacting Particles under Weak Driving Force around a Fixed Point Scatterer
72.10
(b) Vol.158,No.2, 653-664

E.M. KOGAN
Anderson Localization in Ferromagnetic Semiconductors Due to Spin Disorder. III. Selfconsistent Localization Theory ...
72.20; 72.10
(b) Vol.158,No.2, 665-668

G.P. TRIBERIS
On the Importance of Correlations in the Study of the Conductivity Due to Small Polaron Hopping Motion in Amorphous ...
72.10; 71.38
(b) Vol.158,No.2, K149-K154

B.M. ASKEROV, G.I. GUSEINOV, B.I. KULIEV, and S.R. FIGAROVA
Influence of the Surface Scattering upon the Conductivity in Films with a Quasi-Two Dimensional Electron Spectrum
72.10; 73.60
(b) Vol.158,No.2, K179-K184

M. HARTMANN, H. STOLZ, and R. ZIMMERMANN
Kinetics of Screening in Optically Excited Semiconductors
71.25 and 72.10; 78.20
(b) Vol.159,No.1, 35-42

E. HEINER
Screening of the Coulomb Potential in the Athermal Stage of Highly Excited Direct-Gap Semiconductors
72.10; 72.20
(b) Vol.159,No.1, 43-48

W. SCHAEFER and K. HENNEBERGER
Pulse Propagation and Carrier Kinetics in Laser Excited Semiconductors
72.10 and 78.20; S7.12
(b) Vol.159,No.1, 59-70

H. LEON, F. COMAS, and M. SUHRKE
Influence of Level Broadening in Mobility Calculations for Low-Dimensional Semiconductor Heterostructures
72.10 and 73.40; S7.12
(b) Vol.159,No.2, 731-742

I.I. BOIKO and YU.M. SIRENKO
Scattering of Two-Dimensional Electron Gas on the Semibounded Three-Dimensional Electron Gas
72.10 and 73.40
(b) Vol.159,No.2, 805-816

M. DEY (BOSE) and S. GHOSH
Amplification of Acoustohelicon Waves in Piezoelectric Semiconductors. Effect of Nonlinearity in Electron ...
72.10 and 72.50; 71.25; S7.13
(b) Vol.159,No.2, 817-826

P. KUIVALAINEN and K. LINDBERG
Hot Electron Transport in Submicron Semiconductor Devices
72.10 and 73.40; 85; S7.1
(b) Vol.159,No.2, 827-836

S.YA. GOROSHCHENKO
Integer Quantum Hall Effect in Magnetic and Electric Fields of Arbitrary Orientations
72.10 and 73.40
(b) Vol.159,No.2, K55-K58

V.K. ADAMCHUK, A.E. KUCHMA, and S.L. MOLODTSOV
Analytical Description of Electric Field Effect on the Electron Transport in Semiconductors and Insulators
72.10
(b) Vol.159,No.2, K77-K82

S. HILDEBRANDT and W. HERGERT
Unified Theoretical Description of the CL, EBIC, PL, and LBIC Contrast Profile Area of an Individual Surface-Parallel ...
78.55 and 78.60; 72.10
(a) Vol.119,No.2, 689-700

H. LEON and F. COMAS
Confined-LO-Phonon-Limited Mobility in a Semiconductor Quantum Well
72.10 and 73.40; S7.12; S7.15
(b) Vol.160,No.1, 105-116

R. ZIMMERMANN and K. KRÁL
LO-Phonon Generation and Electron Energy Collision Broadening in Polar Semiconductors
63.20 and 72.10
(b) Vol.160,No.2, K109-K112

A. URBANIAK-KUCHARCZYK, T. BALCERZAK, J. MIELNICKI, and G. WIATROWSKI
The Magnetic Contribution to Electrical Resistivity in the Ising Ferromagnet with Fluctuating Exchange Integral
72.10 and 75.10
(b) Vol.160,No.2, K129-K132

V.N. KADANTSEV and L.N. LUPICHEV
The Capture of an Extra Electron by Supersonic Acoustic Soliton in a Molecular Chain in Presence of Thermal Vibrations
71.45 and 72.10; S12.1
(b) Vol.161,No.2, 769-776

E.M. LEVIN, B.S. KUZHEL, O.I. BODAK, B.D. BELAN, and I.N. STETS
Concentrated Kondo Systems in Solid Solutions on the Base of Europium Ternary Compounds
72.10; 64.75; 71.25; S1.4
(b) Vol.161,No.2, 783-796

R. LENK
Transport with Microscopic Inhomogeneities. Residual Resistivity Dipole, Landauer Formula and the ...
72.10
(b) Vol.161,No.2, 797-814

R. LENK
Resistive Wires Instead of Ideal Leads: Breakdown of the Channel Equipartition Rule
72.10
(b) Vol.162,No.1, 227-236

A.O. DONGHOL DIALLO, E. MROSAN, and P. ZIESCHE
Model Calculation of the Specific Dislocation Resistivity for Copper
61.70 and 72.10; S1.2
(b) Vol.162,No.1, 237-242

YU.I. MAZUR, YU.G. RUBO, O.V. SNITKO, and M.V. STRIKHA
Auger Recombination via Defects in Tellurium
72.10; 71.25; 72.40; S5
(b) Vol.162,No.2, 531-538

A.YU. SEMCHUK, L.G. GRECHKO, and V.M. OGENKO
Theory of Transport Phenomena in Ferromagnetic Semiconductors with Superlattice Produced by Coherent Light Beams
72.10 and 75.10
(b) Vol.162,No.2, 539-544

72.15. Electronic conduction in metals and alloys

A.K. MISHRA, B.B. SAHAY, and K.K. MUKHERJEE
Structure and Electrical Resistivity of Alkali-Alkali and Lithium-Based Liquid Binary Alloys
61.25; 72.15; S2
(b) Vol.157,No.1, 85-92

E. FLOUDA, N. BOUKOS, and C. PAPASTAIKOUDIS
Hall Coefficient of Dilute CuPd/CuRh/CuZr and CuY Alloys. d-Resonance Scattering
72.15; S1.2
(b) Vol.157,No.1, 351-356

S.J. TZENG, Y.D. YAO, and T.H. CHUANG
Electrical Resistivity of Nickel-Rich Nickel-Indium Alloys between 10 and 800 K
72.15; S1.2
(a) Vol.117,No.1, K47-K52

S. SALEH and M.A. ZAKI EWISS
Studies of the Amorphous-Crystalline Transition in the Metallic Glass $Fe_{32}Ni_{36}Cr_{14}P_{12}B_6$
72.15; S1
(a) Vol.117,No.2, K133-K136

P. CZARNECKI, M. IDZIKOWSKI, and A. WRZECIONO
Magnetoresistance of $Tb_xCo_{70-x}B_{30}$ and $Tm_xCo_{70-x}B_{30}$ Amorphous Ribbons in High Magnetic Fields
72.15; S1.2
(a) Vol.117,No.2, K137-K140

F.M. MUNTYANU, M.I. ONU, and V.G. KISTOL
Quantum Oscillations of Galvanomagnetic Effects in Bismuth Bicrystals
72.15; S4
(b) Vol.158,No.1, 301-306

R.K. ISLAMGALIEV, N.A. AKHMADEEV, R.R. MULYUKOV, and R.Z. VALIEV
Grain Boundary Influence on the Electrical Resistance of Submicron Grained Copper
72.15; S1.2
(a) Vol.118,No.1, K27-K30

F. VOELKLEIN and E. KESSLER
Thermal Conductivity and Thermoelectric Figure of Merit of Thin Antimony Films
72.15 and 73.60; S4
(b) Vol.158,No.2, 521-530

S.KH. KARPENKOV
s-d Spin-Orbit Interaction between Conduction Electrons and d-Spin System
72.15; 71.70; S1.1; S1.2
(b) Vol.158,No.2, K169-K172

Y. ITOH, H. KADOMATSU, J. SAKURAI, and H. FUJIWARA
Transport Properties of $Yb_xIn_{1-x}Cu_2$
72.15; S1.2; S1.4
(a) Vol.118,No.2, 513-518

L. VATSKICHEV, J. MUCHA, ZH. GEORGIEV, and M. VATSKICHEVA
Measurements of the Rotational Hysteresis Losses in Magnetic Films by the Planar Hall Effect
72.15 and 75.70
(a) Vol.118,No.2, K95-K98

J. MOLENDA, A. STOKLOSA, DO THAN, and T. BĄK
Electrical Properties of $Ti_{1+y}S_2$ at Temperatures 77 to 300K
72.15; S1.63
(a) Vol.119,No.1, 221-228

F. SACHSLEHNER
Anisotropic Electron Scattering by Phase Boundaries in a Copper-Beryllium Alloy
72.15; S1.2
(a) Vol.119,No.1, 229-236

A. RAOUFI and S. ARAJS
Anisotropy of Magnetoresistivity in Crystalline and Amorphous Ferromagnets
72.15 and 75.50; S1.1
(a) Vol.119,No.1, K75-K78

W. EICHLER and M. VOIGT
On the Analysis of the Nernst-Ettingshausen Effect
72.15 and 72.20
(b) Vol.159,No.2, K87-K92

J. MOLENDA, A. STOKŁOSA, S. MROWEC, and DO THAN
On the Defect Structure and Electronic Properties of Titanium Disulfide
72.15; 61.70; S1.63
(a) Vol.119,No.2, 571-582

H.K. LACHOWICZ, E. PULIDO, A. SIEMKO, and A. HERNANDO
Stress-Forced Magnetoresistance in Near-Zero Magnetostrictive Metallic Glasses
72.15 and 75.80; 75.60; S1.1; S1.2
(a) Vol.119,No.2, 595-600

J. KIERUL and J. LEDZION
Conductivity of Thin Metallic Films
72.15 and 73.60
(a) Vol.119,No.2, K117-K120

S.P. ZAKHAROV and V.V. USTINOV
The High Frequency Quantum Size Effect in the Impedance of Metal Films
73.60; 72.15; S4
(b) Vol.160,No.1, 97-104

W. SCHILLER and K. WESCHKE
Calculation of the Normal-State Resistivity for High-T_c Superconductors
72.15 and 74.70; S10
(b) Vol.160,No.1, 297-304

C.A. MOYER, S. ARAJS, Y.D. YAO, and Y.Y. CHEN
Electrical Resistivity of Metal Matrix Composites
72.15; S4
(a) Vol.120,No.1, K71-K72

W. MAYR and A. PHILIPP
Charge Density Wave and Contact Effects in $NbSe_3$ Observed with Microwave Hamonic Mixing
72.15; S1.63
(b) Vol.160,No.2, 613-624

V.S. STEPANYUK, A.A. KATSNELSON, O.S. TRUSHIN, P. NAYAK, and A. SZASZ
Electrical Resistivity of Amorphous $Ni_{1-x}P_x$ Alloys Calculated by Molecular Dynamics
72.15; S1.2
(b) Vol.160,No.2, K117-K122

A. PAJA
Concentration and Temperature Dependence of the Resistivity of Amorphous $Fe_{1-x}Zr_x$ Alloys
72.15; S1.1
(b) Vol.160,No.2, K123-K128

V.I. GOSTISHEV, S.E. DEMYANOV, V.R. SOBOL, and D.V. PASHIK
On the Anisotropy of Electron-Dislocation Scattering in Aluminum in Charge Transport
61.70 and 72.15; S4
(b) Vol.160,No.2, K133-K136

F. SACHSLEHNER, C. WUTTE, and F. STANGLER
Precipitation Effects on Thermopower of a Commercial Cu-2wt%Be Alloy
72.15; S1.2
(a) Vol.120,No.2, 581-592

V.G. SHEPELEVICH
Structure and Electrical Properties of Quickly-Quenched Foils of Antimony and Its Alloys
61.55 and 72.15; S4
(a) Vol.120,No.2, K173-K176

R.B. PRABHU and A.F. BARBOSA
Magnetoresistivity of a Crystal Field Split Kondo System
72.15; S1.4
(b) Vol.161,No.1, 309-318

Y.D. YAO, Y.Y. CHEN, S.J. TZENG, and T.H. CHUANG
Electrical Resistivity, Magnetization, and Grain Boundary Precipitates in Ni-Sn Alloys
72.15; 75.60; S1.2
(a) Vol.121,No.1, 213-218

T. LUCIŃSKI
The Resistivity of Amorphous $Fe_{1-x}Ti_x$ Thin Films
73.60; 72.15; S1.1
(a) Vol.122,No.1, 323-330

N. MATHUR and A.N. NIGAM
Temperature Dependence of Electrolytic Hydrogen Diffusion in Nickel: A Thermoelectric Study
66.30 and 72.15; S1.2
(a) Vol.122,No.1, K59-K62

M.A. ZAKI EWISS and S. SALEH
Temperature Dependence of the Resistivity of $Fe_{40}Ni_{40}P_{14}B_6$ Amorphous and Crystalline Alloys
72.15; S1.1
(a) Vol.122,No.1, K63-K68

72.20. Conductivity phenomena in semiconductors and insulators
(for nonelectronic conduction, see 66.70)

B.A. ARONZON and I.M. TSIDILKOVSKII
Magnetic-Field-Induced Localization of Electrons in Fluctuation-Potential Wells of Impurities (Review Article)
71.30; 72.20; 75.40; S7.13; S8.15
(b) Vol.157,No.1, 17-60

DOAN NHAT QUANG
Auger Recombination in Disordered Semiconductors. Effect of the Band-Gap Fluctuation
71.25 and 72.20; S7.1
(b) Vol.157,No.1, 235-246

V.A. KASIYAN, D.D. NEDEOGLO, A.V. SIMASHKEVICH, and I.N. TIMCHENKO
Metal-Dielectric Transition in n-ZnSe Obtained by Doping with Shallow Donor Impurity
72.20; 71.30; S8.12
(b) Vol.157,No.1, 341-350

K.L. SOROKINA
The Meyer-Neldel Rule in a-Si:H and the Correlation Energy
71.20 and 72.20; S5.11
(b) Vol.157,No.1, K43-K46

S. AHMED, A.M. SHAMAH, H.A. KHALIK, A. IBRAHIM, and F. HANNA
Electrical and Thermal Studies of $[(NH_4)_{1-x}K_x]_2SO_4$ Crystals at High Temperature
64.70; 72.20; S11
(a) Vol.117,No.1, 163-168

H. ALEXANDER, S. DIETRICH, M. HUEHNE, M. KOLBE, and G. WEBER
EBIC Microscopy Applied to Glide Dislocations
61.70; 71.55; 72.20; S7.12
(a) Vol.117,No.2, 417-428

P. NOZAR, V. SECHOVSKY, M. SOLANKI-MOSER, E. BAUER, and E. GRATZ
Point-Contact Spectroscopy on the $Ce(Cu_xAl_{1-x})_5$ System
71.20; 72.20; 75.30; S1.2; S1.4
(a) Vol.117,No.2, 501-508

H. WITTE, W. SIEGEL, G. KUEHNEL, T. FLADE, and H.A. SCHNEIDER
Medium-Deep Acceptors in LEC GaAs
71.55; 72.20; S7.12
(a) Vol.117,No.2, 527-534

V.V ZUBRITSKII and R.P. PURLYS
The Relation between the Orientation of Streamer Discharges and Anisotropy of Thermal Vibrations of Atoms in II-VI ...
63.20; 71.55; 72.20; 77.50; S8.11
(b) Vol.158,No.1, 91-102

V.A. KOSTYLEV, B.A. GIZHEVSKII, A.A. SAMOKHVALOV, M.I. AUSLENDER, and N.G. BEBENIN
Anisotropy of Magnetoresistance of the p-Type Ferromagnetic Semiconductor $HgCr_2Se_4$
72.20; S11.2
(b) Vol.158,No.1, 307-318

A. ANAGNOSTOPOULOS, G.L. BLERIS, and I.Y. YANCHEV
Percolation Conductivity in Polycrystalline Silicon at Moderate Electrical Fields
72.20; S5.11
(b) Vol.158,No.1, K43-K48

H. GERSTENBERG and W. GLAESER
Transmutation Doping and Lattice Defects in Degenerate InSb
61.70 and 72.20; 71.55; S7.13
(a) Vol.118,No.1, 241-252

V.N. BERZHANSKII, I.I. EVSTAFEV, and V.I. IVANOV
Charge Carrier Transfer in Ferrite Spinels Affected by Magnetic Ordering
72.20; 75.40; S11.2
(b) Vol.158,No.2, 643-652

E.M. KOGAN
Anderson Localization in Ferromagnetic Semiconductors Due to Spin Disorder. III. Selfconsistent Localization Theory ...
72.20; 72.10
(b) Vol.158,No.2, 665-668

DOAN NHAT QUANG
Auger Recombination in Heavily Doped Semiconductors
71.25 and 72.20; S7.12
(b) Vol.158,No.2, 669-684

J. BĄK-MISIUK, G. KUEHNEL, W. SIEGEL, and U. PIETSCH
The Possible Influence of the Free Carrier Redistribution within the Conduction Band on the Thermal Expansion ...
65.70; 72.20; S7.12
(b) Vol.158,No.2, K111-K114

G.A. ADEGBOYEGA, A. POGGI, and E. SUSI
Some Electrical Characteristics of Fe and Its Annealing Behaviour in FZ Silicon
71.55; 72.20; S5.11
(a) Vol.118,No.2, 491-496

V.N. BRUDNYI, S.I. BORISENKO, and A.I. POTAPOV
Electrical and Optical Properties and Fermi Level Pinning in Electron Irradiated $ZnSnAs_2$
72.20 and 78.20; 71.55; S7.16
(a) Vol.118,No.2, 505-512

E. HEINER
Screening of the Coulomb Potential in the Athermal Stage of Highly Excited Direct-Gap Semiconductors
72.10; 72.20
(b) Vol.159,No.1, 43-48

M.R. MURTI and K.V. REDDY
Grain Boundary Effects on the Carrier Mobility of Polysilicon
72.20; 61.70; S5.11
(a) Vol.119,No.1, 237-240

S. BISWAS and R. BHATTACHARYA
Magnetic Properties of Single Crystals of Bismuth Telluride Doped with 0.2 at% Lead and Its Thermoelectric Power
75.20; 71.25; 72.20; S8
(b) Vol.159,No.2, 851-860

W. EICHLER and M. VOIGT
On the Analysis of the Nernst-Ettingshausen Effect
72.15 and 72.20
(b) Vol.159,No.2, K87-K92

R.A. MORENO, C.X. CARDOSO, and M. YOSHIDA
Hole Injection in Teflon FEP Films
72.20 and 73.60; S12
(a) Vol.119,No.2, 583-588

W. HOERSTEL, A. KLIMAKOV, R. KRAMER, and A. SCHMIEDE
Defect-Induced Conduction Anomalies in n-Type $Hg_{1-x}Cd_xTe$
72.20; S8.15
(a) Vol.119,No.2, 589-594

P.I. BARANSKII, A.E. BELYAEV, and O.P. GORODNICHII
Specific Features of Kinetic Effects Measured in (Cd,Hg)Te in Quantizing Magnetic Fields
72.20; S8.15
(b) Vol.160,No.1, 291-296

E.V. BOGDANOV, D.A. KOMARKOV, and W. KRAAK
Pressure Dependence of the Direct Energy Gap in $Bi_{1-x}Sb_x$ Semiconductor Alloys
71.25 and 72.20; S4
(b) Vol.160,No.1, K33-K36

J. KASPRZAK
The Influence of Electron Hopping on the Quadrupole Relaxation in Cu_2O
76.60; 72.20; S10.1
(b) Vol.160,No.1, K53-K56

M. STORDEUR
Consideration to the Nernst-Ettingshausen (NE) Effect in Thermoelectric p-$(Bi_{1-x}Sb_x)_2Te_3$ Single Crystals
72.20; S8
(b) Vol.160,No.2, K145-K148

W. LUBCZYŃSKI, J. CISOWSKI, and J.C. PORTAL
Shubnikov-de Haas Effect in $Cd_{3-x}Zn_xAs_2$ Alloys
71.25; 72.20; S1.62
(a) Vol.120,No.2, 525-530

A. WOLKENBERG
Exact Resistivity Calculation of Heavily Doped Silicon
72.20; S5.11
(a) Vol.120,No.2, 567-574

M. BHAGAVANTHA REDDY and P. VENUGOPAL REDDY
Electrical Conductivity of Some Mixed Li-Ti Ferrites
72.20; S11.2
(a) Vol.120,No.2, 575-580

G.S. MYAKENKAYA, G.L. GUTSEV, N.P. AFANASEVA, V.A. EVSEEV, and R.F. KONOPLEVA
Study of Lithium Interaction with Lattice Defects in Silicon
61.70 and 61.80; 72.20; S5.11
(b) Vol.161,No.1, 91-104

S. SOEDERHOLM, J. HELLBERG, J. KRZYSTEK, G. OLOVSSON, I. OLOVSSON, and J.U. VON SCHUETZ
Electronic Properties and Crystal Structure of a Methylthio Substituted Naphthalene Cation Radical Salt: $(DMbTN)_3(ClO_4)_2$
61.65 and 72.20; 71.55; 75.30; 76.30; S12.1
(b) Vol.161,No.1, 245-256

G.A. MATVEEV, I.M. TSIDILKOVSKI, and A.T. LONCHAKOV
On the Anisotropy of Scattering in Doped n-Ge at Low-Temperatures
71.30 and 72.20; S5.12
(b) Vol.161,No.1, 303-308

M. DEY (BOSE) and S. GHOSH
Acoustic Second Harmonic Generation in Magnetized High Resistivity Piezoelectric Semiconductors. Effect of ...
72.20 and 72.50; 77.60; S8.11
(b) Vol.161,No.1, 319-328

S. BHOWMIK, C.K. SARKAR, and K. SANTRA
The Effect of the Quantum Screening on the Energy Loss Rates of Hot Electrons in n-$Hg_{0.8}Cd_{0.2}Te$ in the Extreme Quantum .
71.25 and 72.20; S8.15
(b) Vol.161,No.1, 329-334

A.P. TYUTNEV, S.G. BOEV, V.N. ABRAMOV, E.A. KUZMENKOV, and A.V. VANNIKOV
Track Effects in Radiation Induced Conductivity of Polymers
61.80 and 72.20; S12
(a) Vol.121,No.1, 205-212

P. JANDL, M. RICK, and J. ROSENZWEIG
Defect Analysis of Bipolar CdTe Using Photo-Hall and Photoconductivity Measurements in the Temperature Range ...
72.20; S8.13
(a) Vol.121,No.1, 219-226

H.-J. FITTING, G.O. MUELLER, R. MACH, G.U. REINSPERGER, TH. HINGST, and E. SCHREIBER
Vacuum Emission of Hot Electrons from ZnS
72.20 and 79.70; 85; S8.11
(a) Vol.121,No.1, 305-314

U. DILLNER
The Effect of Non-Parabolicity on the Thermoelectric Figure of Merit of Two-Band Semiconductors
72.20
(a) Vol.121,No.1, K77-K80

V.V. BRYKSIN, A.V. GOLTSEV, S.D. KHANIN, A.V. NOVOTELNOVA, and A.N. VASILEV
Nonlinear Current-Voltage Characteristics of Ta_2O_5 and Nb_2O_5 Amorphous Oxides
72.20; S10
(b) Vol.161,No.2, 777-782

V.E. ARKHINCHEEV
Exact Relations and Galvanomagnetic Properties of the Inhomogeneous Two-Dimensional Medium
72.20
(b) Vol.161,No.2, 815-822

M. ISTRATE and S. ISTRATE
Drag Effects in Semiconductors with Position-Dependent Band Structure
72.20; 71.38
(b) Vol.161,No.2, 823-830

M. STORDEUR
The Thermoelectric Figure of Merit in the Mixed Crystal System $p-(Bi_{1-x}Sb_x)_2Te_3$
72.20; 72.90; S1.63; S8.15
(b) Vol.161,No.2, 831-836

A.T. NAGAT, S.A. HUSSEIN, Y.H. GAMEEL, and G.A. GAMAL
Switching Effect with Memory in Thallium Sulfide Single Crystals
72.20; S8.1
(a) Vol.121,No.2, K201-K206

A.I. DMITRIEV, Z.D. KOVALYUK, V.I. LAZORENKO, and G.V. LASHKAREV
Two-Dimensional Character of Electron Gas in Layered InSe Crystals
72.20; 71.25; S8
(b) Vol.162,No.1, 213-226

N. GUSKOS, G.P. TRIBERIS, M. CALAMIOTOU, CH. TRIKALINOS, A. KOUFOUDAKIS, C. MITROS, H. GAMARI-SEALE, and D. NIARCHOS
Temperature Dependence of the EPR Spectra of $GdBa_2Cu_3O_{7-\delta}$ Compounds in Orthorombic and Tetragonal Phase
72.20 and 74.70; 76.30; S10.15
(b) Vol.162,No.1, 243-250

Z. CHEN and L.C. BURTON
Electrical Properties and Their Depth Variation in Poly-Si under AM1 Illumination
72.20; 72.40; S5.11
(a) Vol.122,No.1, 361-370

A. LARGETEAU, J.P. BONNET, P. DORDOR, and J. RAVEZ
Propriétés électriques des joints de grains d'une céramique ferrite Ni-Zn à couches d'arrêt
72.20; S11.2
(a) Vol.122,No.1, 371-376

A. ROGALSKI and K. JÓŹWIKOWSKI
Intrinsic Carrier Concentration and Effective Masses in $Hg_{1-x}Mn_xTe$
72.20; S8.15
(a) Vol.122,No.1, K39-K44

M. SASAKI, G.X. TAI, and M. INOUE
Transient Thermoelectric Effect in Quasi-Two-Dimensional eta-Mo_4O_{11} Crystal
72.20; 71.30; S10
(b) Vol.162,No.2, 553-558

H.-J. FITTING, E. SCHREIBER, and TH. HINGST
Avalanche Measurement in ZnS by Vacuum Emission
72.20; 78.60; 79.70; 85; S8.11
(a) Vol.122,No.2, K165-K168

72.30. High-frequency effects; plasma effects

R. SANGHVI and S. GHOSH
Amplitude Modulation and Demodulation of Electromagnetic Waves in Magnetised Material with Strain-Dependent Dielectric..
72.30; 62.65; S11.1
(a) Vol.118,No.1, 235-240

F. JAHNKE and K. HENNEBERGER
Nonequilibrium Description of the Intense Resonant Laser Excitation of a Semiconductor
71.35 and 72.30; 78.20; S8.11
(b) Vol.159,No.1, 81-90

P.N. GORLEI, T.A. ZAGORODNYAYA, and V.A. SHENDEROVSKII
Plasma-Electric Effect in Anisotropic Semiconductors in Heating Electric Field
72.30
(b) Vol.159,No.2, 837-844

72.40. Photoconduction and photovoltaic effects; photodielectric effects

V.I. MIKLA, D.G. SEMAK, A.V. MATELESHKO, and A.A. BAGANICH
Photosensitized Deep Gap Center Relaxation in Amorphous Selenium Layers Containing As Additions
71.55; 72.40; 73.60; S5
(a) Vol.117,No.1, 241-250

G.I. ROSENMAN, YU.L. CHEPELEV, and E.I. BOIKOVA
Photo-Stimulated Relaxation of Exoemission in $Ba_2NaNb_5O_{15}$
72.40 and 79.60; S11.1
(a) Vol.117,No.1, 259-264

A.A. ZAHAB, M. ABD-LEFDIL, and M. CADENE
Electrical Conductivity, Optical Absorption, and Photoconductivity Measurements of Well Oriented Indium-Telluride Thin Films
73.60 and 78.65; 72.40; S8
(a) Vol.117,No.2, K103-K106

G. KLUGE, M. GRAUROCK, R. STROBEL, M. LENZNER, H. BERGNER, V. BRUECKNER, and P. SUEPTITZ
Picosecond Photoconductivity in Amorphous Germanium Chalcogenides
72.40; S8
(b) Vol.158,No.1, K37-K42

L. SZARO and J. MISIEWICZ
The AC Surface Photovoltage on Silicon under Subbandgap Illumination
73.20 and 73.30; 72.40; S5.11
(a) Vol.118,No.1, 185-188

D.M. PETTY and M.R. WILLIS
The Effect of Current Injection on Photoconduction
72.40
(b) Vol.158,No.2, K165-K168

Z. KUČERA, P. HLIDEK, P. HOESCHL, V. KOUBELE, V. PROSSER, and M. ZVÁRA
Magneto-Optical Spectroscopy of p-Type $Hg_{0.8}Cd_{0.2}Te$
72.40; 71.55; S8.15
(b) Vol.158,No.2, K173-K178

N.N. KHUSNATDINOV, V.F. PETRENKO, and A.N. TURANOV
Intrinsic Photoconductivity of Hexagonal Ice
61.70 and 72.40; S10
(a) Vol.118,No.2, 401-408

M. KANIEWSKA and J. KANIEWSKI
Influence of DX Centers on Photoresponse of GaAs/AlGaAs Double Heterostructures
73.40; 71.55; 72.40; S7.12; S7.15
(a) Vol.118,No.2, 487-490

J.A. BAQUEDANO and J.M. CABRERA
Absorption Bands Associated to Photorefractive Erasure in $Bi_{12}SiO_{20}$
71.55; 72.40; 78.50; S10
(a) Vol.118,No.2, 497-504

D.A. ARONOV, R. KABULOV, and YU.M. YUABOV
The Photoelectric Characteristics of Solar p-i-n Photocells Based on Hydrogenated Amorphous Silicon with Built-in ...
72.40 and 73.40; 85; S5.11
(a) Vol.118,No.2, 577-590

G. AMATO, G. BENEDETTO, F. FIZZOTTI, C. MANFREDOTTI, and R. SPAGNOLO
Photothermal and Photoconductive Measurements of Low Pressure Chemically Vapour Deposited Amorphous Silicon
72.40 and 73.60; 78.65; S5.11
(a) Vol.119,No.1, 169-176

A.A. ZAHAB, M. ABD-LEFDIL, and M. CADENE
Rectifying and Photovoltaic Parameters of Indium Telluride(p)-Cadmium Sulfide(n) Thin Film Heterojunctions
72.40 and 73.40; S8
(a) Vol.119,No.1, K35-K40

A.T. NAGAT, G.A. GAMAL, Y.H. GAMEEL, and N.M. MOHAMED
Phototransport Studies of Thallium Monosulfide Single Crystals
72.40; S8
(a) Vol.119,No.1, K47-K52

XIQI FENG, LIANAN TANG, LIANGYING XU, and XIAOKE JIANG
Photoconductivity of Nd,Mg-Doped $LiNbO_3$ Crystals
72.40; S11.1
(a) Vol.119,No.2, 561-570

J.W. TOMM, K.H. HERRMANN, H. BOETTNER, M. TACKE, and A. LAMBRECHT
A Luminescence Study in the $Pb_{1-x}Eu_xSe$ System
73.40 and 78.55; 72.40; S7.16
(a) Vol.119,No.2, 711-720

YANG-FANG CHEN
Persistent Photoconductivity in Doping Modulated Amorphous Silicon Multilayers
73.60; 72.40; S5.11
(b) Vol.160,No.2, K103-K108

D.M. PETTY and M.R. WILLIS
The Effect of Current Injection on Photoconduction in Organic Semiconductors and Insulators
72.40; S12
(a) Vol.120,No.2, 593-600

XIAOTONG LIN, ZHIPU YOU, and HUIFEN GUO
An Enhancement Phenomenon of the Minority Carrier Lifetime in Annealed Silicon
72.40; 61.70; S5.11
(a) Vol.120,No.2, K177-K180

V.I. BLYNSKII, A.YU. KULIKOV, S.YU. RAKHLEY, and V.I. OSINSKII
Photoelectric Crosstalk between Silicon Photoelements
72.40; 78.40; S5.11
(a) Vol.121,No.1, 227-234

V.YU. RUD, YU.V. RUD, and M. SERGINOV
Natural Photopleochroism of n-In_2O_3-p-$CdSiAs_2$ Structures
72.40 and 73.40; 85; S7.16; S10.1
(a) Vol.121,No.1, K81-K84

U. VAN OLFEN
Light-Induced Transparency in Absorbing Powders
78.35; 72.40; S11.1
(a) Vol.121,No.1, K121-K124

G.M. SHMELEV, S.V. KRYUCHKOV, and G.I. TSURKAN
Anisotropic Photoconductivity in Superlattices
72.40
(b) Vol.161,No.2, K85-K90

T.G. ABDEL-MALIK, A.A. AHMAD, and A.S. RIAD
Dark and Photoelectric Investigation on Thin Films of beta-Cobalt Phthalocyanine Dispersed in Polymers
73.60; 72.40; S12
(a) Vol.121,No.2, 507-514

H. BOETTCHER, U. WEIH, J. OPFERMANN, and R. ACKERMANN
Electrical Properties of Pure and Mixed Evaporated Dye Layers
72.40 and 73.60; S12
(a) Vol.121,No.2, 563-570

V.YU. RUD, YU.V. RUD, and M. SERGINOV
Photoelectric Properties of H_2O-p-$CdSiAs_2$:In Structures
72.40 and 73.40; S7.16
(a) Vol.121,No.2, K171-K176

L.L. GOLIK, M.M. GUTMAN, and V.E. PAKSEEV
Nonlinear Oscillations and Chaos in CdS Single Crystals under Temperature-Electric Instability Conditions
72.40; 72.70; S8.11
(b) Vol.162,No.1, 199-212

A. BAIDULLAEVA, B.K. DAULETMURATOV, V.A. GNATYUK, and P.E. MOZOL
Surface States of Laser-Irradiated Cadmium and Tellurium-Containing Solid Solutions of II-VI Compounds
73.20 and 73.40; 72.40; S8.15
(a) Vol.122,No.1, 243-248

Z. CHEN and L.C. BURTON
Electrical Properties and Their Depth Variation in Poly-Si under AM1 Illumination
72.20; 72.40; S5.11
(a) Vol.122,No.1, 361-370

YU.I. MAZUR, YU.G. RUBO, O.V. SNITKO, and M.V. STRIKHA
Auger Recombination via Defects in Tellurium
72.10; 71.25; 72.40; S5
(b) Vol.162,No.2, 531-538

HAOLIANG LI, S. RENTSCH, JIPING YANG, M. LENZNER, and H. BERGNER
Picosecond Photoconductivity Studies on Phenyl-Substituted Poly(Phenylenevinylene)s Derivatives
72.40; S12
(b) Vol.162,No.2, 545-552

A.S. POPOV and A.Y. BAHNEV
Photoconductivity Spectra of Fe-Doped GaInP Epitaxial Layers
72.40 and 73.40; S7.16
(a) Vol.122,No.2, 569-574

J. TOUŠEK
Photovoltaic Effect in Schottky Diodes. Contribution to Theory and Experiment
72.40 and 73.30; S8.13
(a) Vol.122,No.2, 757-764

G. GRUMMT, R. PICKENHAIN, and G. OELGART
Photocurrent Investigation on a Graded Heterojunction
73.40; 72.40; S7.12
(a) Vol.122,No.2, 765-772

72.50. Acoustoelectric effects

M. DEY (BOSE) and S. GHOSH
Amplification of Acoustic Waves in Magnetised High Resistivity Piezoelectric Semiconductors. Effect of Nonuniform ...
62.65 and 72.50; S8.11
(b) Vol.157,No.1, 159-166

M. DEY (BOSE) and S. GHOSH
Amplification of Acoustohelicon Waves in Piezoelectric Semiconductors. Effect of Nonlinearity in Electron ...
72.10 and 72.50; 71.25; S7.13
(b) Vol.159,No.2, 817-826

M. DEY (BOSE) and S. GHOSH
Acoustic Second Harmonic Generation in Magnetized High Resistivity Piezoelectric Semiconductors. Effect of ...
72.20 and 72.50; 77.60; S8.11
(b) Vol.161,No.1, 319-328

72.60. Mixed conductivity and conductivity transitions

V.A. KASIYAN, D.D. NEDEOGLO, A.V. SIMASHKEVICH, and I.N. TIMCHENKO
Coulomb Gap in n-ZnSe and Its Critical Behaviour in the Region of the Metal-Dielectric Transition
71.30 and 72.60; S8.12
(b) Vol.158,No.1, 293-300

N.T. MAMEDOV and N.K. MOROZ
Tl-Me Bond and Semiconductor-to-Metal Transition in $TlMeX_2$ Low-Dimensional Crystals
72.60; S8
(b) Vol.159,No.2, K83-K86

72.70. Noise processes and phenomena

I.S. BAKSHEE, L.A. KARACHEVTSEVA, A.V. LYUBCHENKO, V.A. PETRYAKOV, E.A. SALKOV, and B.I. KHIZHNYAK
Superposition of Burst Noises as the Origin of 1/f Noise Due to Grain Boundaries in $Cd_xHg_{1-x}Te$
72.70; 61.70; S8.15
(a) Vol.117,No.1, K37-K42

L.L. GOLIK, M.M. GUTMAN, and V.E. PAKSEEV
Nonlinear Oscillations and Chaos in CdS Single Crystals under Temperature-Electric Instability Conditions
72.40; 72.70; S8.11
(b) Vol.162,No.1, 199-212

72.80. Conductivity of specific semiconductors and insulators

G.P. TRIBERIS and N. GUSKOS
Electronic Conduction in $CdF_2{:}Gd^{3+}$ Single Crystals before and after Electrical Conversion
72.80; S9.12
(b) Vol.158,No.1, K17-K20

F. EL-AKKAD and A. AL-SHAHRANY
Electron Diffusion Length in ZnTe
72.80; S8.13
(a) Vol.118,No.2, K79-K82

T. TSVETKOVA
Modification Effect of Ni Ion Implantation on the Electrical Properties of Vitreous As_2Se_3
61.80 and 72.80; S8
(a) Vol.118,No.2, K83-K86

M. ZĄBKOWSKA-WACŁAWEK and W. WOJCIECHOWSKI
Electrical and Magnetic Properties of Copper Phthalocyanine-Carbon Black Mixtures
72.80; 76.30; S12
(a) Vol.118,No.2, K87-K90

G. BEHR, G. KRABBES, J. WERNER, P. DORDOR, and J.-P. DOUMERC
Crystal Growth and Electrical Properties of Sb-Doped SnO_2 Single Crystals
72.80; 61.50; S10.1
(a) Vol.118,No.2, K91-K94

D.M. ZAYACHUK, D.D. IVANCHUK, R.D. IVANCHUK, S.S. MASLYANCHUK, and V.I. MIKITYUK
The Effect of Gadolinium Doping on the Physical Properties of Lead Telluride
71.55; 72.80; 75.50; S8
(a) Vol.119,No.1, 215-220

A.T. NAGAT, G.A. GAMAL, and S.A. HUSSEIN
Growth and Characterization of Ternary Compound $TlGaTe_2$ Single Crystals
72.80; S8.16
(a) Vol.120,No.2, K163-K168

G.A. ADEGBOYEGA and A. POGGI
Intrinsic Gettering of Cr Impurities in p-Type Cz Silicon
71.55; 72.80; 78.30; S5.11
(a) Vol.121,No.1, 181-186

P.N. GORLEI, V.V. KOSHMAN, A.I. SKITSKO, L.S. SOLONCHUK, V.V. KHOMYAK, and V.A. SHENDEROVSKII
Longitudinal Hall Effect in p-Te
72.80; S5
(b) Vol.161,No.2, K91-K94

O.A. GOLIKOVA, N. AMANDZHANOV, M.M. KAZANIN, G.M. KLIMASHIN, and V.V. KUTASOV
Electrical Activity of Impurities in beta-Rhombohedral Boron
71.55; 72.80; 76.80; S5
(a) Vol.121,No.2, 579-586

F. TAKEDA, K. MATSUURA, and S. SETO
Admittance Spectroscopy and Photoluminescence Characterization of High-Quality p-Type Cadmium Telluride
71.55 and 78.55; 72.80; S8.13
(a) Vol.121,No.2, 595-606

C.B. STEELE and J. BEYNON
Non-Activated Conduction in Au/SiO_x Thin Films
72.80; S10.1
(a) Vol.122,No.2, K169-K170

72.90. Other topics in electronic transport in condensed matter

M. STORDEUR
The Thermoelectric Figure of Merit in the Mixed Crystal System $p\text{-}(Bi_{1-x}Sb_x)_2Te_3$
72.20; 72.90; S1.63; S8.15
(b) Vol.161,No.2, 831-836

73. Electronic Structure and Electriical Properties of Surfaces, Interfaces, and Thin Films

73.20. Electronic surface states
(for emission and impact phenomena, see 79)

J.W. STASIAK, J. KIERUL, and L. KUENNE
Influence of Phonons on the Calculations of the Density of States in Thin Films
71.20 and 73.20; S1.2
(b) Vol.157,No.1, 229-234

F. BECHSTEDT and D. REICHARDT
Calculation of Surface-Induced Core-Level Shifts for Covalent Semiconductors C, Si, Ge, and alpha-Sn. II. (100)2x1 Surfaces
68.20 and 73.20; S5; S5.11; S5.12
(b) Vol.157,No.2, 567-576

XIANG-YANG ZHENG, R.H. RITCHIE, and J.R. MANSON
High-Order Corrections to the Binding Energy of Image States
73.20
(b) Vol.157,No.2, K87-K92

L. SZARO and J. MISIEWICZ
The AC Surface Photovoltage on Silicon under Subbandgap Illumination
73.20 and 73.30; 72.40; S5.11
(a) Vol.118,No.1, 185-188

D.B. TRAN THOAI, E. ZEITLER, and DO NAM
Self-Energy of an Electron in the Vacuum Gap between Two Metals
71.45 and 73.20
(b) Vol.158,No.2, 557-564

B. REICHERT and CH. JUNG
Electronic Structure of the Charged Copper(001)/Vacuum Interface
73.20; 73.30; 73.60; S1.2
(b) Vol.158,No.2, K119-K122

A. FEDOTOV, B. EVTODYI, L. FIONOVA, YU. ILYASHUK, E. KATZ, and L. POLYAK
Electrical Activity of Grain Boundaries in Shaped Grown Silicon
61.70 and 68.55; 73.20; 73.40; S5.11
(a) Vol.119,No.2, 523-534

F. RODRÍGUEZ, A. CAMACHO, L. QUIROGA, and R. BAQUERO
Electronic Structure of (001) Semiconducting MTe Surfaces (M=Zn,Cd,Hg)
71.25 and 73.20; S8.13
(b) Vol.160,No.1, 127-136

E.F. SARMENTO and T. KANEYOSHI
Surface Magnetic Properties of a Transverse Ising Model with a Diluted Surface
73.20 and 75.10
(b) Vol.160,No.1, 337-348

L.I. KURKINA and O.V. FARBEROVICH
Electron Structure and Magnetic Properties of Small Iron Particles
71.50; 73.20; 75.25; S1.1
(b) Vol.160,No.1, K37-K42

R.F. SABIRYANOV, D.L. NOVIKOV, A.L. IVANOVSKII, and V.A. GUBANOV
Quantum-Chemical Calculations of Hydrogen Adsorption on the Polar Surface of Titanium Carbide
68.45 and 73.20; 79.20; S1.61
(b) Vol.160,No.2, 459-472

Š. PICK, M. TOMÁŠEK, G.V. GADIYAK, and YU.N. MOROKOV
Interactions on Silicon Surfaces and Theory of Moments
68.25 and 68.40; 73.20; S5.11
(b) Vol.160,No.2, 483-490

H. YAMAMOTO and Y. KANIE
Transmission Coefficient and Resonance Condition in a Symmetrical Rectangular Triple-Barrier Structure
73.20; 73.40
(b) Vol.160,No.2, K97-K102

A.N. NAZAROV, V.S. LYSENKO, S.A. VALIEV, M.M. LOKSHIN, A.S. TKACHENKO, and I.A. KUNITSKII
Flash Lamp Annealing and RF Plasma Annealing of Al-SiO_2-Si Structures
73.40; 73.20; S5.11; S10.1
(a) Vol.120,No.2, 447-456

D.B. TRAN THOAI and E. ZEITLER
Inelastic Scattering of Fast Electrons by Thin Metal Slabs
71.45 and 73.20; 61.80; S4
(a) Vol.120,No.2, 467-474

YA.M. STRELNIKER and O.S. ZINETS
Inelastic Scattering of Low-Energy Electrons by Electrons Localized in Surface Potential Wells
73.20; 61.80
(b) Vol.161,No.2, 597-602

W. HERGERT and P. RENNERT
LCAO Recursion Approach in k-Space Representations for the Treatment of Bulk and Surface Problems
71.10; 71.20; 71.25; 73.20
(b) Vol.161,No.2, 667-676

A.A. O'CONNEL-BRONIN and A.B. TRESHCHALOV
Exciton Reflection Spectra of NaI Crystals. Effect of Surface Irradiation by Photons and Incidence Angle Dependence
71.35 and 78.40; 73.20; S9.11
(b) Vol.162,No.1, 289-296

H.A. KASSIM
Dipole Polarizability of O^{2-} Ions near the (011) and the (111) Surface of Alkaline Earth Oxides
73.20 and 77.30; S10
(b) Vol.162,No.1, 297-302

A. BAIDULLAEVA, B.K. DAULETMURATOV, V.A. GNATYUK, and P.E. MOZOL
Surface States of Laser-Irradiated Cadmium and Tellurium-Containing Solid Solutions of II-VI Compounds
73.20 and 73.40; 72.40; S8.15
(a) Vol.122,No.1, 243-248

I.R. PLUMMER and D.W. TURNER
Imaging and Electron Energy Analysis of Exo-Electron Emission
79.75; 73.20; S9.11; S11.1
(a) Vol.122,No.1, 285-296

S.YU. KASANDROV
Tight-Binding Description of Surface States in Semiconductors
73.20; S7.1
(b) Vol.162,No.2, 435-442

73.25. Surface conductivity

M. CHYBICKI and J. LIWO
keV-Proton Induced Surface Conductivity of Lead-Silicate Glass
73.25; S10.15
(a) Vol.120,No.2, K159-K162

G. SRINIVASAN, GUO-MEI WU, and T.T. SRINIVASAN
Surface Resistance at Microwave Frequencies in Transition Metal Ion Substituted Yttrium Barium Copper Oxide ...
73.25 and 74.10; S10.15
(a) Vol.122,No.1, 355-360

73.30. Surface double layers, Schottky barriers, and work functions

S. PATRA and A. KUMAR
Effect of Lattice Vibrations on the Change in Work Function
68.30 and 73.30; S2
(b) Vol.157,No.1, 209-220

M. KARAI, K. KIDO, H. NAITO, K. KUROSAWA, M. OKUDA, T. FUJINO, and M. KITAGAWA
Deep Trap Levels in Zn-Annealed ZnSe Single Crystals
71.55; 73.30; S8.12
(a) Vol.117,No.2, 515-526

L. SZARO and J. MISIEWICZ
The AC Surface Photovoltage on Silicon under Subbandgap Illumination
73.20 and 73.30; 72.40; S5.11
(a) Vol.118,No.1, 185-188

B. REICHERT and CH. JUNG
Electronic Structure of the Charged Copper(001)/Vacuum Interface
73.20; 73.30; 73.60; S1.2
(b) Vol.158,No.2, K119-K122

P.S. GLADKOV and K.S. VARBLIANSKA
Ta/GaAs Schottky Barriers Produced by in Situ Sputter Etching, RF Magnetron Sputtering of Ta, and Its Thermal Oxidation
73.30; 85; S1; S7.12
(a) Vol.118,No.2, 479-486

P. WEISSBRODT and E. NEBAUER
SNMS Study of the Pt Out-Diffusion in the Au/Pt/Ti-n-GaAs System
73.30 and 79.20; S7.12
(a) Vol.119,No.2, K131-K134

R. SCHLESINGER, S. ROGASCHEWSKI, and P.J. JANIETZ
Electrochemically Deposited Copper Schottky Contacts on n-Type GaAs for Electron-Beam-Induced Current Measurements
73.40 and 85; 73.30; S7.12
(a) Vol.120,No.2, 687-694

G. EFTEKHARI
Electrical Degradation of InP Schottky Barriers Due to Mercury Contamination
73.30; S7.11
(a) Vol.120,No.2, K155-K158

YU.A. NOVIKOV, A.V. RAKOV, and V.P. SHANTAROVICH
Positron Annihilation in Aluminium Films in Electric Contact with Silicon Substrate
73.30; 78.70; S4
(a) Vol.121,No.2, 495-500

J. KOLNÍK and M. OŽVOLD
The Influence of Inversion Surface Layers on the Evaluation of the Interface State Energy Distribution from Schottky...
73.30 and 73.40; 85; S5.11; S7.12
(a) Vol.122,No.2, 583-588

G. EFTEKHARI
Characterization of Rapidly Thermally Annealed GaAs and InP Surfaces Using Schottky Barriers
73.30 and 85; S7.11; S7.12
(a) Vol.122,No.2, 751-756

J. TOUŠEK
Photovoltaic Effect in Schottky Diodes. Contribution to Theory and Experiment
72.40 and 73.30; S8.13
(a) Vol.122,No.2, 757-764

73.40. Interfaces

J. KRUPSKI
Interfacial Capacitance
73.40; S10
(b) Vol.157,No.1, 199-208

S.I. POKUTNYI
Energy Transfer of Electron Excitation. Decay Kinetics of Exciton Luminescence in a Three-Layer System
73.40 and 78.55
(b) Vol.157,No.1, 221-228

K.S.R. KOTESWARA RAO and V. KUMAR
Estimation of Trap Concentration in Linearly Graded Junctions Using DLTS
71.55; 73.40
(a) Vol.117,No.1, 251-258

L. KOŃCZEWICZ, E. LITWIN-STASZEWSKA, A. MASŁOWSKA, R. PIOTRZKOWSKI, J.L. ROBERT, and J.P. ANDRÉ
Investigation of Electron Mobility in a Two-Dimensional Electron Gas. GaAs-$Al_xGa_{1-x}As$ Heterostructures. Application of ...
73.40; S7.12; S7.15
(b) Vol.157,No.2, 593-600

L. PAVESI and F.-K. REINHART
Transfer Matrix Method to Compute Energy Levels of Superlattices
71.25 and 73.40; S7.1
(b) Vol.157,No.2, 615-626

E.G. SALMAN, A.N. KORSHUNOV, and V.N. VERTOPRAKHOV
A DLTS Study of InAs MIS Structures
71.55 and 73.40; S7.12
(a) Vol.117,No.2, 509-514

R.Z. VALIEV, YA.D. VISHNYAKOV, R.R. MULYUKOV, and G.S. FAINSHTEIN
On the Decrease of Curie Temperature in Submicron-Grained Nickel
75.30; 73.40; S1.2
(a) Vol.117,No.2, 549-554

M.A. DEMYANENKO
The Effect of Back Inversion Channel and Electrically Active Defects on the Nonequilibrium Characteristics of SOS MOS...
73.40 and 85; S5.11
(a) Vol.117,No.2, 601-610

H. BOETTCHER, K. SCHREIBER, and B. VAUPEL
Photosensitivity of Evaporated Mixed AgBr Layers
73.40 and 78.65; S9.15
(a) Vol.117,No.2, K165-K170

E.P. POKATILOV, S.I. BERIL, N.N. SEMENOVSKAYA, and M. FAHOOD
Charge Carrier Energy Spectrum in Multilayer Structures and Superlattices in the Field of the Self-Action Potential
71.50 and 73.40
(b) Vol.158,No.1, 165-174

N.A. GUSHCHINA and V.K. NIKULIN
A Tight-Binding Calculation of Band Discontinuities in the $Al_xGa_{1-x}As$/GaAs(AlAs) Heterostructures
71.25 and 73.40; S7.15
(b) Vol.158,No.1, 175-180

B. SUKUMAR and K. NAVANEETHAKRISHNAN
Diamagnetic Susceptibility of a Donor in a GaAs/$Ga_{1-x}Al_xAs$ Quantum Well Heterostructure and Its Pressure Dependence
73.40 and 75.20; 71.55; S7.12; S7.15
(b) Vol.158,No.1, 193-200

S.V. IVANOV, P.S. KOPEV, N.N. LEDENTSOV, B.YA. MELTSER, M.YU. NADTOCHY, V.M. USTINOV, and A.M. VASILEV
Ultrathin Layers and Short-Period Superlattices in Semiconductor Structures
73.40; 78.65; S7.12; S7.15
(a) Vol.118,No.1, 169-178

A.D. BELYAEV, V.E. RODIONOV, and N.S. CHERNAYA
Effect of Ultrasound Oscillations on Electroluminescent Properties of MISIM Structures
78.60; 62.80; 73.40; S8.11
(a) Vol.118,No.1, 327-332

J.J. LIOU
Modeling the Steady-State Collector Current of AlGaAs/GaAs Heterojunction Bipolar Transistors Using a Generalized ...
73.40 and 85; S7.12; S7.15
(a) Vol.118,No.1, K65-K68

H. RODRÍGUEZ-COPPOLA, J.L. MARTÍNEZ-CUELLAR, and R. PÉREZ-ÁLVAREZ
Two-Band Envelope Function Model for Graded Interfaces
71.25 and 73.40; S7.15
(b) Vol.158,No.2, 481-492

S.K. CHANG, H.L. PARK, H.K. KIM, J.S. HWANG, C.H. CHUNG, and W.T. KIM
Co^{2+} in $CuGa_{1-x}Al_xSe_2$ and Valence Band Offset in $CuGa_{1-x}Al_xSe_2/CuGaSe_2$
73.40; S8.16
(b) Vol.158,No.2, K115-K118

A. GHOSHAL, M. MONDAL, and K.P. GHATAK
On the Burstein-Moss Shift in 3D Quantum Dots of Ge
73.40; S5.12
(b) Vol.158,No.2, K123-K128

K.A. AL-SALEH, I.J. JABR, and N.S. SALEH
A Comparison of Ion Beam Induced Atomic Mixing Kinetics of Ti/Si, Fe/Si, and Ni/Si Systems. In Situ Electrical ...
61.80 and 73.40; S1.1; S1.2; S5.11
(a) Vol.118,No.2, 467-472

M. KANIEWSKA and J. KANIEWSKI
Influence of DX Centers on Photoresponse of GaAs/AlGaAs Double Heterostructures
73.40; 71.55; 72.40; S7.12; S7.15
(a) Vol.118,No.2, 487-490

D.A. ARONOV, R. KABULOV, and YU.M. YUABOV
The Photoelectric Characteristics of Solar p-i-n Photocells Based on Hydrogenated Amorphous Silicon with Built-in ...
72.40 and 73.40; 85; S5.11
(a) Vol.118,No.2, 577-590

N.S. NAM, I. ZÓLOMY, and V. SCHILLER
Bipolar Transistor with Tunnel Oxide Emitter Structure (MISET)
73.40 and 85
(a) Vol.118,No.2, 591-596

G. DIMOPOULOS
A New Model of Transconductance for SOI Gated Double Injection Structures
73.40; S5.11
(a) Vol.118,No.2, K73-K78

D.S. CHEMLA, S. SCHMITT-RINK, W.H. KNOX, D.A.B. MILLER, K.W. GOOSSEN, and G. HASNAIN
Ultrafast Nonlinear Optical Effects in Biased Semiconductor Quantum Wells
71.35 and 78.20; 73.40; S7.12
(b) Vol.159,No.1, 11-28

A.V. NURMIKKO and R.L. GUNSHOR
Excitons and Phonons in CdTe/MnTe and ZnTe/MnTe Single Quantum Wells. Aspects of Confinement and Optical Phonon Coupling
71.35 and 78.20; 73.40; S8.13
(b) Vol.159,No.1, 107-116

C. ELL and H. HAUG
Absorption and Optical Gain Spectra and Band Gap Renormalization of Highly Excited Quantum Well Systems
71.25 and 78.20; 73.40; S7.11; S7.15
(b) Vol.159,No.1, 117-124

E. LACH, V.D. KULAKOVSKII, A. FORCHEL, T.L. REINECKE, J. STRAKA, D. GRUETZMACHER, and G. WEIMANN
Single and Many Particle Effects in the Emission Spectra of Laterally Homogeneous 2D Plasmas
71.35 and 78.55; 73.40; S7.12; S7.15
(b) Vol.159,No.1, 125-132

Y. MASUMOTO, S. IDESHITA, and T. WAMURA
Nonlinear Optical Study of Semiconductor Microstructures, CuCl Quantum Dots, and GaAs Quantum Wells
71.35 and 78.20; 73.40; S7.12; S9
(b) Vol.159,No.1, 133-142

L. WENDLER, F. BECHSTEDT, and M. FIEDLER
Quasi-Two-Dimensional Screening of the Electron-Hole Interaction in Modulation-Doped Quantum Wells
71.35 and 73.40; S7.12; S7.15
(b) Vol.159,No.1, 143-154

A.I. BOBRYSHEVA and S.S. RUSSU
Optical Properties of Biexcitons in ZnSe/$Zn_{1-x}Mn_xSe$ Quantum Wells
71.35 and 78.20; 73.40; S8.12; S8.15
(b) Vol.159,No.1, 155-164

L.V. BUTOV, V.D. KULAKOVSKII, and T.G. ANDERSSON
Photoluminescence of a Dense Quasi-2D Electron-Hole System in Modulation Doped Heterostructure with Strained InGaAs QW
71.35 and 78.55; 73.40; S7.15
(b) Vol.159,No.1, 165-172

K.-H. SCHLAAD, CH. WEBER, D.S. CHEMLA, J. CUNNINGHAM, D.A.B. MILLER, C.V. HOOF, G. BORGHS, G. WEIMANN, W. SCHLAPP et al.
Nonlinear Optical Properties of GaAs/(AlGa)As Multiple Quantum Well Structures under Quasistationary High Excitation ...
71.35 and 78.20; 73.40; S7.12; S7.15
(b) Vol.159,No.1, 173-180

D. FROEHLICH, CH. NEUMANN, B. UEBBING, and R. WILLE
Experimental Investigation of Three-Level Optical Stark Effect in Semiconductors
71.35 and 78.20; 73.40; S7.1; S10.1
(b) Vol.159,No.1, 297-308

N.S. AVERKIEV, V.M. ASNIN, A.A. ROGACHEV, A.YU. SILOV, and V.I. STEPANOV
Photoluminescence Studies of Anomalous Diamagnetic Properties of the 2D e-h System on GaAs Surface
75.20 and 78.55; 73.40; S7.12
(b) Vol.159,No.1, 471-476

R. GOLDHAHN, H. ROEPPISCHER, M. GERICKE, and C.-E. RICHTER
Characterization of $Al_xGa_{1-x}As$/GaAs Heterostructures by Electroreflectance Depth Profiling
73.40 and 78.20; S7.15
(a) Vol.119,No.1, 183-190

V.A. SKRYSHEVSKII and P. KUS
Optical and Electro-Physical Properties of $Gd_1Ba_2Cu_3O_{7-\delta}$-p-Si Structures at $T>T_c$
73.40 and 74.70; 78.65; S10.15
(a) Vol.119,No.1, 201-208

J.J. LIOU
The Effects of Neutron Irradiation on the Current Gain of AlGaAs/GaAs Heterojunction Bipolar Transistors
73.40 and 85; S7.12; S7.15
(a) Vol.119,No.1, 337-342

I.M. HAFEZ, G. GHIBAUDO, and F. BALESTRA
Numerical and Analytical Modelling of Depletion-Mode MOS Transistors in Ohmic Region at Liquid Helium Temperature
73.40 and 85
(a) Vol.119,No.1, 343-352

A.A. ZAHAB, M. ABD-LEFDIL, and M. CADENE
Rectifying and Photovoltaic Parameters of Indium Telluride(p)-Cadmium Sulfide(n) Thin Film Heterojunctions
72.40 and 73.40; S8
(a) Vol.119,No.1, K35-K40

H. LEON, F. COMAS, and M. SUHRKE
Influence of Level Broadening in Mobility Calculations for Low-Dimensional Semiconductor Heterostructures
72.10 and 73.40; S7.12
(b) Vol.159,No.2, 731-742

I.I. BOIKO and YU.M. SIRENKO
Scattering of Two-Dimensional Electron Gas on the Semibounded Three-Dimensional Electron Gas
72.10 and 73.40
(b) Vol.159,No.2, 805-816

P. KUIVALAINEN and K. LINDBERG
Hot Electron Transport in Submicron Semiconductor Devices
72.10 and 73.40; 85; S7.1
(b) Vol.159,No.2, 827-836

S.YA. GOROSHCHENKO
Integer Quantum Hall Effect in Magnetic and Electric Fields of Arbitrary Orientations
72.10 and 73.40
(b) Vol.159,No.2, K55-K58

K.-H. BAETHER and G. ZIES
Film Thickness Effects and Their Influence on the Stability of the Cr-Si-O/V-O/Al Film System
68.55 and 73.40; S1.61; S4
(a) Vol.119,No.2, 503-510

A. FEDOTOV, B. EVTODYI, L. FIONOVA, YU. ILYASHUK, E. KATZ, and L. POLYAK
Electrical Activity of Grain Boundaries in Shaped Grown Silicon
61.70 and 68.55; 73.20; 73.40; S5.11
(a) Vol.119,No.2, 523-534

I. THURZO and E. PINČÍK
Attempts to Probe the Semi-Insulating GaAs-Plasma Oxide Interface by Small-Signal Charge DLTS
73.40 and 85; S7.12
(a) Vol.119,No.2, 701-710

J.W. TOMM, K.H. HERRMANN, H. BOETTNER, M. TACKE, and A. LAMBRECHT
A Luminescence Study in the $Pb_{1-x}Eu_xSe$ System
73.40 and 78.55; 72.40; S7.16
(a) Vol.119,No.2, 711-720

J. KUZMIK and T. LALINSKY
Correlation between the Backgating Effect and the Gate Length Shortening on GaAs MESFETs
73.40 and 85; S7.12
(a) Vol.119,No.2, K185-K188

H. LEON and F. COMAS
Confined-LO-Phonon-Limited Mobility in a Semiconductor Quantum Well
72.10 and 73.40; S7.12; S7.15
(b) Vol.160,No.1, 105-116

H. RUECKER, M. MORA-RAMOS, and F. COMAS
Magnetopolaron in a Quantum Well. LO-Phonon Confinement Effects
71.38 and 73.40; 63.20; S7.12
(b) Vol.160,No.1, 117-126

B. THOMAS, M.W. AHKTAR, D.V. MORGAN, M.A. MOHAMMED, and D.E. DAVIES
A Low-Cost Ion Assisted Evaporation Technique for AuGe/n-GaAs Contact Fabrication
73.40; S7.12
(a) Vol.120,No.1, K45-K48

W. PITTROFF
C-V Measurements of Isotype Heterojunctions with Deep Rechargeable Interface States
73.40
(a) Vol.120,No.1, K53-K56

L.G. GERCHIKOV and A.V. SUBASHIEV
Interface States in Subband Structure of Semiconductor Quantum Wells
71.25 and 73.40; S7.12; S7.15
(b) Vol.160,No.2, 443-458

D. AGASSI and T.K. CHU
Strain-Induced Superconductivity in Lead Salt Superlattices
74.70; 73.40; S8.1
(b) Vol.160,No.2, 601-612

H. YAMAMOTO and Y. KANIE
Transmission Coefficient and Resonance Condition in a Symmetrical Rectangular Triple-Barrier Structure
73.20; 73.40
(b) Vol.160,No.2, K97-K102

A.N. NAZAROV, V.S. LYSENKO, S.A. VALIEV, M.M. LOKSHIN, A.S. TKACHENKO, and I.A. KUNITSKII
Flash Lamp Annealing and RF Plasma Annealing of Al-SiO_2-Si Structures
73.40; 73.20; S5.11; S10.1
(a) Vol.120,No.2, 447-456

H. HEUKENKAMP and W. HOERSTEL
Trap Tunneling in HgCdTe n^{+}-p Junctions Fabricated by Ion Implantation
73.40; 71.55; S8.15
(a) Vol.120,No.2, 485-496

J. KANIEWSKI, M. KANIEWSKA, W. JUNG, and T. PIOTROWSKI
Iron-Related Deep States in Chemical Vapour Deposited Si
71.55 and 73.40; S5.11
(a) Vol.120,No.2, 531-538

H. WENZEL and H.-J. WUENSCHE
A Model for the Calculation of the Threshold Current of SCH-MQW-SAS Lasers
73.40 and 85; S7.12; S7.15
(a) Vol.120,No.2, 661-674

G. KADEN
Zum Einfluss des Oberflaechenpotentials auf die Frequenzdispersion der Admittanz von GaAs-MIS-Teststrukturen ...
73.40 and 85; S7.12
(a) Vol.120,No.2, 675-686

R. SCHLESINGER, S. ROGASCHEWSKI, and P.J. JANIETZ
Electrochemically Deposited Copper Schottky Contacts on n-Type GaAs for Electron-Beam-Induced Current Measurements
73.40 and 85; 73.30; S7.12
(a) Vol.120,No.2, 687-694

E.L. IVCHENKO, V.P. KOCHERESHKO, I.N. URALTSEV, and D.R. YAKOVLEV
Reflectivity and Photoreflectivity in Superlattices and Quantum Wells
73.40 and 78.20; 71.35; S7.12; S7.15
(b) Vol.161,No.1, 217-222

J. LI and J. BEYNON
Fine-Structure in the I-U Characteristics of Cu-SiO Thin-Film Sandwich Devices
73.40 and 73.60; S10
(a) Vol.121,No.1, 163-172

G. KADEN
Zum Einfluss des Oberflaechenpotentials auf die Frequenzdispersion der Admittanz von GaAs-MIT-Teststrukturen ... (II)
73.40 and 85; S7.12
(a) Vol.121,No.1, 315-326

G. EFTEKHARI
Effect of Rapid Thermal Annealing on Anodic Oxides on InP
73.40 and 85; S7.11
(a) Vol.121,No.1, 327-332

A. MZERD, D. SAYAH, P.C. RUSSELL, and B.K. JONES
Capacitance Measurements on Au-$Cd_{1-y}Zn_yS$ Polycrystalline Schottky Diodes
73.40 and 85; S8.15
(a) Vol.121,No.1, 333-340

A. VOHRA and K.K. SRIVASTAVA
Microwave Switching and AC Conductivity of Memory Glass $As_{10}Ge_{15}Te_{75}$
73.40 and 85; 61.40; S8
(a) Vol.121,No.1, 341-348

V.YU. RUD, YU.V. RUD, and M. SERGINOV
Natural Photopleochroism of n-In_2O_3-p-$CdSiAs_2$ Structures
72.40 and 73.40; 85; S7.16; S10.1
(a) Vol.121,No.1, K81-K84

W. KRAAK, J. KALDASCH, P. GILLE, TH. SCHURIG, and R. HERRMANN
Magnetotransport Properties and Subband Structure of the Two-Dimensional Electron Gas in the Inversion Layer of ...
73.40; S8.15
(b) Vol.161,No.2, 613-628

A. MZERD, D. SAYAH, P.C. RUSSELL, and B.K. JONES
Characterization of Electron Traps in Au-$Cd_{1-y}Zn_yS$ Polycrystalline Schottky Diodes by Capacitance Transient ...
71.55 and 73.40; 85; S8.15
(a) Vol.121,No.2, 587-594

V.YU. RUD, YU.V. RUD, and M. SERGINOV
Photoelectric Properties of H_2O-p-$CdSiAs_2$:In Structures
72.40 and 73.40; S7.16
(a) Vol.121,No.2, K171-K176

V.A. GURTOV and P.A. RAIKERUS
Radiation Induced Current in Silicon Dioxide Considering Diffusion Currents
73.40; S5.11
(a) Vol.121,No.2, K199-K200

R. KASCHNER and TH. VOJTA
Accurate Numerical Calculation of the Electron Density at a Bijellic Interface within the Gradient Method
68.48 and 73.40; S2
(b) Vol.162,No.1, 141-146

G. PAASCH, P.H. NGUYEN, and G. GOBSCH
Envelope Equation and Wave Function Matching for Narrow-Gap Semiconductors
71.25 and 73.40
(b) Vol.162,No.1, 155-164

H. YAMAMOTO, Y. KANIE, and K. TANIGUCHI
Resonant Tunneling in a Symmetrical Rectangular Quadruple-Barrier Structure
73.40
(b) Vol.162,No.1, K25-K28

A. BAIDULLAEVA, B.K. DAULETMURATOV, V.A. GNATYUK, and P.E. MOZOL
Surface States of Laser-Irradiated Cadmium and Tellurium-Containing Solid Solutions of II-VI Compounds
73.20 and 73.40; 72.40; S8.15
(a) Vol.122,No.1, 243-248

I. THURZO, V. NÁDAŽDY, and E. PINČÍK
Electric-Field-Enhanced Emission from a Discrete Energy Level at the GaAs-Oxide Interface
73.40; 71.55; S7.12
(a) Vol.122,No.1, 275-284

H.-J. FITTING, P. MAGDANZ, W. MEHNERT, D. HECHT, and TH. HINGST
Charge Trap Spectroscopy in Single and Multiple Layer Dielectrics
73.40; S10.1
(a) Vol.122,No.1, 297-310

A. REIF, P. STREUBEL, A. MEISEL, and D. ZEISSIG
Characterization of the Diffusion and Reaction Behaviour of Ti/Pt/Au Layer Contacts on GaAs by Means of Auger Electron..
68.48 and 73.40; 79.20; S7.12
(a) Vol.122,No.1, 331-340

V.K. ADAMCHUK, V.V. AFANASEV, and A.V. AKULOV
The Electron Trap Activation in Thermal SiO_2
71.55; 73.40; S10.1
(a) Vol.122,No.1, 347-354

P. CHAKRABARTI, M. PURI, A. SINGHAL, and A. MISHRA
Optically Controlled Characteristics of TEGFET
73.40 and 85; S7.12; S7.16
(a) Vol.122,No.1, 405-412

A. SCHENK
Spatially Variable Drift Mobility Model for $Hg_{1-x}Cd_xTe$ Diodes. I. Analytical Base and Fit to Hall Data
73.40 and 85; S8.15
(a) Vol.122,No.1, 413-426

T. ASAHI, M. YUDASAKA, K. NAKANISHI, S. IWASHIMA, J. AOKI, N. SATO, and H. INOKUCHI
Rectification Properties of In/Tetrabenzo[de,hi,op,st]Pentacene/Au Devices
73.40 and 85; S12
(a) Vol.122,No.1, K97-K100

A.G. LIPSON, V.A. KUZNETSOV, D.M. SAKOV, and YU.P. TOPOROV
Electrical and Magnetic Properties of the Polymer-Ceramics Produced on the Basis of Lithium Fluoride
73.40 and 75.50; S9.11; S12
(a) Vol.122,No.2, 563-568

A.S. POPOV and A.Y. BAHNEV
Photoconductivity Spectra of Fe-Doped GaInP Epitaxial Layers
72.40 and 73.40; S7.16
(a) Vol.122,No.2, 569-574

J. KOLNÍK and M. OŽVOLD
The Influence of Inversion Surface Layers on the Evaluation of the Interface State Energy Distribution from Schottky...
73.30 and 73.40; 85; S5.11; S7.12
(a) Vol.122,No.2, 583-588

O. EIBL, J.M. TRISCONE, and O. FISCHER
Imaging of $YBa_2Cu_3O_7/PrBa_2Cu_3O_7$ Superlattices by TEM
68.55 and 73.40; 64.80; S10.15
(a) Vol.122,No.2, 589-598

J. BARTOŠ and L. TESAŘ
Some Electrical and Optical Properties of Nickel-Related Deep Levels in Silicon
61.70 and 71.55; 73.40; S5.11
(a) Vol.122,No.2, 607-616

H. HAEFNER, L. LEHMANN, R. MITDANK, G. OELGART, and E. SCHULZE
Luminescence Characterization of (AlGa)As Single Quantum Wells
73.40 and 78.55; 78.60; S7.15
(a) Vol.122,No.2, 683-694

J. BARTOŠ, E. PINČÍK, I. THURZO, B. MATÁTKO, and T. LALINSKÝ
Characterization of MAOS Structures on GaAs
73.40 and 85; S7.12
(a) Vol.122,No.2, 715-722

A. SCHENK
Spatially Variable Drift Mobility Model for $Hg_{1-x}Cd_xTe$-Diodes. II. Application to Device Simulation
73.40 and 85; S8.15
(a) Vol.122,No.2, 723-730

M. BRAUER
Effect of Surface Charge and Coating on the Field Distribution of Si High-Voltage Thyristor-Like Structures with ...
73.40 and 85; S5.11
(a) Vol.122,No.2, 741-750

G. GRUMMT, R. PICKENHAIN, and G. OELGART
Photocurrent Investigation on a Graded Heterojunction
73.40; 72.40; S7.12
(a) Vol.122,No.2, 765-772

73.60. Electronic properties of thin films

E. GERLACH
A Contribution to DC Transport in Thin Bi Films
73.60; S4
(b) Vol.157,No.1, 189-198

A. BEL HADJ, B. EL JANI, M. GUITTARD, R. BENNACEUR, and P. GIBART
Photoluminescence of Heavily Doped Sn-GaAs Grown by Metal Organics Vapor Phase Epitaxy
61.70 and 78.65; 73.60; S7.12
(a) Vol.117,No.1, 169-176

V.I. MIKLA, D.G. SEMAK, A.V. MATELESHKO, and A.A. BAGANICH
Photosensitized Deep Gap Center Relaxation in Amorphous Selenium Layers Containing As Additions
71.55; 72.40; 73.60; S5
(a) Vol.117,No.1, 241-250

R. MITDANK, A. STURM, and B. JACOBS
Investigation of Excess Carrier Diffusion in $Ga_{0.35}Al_{0.65}As$ by Cathodoluminescence
73.60; 66.30; S7.15
(a) Vol.117,No.2, 485-494

R. SATHYAMOORTHY, SA.K. NARAYANDASS, C. BALASUBRAMANIAN, and D. MANGALARAJ
Structure, Dielectric, and AC Conduction Studies on Yttrium Fluoride Thin Films
68.55 and 73.60; 77.55; S9
(a) Vol.117,No.2, 495-500

R. SALAM
Trapping Parameters of Electronic Defect States in Indium Tin Oxide from AC Conductivity
73.60; 71.55; S10
(a) Vol.117,No.2, 535-540

A.A. ZAHAB, M. ABD-LEFDIL, and M. CADENE
Electrical Conductivity, Optical Absorption, and Photoconductivity Measurements of Well Oriented Indium-Telluride Thin Films
73.60 and 78.65; 72.40; S8
(a) Vol.117,No.2, K103-K106

C.S. WU, C.S. PAI, S.J. PEARTON, F. REN, E. LANE,
and D.M. SCHLEICH
Schottky Barrier Enhancement on n-Type GaAs by As Implantation
73.60; S7.12
(a) Vol.117,No.2, K107-K110

TENG-GE MA, XIAN-JUN YANG, JI-SHENG ZHANG, and PEI-HSIN TSIEN
Localization of the Boundaries in Laser Recrystallization of SOI with Antireflective Stripe
61.80 and 68.55; 73.60; S5.11
(a) Vol.118,No.1, 179-184

H. RATAJCZAK, I. GOŚCIAŃSKA, and T. LUCIŃSKI
Influence of Annealing on Saturation Magnetization and Hall Resistivity in Amorphous $Nd_{12}Co_{82}B_6$ Thin Films
61.40 and 68.55; 73.60; 75.70; S1.2; S1.4
(a) Vol.118,No.1, 189-196

W.A. BADAWY, R.S. MOMTAZ, and E.M. ELGIAR
Solid State Characteristics of Indium-Incorporated TiO_2 Thin Films
68.55 and 78.65; 73.60; S10.1
(a) Vol.118,No.1, 197-202

F. VOELKLEIN and E. KESSLER
Thermal Conductivity and Thermoelectric Figure of Merit of Thin Antimony Films
72.15 and 73.60; S4
(b) Vol.158,No.2, 521-530

E. GERLACH
A Contribution to Reversed Transport
73.60
(b) Vol.158,No.2, 531-538

B. REICHERT and CH. JUNG
Electronic Structure of the Charged Copper(001)/Vacuum Interface
73.20; 73.30; 73.60; S1.2
(b) Vol.158,No.2, K119-K122

B.M. ASKEROV, G.I. GUSEINOV, B.I. KULIEV, and S.R. FIGAROVA
Influence of the Surface Scattering upon the Conductivity in Films with a Quasi-Two Dimensional Electron Spectrum
72.10; 73.60
(b) Vol.158,No.2, K179-K184

T. TSVETKOVA, B. AMOV, E. VATEVA, and V. AVERYANOV
Ion Implantation Induced Modification of Vitreous Chalcogenides
73.60; 61.80; 78.65; S8.12
(a) Vol.119,No.1, 107-112

B. THOMAS and T.R.N. KUTTY
Formation of Single-Phase Indium Selenide Thin Films by Elemental Evaporation
68.55; 73.60; 78.65; S8.12
(a) Vol.119,No.1, 127-138

G. AMATO, G. BENEDETTO, F. FIZZOTTI, C. MANFREDOTTI, and R. SPAGNOLO
Photothermal and Photoconductive Measurements of Low Pressure Chemically Vapour Deposited Amorphous Silicon
72.40 and 73.60; 78.65; S5.11
(a) Vol.119,No.1, 169-176

A.E. TSURKAN, T.D. SHEMYAKOVA, S.P. MEDVETSKII, and L.A. NAZARENKO
Spectral Response and Detectivity of InAs Hybrid Structures
68.48; 73.60; S7.12; S10.1
(a) Vol.119,No.1, 191-200

A. SARKAR, S. CHAUDHURI, and A.K. PAL
Electrical and Optical Properties of Transparent Conducting ZnO Films
73.60 and 78.65; S10.1
(a) Vol.119,No.1, K21-K26

T. DITTRICH
Correlation between Photoreflectance and CV Measurements of n-GaAs Epitaxial Layers
73.60 and 78.65; S7.12
(a) Vol.119,No.2, 479-486

A. MZERD, D. SAYAH, I.J. SAUNDERS, and B.K. JONES
Electrical Properties of $Cd_{1-y}Zn_yS$ and CdS(In) Thin Films Prepared by Pyrolytic Spray Technique
73.60; 71.55; S8.15
(a) Vol.119,No.2, 487-494

R.A. MORENO, C.X. CARDOSO, and M. YOSHIDA
Hole Injection in Teflon FEP Films
72.20 and 73.60; S12
(a) Vol.119,No.2, 583-588

J. KIERUL and J. LEDZION
Conductivity of Thin Metallic Films
72.15 and 73.60
(a) Vol.119,No.2, K117-K120

R. TURCU
AC Conductivity of ClO_4^- Doped Polythiophene Films
73.60; S12.1
(a) Vol.119,No.2, K121-K126

S.P. ZAKHAROV and V.V. USTINOV
The High Frequency Quantum Size Effect in the Impedance of Metal Films
73.60; 72.15; S4
(b) Vol.160,No.1, 97-104

A.G. KAZANSKII, W. FUHS, and H. MELL
On the Fermi-Level Position in a-Si:H(P)
73.60; S5.11
(b) Vol.160,No.1, K25-K28

J. GUTOWSKI, N. PRESSER, and G. KUDLEK
Optical Properties of ZnSe Epilayers and Films (Review Article)
78.65; 68.55; 71.35; 73.60; 78.55; S8.12
(a) Vol.120,No.1, 11-60

J.M. DUSSEAU, J.L. ROBERT, and S. ABDALLA
Quantitative Interpretation of Shubnikov-de Haas and Hall Effects in Amorphous Thin Films of Cd_3As_2
73.60; S7
(a) Vol.120,No.1, 151-158

N.S. SALEH, K.A. AL-SALEH, and A.M. AL-SAIE
Kinetics of Ion Beam Induced Mixing in Au/Si System
61.80 and 68.55; 73.60; S1.3; S5.11
(a) Vol.120,No.1, 169-174

R.D. GOULD and R.I.R. BLYTH
Conductivity and Absorption Measurement in Nickel Phthalocyanine Thin Films
73.60 and 78.65; S12
(a) Vol.120,No.1, K57-K62

A.K. MEIKAP, A.R. JANA, S.K. DE, and S. CHATTERJEE
Weak Localization and Interaction Effects in Thin Disordered Zinc Films
71.38 and 73.60; S1.2
(b) Vol.160,No.2, 473-482

YANG-FANG CHEN
Persistent Photoconductivity in Doping Modulated Amorphous Silicon Multilayers
73.60; 72.40; S5.11
(b) Vol.160,No.2, K103-K108

J.C. BUREAU, A. GLACHANT, and B. BALLAND
Thermal Nitridation of SiO_2 Thin Films in NH_3 + H_2O (or O_2) Mixtures
68.55 and 73.60; 78.65; S10.1
(a) Vol.120,No.2, 507-518

M. DANZIGER and G. ANDRASSY
Experimental and Theoretical Study of Heavy Ion Irradiation Processes in Polymer Materials
61.80 and 68.60; 73.60; S12
(b) Vol.161,No.1, 223-232

E.YU. TURAEV, P.P. SEREGIN, and F.S. NASREDINOV
Tin and Iron Impurity Atom States in Modified As_2Se_3 Films
71.55 and 73.60; 78.65; S8
(b) Vol.161,No.1, 287-292

J. LI and J. BEYNON
Fine-Structure in the I-U Characteristics of Cu-SiO Thin-Film Sandwich Devices
73.40 and 73.60; S10
(a) Vol.121,No.1, 163-172

T. BERGUNDE, M. WIENECKE, and B. THOMAS
High Mobility HgTe and High Resistivity CdTe Grown by MOVPE on GaAs(001)
68.48 and 73.60; S8.13
(a) Vol.121,No.1, K55-K60

O.I. DUBROVSKII, S.I. KURGANSKII, L.A. RUBTSOVA, and G.W. WOLF
Non-Relativistic and Scalar-Relativistic LAPW Electronic Energy Bands of (001)Cu Films
71.25 and 73.60; 71.10; S1.2
(b) Vol.161,No.2, 697-704

M. ZĄBKOWSKA-WACŁAWEK, P. TALIK, and W. WACŁAWEK
On the Compensation Behaviour of Copper Phthalocyanine Films in NO_x Ambient
73.60; S12
(a) Vol.121,No.2, 489-494

T.G. ABDEL-MALIK, A.A. AHMAD, and A.S. RIAD
Dark and Photoelectric Investigation on Thin Films of beta-Cobalt Phthalocyanine Dispersed in Polymers
73.60; 72.40; S12
(a) Vol.121,No.2, 507-514

M.D. KANNAN, SA.K. NARAYANDASS, C. BALASUBRAMANIAN, and D. MANGALARAJ
Aging, Annealing, and Dielectric Properties of Neodymium Oxide Thin Films
73.60; 77.55; S10
(a) Vol.121,No.2, 515-522

H. BOETTCHER, U. WEIH, J. OPFERMANN, and R. ACKERMANN
Electrical Properties of Pure and Mixed Evaporated Dye Layers
72.40 and 73.60; S12
(a) Vol.121,No.2, 563-570

YINYUE WANG, FANGQING ZHANG, and GUANGHUA CHEN
Behavior of the Thermal Equilibrium in Reactively Sputtered $a\text{-}Si_{1-x}Ge_x{:}H$ Alloy Films Studied by Conductivity and...
73.60; 68.55; S5.11; S5.12
(a) Vol.121,No.2, K177-K180

L.I. BEREZHINSKII, S.I. VLASKINA, G.I. LYASHENKO, and V.E. RODIONOV
On the Possibility of Obtaining the Crystalline Phase in Amorphous Films of Silicon Carbide on Silicon at Low Temperature
68.55; 73.60; 78.55; S6
(a) Vol.121,No.2, K193-K198

F. VOELKLEIN and U. DILLNER
Mobilities and Concentrations of Charge Carriers in Polycrystalline $Bi_{0.87}Sb_{0.13}$ Films
73.60; S4
(b) Vol.162,No.1, 147-154

YIFAN HU and RONGCHANG YIN
The Gas-Sensing Characteristics of Bismuth Chromate Material
68.45 and 73.60; S11.1
(a) Vol.122,No.1, 249-254

T. LUCIŃSKI
The Resistivity of Amorphous $Fe_{1-x}Ti_x$ Thin Films
73.60; 72.15; S1.1
(a) Vol.122,No.1, 323-330

I. GOŚCIAŃSKA, H. RATAJCZAK, P. SOVÁK, P. SPIŠÁK, and M. KONČ
The Structure of Fe-Rich Vapor Quenched Fe-Ti Thin Films
68.55; 61.40; 73.60; S1.1; S1.2
(a) Vol.122,No.2, 575-582

73.90. Other topics in electrical properties of surfaces, interfaces,and thin films

M. TROTT, S. TROTT, G. TEICHMANN, and CH. SCHNITTLER
Modified Thomas-Fermi Treatment of delta-Doping Layer-Like Structures
71.10 and 73.90; 71.20; 71.50
(b) Vol.161,No.1, K15-K18

M. ZAŁUŻNY
Coupling between Intersubband Modes in Wide Quantum Wells
71.25 and 73.90; S7.1
(b) Vol.161,No.1, K19-K24

ZHANG-LONG XU, KA-DI ZHU, and SHI-WEI GU
A Quantum Mechanical Treatment of Surface Polaritons in
Semi-Infinite Dielectric Systems
71.36 and 73.90; S7.11
(b) Vol.161,No.1, K25-K28

74. Superconductivity

74.10. Occurence, critical temperature

R.A. EVARESTOV and V.A. VERYAZOV
The Electronic Structure of Copper Oxide Crystalline Compounds. I. LUC-CNDO Approach to the Electronic Structure of Cu_2O and ...
71.10 and 71.25; 74.10; S1.2; S10.1
(b) Vol.157,No.1, 281-292

R.A. EVARESTOV and V.A. VERYAZOV
The Electronic Structure of Copper Oxide Crystalline Compounds II. Chemical Bonding in Copper-Oxygen Crystals
71.20 and 71.25; 74.10; S1.2; S10
(b) Vol.158,No.1, 201-212

J.S. ARELLANO and G.J. VÁZQUEZ
A First Principles Calculation of the Critical Temperature of Superconducting Metallic Hydrogen
74.10; S5.2
(b) Vol.158,No.2, K161-K164

G. FUCHS, A. GLADUN, S.-L. DRECHSLER, R. MUELLER, M. RITSCHEL, G. KRABBES, N. KOPPETZKI, and A. GILEWSKI
Effect of Oxygen Concentration on the Resistive Superconducting Transition of $YBa_2Cu_3O_{7-x}$ in High Magnetic Field
74.10 and 74.60; 74.40; S10.15
(b) Vol.159,No.2, 797-804

G. SRINIVASAN, GUO-MEI WU, and T.T. SRINIVASAN
Surface Resistance at Microwave Frequencies in Transition Metal Ion Substituted Yttrium Barium Copper Oxide ...
73.25 and 74.10; S10.15
(a) Vol.122,No.1, 355-360

V.V. PARASHCHUK
The Influence of Electric Discharge on the Properties of High-Temperature Superconductors
74.10 and 74.30
(a) Vol.122,No.2, 631-636

74.20. Theory

M. FUKUHARA and H.A. MCKINSTRY
Intralattice Strain Analysis in Superconducting Crystals and Superconducting Electrons
74.20; 74.70
(b) Vol.157,No.1, 357-378

A. TEUBEL, E. KOLLEY, and W. KOLLEY
Functional Approaches to Superconductivity within an Extended Attractive Hubbard Model
74.20
(b) Vol.157,No.1, 389-398

E. KOLLEY and W. KOLLEY
Schrieffer-Wolff Transformation of the Emery Model for Cu-O Superconductors
74.20
(b) Vol.157,No.1, 399-410

L.S. BRIZHIK, A.S. DAVYDOV, and V.N. ERMAKOV
The Magnetic Field Influence on a Superconducting Bisoliton Condensate
74.20; 71.45; S12.1
(b) Vol.157,No.1, 417-424

G. HORVÁTH and J. BÁNKUTI
Cell Percolation Model for Electrical Conduction of Granular Superconducting Composites. II. Percolation Conductivity...
74.20 and 74.50
(a) Vol.117,No.1, 273-282

L.S. BRIZHIK
Bisoliton in Constant Magnetic Field
74.20; 71.45; S12.1
(b) Vol.157,No.2, 649-656

L.J. DUNNE and A. TANG KAI
Excited-State Molecular Superconductivity
74.20; S12
(b) Vol.157,No.2, 657-666

E. KOLLEY and W. KOLLEY
Mean-Field Equations of the $\tilde{t}_{pp}$-J_{pd}-J_{dd} Model for CuO_2-Based Superconductors
74.20
(b) Vol.157,No.2, K113-K116

A. TEUBEL, E. KOLLEY, and W. KOLLEY
Influence of the Fock Parameter on T_c in an Extended Attractive Hubbard Model
74.20
(b) Vol.157,No.2, K117-K122

A.S. DAVYDOV
Foundations of the Quasi-One-Dimensional Bisoliton Model of Superconductivity of Ceramic Oxides
71.45 and 74.20; S12.1
(b) Vol.160,No.1, 305-312

GANG SU
Supersymmetry and Superconductivity in the eta-Pairing Hubbard Model
74.20
(b) Vol.160,No.1, K49-K52

S.A. SERGEENKOV
On Logarithmic CVC of High-T_c Superconductivity Glass Model
74.20; 61.40; S10.15
(b) Vol.160,No.2, K137-K140

A.M. GABOVICH, D.P. MOISEEV, A.S. SHPIGEL, and A.I. VOITENKO
Josephson Tunneling Critical Current between Superconductors with Charge- or Spin-Density Waves
74.50; 74.20; S10.15
(b) Vol.161,No.1, 293-302

A.A. KISELEV and A.V. LIAPZEV
Phase Transitions in Conductive Jahn-Teller Crystals
74.20
(b) Vol.161,No.1, K35-K40

A.A. KISELEV and A.V. LIAPTSEV
A Model Theory of Structural and Superconducting Phase Transitions in Chain-Like Systems
74.20; 64.60; S12.1
(b) Vol.161,No.2, 717-730

V.M. LOKTEV
On the Carrier Spectrum of CuO_2 Planes in High-T_c Superconductors and Its Dependence on Stoichiometry
74.20; 71.45; S12.1
(b) Vol.161,No.2, 731-744

V.N. ERMAKOV, S.P. KRUCHININ, and E.A. PONEZHA
Superconducting Current in a Bisoliton Superconductivity Model
74.20; 71.45
(b) Vol.161,No.2, 745-754

J. FOLTIN
Dependence of T_c on the Number of CuO_2 Layers in High-T_c Superconductors
74.20 and 74.70
(b) Vol.161,No.2, K95-K98

W. WELLER
On Functional Integral Representations Satisfying the Constraints in the Correlation Models of High-Temperature ...
74.20
(b) Vol.162,No.1, 251-264

74.30. General properties

C. LUCCHINI, C. GIOVANNELLA, R. MESSI, B. LECUYER, L. FRUCHTER, and M. IANNUZZI
AC Susceptibility and Non-Linear Dynamics in Granular Superconductors
74.30; S10.15
(b) Vol.157,No.2, K123-K126

J. MUCHA and A. JEŻOWSKI
Evidence of Phase Transition in $La_{1.85}CuO_{4+\delta}$ by Thermal Conductivity
74.30; S10.15
(a) Vol.119,No.1, K53-K58

N. FLOROS, L.J. DUNNE, and J.N. MURRELL
The Relative Energies of Ionic Structures for the High-T_c Superconductor $YBa_2Cu_3O_7$ and Implications for the ...
74.30; S10.15
(b) Vol.159,No.2, 781-786

W. SCHILLER and K. WESCHKE
Calculation of the Normal-State Thermal Conductivity of the Lattice for High-T_c Superconductors
74.30 and 74.70; S10.15
(b) Vol.161,No.2, 755-768

F.J. OWENS
Non-Resonant Microwave Absorption in N=3 Phase of Bi-Pb-Sr-Ca-Cu-O High-Temperature Superconductor
74.30; 76.30; S10.15
(b) Vol.162,No.2, 565-574

M.V. SATARIĆ, Z. SHEMSEDINI, S. STAMENKOVIĆ, and R.B. ŽAKULA
The Meissner Effect in Superconductors with Bisolitonic Mechanism of Superconductivity
74.30
(b) Vol.162,No.2, K95-K100

V.V. PARASHCHUK
The Influence of Electric Discharge on the Properties of High-Temperature Superconductors
74.10 and 74.30
(a) Vol.122,No.2, 631-636

74.40. Fluctuations and critical effects

LING CHEN, GUANGHOU WANG, and MINGKANG TENG
Charge Fluctuation in the High T_c Superconductor YBaCuO in the Transition Temperature Region
74.40; 78.70; S10.15
(b) Vol.157,No.1, 411-416

J. BLAESING, T. HOLZ, and J. SCHREIBER
Hysteresis, Period Doubling, and Magnetic Chaos in Josephson Junction Arrays
74.40 and 74.50
(b) Vol.157,No.2, K109-K112

G. FUCHS, A. GLADUN, S.-L. DRECHSLER, R. MUELLER, M. RITSCHEL, G. KRABBES, N. KOPPETZKI, and A. GILEWSKI
Effect of Oxygen Concentration on the Resistive Superconducting Transition of $YBa_2Cu_3O_{7-x}$ in High Magnetic Field
74.10 and 74.60; 74.40; S10.15
(b) Vol.159,No.2, 797-804

74.50. Tunneling phenomena, Josephson effect, and proximity effects

M. KUPKA
A Simple Model for Point Contact Spectra of Normal Metal-Heavy Fermion Metal
74.50
(b) Vol.157,No.1, K39-K42

J. BÁNKUTI and G. HORVÁTH
Cell Percolation Model for Electrical Conduction of Granular Superconducting Composites. I.The Resistivity of an ...
74.50; 74.70
(a) Vol.117,No.1, 265-272

G. HORVÁTH and J. BÁNKUTI
Cell Percolation Model for Electrical Conduction of Granular Superconducting Composites. II. Percolation Conductivity...
74.20 and 74.50
(a) Vol.117,No.1, 273-282

J. BLAESING, T. HOLZ, and J. SCHREIBER
Hysteresis, Period Doubling, and Magnetic Chaos in Josephson Junction Arrays
74.40 and 74.50
(b) Vol.157,No.2, K109-K112

XICHUNNG JIN, LINJIANG SHEN, ZUGANG LI, HUIMIN SHAO, HUAQIN WANG, YITONG ZHANG, LONG HAI, and MEISHENG HU
Microdomain Structure and Temperature Behaviour of $YBa_2(Cu_{1-x}Fe_x)_3O_{7-y}$
61.50; 74.50; S10.15
(a) Vol.119,No.1, K5-K8

A.M. GABOVICH, D.P. MOISEEV, A.S. SHPIGEL, and A.I. VOITENKO
Josephson Tunneling Critical Current between Superconductors with Charge- or Spin-Density Waves
74.50; 74.20; S10.15
(b) Vol.161,No.1, 293-302

F. MUELLER, H.-J. KOEHLER, P. WEBER, K. BLUETHNER, and H.-G. MEYER
A Microwave Circuit for a 1 V Josephson Voltage Standard Operated at 35 GHz
74.50; S4
(a) Vol.121,No.1, K73-K76

J. DRYZEK and M. ROMEK
Positron Annihilation Study of Oxygen Vacancies in the High Temperature Superconductor Y-Ba-Cu-O
61.70 and 78.70; 74.50; S10.15
(a) Vol.121,No.2, K159-K162

P. SEIDEL, M. TURTENWALD, and T. BUESCHEL
Temperature Dependence of the Structures in the Tunneling Characteristics of High-T_c Junctions
74.70; 74.50
(a) Vol.122,No.2, 645-650

74.60. Type-II superconductivity

C. YU, C. LAI, S.A. MARSHALL, D.R. YODER-SHORT, and Y.N. ZHANG
Electron Paramagnetic Resonance Spectra Observed in Single Crystals of $YBa_2Cu_3O_{7-\delta}$
74.60; 76.30; S10.15
(b) Vol.157,No.1, 379-388

T. KRYSZTOŃ
On the Flux Penetration in the Anisotropic Uniaxial Superconductor
74.60; S10.15
(b) Vol.158,No.1, K21-K26

XIAO-GUANG LI, XIANGLEI MAO, LIN CHEN, and YUHENG ZHANG
Study on the Formation Kinetics of Flux Lines in High-T_c Superconductors
74.60; 74.70; S10.15
(b) Vol.158,No.1, K27-K32

P. TEKIEL
Longitudinal Current Distribution in the Mixed State of an Anisotropic Type-II Superconductor
74.60
(b) Vol.158,No.2, K155-K160

R. HERGT, W. ANDRAE, K. FISCHER, N.M. TCHEBOTAEV, and S.L. TOWN
Pinning in Twinned Single Crystals of Superconducting $REBa_2Cu_3O_{7-x}$ Investigated by Torque Magnetometry
74.60 and 74.70; 61.70; S10.15
(a) Vol.119,No.1, 241-250

K.P. SINHA and M. SINGH
Correlated Pair-Transfer Mechanism for High Temperature Superconductivity. The Case of $YBa_2Cu_3O_{6+x}$
74.60; S10.15
(b) Vol.159,No.2, 787-796

G. FUCHS, A. GLADUN, S.-L. DRECHSLER, R. MUELLER, M. RITSCHEL, G. KRABBES, N. KOPPETZKI, and A. GILEWSKI
Effect of Oxygen Concentration on the Resistive Superconducting Transition of $YBa_2Cu_3O_{7-x}$ in High Magnetic Field
74.10 and 74.60; 74.40; S10.15
(b) Vol.159,No.2, 797-804

R. HERGT, W. ANDRAE, and W. SCHUEPPEL
Pinning of Vortices at Twin Boundaries in $YBa_2Cu_3O_{7-x}$ Single Crystals
74.60 and 74.70; 61.70; S10.15
(a) Vol.122,No.1, K51-K54

J. SHA, C.F. ZHU, Z.H. HE, X.-G. LI, Z.P. SU, L.Z. CAO, Q.R. ZHANG, H.R. YI, and H.C. LI
Nonlinear I-V Characteristics of $GdBa_2Cu_3O_y$ Epitaxial Films Related with the Flux Creep
74.60 and 74.70; S10.15
(a) Vol.122,No.1, K55-K58

A.G. SAIF and T.F. REFAI
On the Critical State and Surface Barrier of a Type II Superconducting Long Cylinder
74.60
(b) Vol.162,No.2, 559-564

W. ANDRAE, J. BETZ, B. BRUNNER, R. HERGT, H. LENGFELLNER, K.F. RENK, W. SCHUEPPEL, and K. STEENBECK
In-Plane Rotational Magnetic Losses in a Laser Deposited $YBa_2Cu_3O_{7-x}$ Film
74.60; 74.70; S10.15
(a) Vol.122,No.2, K161-K164

74.70. Superconducting materials

M. FUKUHARA and H.A. MCKINSTRY
Intralattice Strain Analysis in Superconducting Crystals and Superconducting Electrons
74.20; 74.70
(b) Vol.157,No.1, 357-378

H. MIZUBAYASHI, H. AKINAGA, and K. TAKITA
Low-Temperature Elastic Properties of $Nd_{1+x}Ba_{2-x}Cu_3O_{7-\delta}$
62.40; 74.70; S10.15
(a) Vol.117,No.1, 125-134

J. BÁNKUTI and G. HORVÁTH
Cell Percolation Model for Electrical Conduction of Granular Superconducting Composites. I.The Resistivity of an ...
74.50; 74.70
(a) Vol.117,No.1, 265-272

N. KRISTOFFEL and P. RUBIN
Localized One-Electron States of Impurities on Cu Sites in the CuO_2 Planes of Oxide Superconductors
71.55 and 74.70; S10.15
(b) Vol.157,No.2, K93-K98

XI CHEN, CHAO DING, YU WANG, XIAN-TING ZEN, and BAO-SHENG LUO
The Structures of $YBa_2Cu_3O_{6+\varepsilon}$ Superconducting and Nonsuperconducting Phases
61.60 and 74.70; S10.1
(a) Vol.117,No.2, 335-340

M. MANZEL, L. ILLGEN, and R. HERGT
Superconducting Bi-Sr-Ca-Cu-O Thin Films on YAG
74.70; S10.15
(a) Vol.117,No.2, K119-K120

E.M. GOLOLOBOV, N.A. PRYTKOVA, ZH.M. TOMILO, D.M. TURTSEVICH, M.S. TSELUEVSKII, and N.M. SHIMANSKAYA
Properties of High T_c Superconducting Metal Oxide $Bi_2Sr_2CaCu_2O_y$ with Boron Nitride Additions
61.50 and 74.70; S10.15
(a) Vol.117,No.2, K121-K126

XIN JIN, XIAOHUI WANG, YI TANG, MU LU, YITONG ZANG, WUMING CHEN, JIRONG JIN, HUIMIN SHAO, and XIXIAN YAO
The Irreversibility of the Critical Currents of Polycrystalline Bulk $EuBa_2Cu_3O_{7-x}$ with Cycles in an Applied Magnetic Field
74.70; S10.15
(a) Vol.117,No.2, K127-K132

XIAO-GUANG LI, XIANGLEI MAO, LIN CHEN, and YUHENG ZHANG
Study on the Formation Kinetics of Flux Lines in High-T_c Superconductors
74.60; 74.70; S10.15
(b) Vol.158,No.1, K27-K32

C. SUŁKOWSKI, K. ROGACKI, Z. BUKOWSKI, R. HORYŃ, and E. TROJNAR
Correlation between Superconducting Transition Temperature, Thermopower, and Resistivity for $YBa_2Cu_3O_{7-x}$
74.70; S10.15
(b) Vol.158,No.1, K33-K36

S. GRANDE, B. LIPPOLD, G. MAYER, and H. SCHMIEDEL
NQR Measurements of ^{63}Cu in $YBa_2Cu_3O_{7-x}$
74.70 and 76.60; S10.15
(b) Vol.158,No.2, 707-714

E.N. IZAKOVICH, N.A. ZHELTOVA, YU.M. SHULGA, I.N. IVLEVA, ZH.D. SOKOLOVSKAYA, A.V. BALUEV, V.V. GANIN, O.S. ROSHUPKINA et al.
Erratum to Study of Ba and Cu(111) Based Inorganic Oxide Compounds (phys. stat. sol. (a) 115, K195 (1989))
74.70; 78.30; 79.20; 79.60
(a) Vol.118,No.2, K139-K140

YIFENG WANG, WENHUI SU, HONGJIAN LIU, and LIPING LI
Orthorhombic Phase and Oxygen Deficiencies in the Electron Type Superconductor $Ln_{2-x}Ce_xCuO_{4-y}$
64.70 and 74.70; S10.15
(a) Vol.119,No.1, 55-62

V.A. SKRYSHEVSKII and P. KUS
Optical and Electro-Physical Properties of $Gd_1Ba_2Cu_3O_{7-\delta}$-p-Si Structures at $T>T_c$
73.40 and 74.70; 78.65; S10.15
(a) Vol.119,No.1, 201-208

R. HERGT, W. ANDRAE, K. FISCHER, N.M. TCHEBOTAEV, and S.L. TOWN
Pinning in Twinned Single Crystals of Superconducting $REBa_2Cu_3O_{7-x}$ Investigated by Torque Magnetometry
74.60 and 74.70; 61.70; S10.15
(a) Vol.119,No.1, 241-250

J. PIECHOTA and A. PAJĄCZKOWSKA
Magnetic Properties of $YBa_2(Cu_{1-x}Ni_x)_3O_{7-\delta}$
75.30; 74.70; S10.15
(a) Vol.119,No.1, 271-278

SHA JIAN, LIU WEI, C.F. ZHU, S.X. WANG, N.L. WANG, X.A. CHENG, L.Z. CAO, Z.J. CHENG, Z.Y. CHENG, and Q.R. ZHANG
Superconductivity Affected by Zr Doping in $Bi_2Sr_2Ca_1Cu_2O_y$ Ceramic
74.70; S10.15
(a) Vol.119,No.1, K59-K62

P.P. SEREGIN, F.S. NASREDINOV, V.F. MASTEROV, and G.T. DARIBAEVA
Electron Structure and Crystalline Field Parameters Determined by the Emission Moessbauer Spectroscopy for the Copper...
76.80; 74.70; S10.15
(b) Vol.159,No.2, K97-K102

YUPING SUN, JIANYI JIANG, FANCHUN ZHENG, HUAQING YIN, and JIAJU DU
Synthetic Condition and Properties of the Bi-Pb-Sr-Ca-Cu-O Superconductor with Single High-T_c Phase
74.70; 64.80; S10.15
(a) Vol.119,No.2, 555-560

XIAO-GUANG LI, XIANDONG WANG, XIANGLEI MAO, LIN CHEN, XU WANG, YU WANG, and YUHENG ZHANG
The Effect of La Doping on the Superconductivity of $Bi_{1.7}Pb_{0.3}Sr_{2-x}La_xCa_2Cu_3O_y$
74.70; S10.15
(a) Vol.119,No.2, K141-K144

X.Y. ZHOU, H. ZHANG, and Q.R. ZHANG
Substitution Effect of Al and Sn for Cu in the Y-Ba-Cu-O System Studied by X-Ray Photoemission Spectroscopy
74.70 and 79.60; S10.15
(a) Vol.119,No.2, K145-K150

S.N. BARILO, V.I. GATALSKAYA, G.V. GATALSKII, A.P. GES, S.A. GURETSKII, F.P. KORSHUNOV, L.A. KUROCHKIN et al.
Electron Irradiation of Single-Crystalline $Bi_2Sr_2CaCu_2O_{8+\delta}$
61.80 and 74.70; S10.15
(a) Vol.119,No.2, K151-K154

Š. CHROMIK, V. ŠTRBIK, Š. BEŇAČKA, R. ADAM, and M. JERGEL
Thin $YBa_2Cu_3O_{7-x}$ Films Prepared by Low Temperature Coevaporation on Si Substrates without Buffer Layer
74.70; 68.55; S10.15
(a) Vol.119,No.2, K155-K158

W. SCHILLER and K. WESCHKE
Calculation of the Normal-State Resistivity for High-T_c Superconductors
72.15 and 74.70; S10
(b) Vol.160,No.1, 297-304

W. KRAAK, A. KRAPF, G. NACHTWEI, H.-U. MUELLER, H. SCHOETZ, H. DWELK, R. HERRMANN, P. VASEK, and P. SVOBODA
Preparation, Structure, and Physical Properties of Bi-Sr-Ca-Cu-O Superconducting Ceramics
74.70; S10.15
(a) Vol.120,No.1, 185-198

N. KAWAHARA, S. KAWABATA, H. ENAMI, T. SHINOHARA, H. HOSHIZAKI, M. HASEGAWA, SH. ASAI, and T. IMURA
Preparation of Highly Oriented Bulky Bi(Pb)-Sr-Ca-Cu-O Superconductor by Magnetized Twin-Roll
74.70; S10.15
(a) Vol.120,No.1, 199-208

P.W. KLAMUT, I. BENZAR, A.J. ZALESKI, and M. DYBIEC
Resistivity of $Nd_{1.85}Ce_{0.15}CuO_{4-x}$ in High Magnetic Fields
74.70; S10.15
(a) Vol.120,No.1, K73-K78

D. AGASSI and T.K. CHU
Strain-Induced Superconductivity in Lead Salt Superlattices
74.70; 73.40; S8.1
(b) Vol.160,No.2, 601-612

N. KRISTOFFEL
On the Vibronic Nature of the Tetragonal-to-Orthorhombic Structural Transition in $La_{2-x}M_xCuO_4$
64.70; 63.20; 74.70; S10.15
(b) Vol.160,No.2, K93-K96

T. KOPTE and E. HEGENBARTH
Thermal Conductivity of High-T_c YBCO-Ceramics in Magnetic Fields up to 10 T
66.70 and 74.70; S10.15
(a) Vol.120,No.2, 547-556

U.-C. BOEHNKE, H. BOERNER, J. HERRMANN, B. LIPPOLD, CH. SEMMELHACK, L. WALDOESTL, and M. WURLITZER
Some Investigations on Praseodymium-Substituted Yttrium-Barium-Copper-Oxide
74.70; 61.60; 75.30; S10.15
(a) Vol.120,No.2, 557-566

E. ALTSHULER, S. GARCÍA, and A. AGUILAR
Hysteretic Critical Currents in Y-Ba-Cu-O Superconductors. A Microstructural Approach
74.70; S10.15
(a) Vol.120,No.2, K169-K172

A.M. SHAMAH, N.K. MINA, and M.K. EL-MOUSLY
Structural Investigation of the High T_c Superconducting Oxide System $La_{1-x}Pr_xCaBaCu_3O_7$
61.55 and 74.70; S10.15
(a) Vol.121,No.1, 235-240

T. KOPTE and E. HEGENBARTH
Long-Time Effects in the Thermal Conductivity of YBCO
66.70 and 74.70; S10.15
(a) Vol.121,No.1, K17-K20

F.P. KORSHUNOV, G.V. GATALSKII, V.I. GATALSKAYA, E.K. STRIBUK, and V.T. TROSHCHINSKII
Low-Temperature Electron Irradiation of Thallium and Bismuth High-T_c Superconducting Ceramics
74.70; S10.15
(a) Vol.121,No.1, K85-K90

W. SCHILLER and K. WESCHKE
Calculation of the Normal-State Thermal Conductivity of the Lattice for High-T_c Superconductors
74.30 and 74.70; S10.15
(b) Vol.161,No.2, 755-768

J. FOLTIN
Dependence of T_c on the Number of CuO_2 Layers in High-T_c Superconductors
74.20 and 74.70
(b) Vol.161,No.2, K95-K98

R. HEIDINGER, H. APPEL, G. THEN, and W.-G. THIES
Observation of Radiation Damage in HfV_2 and Its Hybrides by TDPAC
61.80; 64.80; 74.70; 76.80; S1.2
(a) Vol.121,No.2, 445-454

HONG ZHANG
Positron Trapping in $YBa_2Cu_{3-x}Fe_xO_y$
74.70 and 78.70; S10.15
(a) Vol.121,No.2, K207-K210

N. GUSKOS, G.P. TRIBERIS, M. CALAMIOTOU, CH. TRIKALINOS, A. KOUFOUDAKIS, C. MITROS, H. GAMARI-SEALE, and D. NIARCHOS
Temperature Dependence of the EPR Spectra of $GdBa_2Cu_3O_{7-\delta}$ Compounds in Orthorombic and Tetragonal Phase
72.20 and 74.70; 76.30; S10.15
(b) Vol.162,No.1, 243-250

O.YU. POLYAK, R.KH. TUKHVATULIN, KIM GEN CHAN, E.M. GASANOV, and E.M. IBRAGIMOVA
Effect of gamma-Irradiation on $YBa_2Cu_3O_{7-x}$ Ceramics and Monocrystals in the Superconducting State
74.70; S10.15
(a) Vol.122,No.1, K45-K50

R. HERGT, W. ANDRAE, and W. SCHUEPPEL
Pinning of Vortices at Twin Boundaries in $YBa_2Cu_3O_{7-x}$ Single Crystals
74.60 and 74.70; 61.70; S10.15
(a) Vol.122,No.1, K51-K54

J. SHA, C.F. ZHU, Z.H. HE, X.-G. LI, Z.P. SU, L.Z. CAO, Q.R. ZHANG, H.R. YI, and H.C. LI
Nonlinear I-V Characteristics of $GdBa_2Cu_3O_y$ Epitaxial Films Related with the Flux Creep
74.60 and 74.70; S10.15
(a) Vol.122,No.1, K55-K58

N. GUSKOS, M. CALAMIOTOU, S.M. PARASKEVAS, A. KOUFOUDAKIS, C. MITROS, H. GAMARI-SEALE, J. KURIATA, L. SADLOWSKI, and M. WABIA
On the Influence of Oxygen Deficiency on EPR Spectra of $NdBa_2Cu_3O_{7-\delta}$
76.30; 74.70; S10.15
(b) Vol.162,No.2, K101-K106

E.I. TOCHITSKII, O.G. SVIRIDOVICH, E.M. GOLOLOBOV, and D.M. TURTSEVICH
Phase Transformations in $Y_1Ba_2Cu_3O_{7-x}$ Films Prepared by Pulsed Laser Deposition
64.70 and 68.55; 74.70; S10.15
(a) Vol.122,No.2, 555-562

M. POPESCU, G. ALDICA, C. SÂRBU, P. PĂUSESCU, E. CRUCEANU, F. VASILIU, and M. APOSTOL
Crystalline Phases in Superconducting $Bi:Sr:Ca:Cu_2:O_x$ Ceramic
61.60 and 74.70; S10.15
(a) Vol.122,No.2, 637-644

P. SEIDEL, M. TURTENWALD, and T. BUESCHEL
Temperature Dependence of the Structures in the Tunneling Characteristics of High-T_c Junctions
74.70; 74.50
(a) Vol.122,No.2, 645-650

X.A. CHENG, Z.H. HE, H. ZHANG, X.Y. ZHOU, S.X. WANG, M.H. FANG, J.S. XIA, J. SHA, Z.P. SU, and Q.R. ZHANG
Crystal Structure and Superconductivity of $Y_{1-x}Pr_xBa_2Cu_3O_{7-y}$ Compounds
74.70; S10.15
(a) Vol.122,No.2, K151-K156

S. MOHAN, T. RADJA KOUMAR, and M. GAL
High T_c Superconductivity of $Ba_2ErCu_3O_7$
74.70; S10.15
(a) Vol.122,No.2, K157-K160

W. ANDRAE, J. BETZ, B. BRUNNER, R. HERGT, H. LENGFELLNER, K.F. RENK, W. SCHUEPPEL, and K. STEENBECK
In-Plane Rotational Magnetic Losses in a Laser Deposited $YBa_2Cu_3O_{7-x}$ Film
74.60; 74.70; S10.15
(a) Vol.122,No.2, K161-K164

74.90. Other topics in superconductivity

F. GOLEK, P.W. KLAMUT, B. SUJAK, A.J. ZALESKI, M. CISZEK, and J. OLEJNICZAK
Thermally Stimulated Charge Emission from the High Temperature Superconductor $YBa_2Cu_3O_{7-x}$
74.90 and 79.75; S10.15
(a) Vol.117,No.1, K43-K46

B.D. FAINBERG
Real Time Optical Spectroscopy of Superconducting-Gap Excitations
74.90 and 78.30
(b) Vol.160,No.2, K169-K172

B.D. FAINBERG
Erratum to Real Time Optical Spectroscopy of Superconducting Gap Excitations (phys. stat. sol. (b) 160, K169 (1990))
74.90 and 78.30
(b) Vol.162,No.1, K67-K68

75. Magnetic Properties and Materials

75.10. General theory and models of magnetic ordering
(see also 71.25. Nonlocalized single-particle electronic states, 71.70. Level splitting and interactions)

L. WANG
Two-Dimensional Electron Gas in a Magnetic Field. The Non-Degenerate Limit
71.25; 75.10
(b) Vol.157,No.1, 247-252

V. JANIŠ
Mean-Field Theory for Spin Glasses with Many Macroscopic States
75.10 and 75.40
(b) Vol.157,No.1, 425-430

L.G. GRECHKO, V.M. OGENKO, and A.YU. SEMCHUK
Field Influence on the Parameters of Ferromagnetic Semiconductor Superlattices Formed by Coherent Light Beams
71.25 and 75.10
(b) Vol.157,No.1, 451-458

J.M. WESSELINOWA
On the Influence of the s-f Exchange Interaction on the Relaxation Time in Ferromagnetic Semiconductors
75.10
(b) Vol.157,No.1, K47-K50

P. PAWLICKI
Magnetoelastic Effects in the Exactly Solvable Vaks-Larkin-Ovchinnikov Model of Ising Spins
75.10 and 75.80
(b) Vol.157,No.1, K51-K54

M. KOTRLA and V. DRCHAL
Mean-Field Solution Strongly Correlated Systems Using Hubbard Atomic Operators. Hubbard Model with Infinite U
71.10; 75.10
(b) Vol.157,No.2, 635-640

A. URBANIAK-KUCHARCZYK, T. BALCERZAK, J. MIELNICKI, and G. WIATROWSKI
The Magnetic Contribution to Electrical Resistivity in Diluted Ising Ferromagnets
72.10 and 75.10
(b) Vol.157,No.2, K103-K108

A.P. MIKHAILOVSKII, A.M. POLUBOTKO, V.D. PROCHUKHAN, and YU.V. RUD
Gapless State in $CuFeS_2$
71.25 and 75.10; S8
(b) Vol.158,No.1, 229-238

G. FISCHER and A. HERR
Mean Magnetic Moments of Polycrystalline Ce Compounds in a Tetragonal Crystal Field
75.10; S1.4
(b) Vol.159,No.1, K23-K26

J. ULNER and K. DURCZEWSKI
Stable Phases and Quadrupolar Excitations in an Ordered Phase of J=1 Paramagnet
75.10 and 75.30
(b) Vol.159,No.1, K27-K32

SHAOPING LI and L.E. CROSS
Analysis of Magnetostatic Modes Modified by Exchange and Anisotropy Energies
68.30 and 75.10; 75.70
(b) Vol.159,No.2, 861-872

R. PINTANEL and R. FERRER
Random Magnetic Anisotropy Model with Biquadratic Exchange
75.10
(b) Vol.159,No.2, K93-K96

I.A. VAKARCHUK and V.M. TKACHUK
Energy Spectrum and Elementary Excitation Damping in the Structurally Disordered Ising Model in a Transverse Field
75.10; 75.40
(b) Vol.160,No.1, 321-328

JIA-LIN ZHONG, JIA-LIANG LI, and CHUAN-ZHANG YANG
Amorphization of the Transverse Ising Model with Spin-1
75.10
(b) Vol.160,No.1, 329-336

E.F. SARMENTO and T. KANEYOSHI
Surface Magnetic Properties of a Transverse Ising Model with a Diluted Surface
73.20 and 75.10
(b) Vol.160,No.1, 337-348

P. PAWLICKI
Exactly Solvable Ising Model with Spin-Pseudospin Interactions
75.10
(b) Vol.160,No.1, K57-K60

T. BALCERZAK, J. MIELNICKI, and G. WIATROWSKI
The Magnetic Properties of Thin Films with Fluctuations of the Exchange Integral
75.10; 75.30
(b) Vol.160,No.1, K61-K64

A. URBANIAK-KUCHARCZYK, T. BALCERZAK, J. MIELNICKI, and G. WIATROWSKI
The Magnetic Contribution to Electrical Resistivity in the Ising Ferromagnet with Fluctuating Exchange Integral
72.10 and 75.10
(b) Vol.160,No.2, K129-K132

A.S.T. PIRES and B.V. COSTA
Phenomenological Calculation of the Soliton Density in the One-Dimensional Antiferromagnet in an External Field
75.10; 75.25
(b) Vol.160,No.2, K165-K168

J. MIELNICKI, T. BALCERZAK, and G. WIATROWSKI
Discussion of the Ferromagnetic Simple-Cubic Ising Lattice with Fluctuating Exchange Integral (S=1/2)
75.10
(b) Vol.161,No.1, 335-344

A.S.T. PIRES
Soliton in a Canted One Dimensional Antiferromagnet
75.10; 75.25
(b) Vol.161,No.1, K45-K48

A.YU. SEMCHUK, L.G. GRECHKO, and V.M. OGENKO
Theory of Transport Phenomena in Ferromagnetic Semiconductors with Superlattice Produced by Coherent Light Beams
72.10 and 75.10
(b) Vol.162,No.2, 539-544

I. JAEGER
The Ising Antiferromagnet with Competing Interactions : Locating the Tricritical Point
75.10; 68.45
(b) Vol.162,No.2, 575-584

D. ŠOKČEVIĆ and V. ZLATIĆ
Pertubation Approach to the Periodic Anderson Hamiltonian with k-Dependent Hybridization
71.10; 71.20; 75.10
(b) Vol.162,No.2, K83-K86

C.Z. YANG and W.J. SONG
Mean Field Renormalization Group and Path-Integral Methods for the Disordered Transverse Ising Model
75.10
(b) Vol.162,No.2, K113-K118

75.20. Diamagnetism and paramagnetism

D. BANERJEE/CHAKRAVORTY and R. BHATTACHARYA
Magnetic Properties of Single Crystals of Bismuth Doped with Lead and Tin
75.20; S4
(b) Vol.157,No.1, 443-450

G.B. TAGGART
Field Dependence of Dilute Ising Paramagnets
75.20
(b) Vol.157,No.2, 711-722

B. SUKUMAR and K. NAVANEETHAKRISHNAN
Diamagnetic Susceptibility of a Donor in a GaAs/$Ga_{1-x}Al_xAs$ Quantum Well Heterostructure and Its Pressure Dependence
73.40 and 75.20; 71.55; S7.12; S7.15
(b) Vol.158,No.1, 193-200

N.S. AVERKIEV, V.M. ASNIN, A.A. ROGACHEV, A.YU. SILOV, and V.I. STEPANOV
Photoluminescence Studies of Anomalous Diamagnetic Properties of the 2D e-h System on GaAs Surface
75.20 and 78.55; 73.40; S7.12
(b) Vol.159,No.1, 471-476

S. BISWAS and R. BHATTACHARYA
Magnetic Properties of Single Crystals of Bismuth Telluride Doped with 0.2 at% Lead and Its Thermoelectric Power
75.20; 71.25; 72.20; S8
(b) Vol.159,No.2, 851-860

K.P. GHATAK and M. MONDAL
On a Simplified Analysis of the Magnetic Susceptibilities of Electrons in Ultrathin Films of Non-Parabolic ...
75.20 and 75.70; 71.25; S7.16
(b) Vol.160,No.2, 673-682

75.25. Spin arrangements in magnetically ordered materials (neutron studies, etc.)

K. MELZER, J. SUWALSKI, and M. ŁUKASIAK
Moessbauer Study on the Spin Arrangement in $BaFe_{12}O_{19}$ and $BaCo_{0.8}Ti_{0.8}Fe_{10.4}O_{19}$
75.25 and 76.80; S11.2
(a) Vol.118,No.1, K35-K38

L.I. KURKINA and O.V. FARBEROVICH
Electron Structure and Magnetic Properties of Small Iron Particles
71.50; 73.20; 75.25; S1.1
(b) Vol.160,No.1, K37-K42

A.S.T. PIRES and B.V. COSTA
Phenomenological Calculation of the Soliton Density in the One-Dimensional Antiferromagnet in an External Field
75.10; 75.25
(b) Vol.160,No.2, K165-K168

A.S.T. PIRES
Soliton in a Canted One Dimensional Antiferromagnet
75.10; 75.25
(b) Vol.161,No.1, K45-K48

N.H. LUONG, N.P. THUY, L.T. TAI, and N.V. HOANG
Effect of Y Substitution on the Spin Reorientation in $NdFe_{11}Ti$ Compound
75.25; 75.30; S1.2; S1.4
(a) Vol.121,No.2, 607-610

75.30. Magnetically ordered materials, other intrinsic properties
(for critical-points effects, see 75.40)

L.I. KOROLEVA and T.V. VIROVETS
Critical Behaviour of Semiconductor Spin Glasses $(CuCr_2Se_4)_x(Co_{0.5}Me_{0.5}Cr_2Se_4)_{1-x}$ $(Me=In,Ga,\ 0<x<0.1)$
75.30; S8.16
(b) Vol.157,No.1, 431-442

I.V. MANZHOS and I.E. CHUPIS
Generation of Spin-Waves by Ferroelectric Wall Excitations in a Ferroelectric Ferromagnet
75.30
(b) Vol.157,No.1, K65-K68

YU.N. DEVYATKO, V.N. TRONIN, and V.I. TROYAN
Kinetics of the New Phase Nucleus Growth on a Magnet near the Curie Point
61.50; 75.30; S1.2
(a) Vol.117,No.1, 61-66

G.E. ABROSIMOVA, N.I. NOVOKHATSKAYA, and A.V. SEREBRYAKOV
The Magnetic Field Effect on the Creep-Induced Structural Anisotropy of Amorphous Co-Fe-Ni-Si-B Alloys
75.30; S1.2
(a) Vol.117,No.1, K67-K70

E. TAMURA, R. FEDER, G. WALLER, and U. GRADMANN
Determination of the Surface Magnetization of Fe(110) by Spin-Polarized Low-Energy Electron Diffraction
68.20 and 75.30; 61.14; S1.1
(b) Vol.157,No.2, 627-634

TIESONG ZHAO, ZHONGZHI FAN, and HANMIN JIN
Magnetization Processes of the Rare-Earth and Fe Sublattices in $R_2Fe_{14}B$ (R=Tb,Dy,Ho,Er,and Tm)
75.30; 75.60; S1.1; S1.4
(b) Vol.157,No.2, 677-684

YOU XU, GUI-LING YANG, DA-PING CHU, and HONG-RU ZHAI
Theory of the Single Ion Magnetocrystalline Anisotropy of 3d Ions
75.30; S11.2
(b) Vol.157,No.2, 685-694

D. SKRZYPEK
Critical Behaviour of the Magnetic Resonance in $KNi_xMn_{1-x}F_3$
75.30 and 76.30; S9
(b) Vol.157,No.2, 695-700

P. NOZAR, V. SECHOVSKY, M. SOLANKI-MOSER, E. BAUER, and E. GRATZ
Point-Contact Spectroscopy on the $Ce(Cu_xAl_{1-x})_5$ System
71.20; 72.20; 75.30; S1.2; S1.4
(a) Vol.117,No.2, 501-508

R.Z. VALIEV, YA.D. VISHNYAKOV, R.R. MULYUKOV, and G.S. FAINSHTEIN
On the Decrease of Curie Temperature in Submicron-Grained Nickel
75.30; 73.40; S1.2
(a) Vol.117,No.2, 549-554

M. HIDAKA, H. FUJII, M. NISHI, and B.M. WANKLYN
Magnetic Phase Transition in the Layer Compound $CsVF_4$
75.30; S9.16
(a) Vol.117,No.2, 563-570

N. SCHELUDKO and M. MIKHOV
Investigation of Magnetocrystalline Anisotropy of $Mm_2Co_{17-x}Zr_x$ Compounds
75.30; S1.2; S1.4
(a) Vol.117,No.2, 571-576

T. PALEWSKI
Metamagnetism in UAs-YSe Solid Solution
75.30; S1.5
(a) Vol.117,No.2, K151-K154

A.Z. MENSHIKOV, A.P. VOKHMYANIN, and YU.A. DOROFEEV
Magnetic Structure and Phase Transformations in Mn_5Si_3
75.30; 61.55; S1.61
(b) Vol.158,No.1, 319-328

T. IWASHITA and N. URYÛ
Magnetic Susceptibility of an Ising Ferromagnet with Biquadratic Exchange and Uniaxial Anisotropy
75.30
(b) Vol.158,No.1, 347-358

A. URBANIAK-KUCHARCZYK and H. ZEYADA
The Hyperfine Field Distribution in Diluted Ising Ferromagnet
75.30; 75.70
(b) Vol.158,No.1, K59-K62

K.G. NIKIFOROV, M. BARAN, V.K. BELYAEV, L.YA. PASENKO, S.I. RADAUTSAN, and A. WISNIEWSKI
Magnetization and Chromium Ions with Minority Valence in $CdCr_2S_4$ and $HgCr_2Se_4$ Magnetic Semiconductors
75.30; 75.60; S11.2
(b) Vol.158,No.1, K63-K68

R.M. CHORBADJIAN and J. POZO
Elementary Excitation in Amorphous Ferromagnets
75.30
(b) Vol.158,No.2, K191-K194

Z.A. MATYSINA, M.I. MILYAN, D.R. RIZDVYANETSKII, and S.YU. ZAGINAICHENKO
The Surface Atomic and Magnetic Order in Crystals
68.20 and 75.30
(b) Vol.159,No.1, K7-K12

J. ULNER and K. DURCZEWSKI
Stable Phases and Quadrupolar Excitations in an Ordered Phase of J=1 Paramagnet
75.10 and 75.30
(b) Vol.159,No.1, K27-K32

J. PIECHOTA and A. PAJACZKOWSKA
Magnetic Properties of $YBa_2(Cu_{1-x}Ni_x)_3O_{7-\delta}$
75.30; 74.70; S10.15
(a) Vol.119,No.1, 271-278

K.M. MUKIMOV, B.YU. SOKOLOV, and U.V. VALIEV
The Faraday Effect of Rare-Earth Ions in Garnets
78.20; 75.30; S11.2
(a) Vol.119,No.1, 307-316

N.P. KOLMAKOVA, I.B. KRYNETSKII, M.M. LUKINA, and A.A. MUKHIN
Crystal Field and Metamagnetic Behavior of Rare Earth Orthoaluminates: $DyAlO_3$
75.30; 71.70; S10
(b) Vol.159,No.2, 845-850

O.M. TOLKACHEV
Spin Fluctuation Mechanism for the Temperature-Induced Magnetic Moment (TIMM) in the Electron Liquid of Y_2Ni_7 Alloy
75.30; S1.2
(b) Vol.159,No.2, K115-K118

M. AHMAD and S.J. GLADYSZ
The Dynamic Spin Susceptibility and l=0.1 Spin Waves in a Charged Fermi Liquid of Alkali Metals in a DC Magnetic Field
75.30; 67.50; S2
(b) Vol.159,No.2, K119-K124

T. BALCERZAK, J. MIELNICKI, and G. WIATROWSKI
The Magnetic Properties of Thin Films with Fluctuations of the Exchange Integral
75.10; 75.30
(b) Vol.160,No.1, K61-K64

M. AHMAD
Spin Waves with Arbitary l in Two-Dimensional Charged Fermi Liquid Immersed in a DC Magnetic Field
67.50 and 75.30
(b) Vol.160,No.1, K65-K70

S.I. KLOKISHNER, A.V. KORYACHENKO, and B.S. TSUKERBLAT
Charge Ordering of Multielectronic Clusters in Mixed-Valence Molecular Crystals
71.70 and 75.30; 64.60; S10
(b) Vol.160,No.2, 641-648

MAI SUAN LI
Phase Diagram and Susceptibility of a Frustrated Ising Ferrimagnet in Zero Magnetic Field
75.30 and 75.40
(b) Vol.160,No.2, K159-K164

U.-C. BOEHNKE, H. BOERNER, J. HERRMANN, B. LIPPOLD, CH. SEMMELHACK, L. WALDOESTL, and M. WURLITZER
Some Investigations on Praseodymium-Substituted Yttrium-Barium-Copper-Oxide
74.70; 61.60; 75.30; S10.15
(a) Vol.120,No.2, 557-566

S. SOEDERHOLM, J. HELLBERG, J. KRZYSTEK, G. OLOVSSON, I. OLOVSSON, and J.U. VON SCHUETZ
Electronic Properties and Crystal Structure of a Methylthio Substituted Naphthalene Cation Radical Salt: $(DMbTN)_3(ClO_4)_2$
61.65 and 72.20; 71.55; 75.30; 76.30; S12.1
(b) Vol.161,No.1, 245-256

MAI SUAN LI
Study of Frustrated Ising Ferrimagnets
75.30 and 75.40
(b) Vol.161,No.1, 345-356

T. IWASHITA and N. URYÛ
Ising Ferromagnet with Three-Site Four-Spin Interaction
75.30
(b) Vol.161,No.1, 367-378

Y. ENDO
Phase Transitions of Heisenberg Ferromagnetic Superlattices
75.30; 75.50; 75.70
(b) Vol.161,No.1, 379-386

ZHI-DONG ZHANG, X.K. SUN, ZHUONG-WU ZHANG, Y.C. CHUANG, and F.R. DE BOER
Effects of Partial Co Substitution on Magnetic Properties of $(Nd,Gd)_2Fe_{14}B$
75.30; 75.60; S1.1
(a) Vol.121,No.1, 241-248

B. LIPPOLD, J. HERRMANN, W. REICHELT, and H. OPPERMANN
Preparation and Magnetic Investigations on $Mn_2Mo_3O_8$ Single Crystals
75.30; 75.50; S1.2; S10.15
(a) Vol.121,No.1, K91-K94

BAO-GEN SHEN, HUI-QUN GUO, LIN-YUAN YANG, JUN-XIAN ZHANG, and JIAN-GAO ZHAO
Crystallization of Amorphous $Co_{84-x}B_xZr_{16}$ Alloys and Its Influence on Hard Magnetic Properties
75.30; 61.40; 75.60; S1.2
(a) Vol.121,No.1, K105-K110

G.S. CHADDHA and J. SINGH
Spin-1 Exchange Interaction Model of Ferromagnetism
75.30
(b) Vol.161,No.2, 837-842

N.H. LUONG, N.P. THUY, L.T. TAI, and N.V. HOANG
Effect of Y Substitution on the Spin Reorientation in $NdFe_{11}Ti$ Compound
75.25; 75.30; S1.2; S1.4
(a) Vol.121,No.2, 607-610

ZHI-DONG ZHANG, X.K. SUN, Y.C. CHUANG, and R.J. RADWAŃSKI
A Model Based on Single-Ion Theory. I. Theoretical Outline
75.50; 75.30; S1.2; S1.4
(a) Vol.121,No.2, 611-616

ZHI-DONG ZHANG, X.K. SUN, Y.C. CHUANG, F.R. DE BOER, and R.J. RADWAŃSKI
A Model Based on Single-Ion Theory. II. Preferential Substitution and Local-Environment Effects at Different ...
75.50; 75.30; S1.1; S1.2
(a) Vol.121,No.2, 617-626

N.S. ALY, I. MAHMOUD, M.E. BASSIOUNI, H.A. MOTAWEH, and A.S. ILYUSHIN
Magnetic Phase Diagram of the $Tb_{1-x}Nd_xFe_2$ Quasibinary System
75.30; S1.1
(a) Vol.121,No.2, K219-K222

H. WADA, T. INOUE, M. HADA, M. SHIGA, and Y. NAKAMURA
Effect of Magnetic Phase Transition on the Electronic Specific Heat Coefficient of $Lu(Co_{1-x}Al_x)_2$
65.40; 75.30; S1.2; S1.4
(b) Vol.162,No.2, 407-412

JUN-XIAN ZHANG, BAO-GEN SHEN, and WEN-SHAN ZHAN
Magnetic Properties of Amorphous $Nd_4Fe_{96-x}B_x$ Alloys and Influences of Heat Treatment on Its Hard Magnetic Properties
75.30; 75.60; S1.1
(a) Vol.122,No.2, 651-656

A. GRUSKOVÁ, J. LIPKA, J. SLÁMA, I. TÓTH, M. SEBERÍNI, and P. KABOŠ
Moessbauer Effect and Its Application in the Investigation of the Substituted Li Ferrites Obtained from the Liquid Phase
75.30 and 75.50; 76.80; S11.2
(a) Vol.122,No.2, K171-K176

75.40. Critical-point effects, specific heats, short-range order
(see also 65.40. Heat capacities)

B.A. ARONZON and I.M. TSIDILKOVSKII
Magnetic-Field-Induced Localization of Electrons in Fluctuation-Potential Wells of Impurities (Review Article)
71.30; 72.20; 75.40; S7.13; S8.15
(b) Vol.157,No.1, 17-60

M.L. MARÍN, M. ORTUÑO, A. HERNÁNDEZ, and J. ABELLÁN
Percolative Treatment of the Verwey Transition in Cobalt-Iron and Nickel-Iron Ferrites
71.30 and 75.40; S11.2
(b) Vol.157,No.1, 275-280

V. JANIŠ
Mean-Field Theory for Spin Glasses with Many Macroscopic States
75.10 and 75.40
(b) Vol.157,No.1, 425-430

S.K. PAL and ANIL KUMAR
Influence of s-d Interaction on the Curie Temperature
75.40 and 75.60
(b) Vol.157,No.2, K131-K136

V.N. BERZHANSKII, I.I. EVSTAFEV, and V.I. IVANOV
Charge Carrier Transfer in Ferrite Spinels Affected by Magnetic Ordering
72.20; 75.40; S11.2
(b) Vol.158,No.2, 643-652

M. BŁASZAK and B. FECHNER
Stochastic Solitons in a Damped Heisenberg Chain
75.40
(b) Vol.159,No.2, K103-K106

M.A. ALDZHANOV, N.G. GUSEINOV, G.D. SULTANOV, and M.D. NADZHAFZADE
Magnetic Heat Capacity and Susceptibility of the Pseudo-One-Dimensional Magnetic Systems $TlFeS_2$ and $TlFeSe_2$
75.40; 75.50; S8
(b) Vol.159,No.2, K107-K110

I.A. VAKARCHUK and V.M. TKACHUK
Energy Spectrum and Elementary Excitation Damping in the Structurally Disordered Ising Model in a Transverse Field
75.10; 75.40
(b) Vol.160,No.1, 321-328

S. KAHL, R. NEUHAUS, H. PINKVOS, and CH. SCHWINK
The Influence of Tensile Deformation and Atomic Short-Range Order on the Susceptibility of CuMn Spin Glasses
62.20 and 75.40; 61.70; S1.2
(a) Vol.120,No.1, 221-232

MAI SUAN LI
Phase Diagram and Susceptibility of a Frustrated Ising Ferrimagnet in Zero Magnetic Field
75.30 and 75.40
(b) Vol.160,No.2, K159-K164

MAI SUAN LI
Study of Frustrated Ising Ferrimagnets
75.30 and 75.40
(b) Vol.161,No.1, 345-356

D.A. GUSEINOV and M.A. ALYANOV
Heat Capacity and Thermal Expansion of $TiFe_2Se_4$
65.40; 75.40; S11.2
(a) Vol.121,No.1, K51-K54

I. ŚLEDZIŃSKA, A. MURASIK, and A. WIŚNIEWSKI
Linear-Chain Antiferromagnetism in alpha-NiC_2O_4 * 2 D_2O
61.60 and 75.40; S10
(b) Vol.161,No.2, K99-K102

V. ILKOVIČ
The Temperature of the Magnetic Phase Transition in Weakly Coupled Magnetic Layers
75.40
(b) Vol.162,No.2, K119-K122

W.A. KACZMAREK, R. BRAMLEY, and A. CALKA
ESR Study of Amorphous Metallic Alloys below the Magnetic Multicritical Point
75.50 and 76.30; 75.40; S1.1; S1.2
(a) Vol.122,No.2, 663-676

75.50. Studies of specific magnetic materials

B.X. GU, B.G. SHEN, and H.R. ZHAI
Magnetic Properties and Crystallization of Amorphous $Nd_{14}Fe_{81}B_5$ Alloys
75.50; 61.40; S1.1
(a) Vol.117,No.1, 283-290

E.P. ELSUKOV, YU.N. VOROBEV, A.V. TRUBACHEV, and V.A. BARINOV
Structure and Magnetic Properties of Fe-P Electrodeposited Alloys
61.55 and 75.50; 76.80; S1.1
(a) Vol.117,No.1, 291-298

M. JURCZYK and O.D. CHISTYAKOV
Magnetic Properties of Substituted $Nd_{2-x}R_x(Fe,Re,Co)_{14}B$ Compounds (R=Tb or Dy)
75.50; S1.1; S1.2
(a) Vol.117,No.1, 299-304

V.V. SINGER, A.E. VAFIN, L.M. SANDRATSKII, L.P. MOKHRACHEVA, S.V. KONOVALOV, I.Z. RADOVSKII, and P.V. GELD
Magnetic Susceptibility and Valency of Ce and Pr in Their Liquid Binary Alloys with Metals of the Iron Subgroup
75.50; S1.1; S1.2; S1.4
(a) Vol.117,No.2, 577-586

W.A. KACZMAREK, R. BRAMLEY, and A. CALKA
Itinerant Electron Contribution to the Magnetism of Ni-TM Glassy Metallic Magnets
75.50 and 76.30; S1.1; S1.2
(a) Vol.117,No.2, K141-K146

A.M. MANSANARES, F.C.G. GANDRA, E.C. DA SILVA, H. VARGAS, M. DO CARMO, R.S. VARELLA, and F. GALEMBECK
Photoacoustic Detection of Magnetite-Crystal Formation from Iron(III) Hydroxide Acetate
75.50 and 76.50; S12
(a) Vol.117,No.2, K155-K160

G. SERFŐZŐ, E. KISDI-KOSZÓ, A. SLAWSKA-WANIEWSKA, L. POTOCKY, S. JONELIUNAS, and M. BARAN
Study of Metallic Glasses Rapidly Quenched in Magnetic Field
75.50; 76.50; S1.1
(a) Vol.118,No.1, 307-310

A. GRAF, R. OBERMAIER, and W. GEBHARDT
Spin Glass Properties of $Zn_{1-x}Mn_xSe$
75.50; S8.15
(b) Vol.158,No.2, 695-706

J. ZBROSZCZYK, W. CIURZYŃSKA, J. ŚWIERCZEK, S. SZYMURA, B. WYSŁOCKI, M. BARAN, and A. WISNIEWSKI
Effect of Surface Layer Thinning on Approach to Magnetic Saturation in $Co_{64}Fe_{14}Si_{11}B_{11}$ Amorphous Alloy
75.50; S1.1; S1.2
(a) Vol.118,No.2, 525-530

I.O. TROYANCHUK, S.N. PASTUSHONOK, A.K. BOGUSH, and V.I. PAVLOV
Magnetic Properties of Nonstoichiometric Orthomanganites
75.50; S11.1
(a) Vol.118,No.2, K111-K114

D.M. ZAYACHUK, D.D. IVANCHUK, R.D. IVANCHUK, S.S. MASLYANCHUK, and V.I. MIKITYUK
The Effect of Gadolinium Doping on the Physical Properties of Lead Telluride
71.55; 72.80; 75.50; S8
(a) Vol.119,No.1, 215-220

BAO-GEN SHEN, LIN-YUAN YANG, JUN-XIAN ZHANG, FENG WO, TAI-SHAN NING, JIAN-GAO ZHAO, HUI-QUN GUO, and WEN-SHAN ZHAN
Hard Magnetic Properties of Heat Treated $Fe_{72.8}Co_{4.7}Nd_4B_{18.5}$ Alloy
75.50; 75.60; S1.1
(a) Vol.119,No.1, K63-K66

P. SVOBODA, M. DIVIŠ, J. BISCHOF, Z. SMETANA, R. ČERNÝ, and J. BURIÁNEK
The Magnetic Properties of $SmCu_5$
75.50; S1.4
(a) Vol.119,No.1, K67-K70

K. IWAUCHI and Y. IKEDA
Magnetic Properties of Sintered Fe_2TiO_5
75.50; S10
(a) Vol.119,No.1, K71-K74

A. RAOUFI and S. ARAJS
Anisotropy of Magnetoresistivity in Crystalline and Amorphous Ferromagnets
72.15 and 75.50; S1.1
(a) Vol.119,No.1, K75-K78

M.A. ALDZHANOV, N.G. GUSEINOV, G.D. SULTANOV, and M.D. NADZHAFZADE
Magnetic Heat Capacity and Susceptibility of the Pseudo-One-Dimensional Magnetic Systems $TlFeS_2$ and $TlFeSe_2$
75.40; 75.50; S8
(b) Vol.159,No.2, K107-K110

M. HRABČÁK and E. KISDI-KOSZÓ
Modification of Magnetic Properties of Amorphous Fe-Si-B Ribbons by Flash Annealing
75.50; 75.60; S1.1
(a) Vol.119,No.2, 601-606

W.A. KACZMAREK
Magnetic Structure Rearrangement during Crystallization in $Ni_{73}Mn_5Si_{10}B_{12}$ Ribbons
75.50 and 76.30; S1.2
(a) Vol.119,No.2, K159-K162

E.P. NAIDEN, V.I. MALTSEV, and G.I. RYABTSEV
Magnetic Structure and Spin-Orientational Transitions of Hexaferrites of the $BaCo_{2-x}Zn_xFe_{16}O_{27}$ System
75.50; S11.2
(a) Vol.120,No.1, 209-220

W.A. KACZMAREK
Magnetic Hyperfine Interactions in $BaIn_{1.5}Fe_{10.5}O_{19}$ Ferrite
76.80; 75.50; S11.2
(a) Vol.120,No.1, K79-K84

G.D. MAKSIMOVIĆ, Z.S. POPOVIĆ, F.R. VUKAJLOVIĆ, and K. VULETIĆ
Magnetic Properties of Pd_3Fe
75.50; S1.1
(b) Vol.160,No.2, 635-640

W. WALZ and H. KRONMUELLER
Point Defects in Electron-Irradiated Stoichiometric Magnetite
61.70 and 75.50; S10
(b) Vol.160,No.2, 661-672

A.YA. FISHMAN, M.A. IVANOV, V.YA. MITROFANOV, and A.A. SHEMYAKOV
Hyperfine Interactions of Jahn-Teller Ion Mn^{3+} in Cubic Magnets with Spinel and Garnet Structures
75.50; 76.20; S11.2
(b) Vol.160,No.2, K153-K158

G.F. ZHOU, X.K. SUN, S.Y. FU, Y.C. CHUANG, R. GROESSINGER, and H.R. KIRCHMAYR
Relationship between Microstructure and Coercivity in (Nd,Dy)-(Fe,Co)-B Based Sintered Magnets with Minor Nb and Al ...
75.50; 64.80; S1.1
(a) Vol.120,No.2, 627-634

U.C. JOHRI, R.M. SINGRU, and D. BAHADUR
Magnetic and Moessbauer Studies of the System $LaCo_{1-x}Zr_xO_3$ (x=0.1,0.3,0.5)
76.80; 75.50; S10.15
(b) Vol.161,No.1, 357-366

Y. ENDO
Phase Transitions of Heisenberg Ferromagnetic Superlattices
75.30; 75.50; 75.70
(b) Vol.161,No.1, 379-386

D. VLADIKOVA, L. ILKOV, and S. KARBANOV
Optimal Prefiring Conditions of Substituted Nickel Ferrites for Microwave Applications
64.75 and 75.50; S11.2
(a) Vol.121,No.1, 249-256

S. UNNIKRISHNAN and D.K. CHAKRABARTY
Magnetic Properties and Domain State of the Magnesium Ferrite-Magnesium Titanate Solid Solutions
75.50; 76.80; S11.2
(a) Vol.121,No.1, 265-272

B. LIPPOLD, J. HERRMANN, W. REICHELT, and H. OPPERMANN
Preparation and Magnetic Investigations on $Mn_2Mo_3O_8$ Single Crystals
75.30; 75.50; S1.2; S10.15
(a) Vol.121,No.1, K91-K94

L. SADŁOWSKI, J. KURIATA, B. BOJANOWSKI, J. WALCZAK, M. KURZAWA, M.L. FALIN, and V.V. IZOTOV
EPR Study of Antiferromagnetic Phase Transition in $Fe_4V_2Mo_3O_{20}$
76.30; 75.50; S11.1
(a) Vol.121,No.1, K95-K98

B.X. GU, B.G. SHEN, and H.R. ZHAI
Crystallization and Magnetic Properties of Amorphous $Nd_{0.19}Fe_{0.81}$ Alloy
75.50; 61.40; S1.1; S1.4
(a) Vol.121,No.1, K99-K104

A.G . POPOV, YE.V. BELOZEROV, A.G. KUCHIN, A.S. ERMOLENKO, G.M. MAKAROVA, V.S. GAVIKO, and V.I. KHRABROV
Magnetic Properties of $Sm_2(Fe,M)_{17}C_x$ Compounds, M = Co,Al,Ga
75.50; S1
(a) Vol.121,No.1, K111-K116

A.A. SOROKIN, D.D. PERLOV, R.Z. LEVITIN, N.P. KOLMAKOVA, and A.P. DODOKIN
Determination of Ground Quasidoublet Parameters for Ho^{3+} Ion in $(Ho_xY_{1-x})_3Al_5O_{12}$ from Low-Temperature Magnetic ...
71.70 and 75.50; S11.2
(b) Vol.161,No.2, K103-K106

ZHI-DONG ZHANG, X.K. SUN, Y.C. CHUANG, and R.J. RADWAŃSKI
A Model Based on Single-Ion Theory. I. Theoretical Outline
75.50; 75.30; S1.2; S1.4
(a) Vol.121,No.2, 611-616

ZHI-DONG ZHANG, X.K. SUN, Y.C. CHUANG, F.R. DE BOER,
and R.J. RADWAŃSKI
A Model Based on Single-Ion Theory. II. Preferential Substitution and Local-Environment Effects at Different ...
75.50; 75.30; S1.1; S1.2
(a) Vol.121,No.2, 617-626

L.I. VERSHININA, S.Z. SKLUEV, V.S. ZHIGALOV, A.G. STEPANOV,
G.I. FROLOV, A.S. AVILOV, and S.V. OREKHOV
The Structure of Amorphous Dy-Co Films with Magnetic Anisotropy
61.40 and 75.50; S1.2
(a) Vol.121,No.2, K145-K148

BAO-GEN SHEN and S. METHFESSEL
Crystallization of the Amorphous $Fe_{83}Nd_7B_{10}$ Alloy and the Formation Process of $Nd_2Fe_{23}B_3$ Phase
75.50; S1.1
(a) Vol.121,No.2, K211-K214

Y. KHAN
Phases Obtained by Continuous Heating of the Amorphous Permanent Magnet Alloy $Nd_{4.4}Fe_{77.8}B_{17.8}$
61.55 and 75.50; S1.1
(a) Vol.122,No.1, K1-K6

X.K. SUN, ZHI-GANG ZHAO, LIU WEI, and Y.C. CHUANG
Structure and Magnetic Properties of $Pr_2Fe_{14-x}Co_xB_{1-y}C_y$ Compounds
75.50; S1.1
(a) Vol.122,No.1, K69-K72

A.G. LIPSON, V.A. KUZNETSOV, D.M. SAKOV, and YU.P. TOPOROV
Electrical and Magnetic Properties of the Polymer-Ceramics Produced on the Basis of Lithium Fluoride
73.40 and 75.50; S9.11; S12
(a) Vol.122,No.2, 563-568

W.A. KACZMAREK, R. BRAMLEY, and A. CALKA
ESR Study of Amorphous Metallic Alloys below the Magnetic Multicritical Point
75.50 and 76.30; 75.40; S1.1; S1.2
(a) Vol.122,No.2, 663-676

A. GRUSKOVÁ, J. LIPKA, J. SLÁMA, I. TÓTH, M. SEBERÍNI,
and P. KABOŠ
Moessbauer Effect and Its Application in the Investigation of the Substituted Li Ferrites Obtained from the Liquid Phase
75.30 and 75.50; 76.80; S11.2
(a) Vol.122,No.2, K171-K176

75.60. Domain effects, magnetization curves, and hysteresis

A.G. SHISHKOV, E.N. ILICHEVA, E.V. KOCHETKOVA, and YU.N. FEDYUNIN
On the Choice of the Pulse Waveform for Reading without Distortion of the Vertical Bloch Line Information
75.70; 75.60; S11.2
(a) Vol.117,No.1, K53-K56

A.H. WAFIK and A.A. SATTAR
Dependence of the Irreversible Magnetization in $Co_{1-a}Cd_aFe_2O_4$ on the Magnetizing Field
75.60; S10
(a) Vol.117,No.1, K61-K66

TIESONG ZHAO, ZHONGZHI FAN, and HANMIN JIN
Magnetization Processes of the Rare-Earth and Fe Sublattices in $R_2Fe_{14}B$ (R=Tb,Dy,Ho,Er,and Tm)
75.30; 75.60; S1.1; S1.4
(b) Vol.157,No.2, 677-684

A. TARI and J.S. HWANG
Magnetisation and ESR Studies of $Gd(Ni_{1-x}Rh_x)_5$ Compounds
75.60 and 76.30; S1.4
(b) Vol.157,No.2, 701-710

S.K. PAL and ANIL KUMAR
Influence of s-d Interaction on the Curie Temperature
75.40 and 75.60
(b) Vol.157,No.2, K131-K136

D. PŁUSA, R. PFRANGER, and B. WYSŁOCKI
Magnetic Domain Structure of Sintered $Nd_{15}Fe_{77}B_8$ Permanent Magnet in Rotational Magnetic Field
75.60; S1.1
(a) Vol.117,No.2, 541-548

K.G. NIKIFOROV, M. BARAN, V.K. BELYAEV, L.YA. PASENKO, S.I. RADAUTSAN, and A. WISNIEWSKI
Magnetization and Chromium Ions with Minority Valence in $CdCr_2S_4$ and $HgCr_2Se_4$ Magnetic Semiconductors
75.30; 75.60; S11.2
(b) Vol.158,No.1, K63-K68

K. MAKI and H. KRONMUELLER
Investigation of Demagnetization Processes in $Fe_{14}Nd_2B$-Type Sintered Magnets
75.60; S1.1; S1.4
(a) Vol.118,No.1, 253-270

R. SCHAEFER and A. HUBERT
A New Magnetooptic Effect Related to Non-Uniform Magnetization on the Surface of a Ferromagnet
75.60; 78.20; S1.1; S1.2; S1.4; S11.2
(a) Vol.118,No.1, 271-288

O. POPOV and M. MIKHOV
Temperature and Magnetic Field Dependence of the Critical Single Domain Size
75.60
(a) Vol.118,No.1, 289-294

H. PFEIFFER
Determination of Anisotropy Field Distribution in Particle Assemblies Taking into Account Thermal Fluctuations
75.60
(a) Vol.118,No.1, 295-306

D. MARTÍNEZ, C. DE FRANCISCO, J. IÑIGUEZ, J. RIVAS, and F. WALZ
Influence of Ba Substitutions on Magnetic After-Effects in Magnetite Containing Vacancies
75.60; S11.2
(a) Vol.118,No.1, K31-K34

T. TYMOSZ, K. ŚWIDERCZAK, and K. BRAŃSKA
Lorentz Microscope Study of Obliquely Evaporated CoCr Thin Films
75.60 and 75.70; S1.2
(a) Vol.118,No.2, 519-524

U. HEIBER
On the Domain Configuration of Finite Uniaxial Crystals
75.60; S10
(a) Vol.118,No.2, 531-538

V.P. GOGIN, V.A. MATVEEV, V.G. TASHIROV, YU.G. RAYZANOV, and A.A. ROMANYUKHA
Compensation Mechanism of the Temperature Stabilization of the Bubble Collapse Field in Eu-Tm Garnet
75.70; 75.60; S11.2
(a) Vol.118,No.2, K107-K110

H. PFEIFFER and W. SCHUEPPEL
Investigation of Magnetic Properties of Barium Ferrite Powders by Remanence Curves
75.60; S11.2
(a) Vol.119,No.1, 259-270

BAO-GEN SHEN, LIN-YUAN YANG, JUN-XIAN ZHANG, FENG WO, TAI-SHAN NING, JIAN-GAO ZHAO, HUI-QUN GUO, and WEN-SHAN ZHAN
Hard Magnetic Properties of Heat Treated $Fe_{72.8}Co_{4.7}Nd_4B_{18.5}$ Alloy
75.50; 75.60; S1.1
(a) Vol.119,No.1, K63-K66

K. ELK and V. CHRISTOPH
Dipole-Dipole Interaction in Fine Particle Magnets
75.60
(b) Vol.159,No.2, K111-K114

H.K. LACHOWICZ, E. PULIDO, A. SIEMKO, and A. HERNANDO
Stress-Forced Magnetoresistance in Near-Zero Magnetostrictive Metallic Glasses
72.15 and 75.80; 75.60; S1.1; S1.2
(a) Vol.119,No.2, 595-600

M. HRABCÁK and E. KISDI-KOSZÓ
Modification of Magnetic Properties of Amorphous Fe-Si-B Ribbons by Flash Annealing
75.50; 75.60; S1.1
(a) Vol.119,No.2, 601-606

A.V. KOROLEV, V.S. GAVIKO, and N.V. MUSHNIKOV
The Magnetic Annealing Effect in Hydrogen Containing Intermetallic $Sm(Fe,Co)_2$ Compounds
75.60 and 75.80; S1.2
(a) Vol.119,No.2, K163-K166

H. PFEIFFER
Relaxation Behaviour of Magnetic Particle Assemblies Due to Thermal Fluctuations
75.60
(a) Vol.120,No.1, 233-246

H. PFEIFFER, W. SCHUEPPEL, and TH. KLUPSCH
On the Magnetic Properties of Planar Hexaferrite Powders
75.60; S11.2
(a) Vol.120,No.1, K89-K92

A.H. WAFIK and A.Z. MOHAMED
Barkhausen Jumps and Microstructure in Molybdenum Eutectoid Steel
75.60; 64.80; S1.1
(a) Vol.120,No.2, 601-608

R.L. SOMMER and F.P. LIVI
Barkhausen Noise Measurements in Small Samples of (110)[001] Silicon-Iron
75.60; S1.61
(a) Vol.120,No.2, 609-616

J. CHEN and H. KRONMUELLER
Properties of Nd-Fe-B Magnets Processed by the Hydrogen Decrepitation Technique
75.60; S1.1
(a) Vol.120,No.2, 617-626

K. ELK and V. CHRISTOPH
On the Wohlfarth Remanence Relations in Magnetic Particle Ensembles
75.60
(b) Vol.161,No.1, K41-K44

Y.D. YAO, Y.Y. CHEN, S.J. TZENG, and T.H. CHUANG
Electrical Resistivity, Magnetization, and Grain Boundary Precipitates in Ni-Sn Alloys
72.15; 75.60; S1.2
(a) Vol.121,No.1, 213-218

ZHI-DONG ZHANG, X.K. SUN, ZHUONG-WU ZHANG, Y.C. CHUANG, and F.R. DE BOER
Effects of Partial Co Substitution on Magnetic Properties of $(Nd,Gd)_2Fe_{14}B$
75.30; 75.60; S1.1
(a) Vol.121,No.1, 241-248

G.F. ZHOU, S.Y. FU, X.K. SUN, and Y.C. CHUANG
Influence of Annealing on the Magnetic Properties and Microstructure of Nd-Fe-B Based Magnets
75.60; S1.1
(a) Vol.121,No.1, 257-264

BAO-GEN SHEN, HUI-QUN GUO, LIN-YUAN YANG, JUN-XIAN ZHANG, and JIAN-GAO ZHAO
Crystallization of Amorphous $Co_{84-x}B_xZr_{16}$ Alloys and Its Influence on Hard Magnetic Properties
75.30; 61.40; 75.60; S1.2
(a) Vol.121,No.1, K105-K110

H. PFEIFFER, W. SCHUEPPEL, and A. SAECKL
On the Determination of the Volume and the Anisotropy Field of Small Magnetic Particles by Coercivity Measurements
75.60
(a) Vol.121,No.1, K117-K120

L. KRAUS
Bloch Walls in Magnetostrictive Metallic Glasses
75.60; 75.80
(b) Vol.161,No.2, 853-860

P. NOVOTNÝ and R. GEMPERLE
Wall Hysteresis Loops Exhibiting Asymmetry
75.60; S11.2
(a) Vol.121,No.2, K215-K218

H. PFEIFFER
Influence of Thermal Fluctuations on the Magnetic Properties of Particle Assemblies
75.60
(a) Vol.122,No.1, 377-390

JUN-XIAN ZHANG, BAO-GEN SHEN, and WEN-SHAN ZHAN
Magnetic Properties of Amorphous $Nd_4Fe_{96-x}B_x$ Alloys and Influences of Heat Treatment on Its Hard Magnetic Properties
75.30; 75.60; S1.1
(a) Vol.122,No.2, 651-656

A.H. WAFIK
Dependence of the Irreversible Magnetization in $Co_{1-a}Cu_aFe_2O_4$ on the Magnetizing Field
75.60; S10
(a) Vol.122,No.2, 657-662

75.70. Magnetic films and plates

A.G. SHISHKOV, E.N. ILICHEVA, E.V. KOCHETKOVA, and YU.N. FEDYUNIN
On the Choice of the Pulse Waveform for Reading without Distortion of the Vertical Bloch Line Information
75.70; 75.60; S11.2
(a) Vol.117,No.1, K53-K56

A. URBANIAK-KUCHARCZYK and H. ZEYADA
The Hyperfine Field Distribution in Diluted Ising Ferromagnet
75.30; 75.70
(b) Vol.158,No.1, K59-K62

H. RATAJCZAK, I. GOŚCIAŃSKA, and T. LUCIŃSKI
Influence of Annealing on Saturation Magnetization and Hall Resistivity in Amorphous $Nd_{12}Co_{82}B_6$ Thin Films
61.40 and 68.55; 73.60; 75.70; S1.2; S1.4
(a) Vol.118,No.1, 189-196

T. TYMOSZ, K. ŚWIDERCZAK, and K. BRAŃSKA
Lorentz Microscope Study of Obliquely Evaporated CoCr Thin Films
75.60 and 75.70; S1.2
(a) Vol.118,No.2, 519-524

L. VATSKICHEV, J. MUCHA, ZH. GEORGIEV, and M. VATSKICHEVA
Measurements of the Rotational Hysteresis Losses in Magnetic Films by the Planar Hall Effect
72.15 and 75.70
(a) Vol.118,No.2, K95-K98

V.P. GOGIN, V.A. MATVEEV, V.G. TASHIROV, YU.G. RAYZANOV, and A.A. ROMANYUKHA
Compensation Mechanism of the Temperature Stabilization of the Bubble Collapse Field in Eu-Tm Garnet
75.70; 75.60; S11.2
(a) Vol.118,No.2, K107-K110

M. MANZEL, E. STEINBEISS, and W. SCHUEPPEL
Magnetic Properties of Sputter-Synthesized CoN_x-Films
75.70; S1.62
(a) Vol.119,No.1, 279-284

SHAOPING LI and L.E. CROSS
Analysis of Magnetostatic Modes Modified by Exchange and Anisotropy Energies
68.30 and 75.10; 75.70
(b) Vol.159,No.2, 861-872

B.X. GU, H. HOMBURG, S. METHFESSEL, and H.R. ZHAI
Crystallization Behavior and Magnetic Properties of Amorphous Nd-Fe-B Thin Films
68.55 and 75.70; 76.80; S1.1
(a) Vol.120,No.1, 159-168

K.P. GHATAK and M. MONDAL
On a Simplified Analysis of the Magnetic Susceptibilities of Electrons in Ultrathin Films of Non-Parabolic ...
75.20 and 75.70; 71.25; S7.16
(b) Vol.160,No.2, 673-682

N.V. EDNERAL, V.G. KUTILIN, YU.A. LISOVSKII, M.L. SOKOLOVA, A.M. UMPELEV, A.O. TITOV, and L.K. FIONOVA
Thin Film Structures for Magneto-Optical Information Recording
68.55; 75.70; 78.20; S1.1
(a) Vol.120,No.2, 457-466

Y. ENDO
Phase Transitions of Heisenberg Ferromagnetic Superlattices
75.30; 75.50; 75.70
(b) Vol.161,No.1, 379-386

75.80. Magnetomechanical and magnetoelectric effects, magnetostriction

For galvanomagnetic effects, see 72.15 and 72.20
For magneto-optical effects, see 78.20

P. PAWLICKI
Magnetoelastic Effects in the Exactly Solvable Vaks-Larkin-Ovchinnikov Model of Ising Spins
75.10 and 75.80
(b) Vol.157,No.1, K51-K54

M. FAEHNLE and J. FURTHMUELLER
On the Change of Magnetostriction by Field Annealing of Amorphous Ferromagnetic Alloys
75.80; S1.2
(a) Vol.117,No.1, K71-K76

H.K. LACHOWICZ, E. PULIDO, A. SIEMKO, and A. HERNANDO
Stress-Forced Magnetoresistance in Near-Zero Magnetostrictive Metallic Glasses
72.15 and 75.80; 75.60; S1.1; S1.2
(a) Vol.119,No.2, 595-600

A.V. KOROLEV, V.S. GAVIKO, and N.V. MUSHNIKOV
The Magnetic Annealing Effect in Hydrogen Containing Intermetallic $Sm(Fe,Co)_2$ Compounds
75.60 and 75.80; S1.2
(a) Vol.119,No.2, K163-K166

L. KRAUS
Bloch Walls in Magnetostrictive Metallic Glasses
75.60; 75.80
(b) Vol.161,No.2, 853-860

76. Magnetic Resonances and Relaxation in Condensed Matter; Moessbauer Effect

76.20. General theory of resonances and relaxation

E.I. NEIMARK, I.N. NURUTDINOVA, and A.B. ROITSIN
A Method for Calculating the Inhomogeneously Broadened Resonance Line Shape
76.20
(b) Vol.158,No.1, K55-K58

A.YA. FISHMAN, M.A. IVANOV, V.YA. MITROFANOV, and A.A. SHEMYAKOV
Hyperfine Interactions of Jahn-Teller Ion Mn^{3+} in Cubic Magnets with Spinel and Garnet Structures
75.50; 76.20; S11.2
(b) Vol.160,No.2, K153-K158

M.I. KAGANOV and T.I. SHALAEVA
Resonance Polaritons
71.36; 76.20
(b) Vol.162,No.2, 469-476

76.30. Electron paramagnetic resonance and relaxation

C. YU, C. LAI, S.A. MARSHALL, D.R. YODER-SHORT, and Y.N. ZHANG
Electron Paramagnetic Resonance Spectra Observed in Single Crystals of $YBa_2Cu_3O_{7-\delta}$
74.60; 76.30; S10.15
(b) Vol.157,No.1, 379-388

WEN-CHEN ZHENG
Spin-Lattice Coupling Coefficients F_{ij} for $SrTiO_3:Cr^{3+}$ Crystal
71.70; 76.30; S11.1
(b) Vol.157,No.1, K33-K38

KH.B. GEZALOV, A.M. GASANOV, A.A. GARIBOV, M.M. ALIEV, and N.I. MUSAEV
The Nature of Paramagnetic Centers in gamma-Irradiated Boron Oxides
61.80 and 76.30; S10.1
(a) Vol.117,No.1, K57-K60

D. SKRZYPEK
Critical Behaviour of the Magnetic Resonance in $KNi_xMn_{1-x}F_3$
75.30 and 76.30; S9
(b) Vol.157,No.2, 695-700

A. TARI and J.S. HWANG
Magnetisation and ESR Studies of $Gd(Ni_{1-x}Rh_x)_5$ Compounds
75.60 and 76.30; S1.4
(b) Vol.157,No.2, 701-710

S. WAPLAK and N.N. KOLPAKOVA
Phase Transitions in $Cd_2Nb_2O_7$ Studied by EPR of Cr^{3+} Ion
64.70; 76.30; S10.15
(a) Vol.117,No.2, 461-466

W.A. KACZMAREK, R. BRAMLEY, and A. CALKA
Itinerant Electron Contribution to the Magnetism of Ni-TM Glassy Metallic Magnets
75.50 and 76.30; S1.1; S1.2
(a) Vol.117,No.2, K141-K146

E.A. EIVAZOV, A.F. SAFAROV, S.M. ATAKISHIEV, and YA.M. ABASOV
The EPR Spectrum of $Co_{0.7}Cu_{0.3}Cr_2S_{4-x}Se_x$ System
76.30; S8.16
(a) Vol.117,No.2, K147-K150

V.K. VORONKOVA, L.V. MOSINA, A.E. USACHEV, and YU.V. YABLOKOV
Weak Pair Exchange Interactions between Cobalt Ions in the Quasi-One-Dimensional Antiferromagnet $CsCoCl_3$
76.30; 71.70; S9.15
(b) Vol.158,No.1, 337-346

WAN-LUN YU
Crystal Field Effect on the g-Factors of ^6S-State Ions
71.70; 76.30; S8.11; S10.1; S11.1
(b) Vol.158,No.1, K13-K16

M. WATTENBACH, C. KISIELOWSKI-KEMMERICH, H. ALEXANDER, V.V. KVEDER, T.R. MCHEDLIDZE, and YU.A. OSIPYAN
Electric-Dipole Spin Resonance of Dislocations in Plastically Deformed p-Type Silicon
71.55 and 76.30; S5.11
(b) Vol.158,No.1, K49-K54

S. FUJITA and S. WATANABE
On the Orientation Dependence of the Cyclotron Resonance Peaks for Holes in Ge and Si
71.25 and 76.30; S5.11; S5.12
(b) Vol.158,No.1, K69-K74

DEYAN HE, FANGQING ZHANG, and GUANGHUA CHEN
A Direct Measurement of Thermal Equilibrium Defects in Hydrogenated Amorphous Silicon-Nitrogen Alloy Films
61.70 and 76.30; S5.11
(a) Vol.118,No.1, K39-K42

L.K. AMINOV and YU.YU. KOSTETSKII
Temperature Dependence of EPR Linewidth of Impurities in Van-Vleck Paramagnets
71.70 and 76.30; S9.16
(b) Vol.158,No.2, 595-602

K. ZINK, A. KROST, H. NELKOWSKI, J. SAHM, and H. STUTENBECKER
Energy Transfer Mechanisms in $Zn_{1-x}Mn_xSe$ (0.005<x<0.017) Detected by ODMR
71.70; 76.30; 78.55; S8.15
(b) Vol.158,No.2, 603-608

WEN-CHEN ZHENG
An Estimate of the Spin-Lattice Coupling Coefficients for Trigonal $CdS:Mn^{2+}$ Crystals
71.70; 76.30; S8.11
(b) Vol.158,No.2, K137-K142

GUOYIN SHEN, CHANGQING XU, and GUIRU BAI
Cu^{2+}-Cu^{2+} Pair Effect on EPR Parameters in C_{2v} Symmetry Crystals
71.70 and 76.30; S11
(b) Vol.158,No.2, K185-K190

D.L. DZHAPARIDZE, S.V. ALCHANGYAN, D.M. DARASELIA, and T.I. SANADZE
EPR and RFDS Study of Gd^{3+} in alpha-$LiIO_3$ Single Crystals
76.30; S11.1
(b) Vol.158,No.2, K195-K196

W. ZAPART
Possibility of Simultaneously Incommensurate and Ferroelastic Phase in $RbIn(MoO_4)_2$ by EPR of Cr^{3+} Ion
64.70 and 76.30; S11.1
(a) Vol.118,No.2, 447-454

M. ZABKOWSKA-WACŁAWEK and W. WOJCIECHOWSKI
Electrical and Magnetic Properties of Copper Phthalocyanine-Carbon Black Mixtures
72.80; 76.30; S12
(a) Vol.118,No.2, K87-K90

P. SIVAPRASAD, K. RAMESH, and Y.P. REDDY
Optical and EPR Studies of Cu^{2+} in $MgNH_4PO_4$ * 6 H_2O
76.30 and 78.50; 71.55; S11.1
(a) Vol.118,No.2, K103-K106

GUANGHUA CHEN, WEI JIA, and FANGQING ZHANG
ESR Study on Annealing Behavior of Reactively Sputtered a-GeN_x:H Films
68.60 and 76.30; S6
(a) Vol.118,No.2, K115-K118

H. HIKITA, K. TAKEDA, and Y. KIMURA
EPR of New Cr^{3+} Centers in SnO_2 Crystals
76.30; S10.1
(a) Vol.119,No.1, 251-258

V.K. JAIN and G. LEHMANN
Electron Paramagnetic Resonance of Mn^{2+} in Orthorhombic and Higher Symmetry Crystals (Review Article)
76.30; 64.60; S11
(b) Vol.159,No.2, 495-544

T. KAMIKAWA
A New Hole Centre in Neutron Irradiated LiF Crystals
61.70; 76.30; S9.11
(b) Vol.159,No.2, 571-576

WEN-CHEN ZHENG
Spin-Lattice Coupling Coefficients F_{ij} for d^8 Ions in Cubic Symmetry
71.70; 76.30; S10.1
(b) Vol.159,No.2, K67-K72

W.A. KACZMAREK
Magnetic Structure Rearrangement during Crystallization in $Ni_{73}Mn_5Si_{10}B_{12}$ Ribbons
75.50 and 76.30; S1.2
(a) Vol.119,No.2, K159-K162

H.A. BUCKMASTER and K. KUMAR
The C_{3h} Symmetry EPR Spectral Parameters for $(0.005Gd,0.995La)(C_2H_5SO_4)_3$ * 9 H_2O and 9 D_2O
76.30; 71.70; S11
(b) Vol.160,No.1, 349-356

WEN-CHEN ZHENG
Spin-Lattice Coupling Coefficients for $CaO:Ni^{2+}$ Crystals
71.70; 76.30; S10.1
(b) Vol.160,No.1, K43-K48

A.A. ALYBAKOV, O.M. ARBOTOEV, V.A. GUBANOVA, and I.K. TURDALIEV
Paramagnetic Centers in X-Irradiated LiH_2PO_4 Single Crystals
76.30; S11.1
(a) Vol.120,No.1, K85-K88

S.A. MARSHALL, D.R. YODER-SHORT, C. YU, Y.N. ZHANG, and J.K. FURDYNA
Super Hyperfine Structure in the EPR Spectrum of Divalent Manganese in Single Crystal Zinc Selenide
71.70 and 76.30; S8.12
(b) Vol.160,No.2, 591-600

A. GOLTZENÉ and C. SCHWAB
Qualitative and Quantitative Differentiation of Paramagnetic Anion-Antisite-Related Spectra in GaAs
71.70 and 76.30; S7.12
(b) Vol.160,No.2, 649-660

S.M. AKHMIN, V.P. MEIKLYAR, A.E. USACHEV, and YU.V. YABLOKOV
$^{59}Co^{2+}$ ENDOR in the Tetragonal Phase of $CsCaCl_3$
76.30 and 76.70; 64.60; S9.16
(b) Vol.160,No.2, K149-K152

C. KISIELOWSKI-KEMMERICH
Vacancies and Their Complexes in the Core of Screw Dislocations. Models which Account for ESR-Investigation ... (Review Article)
61.70; 76.30; S5.11
(b) Vol.161,No.1, 11-42

S. SOEDERHOLM, J. HELLBERG, J. KRZYSTEK, G. OLOVSSON, I. OLOVSSON, and J.U. VON SCHUETZ
Electronic Properties and Crystal Structure of a Methylthio Substituted Naphthalene Cation Radical Salt: $(DMbTN)_3(ClO_4)_2$
61.65 and 72.20; 71.55; 75.30; 76.30; S12.1
(b) Vol.161,No.1, 245-256

F. MAES, F. CALLENS, P. MATTHYS, and E. BOESMAN
EPR of S_2^- in NaBr
61.70 and 76.30; S9.11
(b) Vol.161,No.1, K1-K4

FU-ZHEN LI and ZHAO-MIN LI
The Eight-Order Perturbation Formula of Zero-Field Splitting for $3d^5$ Ions in D_{3d} Site and Its Application to $CaCO_3:Mn^{2+}$
71.70; 76.30; S11.1
(b) Vol.161,No.1, K29-K34

E. POSSENRIEDE, O.F. SCHIRMER, and G. GODEFROY
ESR of Nd^{3+} in $BaTiO_3$
76.30; S11.1
(b) Vol.161,No.1, K55-K58

D.P. ERCHAK, R.B. GELFAND, N.M. PENINA, V.F. STELMAKH, V.P. TOLSTYKH, A.G. ULYASHIN, V.S. VARICHENKO, and A.M. ZAITSEV
Point Paramagnetic Defects in Diamond Irradiated by High-Energy Ions
61.70; 76.30; 78.60; S5
(a) Vol.121,No.1, 63-72

A. POEPPL and G. VOELKEL
ESR and Photo-ESR Investigations of the V_0^+ Centre in ZnO Raw Material and Li-Doped ZnO Ceramic Powder
71.55; 76.30; S10.1
(a) Vol.121,No.1, 195-204

W. ZAPART and M.B. ZAPART
Incommensurate Phase in $KSc(MoO_4)_2$ by EPR of Cr^{3+}
64.70 and 76.30; S11.1
(a) Vol.121,No.1, K43-K46

A. POEPPL and G. VOELKEL
ESR and Photo-ESR Investigations of the Pb^{3+} Centre in Undoped ZnO Ceramics
71.55; 76.30; S10.1
(a) Vol.121,No.1, K69-K72

L. SADŁOWSKI, J. KURIATA, B. BOJANOWSKI, J. WALCZAK, M. KURZAWA, M.L. FALIN, and V.V. IZOTOV
EPR Study of Antiferromagnetic Phase Transition in $Fe_4V_2Mo_3O_{20}$
76.30; 75.50; S11.1
(a) Vol.121,No.1, K95-K98

A.CH. LUSHCHIK and A.G. FRORIP
Thermalized and Hot Interstitial Halogen Ions in Alkali Halides
61.70 and 78.55; 61.80; 76.30; S9.11
(b) Vol.161,No.2, 525-536

WEN-CHEN ZHENG
Spin-Lattice Coupling Coefficients G_{ij} for Trigonal $Al_2O_3:Fe^{3+}$ and $ZnSiF_6 * 6 H_2O:Ni^{2+}$ Crystals
71.70; 76.30; S9.16; S10
(b) Vol.161,No.2, 711-716

A.P. PECHONII, T.V. ANTIMIROVA, M.D. GLINCHUK, and I.M. SMOLYANINOV
An ESR Study of Li^+ Impurity Ion Deformational Fields in $KTaO_3$
76.30; 61.70; S11
(b) Vol.161,No.2, 843-852

T.L. CHOY and Y.Y. YEUNG
Local Distortion and Crystal Field Parameters for Cr^{3+} at Orthorhombic Sites in MgO
76.30; S10.1
(b) Vol.161,No.2, K107-K110

GUANGHUA CHEN, JINZHANG XU, and FANGQING ZHANG
The ESR Studies of B-Doped a-Si:H Thin Films Prepared by Reactive-Sputtering
61.40 and 76.30; 68.55; S5.11
(a) Vol.121,No.2, 501-506

KANG-WEI ZHOU, SANG-BO ZHAO, PING-FENG WU, and JUN-KAI XIE
EPR Parameters and Spectral Fine Structure of Ni^{2+} in $LiNbO_3$
71.70 and 76.30; 78.50; S11.1
(b) Vol.162,No.1, 193-198

N. GUSKOS, G.P. TRIBERIS, M. CALAMIOTOU, CH. TRIKALINOS,
A. KOUFOUDAKIS, C. MITROS, H. GAMARI-SEALE, and D. NIARCHOS
Temperature Dependence of the EPR Spectra of $GdBa_2Cu_3O_{7-\delta}$ Compounds in Orthorombic and Tetragonal Phase
72.20 and 74.70; 76.30; S10.15
(b) Vol.162,No.1, 243-250

S.K. MISRA and JIANSHENG SUN
EPR of a Gd^{3+}-Doped $NH_4Pr(SO_4)_2$*$4H_2O$ Single Crystal. Study of Phase Transition and Superposition-Model Calculation of ..
76.30; S11
(b) Vol.162,No.1, 265-274

R. HRABAŃSKI and J. LECH
Optimization of Spin Hamiltonian Parameters by the Method of Non-Linear Least Squares Fitting
76.30; S11
(b) Vol.162,No.1, 275-280

M.D. GLINCHUK, I.P. BYKOV, V.M. KURLIAND, M. BOUDYŠ, and K. NEJEZCHLEB
Valency States and Distribution of Manganese Ions in PZT Ceramics Simultaneously Doped with Mn and Nb
71.55; 76.30; S10.15
(a) Vol.122,No.1, 341-346

H. METZ, G. VOELKEL, and W. WINDSCH
A Simple Method for Fitting of Inhomogenous ESR Saturation Curves
76.30
(a) Vol.122,No.1, K73-K76

S. SWAMI, V.K. AGARWAL, S.S. SHARMA, and J. VARMA
Moessbauer and ESR Studies of Polymetallic Nodules
76.30 and 76.80; S1.2
(a) Vol.122,No.1, K77-K82

F.J. OWENS
Non-Resonant Microwave Absorption in N=3 Phase of Bi-Pb-Sr-Ca-Cu-O High-Temperature Superconductor
74.30; 76.30; S10.15
(b) Vol.162,No.2, 565-574

S.K. MISRA, JIANSHENG SUN, and U. ORHUN
EPR of a VO^{2+}-Doped $Fe(NH_4)_2(SO_4)_2$*$6H_2O$ Single Crystal: VO^{2+}-Fe^{2+} Exchange Interaction and Spin-Lattice Relaxation ...
76.30; S11
(b) Vol.162,No.2, 585-596

N. GUSKOS, M. CALAMIOTOU, S.M. PARASKEVAS, A. KOUFOUDAKIS, C. MITROS, H. GAMARI-SEALE, J. KURIATA, L. SADLOWSKI, and M. WABIA
On the Influence of Oxygen Deficiency on EPR Spectra of $NdBa_2Cu_3O_{7-\delta}$
76.30; 74.70; S10.15
(b) Vol.162,No.2, K101-K106

W.A. KACZMAREK, R. BRAMLEY, and A. CALKA
ESR Study of Amorphous Metallic Alloys below the Magnetic Multicritical Point
75.50 and 76.30; 75.40; S1.1; S1.2
(a) Vol.122,No.2, 663-676

76.50. Ferromagnetic, antiferromagnetic, and ferrimagnetic resonances; spin-wave resonance
(see also 75.30. Spin waves)

A.M. MANSANARES, F.C.G. GANDRA, E.C. DA SILVA, H. VARGAS, M. DO CARMO, R.S. VARELLA, and F. GALEMBECK
Photoacoustic Detection of Magnetite-Crystal Formation from Iron(III) Hydroxide Acetate
75.50 and 76.50; S12
(a) Vol.117,No.2, K155-K160

G. SERFŐZŐ, E. KISDI-KOSZÓ, A. SLAWSKA-WANIEWSKA, L. POTOCKY, S. JONELIUNAS, and M. BARAN
Study of Metallic Glasses Rapidly Quenched in Magnetic Field
75.50; 76.50; S1.1
(a) Vol.118,No.1, 307-310

76.60. Nuclear magnetic resonance and relaxation

K.D. TOVSTYUK, I.O. POLYAKOV, E.I. SLINKO, N.K. TOVSTYUK, and A.G. KHANDOZHKO
Lithium Impurity States in PbTe
64.60; 61.70; 76.60; S8
(b) Vol.157,No.1, 151-158

N.T. MAMEDOV and A.M. PANICH
Phase Transition in $TlGaTe_2$
64.70 and 76.60; S8.16
(a) Vol.117,No.1, K15-K18

XIQI FENG, DONGSHENG WANG, and JIZHOU ZHANG
NMR Spectra of Mg Nuclei in Mg-Doped $LiNbO_3$ Crystals
76.60; S11.1
(b) Vol.157,No.2, K127-K130

J. CZAPLICKI, N. WEIDEN, and A. WEISS
Dynamics of Protons in $Cd(NH_3)_6X_2$ Type Compounds
76.60; S9.16
(a) Vol.117,No.2, 555-562

S.G. KOZLOVA, YU.G. KRIGER, and S.P. GABUDA
Anisotropic Reorientation of Octahedral Ions According to NMR Relaxation in Solid State
76.60; 71.70; S11
(b) Vol.158,No.1, 329-336

E.K. SADYKOV and A.I. SKVORTSOV
Theory of R.F. Moessbauer Spectra of the Magnetics with Magnetic Anisotropy
76.60 and 76.80; S11.1
(b) Vol.158,No.2, 685-694

S. GRANDE, B. LIPPOLD, G. MAYER, and H. SCHMIEDEL
NQR Measurements of ^{63}Cu in $YBa_2Cu_3O_{7-x}$
74.70 and 76.60; S10.15
(b) Vol.158,No.2, 707-714

R. PIRLOT, M. CYAMUKUNGU, L. GRENACS, J. LEHMANN, and K. TOMPA
Relaxation Time Measurements and Hyperfine Fields at ^{12}B in Transition Metals
76.60; S1
(b) Vol.158,No.2, K197-K200

M. MAĆKOWIAK, N. WEIDEN, and A. WEISS
Phase Transitions and Molecular Dynamics in Tris(Methylammonium) Nonabromodiantimonate(III) and Its ...
63.20 and 64.70; 76.60; S12
(a) Vol.119,No.1, 77-86

P.K. KAHOL
Evaluation of Moments from the NMR Solid Echoes in Dipolar Solids
76.60
(b) Vol.159,No.2, 873-882

J. KASPRZAK
The Influence of Electron Hopping on the Quadrupole Relaxation in Cu_2O
76.60; 72.20; S10.1
(b) Vol.160,No.1, K53-K56

J.E. GARBARCZYK
Study of Protonic Diffusion in $H_{4.5}UO_2(IO_6)_{1.3} * m\ H_2O$ Using PFG NMR Method
66.30; 76.60; S11
(a) Vol.120,No.2, 387-390

A.A. BOGUSLAVSKII, YU.N. IVANOV, A.I. KRIEGER, A.K. MOSKALEV, V.I. PAKHOMOV, and R.SH. LOTFULLIN
NMR Investigation of Quadrupole Interaction and Chemical Shift ^{133}Cs in Cs_2HgBr_4 Crystal
76.60 and 77.80; S9.16
(b) Vol.161,No.1, K49-K54

T.SH. ABESADZE and R.A. EL EGAIMI
Subharmonic Excitation of Multiple Spin Echo Signals in Systems with Quadrupole Interaction
76.60
(b) Vol.161,No.2, K111-K112

76.70. Magnetic double resonances and cross effets

S.M. AKHMIN, V.P. MEIKLYAR, A.E. USACHEV, and YU.V. YABLOKOV
$^{59}Co^{2+}$ ENDOR in the Tetragonal Phase of $CsCaCl_3$
76.30 and 76.70; 64.60; S9.16
(b) Vol.160,No.2, K149-K152

S.M. AKHMIN, V.P. MEIKLYAR, A.E. USACHEV, YU.V. YABLOKOV, and S.V. YURTAEVA
Electron-Nuclear Interactions of the Ce^{3+} Ion in $Cs_2NaInCl_6$
76.70; S9.16
(b) Vol.162,No.2, K107-K112

76.80. Moessbauer effect; other gamma-ray spectroscopy

A.K. ZHETBAEV, K.M. DONBAEV, and M.B. ABLANOV
Spin Reorientation in Irradiated Hematite at 7 to 60 K
61.70 and 76.80; S10.1
(b) Vol.157,No.1, K55-K60

E.A. VASILEV and T.M. TKACHENKO
The ^{119}Sn Atom Dynamics in NiSb-$Ni_{1.5}$ Solid Solutions
76.80; S1.2
(b) Vol.157,No.1, K61-K64

E.P. ELSUKOV, YU.N. VOROBEV, A.V. TRUBACHEV, and V.A. BARINOV
Structure and Magnetic Properties of Fe-P Electrodeposited Alloys
61.55 and 75.50; 76.80; S1.1
(a) Vol.117,No.1, 291-298

J.J. BARA, B.F. BOGACZ, A. POLACZEK, M. PĘKAŁA, and A. GRODZIŃSKI
The Temperature Induced Transition in the theta-Type Vanadium Oxide Bronzes
61.60; 64.70; 76.80; S10
(a) Vol.118,No.1, 41-52

K. MELZER, J. SUWALSKI, and M. ŁUKASIAK
Moessbauer Study on the Spin Arrangement in $BaFe_{12}O_{19}$ and $BaCo_{0.8}Ti_{0.8}Fe_{10.4}O_{19}$
75.25 and 76.80; S11.2
(a) Vol.118,No.1, K35-K38

E.K. SADYKOV and A.I. SKVORTSOV
Theory of R.F. Moessbauer Spectra of the Magnetics with Magnetic Anisotropy
76.60 and 76.80; S11.1
(b) Vol.158,No.2, 685-694

V.V. BOGDANOV, V.V. ZAKURKIN, D.V. PETROV, and YU.P. PENKOV
Investigation of Proton-Irradiated Platinum by Moessbauer Spectroscopy
61.80 and 76.80; S1.3
(b) Vol.158,No.2, K105-K110

H. REUTHER, E. RICHTER, and I.S. SZUECS
CEMS Measurements on Ion Implanted Tool Steels
61.80 and 76.80; S1.1
(a) Vol.118,No.2, 409-414

P.H. DIONISIO, B.A.S. DE BARROS, JR., and I.J.R. BAUMVOL
A Moessbauer Spectroscopic Study of the High-Temperature Oxidation Behaviour of an Austenitic Stainless Steel ...
61.80 and 68.20; 76.80; S1.1
(a) Vol.118,No.2, 473-478

Z.T. SERIKBAEVA, T.M. ZHANTIKIN, and A.K. ZHETBAEV
Moessbauer Study of 30 MeV Proton Dose Dependence of Radiation Damage in Disordered Ni_3Fe
61.80 and 76.80; S1.1
(a) Vol.118,No.2, K99-K102

D. BALTRUNAS, R. MIKAITIS, V.YU. SLIVKA, and YU.M. VYSOCHANSKII
Manifestation of the Ferroelectric Phase Transition in the Moessbauer Parameters of the System $(Sn_xPb_{1-x})_2P_2S_6$
64.70 and 76.80; S8.16
(a) Vol.119,No.1, 71-76

J. CHOJCAN
The Moessbauer Effect and Positron Annihilation Studies of Iron-Cobalt Alloys
71.20 and 76.80; S1.1
(b) Vol.159,No.2, K59-K62

P.P. SEREGIN, F.S. NASREDINOV, V.F. MASTEROV, and G.T. DARIBAEVA
Electron Structure and Crystalline Field Parameters Determined by the Emission Moessbauer Spectroscopy for the Copper...
76.80; 74.70; S10.15
(b) Vol.159,No.2, K97-K102

Z.Z. AKSELROD, B.A. KOMISSAROVA, L.N. KRYUKOVA, G.K. RYASNYI, L.G. SHPINKOVA, and A.A. SOROKIN
Electron Quadrupole Interaction of ^{181}Ta in Hf_2Fe Remeasured
71.70; 76.80; S1.1; S4
(b) Vol.160,No.1, 255-262

S.K. KULSHRESHTHA
On Moessbauer Spectrum Analysis for Spin Orientation in Fe_2As
76.80; S1.62
(b) Vol.160,No.1, 313-320

B.X. GU, H. HOMBURG, S. METHFESSEL, and H.R. ZHAI
Crystallization Behavior and Magnetic Properties of Amorphous Nd-Fe-B Thin Films
68.55 and 75.70; 76.80; S1.1
(a) Vol.120,No.1, 159-168

W.A. KACZMAREK
Magnetic Hyperfine Interactions in $BaIn_{1.5}Fe_{10.5}O_{19}$ Ferrite
76.80; 75.50; S11.2
(a) Vol.120,No.1, K79-K84

E.V. VORONINA, N.V. ERSHOV, A.L. AGEEV, and YU.A. BABANOV
Regular Algorithm for the Solution of the Inverse Problem in Moessbauer Spectroscopy
76.80
(b) Vol.160,No.2, 625-634

Z.T. SERIKBAEVA, T.M. ZHANTIKIN, A.K. ZHETBAEV, and K.G. BINNATOV
Moessbauer Investigations of Ordering of the $Ni_3(Fe_{1-x}Mn_x)$ Alloy System
61.55; 76.80; S1.1; S1.2
(a) Vol.120,No.2, 305-314

D. SUEVA, V. SPASSOV, N. CHIKOV, E.I. VAPIREV, and I. IVANOV
Silicon Avalanche Detectors for X-Ray Analysis
76.80 and 85; S5.11
(a) Vol.120,No.2, K185-K188

U.C. JOHRI, R.M. SINGRU, and D. BAHADUR
Magnetic and Moessbauer Studies of the System $LaCo_{1-x}Zr_xO_3$ (x=0.1,0.3,0.5)
76.80; 75.50; S10.15
(b) Vol.161,No.1, 357-366

S. UNNIKRISHNAN and D.K. CHAKRABARTY
Magnetic Properties and Domain State of the Magnesium Ferrite-Magnesium Titanate Solid Solutions
75.50; 76.80; S11.2
(a) Vol.121,No.1, 265-272

R. HEIDINGER, H. APPEL, G. THEN, and W.-G. THIES
Observation of Radiation Damage in HfV_2 and Its Hybrides by TDPAC
61.80; 64.80; 74.70; 76.80; S1.2
(a) Vol.121,No.2, 445-454

F.S. NASREDINOV, E.YU. TURAEV, P.P. SEREGIN, H.B. RAKHMATULLAEV, and M.K. BAKHADIRKHANOV
Mechanism of the Two-Electron Exchange between Neutral and Ionized Centres of Tin in the $PbS_{1-x}Se_x$ Solid Solutions
61.70 and 71.55; 76.80; S8.15
(a) Vol.121,No.2, 571-578

O.A. GOLIKOVA, N. AMANDZHANOV, M.M. KAZANIN, G.M. KLIMASHIN, and V.V. KUTASOV
Electrical Activity of Impurities in beta-Rhombohedral Boron
71.55; 72.80; 76.80; S5
(a) Vol.121,No.2, 579-586

S. SWAMI, V.K. AGARWAL, S.S. SHARMA, and J. VARMA
Moessbauer and ESR Studies of Polymetallic Nodules
76.30 and 76.80; S1.2
(a) Vol.122,No.1, K77-K82

A. GRUSKOVÁ, J. LIPKA, J. SLÁMA, I. TÓTH, M. SEBERÍNI, and P. KABOŠ
Moessbauer Effect and Its Application in the Investigation of the Substituted Li Ferrites Obtained from the Liquid Phase
75.30 and 75.50; 76.80; S11.2
(a) Vol.122,No.2, K171-K176

77. Dielectric Properties and Materials

(for conductivity phenomena, see 72.20 and 72.80)

77.20. Permittivity

J. VON CIEMINSKI, H.TH. LANGHAMMER, and H.-P. ABICHT
Peculiar Electromechanical Properties of Some $Ba(Ti,Sn)O_3$ Ceramics
77.60; 77.20; 77.30; S10.15
(a) Vol.120,No.1, 285-294

K. KUMAZAKI
Dielectric Constant in $Zn_xHg_{1-x}Se$ Determined by Raman Scattering
77.20; S8.15
(b) Vol.160,No.2, K173-K176

R. CACH and Z. CZAPLA
Dielectric State Equation Parameters of Monoclinic Telluric Acid Ammonium Phosphate
77.80; 77.20; S11
(a) Vol.121,No.1, 299-304

A. LARGETEAU, J.M. REAU, and J. RAVEZ
Dispersions diélectrique et d'impédance et énergies d'activation de céramiques à couches d'arrêt de type ...
77.20; 64.80; 77.40; S11.2
(a) Vol.121,No.2, 627-634

77.30. Polarization and depolarization effects

J. MANGIN, G. JEANDEL, and G. MARNIER
Temperature Dependence of Polarization in $KTiOPO_4$ Single Crystals
77.30; 77.70; S11.1
(a) Vol.117,No.1, 319-324

Z. ZIKMUND and J. FOUSEK
Reduction of the Number of Polarization Vectors in Multiaxial Ferroelectrics by Means of Chiral Dopants. Macroscopic...
77.30 and 77.80
(a) Vol.118,No.2, 539-552

S.S. BAWA, K. SAXENA, A.M. BIRADAR, K.K. SAINI, and S. CHANDRA
Bistable Switching and Molecular Reorientation of Ferroelectric Liquid Crystals by Current Analysis
77.30 and 77.80; S12
(a) Vol.118,No.2, 561-566

J. VON CIEMINSKI, H.TH. LANGHAMMER, and H.-P. ABICHT
Peculiar Electromechanical Properties of Some $Ba(Ti,Sn)O_3$ Ceramics
77.60; 77.20; 77.30; S10.15
(a) Vol.120,No.1, 285-294

V.A. KOSOBUKIN
Localization and Resonance Transmission of Light Waves in a One-Dimensionally Disordered Dielectric. Polarization Effects
78.20; 77.30
(b) Vol.161,No.1, 405-418

W. KUCHARCZYK
Correlation between Electronic Polarizabilities of Ions and Ratio of Their Radii in Alkali Halides
77.30; S9.11
(b) Vol.161,No.1, K59-K62

M. BUIVYDAS, O. ADOMENIENE, and A. BEGANSKIENE
Model of Polarization Reversal Kinetics after an Excitation by an Asymmetric Rectangular Shaped Voltage Pulse
77.30; S12
(b) Vol.161,No.2, K113-K118

E. LAREDO, A. BELLO, N. SUAREZ, and M. DÍAZ
Comparison of Analyzing Procedures for Determining the Relaxation Parameters from ITC Curves
77.30; S9.12
(a) Vol.121,No.2, K235-K238

H.A. KASSIM
Dipole Polarizability of O^{2-} Ions near the (011) and the (111) Surface of Alkaline Earth Oxides
73.20 and 77.30; S10
(b) Vol.162,No.1, 297-302

M. FUKUHARA, A. BHALLA, and R.E. NEWNHAM
Morphotropic Phase Boundary in the $Pb(Zr_xTi_{1-x})O_3$ System
77.30 and 77.90; S10.15
(a) Vol.122,No.2, 677-682

77.40. Dielectric loss and relaxation

H. KRESSE and F. KREMER
Dielectric Relaxation of a Swallow-Tailed Compound at the Phase Transition from the Nematic into the Isotropic Phase
77.40; 64.70; S12
(a) Vol.118,No.2, K119-K122

V. LHOTSKÁ, J. FOUSEK, and N. NEUMANN
Possible Origin of Dielectric Loss Behaviour in Samples of TGS-Type Crystals
77.40; 77.80; S12
(a) Vol.120,No.1, 273-284

A. LARGETEAU, J.M. REAU, and J. RAVEZ
Dispersions diélectrique et d'impédance et énergies d'activation de céramiques à couches d'arrêt de type ...
77.20; 64.80; 77.40; S11.2
(a) Vol.121,No.2, 627-634

K. IWAUCHI
Ionic Conductivity and Dielectric Relaxation of beta-Al_2O_3 Fiber Superionic Conductor
66.30; 77.40; S10
(a) Vol.122,No.2, 503-510

77.50. Dielectric breakdown and space-charge effects

V.V ZUBRITSKII and R.P. PURLYS
The Relation between the Orientation of Streamer Discharges and Anisotropy of Thermal Vibrations of Atoms in II-VI ...
63.20; 71.55; 72.20; 77.50; S8.11
(b) Vol.158,No.1, 91-102

77.55. Dielectric thin films

R. SATHYAMOORTHY, SA.K. NARAYANDASS, C. BALASUBRAMANIAN, and D. MANGALARAJ
Structure, Dielectric, and AC Conduction Studies on Yttrium Fluoride Thin Films
68.55 and 73.60; 77.55; S9
(a) Vol.117,No.2, 495-500

M.D. KANNAN, SA.K. NARAYANDASS, C. BALASUBRAMANIAN, and D. MANGALARAJ
Aging, Annealing, and Dielectric Properties of Neodymium Oxide Thin Films
73.60; 77.55; S10
(a) Vol.121,No.2, 515-522

77.60. Piezoelectricity and electrostriction
(for piezo-optical effects, see 78.20)

E. KLIER
Direct Piezoelectric Effect in Semi-Insulating InP:Fe
77.60; S7.11
(a) Vol.119,No.1, K79-K82

J. VON CIEMINSKI, H.TH. LANGHAMMER, and H.-P. ABICHT
Peculiar Electromechanical Properties of Some Ba(Ti,Sn)O_3 Ceramics
77.60; 77.20; 77.30; S10.15
(a) Vol.120,No.1, 285-294

M. DEY (BOSE) and S. GHOSH
Acoustic Second Harmonic Generation in Magnetized High Resistivity Piezoelectric Semiconductors. Effect of ...
72.20 and 72.50; 77.60; S8.11
(b) Vol.161,No.1, 319-328

77.70. Pyroelectric and electrocaloric effects

J. MANGIN, G. JEANDEL, and G. MARNIER
Temperature Dependence of Polarization in $KTiOPO_4$ Single Crystals
77.30; 77.70; S11.1
(a) Vol.117,No.1, 319-324

S.R. STOYANOV
Heat Pretreatment Effect on the Pyroelectric Current of Pure and Sb^{3+} Doped Triglycine Sulphate Monocrystals
77.70; S12
(a) Vol.117,No.2, K161-K164

P. ROTH and E. HEGENBARTH
How Universal Are the Glassy States in Solids with a Structural Instability?
61.40; 77.70; 77.80
(b) Vol.159,No.1, K1-K6

P. KOZIOL and R. JAKUBAS
Effect of Hydrostatic Pressure on the Phase Transitions in $(CH_3NH_3)_3Sb_2Br_9$
77.70; S12
(a) Vol.120,No.1, K93-K94

D. DĂDÂRLAT, M. CHIRTOC, C. NEAMŢU, R.M. CÂNDEA, and D. BICANIC
Inverse Photopyroelectric Detection Method
65.90 and 77.70; S1.1; S1.2
(a) Vol.121,No.2, K231-K234

77.80. Ferroelectricity and antiferroelectricity

B.E. VUGMEISTER and T.V. ANTIMIROVA
Simple Model for Ferroelectric Phase Transition in $KTa_{1-f}Nb_fO_3$
64.60 and 77.80; S10.15
(b) Vol.157,No.1, 183-188

K. BIEDRZYCKI
Thermally Stimulated Electron Emission from Triglycine Sulphate Crystals with Well-Defined Initial Macroscopic ...
77.80 and 79.40; S12
(a) Vol.117,No.1, 313-318

T. FUKAMI
Refinement of the Crystal Structure of KH_2PO_4 in the Ferroelectric Phase
61.50 and 77.80; S11.1
(a) Vol.117,No.2, K93-K96

J.M. WESSELINOWA
On the Theory of the Central Peak in CsH_2PO_4
77.80
(b) Vol.158,No.1, K95-K100

F.KH. ABDULLAEV and S.A. DARMANYAN
Dynamical Chaos of Domain Wall Motion in the Ferroelectric under Action of an Oscillating Field
77.80
(b) Vol.158,No.2, K201-K204

P.S. GOYAL, R. CHAKRAVARTHY, B.A. DASANNACHARYA, and C.J. CARLILE
Reorientations of NH^{4+} Ions in the Ferroelectric Phase of $(NH_4)_2SO_4$
63.90; 64.70; 77.80; S11.1
(a) Vol.118,No.2, 425-430

Z. ZIKMUND and J. FOUSEK
Reduction of the Number of Polarization Vectors in Multiaxial Ferroelectrics by Means of Chiral Dopants. Macroscopic...
77.30 and 77.80
(a) Vol.118,No.2, 539-552

S.S. BAWA, K. SAXENA, A.M. BIRADAR, K.K. SAINI, and S. CHANDRA
Bistable Switching and Molecular Reorientation of Ferroelectric Liquid Crystals by Current Analysis
77.30 and 77.80; S12
(a) Vol.118,No.2, 561-566

P. ROTH and E. HEGENBARTH
How Universal Are the Glassy States in Solids with a Structural Instability?
61.40; 77.70; 77.80
(b) Vol.159,No.1, K1-K6

H. KRESSE, K. JAHN, and D. DEMUS
Investigation of the Temperature Dependence of Ferroelectricity in Smectic C Liquid Crystals
77.80; S12
(a) Vol.119,No.1, K83-K86

H.L. PARK, G.C. KIM, HYUN KUK KIM, and S.W . SHIN
Study on the Temperature Dependent Dielectric Constant of the $BaTiO_3$-$Cd_5(PO_4)_3Cl$ System
77.80; S11
(a) Vol.119,No.2, K177-K180

V. LHOTSKÁ, J. FOUSEK, and N. NEUMANN
Possible Origin of Dielectric Loss Behaviour in Samples of TGS-Type Crystals
77.40; 77.80; S12
(a) Vol.120,No.1, 273-284

J.M. WESSELINOWA
The Dielectric Function and Reflectivity in KDP-Type Ferroelectrics
77.80; S11.1
(b) Vol.160,No.2, 697-704

V.M. BURLAKOV, S.A. KISELEV, V.N. PETUKHOV, and A.A. ANIKIEV
The Dynamics of the Domain Boundaries and Soft Lattice Mode in SbSI Crystals
77.80; 65.40; S9.16
(b) Vol.161,No.1, 387-394

A.A. BOGUSLAVSKII, YU.N. IVANOV, A.I. KRIEGER, A.K. MOSKALEV, V.I. PAKHOMOV, and R.SH. LOTFULLIN
NMR Investigation of Quadrupole Interaction and Chemical Shift ^{133}Cs in Cs_2HgBr_4 Crystal
76.60 and 77.80; S9.16
(b) Vol.161,No.1, K49-K54

E.V. SVIRIDOV, V.A. ALYOSHIN, YU.I. GOLOVKO, I.N. ZAKHARCHENKO, V.M. MUKHORTOV, and V.P. DUDKEVICH
Roentgenographic and Electron-Microscopic Data on Heteroepitaxial $PbTiO_3$/(001)MgO Films
68.55; 77.80; S10
(a) Vol.121,No.1, 157-162

R. CACH and Z. CZAPLA
Dielectric State Equation Parameters of Monoclinic Telluric Acid Ammonium Phosphate
77.80; 77.20; S11
(a) Vol.121,No.1, 299-304

V.M. BURLAKOV, A.A. ANIKIEV, and M.R. YAKHEEV
Evidence for Order Parameter Spatial Fluctuations in $TlGa(Se_{1-x}S_x)_2$ Crystals by Raman Spectroscopy
78.30; 77.80; S8.16
(b) Vol.161,No.2, 883-890

E.S. TSIKHOTSKII, I.N. ZAKHARCHENKO, and P.V. DUDKEVICH
Linear Defects and Ferroelectric Phase Transition in $BaTiO_3$ Crystals
61.70; 77.80; S10.15
(a) Vol.121,No.2, 391-398

R. STYRKOWIEC, W.A. SYSLO, A. WIECZOREK, and Z. CZAPLA
Splitting of the Incommensurate-Commensurate Phase Transition of NH_4HSeO_4 Crystals under Uniaxial Mechanical Stress
77.80; S11.1
(a) Vol.121,No.2, K227-K230

T. FUKAMI
Refinement of the Crystal Structure of RbH_2PO_4 in the Ferroelectric Phase
61.60 and 77.80; S11.1
(a) Vol.122,No.2, K117-K120

77.90. Other topics in dielectric properties and materials

J. D'ALBUQUERQUE E CASTRO, R. BECHARA MUNIZ, and L.E. OLIVEIRA
A Model Calculation of the Dielectric Band Structure of Diamond and Silicon
71.25 and 77.90; S5.11
(b) Vol.158,No.2, 743-752

M. FUKUHARA, A. BHALLA, and R.E. NEWNHAM
Morphotropic Phase Boundary in the $Pb(Zr_xTi_{1-x})O_3$ System
77.30 and 77.90; S10.15
(a) Vol.122,No.2, 677-682

78. Optical Properties and Condensed-Matter Spectroscopy and Other Interactions of Matter with Particles and Radiation

(for phonon spectra, see 63)

78.20. Optical properties and materials

A.A. KORNIENKO, A.A. KAMINSKII, and E.B. DUNINA
Dependence of the Line Strength of f-f Transitions on the Manifold Energy. I. Projector on the Basis Nonorthogonal ...
71.10 and 78.20
(b) Vol.157,No.1, 261-266

A.A. KORNIENKO, A.A. KAMINSKII, and E.B. DUNINA
Dependence of the Line Strength of f-f Transitions on the Manifold Energy. II. Analysis of Pr^{3+} in $KPrP_4O_{12}$
71.10 and 78.20; 71.70; S10
(b) Vol.157,No.1, 267-274

O. KELLER
Moment Expansion of the Optical Second-Harmonic Response Tensor of Condensed Media
78.20
(b) Vol.157,No.1, 459-472

P. MORETTI, P. THEVENARD, G. GODEFROY, R. SOMMERFELDT, P. HERTEL, and E. KRAETZIG
Waveguides in Barium Titanate by Helium Implantation
78.20; 61.80; S11.1
(a) Vol.117,No.1, K85-K88

P. AGHAMKAR and P.K. SEN
Parametric Excitation of Laser in Magnetoactive Doped Piezoelectric Semiconductors
78.20; S7.13
(b) Vol.157,No.2, 735-744

M. KUČERA, V.N. KOLOBANOV, V.V. MIKHAILIN, P.A. OREKHANOV, and V.N. MAKHOV
Reflection Spectra of Some Garnet and Orthoferrite Single Crystals in Vacuum Ultraviolet
78.20; 78.40; S11.2
(b) Vol.157,No.2, 745-752

G. VIELSACK and R. VON BALTZ
Dielectric Function of La_2CuO_4
71.25; 78.20; S11.1
(b) Vol.158,No.1, 249-260

D. FROEHLICH, ST. KIRCHHOFF, P. KOEHLER, and W. NIESWAND
Magnetic Field Splitting and Dispersion of the Lower 1S-Exciton Polariton in KI and RbI
71.35 and 71.70; 71.36; 78.20; S9.11
(b) Vol.158,No.1, 267-276

V.E. GUSEV
Generation of Picosecond Acoustic Pulses by Laser Radiation in Piezosemiconductors
62.30 and 78.20
(b) Vol.158,No.1, 367-382

E.I. SHREDER, I.A. POPOV, E.E. KRASOVSKII, and I.I. SASOVSKAYA
The Optical Properties and Electronic Structure of AgZn Alloy Undergoing $\beta' \rightleftharpoons \zeta$ Structural Transformation
64.70 and 78.20; 71.25; S1.2; S1.3
(a) Vol.118,No.1, 147-152

R. SCHAEFER and A. HUBERT
A New Magnetooptic Effect Related to Non-Uniform Magnetization on the Surface of a Ferromagnet
75.60; 78.20; S1.1; S1.2; S1.4; S11.2
(a) Vol.118,No.1, 271-288

A.A. KAMINSKII, B.V. MILL, and A.V. BUTASHIN
Stimulated Emission from Nd^{3+} Ions in Acentric $LaBGeO_5$ Crystals
78.20; 78.30; 78.45; 78.55; S10
(a) Vol.118,No.1, K59-K64

T.D. SHERMERGOR, A.G. FOKIN, and A.V. DIKAREV
Scattering and Velocity Dispersion of Electromagnetic Waves in Multiphase Dielectric Materials
78.20
(b) Vol.158,No.2, 715-724

I. RUECKMANN, J. KORNACK, J. KOLENDA, and M. PETRAUSKAS
Transient Energy Transfer in ps Two-Photon Absorption Experiments in CdTe at Room Temperature
78.20; S8.13
(b) Vol.158,No.2, 769-780

M.P. PETROV and P.D. SIMOVA
Textural and Electrooptical Investigations of the HOBA and OOBA Cholesteric Phases Induced by Chiral Dopants
61.40; 78.20; S12
(a) Vol.118,No.2, 357-364

V.N. BRUDNYI, S.I. BORISENKO, and A.I. POTAPOV
Electrical and Optical Properties and Fermi Level Pinning in Electron Irradiated $ZnSnAs_2$
72.20 and 78.20; 71.55; S7.16
(a) Vol.118,No.2, 505-512

D.S. CHEMLA, S. SCHMITT-RINK, W.H. KNOX, D.A.B. MILLER, K.W. GOOSSEN, and G. HASNAIN
Ultrafast Nonlinear Optical Effects in Biased Semiconductor Quantum Wells
71.35 and 78.20; 73.40; S7.12
(b) Vol.159,No.1, 11-28

I. ABRAM
Photon Echoes and the Structure of Wannier Excitons
71.35 and 78.20; S7.12; S7.15
(b) Vol.159,No.1, 29-34

M. HARTMANN, H. STOLZ, and R. ZIMMERMANN
Kinetics of Screening in Optically Excited Semiconductors
71.25 and 72.10; 78.20
(b) Vol.159,No.1, 35-42

W. RUDOLPH, J. PULS, F. HENNEBERGER, and D. LAP
Femtosecond Studies of Transient Nonlinearities in Wide-Gap II-VI Semiconductor Compounds
71.35 and 78.20; S8.11; S8.12; S8.15
(b) Vol.159,No.1, 49-58

W. SCHAEFER and K. HENNEBERGER
Pulse Propagation and Carrier Kinetics in Laser Excited Semiconductors
72.10 and 78.20; S7.12
(b) Vol.159,No.1, 59-70

N.A. GIPPIUS, A.L. IVANOV, S.G. TIKHODEEV, and S.G. VYGOVSKII
Phonoriton Transient Phenomena
78.20; 71.35; 71.36; S8.11
(b) Vol.159,No.1, 71-80

F. JAHNKE and K. HENNEBERGER
Nonequilibrium Description of the Intense Resonant Laser Excitation of a Semiconductor
71.35 and 72.30; 78.20; S8.11
(b) Vol.159,No.1, 81-90

I.V. BELOUSSOV and YU.M. SHVERA
Kinetics of Partially Coherent Polaritons in Semiconductors. Schwinger-Dyson Equations
71.36 and 78.20
(b) Vol.159,No.1, 91-100

M.J.M. GOMES, B. KIPPELEN, R. LEVY, J.B. GRUN, and B. HOENERLAGE
Time-Resolved Four-Wave Mixing Experiments in CuCl
71.35 and 78.20; S9
(b) Vol.159,No.1, 101-106

A.V. NURMIKKO and R.L. GUNSHOR
Excitons and Phonons in CdTe/MnTe and ZnTe/MnTe Single Quantum Wells. Aspects of Confinement and Optical Phonon Coupling
71.35 and 78.20; 73.40; S8.13
(b) Vol.159,No.1, 107-116

C. ELL and H. HAUG
Absorption and Optical Gain Spectra and Band Gap Renormalization of Highly Excited Quantum Well Systems
71.25 and 78.20; 73.40; S7.11; S7.15
(b) Vol.159,No.1, 117-124

Y. MASUMOTO, S. IDESHITA, and T. WAMURA
Nonlinear Optical Study of Semiconductor Microstructures, CuCl Quantum Dots, and GaAs Quantum Wells
71.35 and 78.20; 73.40; S7.12; S9
(b) Vol.159,No.1, 133-142

A.I. BOBRYSHEVA and S.S. RUSSU
Optical Properties of Biexcitons in ZnSe/$Zn_{1-x}Mn_xSe$ Quantum Wells
71.35 and 78.20; 73.40; S8.12; S8.15
(b) Vol.159,No.1, 155-164

K.-H. SCHLAAD, CH. WEBER, D.S. CHEMLA, J. CUNNINGHAM, D.A.B. MILLER, C.V. HOOF, G. BORGHS, G. WEIMANN, W. SCHLAPP et al.
Nonlinear Optical Properties of GaAs/(AlGa)As Multiple Quantum Well Structures under Quasistationary High Excitation ...
71.35 and 78.20; 73.40; S7.12; S7.15
(b) Vol.159,No.1, 173-180

J.L. OUDAR, B. SFEZ, R. KUSZELEWICZ, J.C. MICHEL, and R. AZOULAY
Optical Nonlinearities of GaAs-Based Epitaxial Structures for All-Optical Switching
78.20; 68.55; S7.12; S7.15
(b) Vol.159,No.1, 181-190

V.A. KOCHELAP and A.V. KUZNETSOV
Self-Supporting Acoustoelectronic Pulses Driven by Optical Excitation of a Semiconductor in the Transparency Region
78.20
(b) Vol.159,No.1, 191-196

I.L. ALEINER, L.M. CHISTYAKOVA, S.I. KOKHANOVSKII, R.P. SEISYAN, A.A. TOROPOV, and T.V. SHUBINA
Passive and Active Q-Switching in an AlGaAs/GaAs SQW Laser. Giant Stark Shift in the Laser p-n SQW Heterostructure
71.35 and 78.20; S7.12; S7.15
(b) Vol.159,No.1, 197-204

J. HOLLANDT, J. GUTOWSKI, and I. BROSER
A Detailed Study Concerning the Main Aspects Influencing Thermally Induced Optical Bistability of CdS
78.20; S8.11
(b) Vol.159,No.1, 205-212

G. JUNGK
Bistability in Composites of Linear and Nonlinear Media
78.20
(b) Vol.159,No.1, 213-222

A.V. KUZNETSOV
Kinks in Resonatorless Optical Bistability. Stability Limits and Formation Dynamics
78.20
(b) Vol.159,No.1, 223-228

F. ŠROBÁR
Absorptive Optical Bistability. The Case of Fabry-Perot
78.20 and 78.90
(b) Vol.159,No.1, 229-234

M. MAREYEN, F.-J. SCHUETTE, and R. TIEBEL
Dynamics of Carrier-Screened Photon-Exciton Interaction
71.35 and 78.20
(b) Vol.159,No.1, 235-240

V.A. STADNIK
Multistability of Absorption Domains in Bulk CdS Due to Band Edge Nonlinearities
71.35 and 78.20; S8.11
(b) Vol.159,No.1, 241-248

Y.Z. HU, S.W. KOCH, M. LINDBERG, and N. PEYGHAMBARIAN
Theoretical and Experimental Results on Coulomb Effects in Semiconductor Quantum Dots
71.35 and 78.20; S8.12
(b) Vol.159,No.1, 249-258

P. GILLIOT, B. HOENERLAGE, R. LEVY, and J.B. GRUN
Nonlinearities of CuCl Microcrystals
71.35 and 78.20; S9
(b) Vol.159,No.1, 259-266

D. RICARD, P. ROUSSIGNOL, F. HACHE, and CH. FLYTZANIS
Nonlinear Optical Properties of Quantum Confined Semiconductor Microcrystallites
78.20; S8.15
(b) Vol.159,No.1, 275-286

H. ROSSMANN, A. SCHUELZGEN, F. HENNEBERGER, and M. MUELLER
Quantum Confined DC Stark Effect in Microcrystallites
71.35 and 78.20; S8.11
(b) Vol.159,No.1, 287-290

TH. PESCHEL
Possibilities for Intrinsic Optical Bistability in Semiconductor-Doped Glasses
71.35 and 78.20
(b) Vol.159,No.1, 291-296

D. FROEHLICH, CH. NEUMANN, B. UEBBING, and R. WILLE
Experimental Investigation of Three-Level Optical Stark Effect in Semiconductors
71.35 and 78.20; 73.40; S7.1; S10.1
(b) Vol.159,No.1, 297-308

H. HAUG, L. BANYAI, J. LIEBLER, and T. WICHT
The Pulsed Nonresonant Optical Stark Effect and the Urbach Tail in Semiconductors
71.35 and 78.20; S7.12; S8.11; S9
(b) Vol.159,No.1, 309-316

R. ZIMMERMANN
Transverse Relaxation and Polarization Specifics in the Dynamical Stark Effect
71.35 and 78.20; S8.11
(b) Vol.159,No.1, 317-326

A. STAHL
Coupled Two-Level Systems and the Dynamics of Semiconductor Electrons
71.25 and 78.20; 71.35
(b) Vol.159,No.1, 327-338

I. BALSLEV
Induced Absorption and the Dynamical Stark Shift Associated with Biexcitons
71.35 and 78.20; S8.12; S9
(b) Vol.159,No.1, 339-346

M. KUWATA-GONOKAMI, R. SHIMANO, J. IWAMATSU, H. AKIYAMA, T. KUGA, and M. MATSUOKA
Optical Nonlinearity and Dynamics of Exciton-Biexciton System in CuCl in the Weak Excitation Regime
71.35 and 78.20; S9
(b) Vol.159,No.1, 347-352

CH. SPIEGELBERG, J. PULS, and F. HENNEBERGER
Nonlinear Optical Properties of CdS at Room Temperature
71.35 and 78.20; S8.11
(b) Vol.159,No.1, 353-362

T. HOENIG, J. GUTOWSKI, and I. BROSER
Carrier-Induced Pulse Narrowing in Indium-Doped CdS
71.55 and 78.20; 71.36; S8.11
(b) Vol.159,No.1, 363-370

T. ISHIHARA
The Origin of Third-Order Optical Nonlinearity in Exciton Systems
71.35 and 78.20; 78.65; S9
(b) Vol.159,No.1, 371-378

V.P. GRIBKOVSKII, V.A. ZYULKOV, A.E. KAZACHENKO, and V.A. SHAKIN
Second Harmonic Generation on the Grating of Optical Inhomogeneities in ZnSe and ZnS
78.20; S8.11; S8.12
(b) Vol.159,No.1, 379-386

V.D. EGOROV, HOANG XUAN NGUYEN, R. ZIMMERMANN, V.S. DNEPROVSKII, M. KASCHKE, and D.S. KHECHINASHVILI
Mott Transition in CdS Revealed by Picosecond Nonlinear Absorption Measurements
71.35 and 78.20; S8.11
(b) Vol.159,No.1, 403-410

A.L. IVANOV and V.V. PANASHCHENKO
The Exciton-Biexciton Spectrum Renormalization in the Presence of a Polariton Pumping Wave. The New Mechanism, Exciton ...
71.35 and 78.20; 71.36; S8.11
(b) Vol.159,No.1, 417-424

K.-H. PANTKE, H. OVER, and I. BROSER
Dissipation of Excitonic Polaritons in CdS
71.36 and 78.20; S8.11
(b) Vol.159,No.1, 437-442

L.G. ZIMIN, S.V. GAPONENKO, I.E. MALINOVSKII, V.YU. LEBED, P.I. KUZNETSOV, G.G. YAKUSHCHEVA, and A.V. KUZNETSOV
Excitonic Optical Nonlinearity in II-VI MOCVD-Grown Crystals
71.35 and 78.20; S8.1
(b) Vol.159,No.1, 449-456

A. FORCHEL, B.E. MAILE, H. LEIER, G. MAYER, and R. GERMANN
Optical Spectroscopy on One-Dimensional Compound Semiconductor Structures
71.35 and 78.20; S7.12; S7.15
(b) Vol.159,No.1, 457-470

V.A. GANSHIN and YU.N. KORKISHKO
Proton Exchange in Lithium Niobate and Lithium Tantalate Single Crystals: Regularities and Specific Features (Review Article)
61.40 and 68.48; 78.20; S11.1
(a) Vol.119,No.1, 11-26

R. GOLDHAHN, H. ROEPPISCHER, M. GERICKE, and C.-E. RICHTER
Characterization of $Al_xGa_{1-x}As$/GaAs Heterostructures by Electroreflectance Depth Profiling
73.40 and 78.20; S7.15
(a) Vol.119,No.1, 183-190

V.E. ZUBOV, G.S. KRINCHIK, A.S. TABLIN, and A.A. KOSTYURIN
Magnetooptical Techniques in Self-Consistent Determination of Optical and Magnetooptical Constants of Magnetics
78.20; S1.2; S11.2
(a) Vol.119,No.1, 297-306

K.M. MUKIMOV, B.YU. SOKOLOV, and U.V. VALIEV
The Faraday Effect of Rare-Earth Ions in Garnets
78.20; 75.30; S11.2
(a) Vol.119,No.1, 307-316

J.TH. ZETTLER, TH. DITTRICH, and L. SCHROTTKE
Modulation Ellipsometry - a New Technique for the Characterization of Semiconductor Materials and Complex ...
78.20; S7.12
(a) Vol.119,No.1, K91-K96

K.A. KIKOIN, G.P. NIZHNIKOVA, and O.V. FARBEROVICH
Optical Properties of the Mixed-Valent TmSe
71.25; 71.50; 78.20; S1.4; S8
(b) Vol.159,No.2, 743-754

A. BECKMANN, K. DETTMER, and F.R. KESSLER
Determination of the Indirect Exciton Energy Gap of GaP Using Thermomodulation in the Temperature Region 80 to 350 K
71.35; 78.20; S7.11
(b) Vol.159,No.2, 765-770

A.V. MALAKHOVSKII
Magnetooptical Activity of Allowed Electron Transitions in Atoms in Free State and in Crystals
78.20
(b) Vol.159,No.2, 883-892

S.A. OSTANIN, N.A. SHILKOVA, and V.P. SHIROKOVSKII
Optical Absorption in Heavy Transition B.C.C. Metals
78.20; 71.10; S4
(b) Vol.159,No.2, 893-902

A.M. ELKORASHY
Optical Constants of Tin Sulphide Single Crystals Measured by Interference Method
78.20; S8
(b) Vol.159,No.2, 903-916

V.N. BOGDANOV, I.M. BROVCHENKO, L.V. MAKSIMOV, A.R. SILIN, and O.V. YANUSH
Spectroscopic Optical and Acoustic Investigations of Radiation Modified Silica Glasses
61.40 and 78.20; 61.80; 62.65; S10.1
(a) Vol.119,No.2, 621-630

J. MUSILOVÁ
Method of the Exact Determination of Transition Energies from Modulated Piezoreflectance Spectra of Cubic Crystals
78.20; 71.20; S5.12
(a) Vol.119,No.2, 669-676

G.C. BHAR, S. DAS, and P.K. DATTA
Efficient Frequency Doubling of Nd Laser Radiation
78.20; S11.1
(a) Vol.119,No.2, K173-K176

G. BACQUET, J. FRANDON, J. BANDET, F. FABRE, R. TAOUINT, C. FONTAINE, and A. MUNOZ-YAGUE
Optical Orientation in GaAs Grown on Si
78.20; S7.12
(a) Vol.119,No.2, K181-K184

A.R. HASSAN
Three-Photon Interband Process in Solid
71.25; 78.20
(b) Vol.160,No.1, 185-192

U. WENSCHUH and E. HEINER
Optical Properties of Gold in the Nonequilibrium Thermalized Stage
71.10; 65.40; 71.38; 78.20; S1.3
(b) Vol.160,No.1, 225-236

T.A. KUDYKINA
Boundary Conditions in Case of Electromagnetic Wave Absorption
78.20; 71.35
(b) Vol.160,No.1, 365-374

U. WOGGON, M. MUELLER, I. RUECKMANN, J. KOLENDA, and M. PETRAUSKAS
Influence of Hydrogenation of Defects at the Microcrystallite-Glass Interface on the Nonlinear Optics ...
61.70 and 78.20; S8.12
(b) Vol.160,No.1, K79-K84

A.K. ABASS, Z.A. AHMED, and R.M. SAMUEL
Optical Properties of Chemically Deposited FeS_2 Coatings
78.20; S8
(a) Vol.120,No.1, 247-252

J. MANGIN, A. KHODJAOUI, and G. MARNIER
Optical Absorption of KTP Single Crystals
78.20; 78.40; S11.1
(a) Vol.120,No.1, K111-K116

I.V. STASYUK, S.S. KOTSUR, and V.P. TUPYCHAK
On the Theory of the Faraday Effect in Dielectric Ionic Crystals and Crystals with Structural Phase Transition of ...
78.20; S11.1
(b) Vol.160,No.2, 683-696

N.V. EDNERAL, V.G. KUTILIN, YU.A. LISOVSKII, M.L. SOKOLOVA, A.M. UMPELEV, A.O. TITOV, and L.K. FIONOVA
Thin Film Structures for Magneto-Optical Information Recording
68.55; 75.70; 78.20; S1.1
(a) Vol.120,No.2, 457-466

E.L. IVCHENKO, V.P. KOCHERESHKO, I.N. URALTSEV, and D.R. YAKOVLEV
Reflectivity and Photoreflectivity in Superlattices and Quantum Wells
73.40 and 78.20; 71.35; S7.12; S7.15
(b) Vol.161,No.1, 217-222

V.A. KOSOBUKIN
Localization and Resonance Transmission of Light Waves in a One-Dimensionally Disordered Dielectric. Polarization Effects
78.20; 77.30
(b) Vol.161,No.1, 405-418

S.A. STUDENIKIN
Dispersion Properties of Indium Antimonide under a Strong Magnetic Field
78.20; S7.13
(b) Vol.161,No.1, K67-K70

T. IKARI, H. YOKOYAMA, K. MAEDA, and K. FUTAGAMI
Near Band Edge Photoacoustic Spectra of GaAs below Room Temperature
78.20; 71.50; S7.12
(a) Vol.121,No.1, K125-K128

TS.S. VELINOV and K.P. BRANSALOV
Interference Analysis of Photothermal Generation of Waves in Layered Structures
62.30 and 68.20; 78.20
(a) Vol.121,No.2, 555-562

K.H. HERRMANN, K.-P. MOELLMANN, and J.W. TOMM
On the Broadening Mechanisms near the E_0 Transition in Narrow-Gap (Hg,Cd)Te
78.20; 71.25; 78.55; S8.15
(a) Vol.121,No.2, 635-640

P. MALY, A. SVOBODA, and P. BUSA
Picosecond Transient Concentration Grating in Silicon
78.20; S5.11
(a) Vol.121,No.2, K223-K226

H.H. WEBER
A Shorter Wavelength Mode Shift of Ridge-Waveguided 1.3 µm Lasers
78.45; 78.20; S7.15
(a) Vol.121,No.2, K239-K242

P. GÓRSKI and W. KUCHARCZYK
The Fourth-Order Electrooptic Effect in KDP and ADP Crystals
78.20; S11.1
(a) Vol.121,No.2, K243-K246

I.T. BODNAR and I.V. BODNAR
Dispersion and Temperature Dependence of Refractive Indicies of Some $A^{I}B^{III}C_2^{VI}$ Compounds
78.20; S8.16
(a) Vol.121,No.2, K247-K250

U. WENSCHUH and E. HEINER
Nonequilibrium Thermoreflectance Evolution of Laser Irradiated Noble Metals
78.20; 65.40; 71.10; S1.3
(b) Vol.162,No.1, 303-310

S. DEFREGGER, G.F. ENGEL, and P.W. KREMPL
The Temperature Dependence of the Optical Second Harmonic Generation Coefficient d_{11} of Quartz Homeotypic $AlPO_4$
78.20; 64.60; S11.1
(b) Vol.162,No.1, 311-320

V.I. GAVRILENKO, A.V. POSTNIKOV, N.I. KLYUI, and V.G. LITOVCHENKO
Energy Band Structure and Optical Properties of Wurtzite-Structure of Silicon Carbide Crystals
71.25 and 78.20; S6
(b) Vol.162,No.2, 477-488

M. QUINTERO, R. TOVAR, C. BELLABARBA, and J.C. WOOLLEY
Temperature Variation of Optical Energy Gap and Deformation Potentials in $AgInTe_2$
71.25; 78.20; S8
(b) Vol.162,No.2, 517-522

E. LOPEZ-CRUZ
A Comment on the Temperature Dependence of the Band Gap of FeS_2
78.20; S8
(b) Vol.162,No.2, K131-K134

A. KARA, J.J. COUDERC, C. LEVADE, and G. VANDERSCHAEVE
Memory Photoelastic Effects in ZnS Sphalerite
62.20 and 78.20; S8.11
(a) Vol.122,No.2, 545-554

F.R. KESSLER, W. KOEPPEN, and G. KRAUSE
External Stabilization of CO_2 Laser Radiation by Means of i-Semiconductor Absorption
78.20 and 85; 78.30; S5.12; S7.13
(a) Vol.122,No.2, 731-740

78.30. Infrared and Raman spectra and scattering

M.H. BROOKER
A Search for a Phase Transition in $Sr(NO_3)_2$ by Raman Spectroscopy. Evidence for Partial Disorder
64.60 and 78.30; 63.20; 63.50; S11.1
(b) Vol.157,No.1, 167-182

J.P. SRIVASTAVA, ASITA KULSHRESHTHA, and H. RAUH
FIR Reflectivity and Vibrational Spectra of $(CH_3NH_3)_2ZnCl_4$ Crystals
78.30; 63.20; S12
(a) Vol.117,No.1, K77-K80

S. MOCHIZUKI, K. ISHI, and A. JOHGO
Far-Infrared Absorption in Nickel and Copper Microcrystals
78.30; S1.2
(b) Vol.157,No.2, K137-K140

M.N. KHAN
A Study of the Effects of CuO on the Structure and Optical Absorption of Tellurite Glasses
78.30; 61.40; S10.15
(a) Vol.117,No.2, 593-600

J.L. LINDSTROEM, B.G. SVENSSON, W.-X. NI, and M. WILLANDER
Infrared Absorption of Oxygen-Doped Silicon Grown by Molecular Beam Epitaxy
78.30; S5.11
(a) Vol.117,No.2, K171-K176

E. PIGORSCH and W.E. STEGER
Identification of LiO Bands in the Infrared Spectra of the Insertion Compound delta-LiV_2O_5
61.70 and 78.30; S10.1
(a) Vol.117,No.2, K189-K192

R.M. SARDARLY, R.A. ALIEV, E.M. KERIMOVA, N.A. BAKHYSHOV, and I.SH. SADYKHOV
Rearrangement Features of the Vibrational Spectra of $TlInS_{2x}Se_{2(1-x)}$ and $TlGa_xIn_{1-x}Se_2$ Solid Solutions
63.20 and 78.30; 64.75; S8.16
(b) Vol.158,No.1, 73-80

M. SIESKIND, D. AYACHOUR, J.-C. MERLE, and J.-C. BOULOU
Crystal Growth, Raman and Infrared Spectra of BaFI Single Crystals
63.20 and 78.30; S9.16
(b) Vol.158,No.1, 103-112

N.M. GASANLY, S.A. EL-HAKIM, L.G. GASANOVA, and A.Z. MAGOMEDOV
Lattice Vibrations in $CuIn_5S_8$ Crystals
63.20 and 78.30; S8.16
(b) Vol.158,No.1, K1-K4

N.M. GASANLY, A.G. GUSEINOV, E.A. ASLANOV, and S.A. EL-HAMID
Infrared Reflection Spectra of $Cu_3B_5^{III}C_9^{VI}$ Single Crystals
78.30; 63.20; S8.16
(b) Vol.158,No.1, K85-K88

B. GILLOT and A. ROUSSET
On the Limit of Aluminium Substitution in Fe_3O_4 and gamma-Fe_2O_3
61.60; 78.30; S10.15
(a) Vol.118,No.1, K5-K8

A.A. KOLB, I.G. MEGELA, A.P. BUTURLAKIN, and D.B. GOYER
Low-Temperature Annealing of Radiation Defects in Electron-Irradiated Gallium Phosphide
61.80 and 78.30; S7.11
(a) Vol.118,No.1, K9-K12

O. FOLTIN
Simulation of the Oblique Incidence Specular Reflectivity of Uniaxial Semiconductors at Plasma Resonance
78.30; S8
(a) Vol.118,No.1, K43-K46

HONG WANG, JINKE WEN, BIN LI, and HUAFU WANG
Infrared Absorption Study of OH^- in $LiNbO_3$:Mg and $LiNbO_3$:Mg:Fe Crystals
61.70 and 78.30; S11.1
(a) Vol.118,No.1, K47-K50

GUANG-XU CHENG, HUA XIA, KUN-JI CHEN, WEI ZHANG, and XING-KUI ZHANG
Raman Measurement of the Grain Size for Silicon Crystallites
78.30; S5.11
(a) Vol.118,No.1, K51-K54

A.A. KAMINSKII, B.V. MILL, and A.V. BUTASHIN
Stimulated Emission from Nd^{3+} Ions in Acentric $LaBGeO_5$ Crystals
78.20; 78.30; 78.45; 78.55; S10
(a) Vol.118,No.1, K59-K64

I.V. BODNAR, G.F. SMIRNOVA, A.G. KAROZA, and A.P. CHERNYAKOVA
Vibrational Spectra of $CuGaS_2$ and $CuGaSe_2$ Compounds and $CuGaS_{2x}Se_{2(1-x)}$ Solid Solutions
63.20; 64.75; 78.30; S8.16
(b) Vol.158,No.2, 469-474

C. JULIEN, A. CHEVY, and D. SIAPKAS
Optical Properties of In_2Se_3 Phases
78.50; 78.30; S8
(a) Vol.118,No.2, 553-560

E.N. IZAKOVICH, N.A. ZHELTOVA, YU.M. SHULGA, I.N. IVLEVA, ZH.D. SOKOLOVSKAYA, A.V. BALUEV, V.V. GANIN, O.S. ROSHUPKINA et al.
Erratum to Study of Ba and Cu(111) Based Inorganic Oxide Compounds (phys. stat. sol. (a) 115, K195 (1989))
74.70; 78.30; 79.20; 79.60
(a) Vol.118,No.2, K139-K140

S.E. SARKISOV, A.A. KAMINSKII, and A.V. BUTASHIN
Manifestation of Structural Disorder Peculiarities of $Ca_3Ga_2Ge_3O_{12}$-$Ca_3(Nb/Ga)_2Ga_3O_{12}$ and BaF_2-YF_3 Crystalline ...
78.30; S11.2
(a) Vol.119,No.1, 285-296

S. VES
High Pressure Raman Study of the Chain Chalcogenide $TlInTe_2$
63.20 and 78.30; 64.60; S8.16
(b) Vol.159,No.2, 699-706

V.M. FOMIN and E.P. POKATILOV
Optical Properties of Multi-Layer Structures. III. Raman Scattering in Multi-Layer Structures and Superlattices
68.30 and 78.30
(b) Vol.159,No.2, 707-720

U. TESCHNER and K. HUEBNER
IR-Spectroscopic Data of Thin Insulating Films on Semiconductors. New Methods of Interpretation and Analysis
78.30 and 78.65; S10.1
(b) Vol.159,No.2, 917-926

I.KH. AKOPYAN, A.A. KLOCHIKHIN, B.V. NOVIKOV, M.YA. VALAKH, A.P. LITVINCHUK, and I. KOSAZKII
Optical Spectroscopy of the Superionic Crystals (Review Article)
78.30 and 78.40; 64.70; 71.25; 71.35; 78.55; S9
(a) Vol.119,No.2, 363-416

S.E. SARKISOV, A.A. KAMINSKII, A.V. BUTASHIN, YU.E. PERLIN, V.N. ENAKII, and M.G. BLAZHA
Selective Spectroscopy, Luminescence Kinetics of Pr^{3+} Ions, and Phonon Density Calculation in Multicentred and ...
78.55; 63.20; 71.55; 78.30; S11.2
(a) Vol.119,No.2, 607-620

A. LEHMANN, W. HEERDEGEN, G. SCHIRMER, H. MUTSCHKE, W. RICHTER, E. HACKER, and R. DOHLE
Structure Correlated Infrared Properties of Fluoride Films
78.30 and 78.65; S9.12
(a) Vol.119,No.2, 683-688

P. MIKULSKI, K. FABISIAK, S. ORZESZKO, and F. ROZPŁOCH
The Effect of Starting Material on the Properties of Soft a-C:H Films Obtained by the Glow Discharge Technique
68.55; 78.30; S5
(a) Vol.119,No.2, K127-K130

M. KOLESÍK, D. TUNEGA, and B.P. SOBOLEV
A Study of the Disorder in Heavily Doped $Ba_{1-x}La_xF_{2+x}$ by Raman Scattering
78.30; S9.15
(b) Vol.160,No.1, 375-380

A.V. EGRANOV, V.V. OTROSHOK, and B.P. CHERNYAGO
Infrared Absorption Due to U-Centres in Highly Doped $LiF:H^-$ Crystals
61.70 and 78.30; S9.11
(b) Vol.160,No.1, K75-K78

L. KOVÁCS, K. POLGÁR, R. CAPELLETTI, and C. MORA
Diffusion of Hydrogen Isotopes in Pure and Mg-Doped $LiNbO_3$ Crystals
66.30; 78.30; S10.15
(a) Vol.120,No.1, 97-104

B.D. FAINBERG
Real Time Optical Spectroscopy of Superconducting-Gap Excitations
74.90 and 78.30
(b) Vol.160,No.2, K169-K172

V.V. ARTAMONOV, V.S. LYSENKO, A.N. NAZAROV, B.D. NECHIPORUK, V.V. STRELCHUK, and M.YA. VALAKH
Radio-Frequency Plasma Treatment and Thermal Annealing in Implanted Si. Raman Study
61.80 and 68.55; 78.30; S5.11
(a) Vol.120,No.2, 475-484

U. BECKER, P. RUDOLPH, R. BOYN, M. WIENECKE, and I. UTKE
Characterization of p-Type CdTe Bridgman Crystals by Infrared Extinction Spectra
64.75 and 78.30; S8.13
(a) Vol.120,No.2, 653-660

O. SILD
The Model of Rotation of NO_2^- Impurity Molecule in KCl Crystal
61.70; 78.30; 78.55; S9.11
(b) Vol.161,No.1, 105-112

T. DUMELOW, A.R. EL GOHARY, K.A. MASLIN, T.J. PARKER, D.R. TILLEY, and S.N. ERSHOV
Analysis of Far Infrared Spectra Showing Bulk and Surface Phonon-Polaritons in CdTe Epilayers on GaAs Substrates
78.30 and 78.65; 71.36; S8.13
(b) Vol.161,No.1, 233-244

YU.P. GNATENKO, YU.I. ZHIRKO, and Z.D. KOVALYUK
Optical Properties of GaSe Crystals Containing Mn Impurity Atoms. I. Exciton-Phonon Interaction
71.35 and 78.30; S8
(b) Vol.161,No.1, 419-426

H.-B. ABEL
Low Frequency SERS from Ag Colloids in Alkali Halide Crystals
78.30; 61.70; S9.11
(b) Vol.161,No.1, 435-446

G.A. ADEGBOYEGA and A. POGGI
Intrinsic Gettering of Cr Impurities in p-Type Cz Silicon
71.55; 72.80; 78.30; S5.11
(a) Vol.121,No.1, 181-186

E. ANACHKOVA, M. GOSPODINOV, A. NIKOLOV, P. SVESTAROV, N. PETKOV, YU. MARKOV, M. LIMONOV, and G. BRUCHMAN
Fundamental Vibrations in $Pb_5(GeO_4)(VO_4)_2$ Crystals
63.20; 78.30; S10.15
(b) Vol.161,No.2, 575-582

F.M. GASHIMZADE, T.G. ISMAILOV, and R.S. NADIRZADE
Interband Resonance Raman Scattering in p-Type Semimagnetic Semiconductors in Quantizing Magnetic Field
78.30; S8.15
(b) Vol.161,No.2, 861-868

V.M. BURLAKOV, A.A. ANIKIEV, and M.R. YAKHEEV
Evidence for Order Parameter Spatial Fluctuations in $TlGa(Se_{1-x}S_x)_2$ Crystals by Raman Spectroscopy
78.30; 77.80; S8.16
(b) Vol.161,No.2, 883-890

R. ASHOKAN, V. GOPAL, and K.C. CHHABRA
IR Absorption Spectra of SiO_2 Films Grown by Photo-CVD
68.55 and 78.30; S10.1
(a) Vol.121,No.2, 533-538

U. TESCHNER
Infrared Optical Properties of Amorphous Silicon Oxynitride and Their Theoretical Interpretation
78.30 and 78.65; S6
(a) Vol.121,No.2, 641-656

C. BENOIT, M. ROLLAND, and M. PALPACUER
Infrared Properties of Polyacetylene. Experiments and Theory (Review Article)
78.30; S12
(b) Vol.162,No.1, 9-60

H.L. PARK, S.Y. GEE, S.K. CHANG, S. PARK, S.K. LEE, J.Y. LEE, C.S. YOON, J.E. KIM, H.Y. PARK, W.T. KIM, and H. LIM
Solid Solubility of Selenium in $AgGaSe_2$
64.75; 78.30; S8.16
(b) Vol.162,No.1, K9-K14

A. LEHMANN, E. BROESE, and B. SCHROETER
Vibrational Modes of Ge_xSe_{1-x} Films
63.20 and 78.30; S5
(b) Vol.162,No.1, K53-K58

S.I. RADAUTSAN, I.M. TIGINYANU, V.V. URSAKII, V.M. FOMIN, and E.P. POKATILOV
The Peculiarities of the Temperature Broadening of Raman Light Scattering Lines in $Zn(Cd)Ga_2Se_4$ Single Crystals
78.30; S8.16
(b) Vol.162,No.1, K63-K66

B.D. FAINBERG
Erratum to Real Time Optical Spectroscopy of Superconducting Gap Excitations (phys. stat. sol. (b) 160, K169 (1990))
74.90 and 78.30
(b) Vol.162,No.1, K67-K68

P. DUBČEK, B. ETLINGER, K. FURIĆ, and M. KRANJČEC
Raman Spectra of $(Ga_xIn_{1-x})_2Se_3$
78.30; S8.16
(a) Vol.122,No.1, K87-K90

V.I. TYUTYUNNIK and O.I. TYUTYUNNIK
Phonon Structure of Raman Scattering Spectra of LiH Crystals
63.20 and 78.30; S9.11
(b) Vol.162,No.2, 597-604

F.R. KESSLER, W. KOEPPEN, and G. KRAUSE
External Stabilization of CO_2 Laser Radiation by Means of i-Semiconductor Absorption
78.20 and 85; 78.30; S5.12; S7.13
(a) Vol.122,No.2, 731-740

A.N. VEIS, D.D. KODITSA, and D.V. GITSU
Optical Properties of p-$TlSbTe_2$ Single Crystals
78.30; S8.16
(a) Vol.122,No.2, K177-K182

78.35. Brillouin and Rayleigh scattering

S.A. LEE and S.M. LINDSAY
Surface Brillouin Scattering from Graphite
68.30 and 78.35; S5
(b) Vol.157,No.2, K83-K86

F. MICHARD and A. ZAREMBOWITCH
Changes in the Propagation of Ultrasonic Waves in Silica Aerogels
62.65; 78.35; S10
(a) Vol.118,No.2, 461-466

U. VAN OLFEN
Light-Induced Transparency in Absorbing Powders
78.35; 72.40; S11.1
(a) Vol.121,No.1, K121-K124

G.I. BERSUKER
The Origin of the Dynamical Central Peak in Brillouin Scattering
78.35
(b) Vol.162,No.2, K123-K126

78.40. Visible and ultraviolet spectra

YA.O. DOVGYI and I.V. KITYK
Antiresonances in the in the Fundamental Spectra of CdI_2-4H and $CdBr_2$-6R Crystals
71.35 and 78.40; S9
(b) Vol.157,No.1, 473-480

M. KUČERA, V.N. KOLOBANOV, V.V. MIKHAILIN, P.A. OREKHANOV, and V.N. MAKHOV
Reflection Spectra of Some Garnet and Orthoferrite Single Crystals in Vacuum Ultraviolet
78.20; 78.40; S11.2
(b) Vol.157,No.2, 745-752

R. LÉVY, M.J.M. GOMES, B. KIPPELEN, and B. HOENERLAGE
Coherent Signal Generation in CuCl by Light-Induced Grating and Induced Biexciton Decay
78.40; 71.35; S9
(b) Vol.158,No.1, 391-396

V. HIZHNYAKOV
Stimulated Echo and Vibrational Dynamics
78.40; 78.45; 78.50; S9.11
(b) Vol.158,No.2, 725-736

K. NITSCH, M. NIKL, Č. BARTA, D. SCHULTZE, A. TŘÍSKA, and R. UECKER
Preparation and Emission Properties of $NaBi(WO_4)_2$ and $NaBi(WO_4)_2$:Ce Single Crystals
78.40 and 78.55; S11.1
(a) Vol.118,No.2, K133-K138

I.KH. AKOPYAN, A.A. KLOCHIKHIN, B.V. NOVIKOV, M.YA. VALAKH, A.P. LITVINCHUK, and I. KOSAZKII
Optical Spectroscopy of the Superionic Crystals (Review Article)
78.30 and 78.40; 64.70; 71.25; 71.35; 78.55; S9
(a) Vol.119,No.2, 363-416

V.V. HARUTUNIAN, V.A. GEVORKIAN, N.E. GRIGORIAN, and G.N. ERITSIAN
Study of Optical Properties of alpha-Al_2O_3 in the Vacuum Ultraviolet Region of Synchrotron Radiation
78.40; S10.1
(b) Vol.160,No.1, K71-K74

J. MANGIN, A. KHODJAOUI, and G. MARNIER
Optical Absorption of KTP Single Crystals
78.20; 78.40; S11.1
(a) Vol.120,No.1, K111-K116

S. JIANG, S.C. SHEN, and W. GIRIAT
Absorption Edge and New Absorption Peaks of $Cd_{1-x}Mn_xS$ under Hydrostatic Pressure
78.40; S8.15
(b) Vol.161,No.1, K71-K76

V.I. BLYNSKII, A.YU. KULIKOV, S.YU. RAKHLEY, and V.I. OSINSKII
Photoelectric Crosstalk between Silicon Photoelements
72.40; 78.40; S5.11
(a) Vol.121,No.1, 227-234

A.A. O'CONNEL-BRONIN and A.B. TRESHCHALOV
Exciton Reflection Spectra of NaI Crystals. Effect of Surface Irradiation by Photons and Incidence Angle Dependence
71.35 and 78.40; 73.20; S9.11
(b) Vol.162,No.1, 289-296

K.S. CHOI, B.J. KOO, M. JUNG, H.J. SHIN, J.H. LEE, S.B. KIM, S.K. CHANG, and H.L. PARK
(100) CdTe Thin Film Growth by Temperature Gradient Vapor Transport (TGVT)
68.55; 78.40; 78.55; S8.13
(a) Vol.122,No.2, K135-K138

78.45. Stimulated emission

A.A. KAMINSKII, B.V. MILL, and A.V. BUTASHIN
Stimulated Emission from Nd^{3+} Ions in Acentric $LaBGeO_5$ Crystals
78.20; 78.30; 78.45; 78.55; S10
(a) Vol.118,No.1, K59-K64

V. HIZHNYAKOV
Stimulated Echo and Vibrational Dynamics
78.40; 78.45; 78.50; S9.11
(b) Vol.158,No.2, 725-736

H.H. WEBER
A Shorter Wavelength Mode Shift of Ridge-Waveguided 1.3 μm Lasers
78.45; 78.20; S7.15
(a) Vol.121,No.2, K239-K242

J.W. TOMM, K.H. HERRMANN, and A.E. YUNOVICH
Infrared Photoluminescence in Narrow-Gap Semiconductors (Review Article)
78.55; 78.45; S8.1; S8.15
(a) Vol.122,No.1, 11-42

78.50. Impurity and defect absorption in solids

R. CYWIŃSKI, E. MUGEŃSKI, and I. SOKÓLSKA
Eu^{2+}-Mn^{2+} Energy Transfer in NaCl and KCl Crystals
61.70; 78.50; S9.11
(b) Vol.157,No.1, 137-144

A.G. MAKHANEK
On the Mechanism of Two-Photon Absorption by Nd^{3+} Ions in Yttrium Aluminium Garnet and Yittrium Lithium Fluoride in ...
71.70; 78.50; S11.2
(b) Vol.157,No.2, 641-648

R. DEBNATH, J. CHAUDHURY, and S. CHANDRA BERA
Optical Properties and Nature of Coordination of Cu^{+} Ions in Calcium-Metaphosphate Glass
78.50 and 78.55; 71.55; S11.1
(b) Vol.157,No.2, 723-734

L.P. WANG, L.Z. Zhang, W.X. ZHU, X.T. LU, and G.G. QIN
The Behaviour of the Hydrogen-Localized Vibrational Mode in Proton-Implanted GaAs
63.20 and 78.50; S7.12
(b) Vol.158,No.1, 113-118

T. ABRITTA, F.H. BLAK, and R.J.M. DA FONSECA
Photoacoustic Measurements of $LiGa_5O_8:Fe^{3+}$
78.50 and 78.55; S11.2
(b) Vol.158,No.1, K75-K78

V. HIZHNYAKOV
Stimulated Echo and Vibrational Dynamics
78.40; 78.45; 78.50; S9.11
(b) Vol.158,No.2, 725-736

J.A. BAQUEDANO and J.M. CABRERA
Absorption Bands Associated to Photorefractive Erasure in $Bi_{12}SiO_{20}$
71.55; 72.40; 78.50; S10
(a) Vol.118,No.2, 497-504

C. JULIEN, A. CHEVY, and D. SIAPKAS
Optical Properties of In_2Se_3 Phases
78.50; 78.30; S8
(a) Vol.118,No.2, 553-560

P. SIVAPRASAD, K. RAMESH, and Y.P. REDDY
Optical and EPR Studies of Cu^{2+} in $MgNH_4PO_4 * 6 H_2O$
76.30 and 78.50; 71.55; S11.1
(a) Vol.118,No.2, K103-K106

R. KRAIENHEMKE, P. SEMIG, and F. FISCHER
Travelling Solvent Zone Melting Growth and Optical Properties of Lead Activated Calcite Crystals
78.50 and 78.55; 61.50; S11.1
(a) Vol.119,No.1, 327-336

F. PIO, M. GUZZI, G. SPINOLO, and M. MARTINI
Intrinsinc and Impurity-Related Point Defects in Amorphous Silica: A Spectroscopic Study
61.70; 78.50; 78.55; S10.1
(b) Vol.159,No.2, 577-588

J. HUGEL and J.C. PARLEBAS
Optical Spectrum Model for a Substitutional Isoelectronic Impurity in Binary Semiconductors
71.55 and 78.50
(b) Vol.159,No.2, 927-936

I.YA. FUGOL, A.M. RATNER, and E.M. YURTAEVA
On Conditions of Spectroscopic Manifestation of Dynamical Jahn-Teller Effect
71.70; 78.50; S5.2
(b) Vol.160,No.1, 245-254

E.P. SINYAVSKII and E.YU. SAFRONOV
The Optical Properties of Impurity Center Short-Range Potentials in the Magnetic Field
78.50 and 78.55; 71.55; S5.12
(b) Vol.160,No.1, 357-364

A.A. KAMINSKII, B.V. MILL, A.V. BUTASHIN, K. KURBANOV, and L.A. POLYAKOVA
Spectroscopy of Disordered $La_3Ga_5SiO_{14}$ Single Crystals Doped with Pr^{3+}, Ho^{3+}, and Er^{3+} Ions
78.55; 71.55; 78.50; S10
(a) Vol.120,No.1, 253-266

T. HATTORI, T. TAKADA, K. KAGAWA, A. MITSUISHI, and Y. KAMIURA
Copper Related Acceptors in Quenched Germanium
71.55 and 78.50; S5.12
(a) Vol.120,No.1, K63-K66

S.V. NISTOR, GH. STOICESCU, and C.D. MATEESCU
Incorporation and Aggregation Properties of Bismuth Ions in KCl Crystals
61.70; 66.30; 78.50; S9.11
(b) Vol.160,No.2, 423-432

CHANGHE HUANG, ZHENZHONG YU, and DINGYUAN TANG
Defect Related Absorption in p-Type $Hg_{1-x}Cd_xTe$ Alloys
78.50; 61.70; S8.15
(a) Vol.121,No.1, 273-280

P.A. APANASEVICH, V.I. KRUGLOV, and I.E. TRALLE
Absorption of Impurity Centres of a Crystal in the Field of a Strong Monochromatic Wave
71.55 and 78.50; 71.38
(b) Vol.161,No.2, 869-882

A.D. VISHNOI, V.B. SAPRE, and C. MANDE
A Study of the White Lines Associated with the Rare Earth L_{II} and L_{III} Absorption Discontinouities in Some Oxides ...
78.50; 71.55; S11.1
(b) Vol.161,No.2, 891-896

V. SLUSARENKO, Y. BURKI, W. CZAJA, and H. BERGER
Temperature Dependence of the Oxygen Absorption Band in ZnTe:O
78.50; S8.13
(b) Vol.161,No.2, 897-906

KANG-WEI ZHOU, SANG-BO ZHAO, PING-FENG WU, and JUN-KAI XIE
EPR Parameters and Spectral Fine Structure of Ni^{2+} in $LiNbO_3$
71.70 and 76.30; 78.50; S11.1
(b) Vol.162,No.1, 193-198

SH.M. EFENDIEV, V. GAVRYUSHIN, G. RACIUKAITIS, V.E. BAGIEV, V. KAZLAUSKAS, G. PUZONAS, R.A. ALIEV, and E.R. MUSTAFAEV
Induced Absorption Spectroscopy by Lattice Defects in Sillenite-Type Crystals
78.50; 71.55; S11
(b) Vol.162,No.1, 281-288

78.55. Photoluminescence

S.I. POKUTNYI
Energy Transfer of Electron Excitation. Decay Kinetics of Exciton Luminescence in a Three-Layer System
73.40 and 78.55
(b) Vol.157,No.1, 221-228

A. ZEHE and D. DE LA CRUZ
Effect of Fluctuations in delta-Doping Superlattices on Donor-Acceptor Pair Luminescence
71.55; 78.55; S7.12
(b) Vol.157,No.1, 305-310

A. SVOBODA, J. VEČEŘ, C. BÁRTA, and I. PELANT
Kinetics of Two-Photon Excited Luminescence of Hg_2Cl_2 Crystals
78.55; S9
(b) Vol.157,No.1, K69-K72

M. NIKL and K. POLAK
Luminescence Kinetics of PbF_2 Single Crystals
78.55; S9
(a) Vol.117,No.1, K89-K92

R. DEBNATH, J. CHAUDHURY, and S. CHANDRA BERA
Optical Properties and Nature of Coordination of Cu^+ Ions in Calcium-Metaphosphate Glass
78.50 and 78.55; 71.55; S11.1
(b) Vol.157,No.2, 723-734

K.R. MURALI and Y.V.G.S. MURTI
Color Centers in Irradiated $NaF:Mn^{++}$ Crystals
61.70; 78.55; 78.60; S9.11
(a) Vol.117,No.2, 377-386

A.I. BOBRYSHEVA, G.P. GOLUBEV, I.I. DOBÎNDǍ, E.A. ZHUKOV, D.G. LUCHINSKII, and YU.G. SHEKUN
Picosecond Kinetics of Photoluminescence in GaSe
78.55; S8
(a) Vol.117,No.2, 587-592

T. KUROBORI, E. YOSHIDA, S. NAKAMURA, N. TAKEUCHI, K. KAWABATA, and Y. FUKUDA
Optical Properties of a Modified F^+ Center in CaO Crystal Doped with Impurities
78.55; 61.70; S10.1
(a) Vol.117,No.2, K177-K180

T. PIOTROWSKI, J.W. TOMM, and N. PUHLMANN
Photoluminescence Studies of $Hg_{1-x}Mn_xTe$ Single Crystals
78.55; S8.15
(a) Vol.117,No.2, K181-K184

V.M. BABICH, M.YA. VALAKH, V.B. KOVALCHUK, G.YU. RUDKO, and N.I. SHAKHRAYCHUK
Photoluminescence and Electrical Properties of Germanium-Doped and Thermally Annealed Silicon
78.55; S5.11
(a) Vol.117,No.2, K185-K188

T. ABRITTA, F.H. BLAK, and R.J.M. DA FONSECA
Photoacoustic Measurements of $LiGa_5O_8:Fe^{3+}$
78.50 and 78.55; S11.2
(b) Vol.158,No.1, K75-K78

H. ZIMMERMANN, R. BOYN, C. MICHEL, and P. RUDOLPH
Absorption-Calibrated Determination of Impurity Concentrations in CdTe from Excitonic Photoluminescence
71.55; 78.55; S8.13
(a) Vol.118,No.1, 225-234

J. PASTRŇÁK, J. OSWALD, O. PETŘIČEK, and V.N. OVSYUK
The Degradation of Band Edge Luminescence of GaAs under Laser Light Excitation
78.55; 68.20; S7.12
(a) Vol.118,No.1, 317-326

K.-H. PANTKE, H. OVER, and I. BROSER
Excitation of the Impurity Emission Due to the Mixed Mode Exciton Absorption in CdS
78.55; 71.35; S8.11
(a) Vol.118,No.1, 343-348

A.A. KAMINSKII, B.V. MILL, and A.V. BUTASHIN
Stimulated Emission from Nd^{3+} Ions in Acentric $LaBGeO_5$ Crystals
78.20; 78.30; 78.45; 78.55; S10
(a) Vol.118,No.1, K59-K64

K. ZINK, A. KROST, H. NELKOWSKI, J. SAHM, and H. STUTENBECKER
Energy Transfer Mechanisms in $Zn_{1-x}Mn_xSe$ $(0.005<x<0.017)$ Detected by ODMR
71.70; 76.30; 78.55; S8.15
(b) Vol.158,No.2, 603-608

W. LAASCH, H. HAGEDORN, T. KLOIBER, and G. ZIMMERER
Fine Structure of the Luminescence of Solid Neon and Its Relation to Exciton Trapping and Desorption Induced by ...
78.55; 71.35; S5.2
(b) Vol.158,No.2, 753-768

J. PASTRŇÁK, J. OSWALD, M. LÁZNIČKA, A. BOSACCHI, and A. SALOKATVE
The Photoluminescence Spectra of Thin Si-Doped GaAs Layers Grown by MBE
78.55 and 78.65; S7.12
(a) Vol.118,No.2, 567-576

D. SCHIKORA, J.W. TOMM, K. RAKKENUS, and M. PESSA
Growth and Optical Characterization of HgTe-CdTe Superlattices
78.55; S8.16
(a) Vol.118,No.2, K123-K128

K. NITSCH, M. NIKL, Č. BARTA, D. SCHULTZE, A. TŘÍSKA, and R. UECKER
Preparation and Emission Properties of $NaBi(WO_4)_2$ and $NaBi(WO_4)_2$:Ce Single Crystals
78.40 and 78.55; S11.1
(a) Vol.118,No.2, K133-K138

E. LACH, V.D. KULAKOVSKII, A. FORCHEL, T.L. REINECKE, J. STRAKA, D. GRUETZMACHER, and G. WEIMANN
Single and Many Particle Effects in the Emission Spectra of Laterally Homogeneous 2D Plasmas
71.35 and 78.55; 73.40; S7.12; S7.15
(b) Vol.159,No.1, 125-132

L.V. BUTOV, V.D. KULAKOVSKII, and T.G. ANDERSSON
Photoluminescence of a Dense Quasi-2D Electron-Hole System in Modulation Doped Heterostructure with Strained InGaAs QW
71.35 and 78.55; 73.40; S7.15
(b) Vol.159,No.1, 165-172

A. MYSYROWICZ, D.W. SNOKE, and J.P. WOLFE
Progress on Bose-Einstein Condensation in Excitons
71.35 and 78.55; S10.1
(b) Vol.159,No.1, 387-402

H. FIESELER, R. SCHWABE, and J.L. STAEHLI
Electron-Hole Plasma in Cross-Over-Near Ga(As,P)
71.35 and 78.55; S7.15
(b) Vol.159,No.1, 411-416

V.A. KHARCHENKO, G.V. MIKHAILOV, D.K. NELSON, and B.S. RAZBIRIN
Impact Excitation and Auger Decay of Bound Exciton Complexes in a Dense Exciton Gas
71.35 and 78.55; S8.11
(b) Vol.159,No.1, 425-430

U. NEUKIRCH, I. BROSER, and R. RASS
Nonlinear Luminescence of Heavily Doped CdS:In Crystals
78.55; 71.45; S8.11
(b) Vol.159,No.1, 431-436

N. PRESSER, G. KUDLEK, J. GUTOWSKI, S. DURBIN, D.R. MENKE, M. KOBAYASHI, and R.L. GUNSHOR
High-Density Spectroscopy of ZnSe/GaAs Epilayers in the Near-Band-Edge Region
71.35 and 78.65; 78.55; S7.12; S8.12
(b) Vol.159,No.1, 443-448

N.S. AVERKIEV, V.M. ASNIN, A.A. ROGACHEV, A.YU. SILOV, and V.I. STEPANOV
Photoluminescence Studies of Anomalous Diamagnetic Properties of the 2D e-h System on GaAs Surface
75.20 and 78.55; 73.40; S7.12
(b) Vol.159,No.1, 471-476

M. STIEBLER, J. REICHARDT, R. HIRRLE, and S. KEMMLER-SACK
Cathodo- and Photoluminescence in the System $Y_{2-x}Tb_xSiO_5:Eu^{3+}$. A Case with an Unusual Effective Energy Transfer in ...
78.55 and 78.60; 71.55; S10.15
(a) Vol.119,No.1, 317-326

R. KRAIENHEMKE, P. SEMIG, and F. FISCHER
Travelling Solvent Zone Melting Growth and Optical Properties of Lead Activated Calcite Crystals
78.50 and 78.55; 61.50; S11.1
(a) Vol.119,No.1, 327-336

S.R. ROTMAN
Energy Transfer in Iron-Doped Yttrium Aluminum Garnet
78.55; S11.2
(a) Vol.119,No.1, K87-K90

F. PIO, M. GUZZI, G. SPINOLO, and M. MARTINI
Intrinsinc and Impurity-Related Point Defects in Amorphous Silica: A Spectroscopic Study
61.70; 78.50; 78.55; S10.1
(b) Vol.159,No.2, 577-588

I.KH. AKOPYAN, A.A. KLOCHIKHIN, B.V. NOVIKOV, M.YA. VALAKH, A.P. LITVINCHUK, and I. KOSAZKII
Optical Spectroscopy of the Superionic Crystals (Review Article)
78.30 and 78.40; 64.70; 71.25; 71.35; 78.55; S9
(a) Vol.119,No.2, 363-416

S.E. SARKISOV, A.A. KAMINSKII, A.V. BUTASHIN, YU.E. PERLIN, V.N. ENAKII, and M.G. BLAZHA
Selective Spectroscopy, Luminescence Kinetics of Pr^{3+} Ions, and Phonon Density Calculation in Multicentred and ...
78.55; 63.20; 71.55; 78.30; S11.2
(a) Vol.119,No.2, 607-620

J. REICHARDT, M. STIEBLER, R. HIRRLE, and S. KEMMLER-SACK
Cathodo- and Photoluminescence in Oxyorthosilicates of X_1 and X_2 Type: System $Y_{2-x}Gd_xSiO_5:Tb^{3+}$
78.55 and 78.60; 71.55; S10
(a) Vol.119,No.2, 631-642

J. LLOPIS
Luminescence of MgO and CaO Stabilized ZrO_2 Crystals
78.55 and 78.60; S10.1
(a) Vol.119,No.2, 661-668

S. HILDEBRANDT and W. HERGERT
Unified Theoretical Description of the CL, EBIC, PL, and LBIC Contrast Profile Area of an Individual Surface-Parallel ...
78.55 and 78.60; 72.10
(a) Vol.119,No.2, 689-700

J.W. TOMM, K.H. HERRMANN, H. BOETTNER, M. TACKE, and A. LAMBRECHT
A Luminescence Study in the $Pb_{1-x}Eu_xSe$ System
73.40 and 78.55; 72.40; S7.16
(a) Vol.119,No.2, 711-720

I.CH. AKOPYAN, B.V. NOVIKOV, R.V. GRIGOREV, S.R. GRIGOREV, H. ROEPPISCHER, and N. STEIN
Photoluminescence Study of Na and Li Acceptors in ZnSe
78.55; 61.80; S8.12
(a) Vol.119,No.2, K167-K172

E.P. SINYAVSKII and E.YU. SAFRONOV
The Optical Properties of Impurity Center Short-Range Potentials in the Magnetic Field
78.50 and 78.55; 71.55; S5.12
(b) Vol.160,No.1, 357-364

J. GUTOWSKI, N. PRESSER, and G. KUDLEK
Optical Properties of ZnSe Epilayers and Films (Review Article)
78.65; 68.55; 71.35; 73.60; 78.55; S8.12
(a) Vol.120,No.1, 11-60

A.A. KAMINSKII, B.V. MILL, A.V. BUTASHIN, K. KURBANOV, and L.A. POLYAKOVA
Spectroscopy of Disordered $La_3Ga_5SiO_{14}$ Single Crystals Doped with Pr^{3+}, Ho^{3+}, and Er^{3+} Ions
78.55; 71.55; 78.50; S10
(a) Vol.120,No.1, 253-266

G. SCHRAMM
Determination of the Free Exciton Capture Cross Section of As Doped Silicon by Photoluminescence
78.55; 71.35; S5.11
(a) Vol.120,No.1, K101-K104

A. CORNET, A. MIRALLES, O. RUIZ, and J.R. MORANTE
Near Infrared Photoluminescence of ZnO:Co Varistors
78.55; S10.1
(a) Vol.120,No.1, K105-K110

J. ATAEV, O.I. KONKOV, E.I. TERUKOV, V.A. VASSILEV, and A.S. VOLKOV
Influence of Preparation Conditions on the Radiative Recombination in a-Si:H Films
68.55 and 78.55; S5.11
(a) Vol.120,No.2, 519-524

S. BENCI, F. FERMI, and A.A. NAGORNYI
Annealing of Eu^{2+} Centres in CdF_2 Crystals
78.55 and 78.60; S9.12
(a) Vol.120,No.2, 635-642

R. HIRRLE, J. WIEHL, W. WISCHERT, and S. KEMMLER-SACK
Cathodo- and Photoluminescence in Nd^{3+} and Tb^{3+} Activated Garnets of Type $Cd_{3-x}Ca_xGa_2Ge_3O_{12}$
78.55 and 78.60; 71.55; S11.2
(a) Vol.120,No.2, 643-652

G. SCHRAMM
Quantitative Analysis of As Doped Silicon by Photoluminescence
78.55; S5.11
(a) Vol.120,No.2, K181-K184

A. KALNIŇŠ, I. PLAVINA, A.I. POPOV, and A. TALE
Determination of the Effective Absorption Cross-Section of F-Centres in KBr-In by Photostimulated Luminescence
61.70; 78.55; S9.11
(b) Vol.161,No.1, 85-90

O. SILD
The Model of Rotation of NO_2^- Impurity Molecule in KCl Crystal
61.70; 78.30; 78.55; S9.11
(b) Vol.161,No.1, 105-112

M.A. JAKOBSON, V.D. KAGAN, R. KATILIUS, and G.O. MUELLER
Impact Ionization of Free and Bound Excitons by Free Carriers in an Electric Field in CdS: Investigation by Optical ...
78.55; 71.35; S8.11
(b) Vol.161,No.1, 395-404

YU.P. GNATENKO, YU.I. ZHIRKO, and Z.D. KOVALYUK
Optical Properties of GaSe Crystals Containing Mn Impurity Atoms. II. Excitonic and Impurity Emission
78.55; 71.35; 71.55; S8
(b) Vol.161,No.1, 427-434

R. HIRRLE, W. WISCHERT, and S. KEMMLER-SACK
Cathodo- and Photoluminescence in the Mn^{2+} Activated Garnets of Type $Cd_{3-x}Ca_xGa_2Ge_3O_{12}$
78.55 and 78.60; 71.55; S11.2
(a) Vol.121,No.1, 289-298

N.S. AVERKIEV, B.N. KALININ, A.V. LOSEV, A.A. ROGACHEV, and A.S. FILIPCHENKO
Recombination Radiation from Heavily-Doped n-Type Indium Antimonide Crystals
78.55; S7.13
(a) Vol.121,No.1, K129-K134

A.CH. LUSHCHIK and A.G. FRORIP
Thermalized and Hot Interstitial Halogen Ions in Alkali Halides
61.70 and 78.55; 61.80; 76.30; S9.11
(b) Vol.161,No.2, 525-536

A.N. GRUZINTSEV, S.V. DEMISHEV, and T.V. NIKIFOROVA
The Influence of Complex Luminescence Centres on the Luminescent Properties of CaS:Eu
71.55 and 78.55; 78.60; S8
(b) Vol.161,No.2, 705-710

H. OPYRCHAŁ and M. MANFREDI
Spectroscopic Evidences of Eu-V Dipole Precipitation in KCl:Eu Crystals
61.70; 71.55; 78.55; S9.11
(a) Vol.121,No.2, 407-414

F. TAKEDA, K. MATSUURA, and S. SETO
Admittance Spectroscopy and Photoluminescence Characterization of High-Quality p-Type Cadmium Telluride
71.55 and 78.55; 72.80; S8.13
(a) Vol.121,No.2, 595-606

K.H. HERRMANN, K.-P. MOELLMANN, and J.W. TOMM
On the Broadening Mechanisms near the E_0 Transition in Narrow-Gap (Hg,Cd)Te
78.20; 71.25; 78.55; S8.15
(a) Vol.121,No.2, 635-640

L.I. BEREZHINSKII, S.I. VLASKINA, G.I. LYASHENKO, and V.E. RODIONOV
On the Possibility of Obtaining the Crystalline Phase in Amorphous Films of Silicon Carbide on Silicon at Low Temperature
68.55; 73.60; 78.55; S6
(a) Vol.121,No.2, K193-K198

T. TSUBOI and P. SILFSTEN
Detection of Structural Phase Transition in $BaMnF_4$ by Luminiscence Lifetime
64.70 and 78.55; S9.16
(b) Vol.162,No.1, K15-K20

R.J.M. DA FONSECA and T. ABRITTA
Photoluminescence of Octahedrally Coordinated Mn^{2+} in Lithium Gallate
78.55; S11.2
(b) Vol.162,No.1, K59-K62

J.W. TOMM, K.H. HERRMANN, and A.E. YUNOVICH
Infrared Photoluminescence in Narrow-Gap Semiconductors (Review Article)
78.55; 78.45; S8.1; S8.15
(a) Vol.122,No.1, 11-42

T. KUROBORI and N. TAKEUCHI
On the Lifetime in Magnesium-Doped Calcium Oxide Crystals
78.55; 61.70; 71.55; S10.1
(a) Vol.122,No.1, K91-K96

A. KANOUN, S. ALAYA, and H. MAAREF
Spectroscopic Properties of Pr^{3+} and Nd^{3+} Ions in Zinc Tellurite Glasses
71.55; 78.55; S8.13
(b) Vol.162,No.2, 523-530

V. HIZHNYAKOV, M. KARKLINYA, V. NAGIRNYI, L. NAGLI, and S. ZAZUBOVICH
On the Reasons of the Off-Centre Position of Excited Ga^+ and In^+ Ions in Alkali Halides
71.55 and 78.55; S9.11
(b) Vol.162,No.2, K91-K94

H. HAEFNER, L. LEHMANN, R. MITDANK, G. OELGART, and E. SCHULZE
Luminescence Characterization of (AlGa)As Single Quantum Wells
73.40 and 78.55; 78.60; S7.15
(a) Vol.122,No.2, 683-694

N. PUHLMANN and G. OELGART
Semiconductor Characterization by Means of EBIC, Cathodo-, and Photoluminescence
78.60; 78.55; S7.12
(a) Vol.122,No.2, 705-714

K.S. CHOI, B.J. KOO, M. JUNG, H.J. SHIN, J.H. LEE, S.B. KIM, S.K. CHANG, and H.L. PARK
(100) CdTe Thin Film Growth by Temperature Gradient Vapor Transport (TGVT)
68.55; 78.40; 78.55; S8.13
(a) Vol.122,No.2, K135-K138

78.60. Other luminescence and radiative recombination
For photoconduction and photovoltaic effects, see 72.40

R.J. TARENTO
Injection Effects on Cathodoluminescence Experiments. Dember-Electric Field
78.60; S5.11
(a) Vol.117,No.1, 305-312

K. SOMAIAH, M. VENKATA NARAYANA, and L.H. BRIXNER
Thermally Stimulated Luminescence of LaOBr:Tm
78.60; S9
(a) Vol.117,No.1, K81-K84

K.R. MURALI and Y.V.G.S. MURTI
Color Centers in Irradiated NaF:Mn^{++} Crystals
61.70; 78.55; 78.60; S9.11
(a) Vol.117,No.2, 377-386

A. TOMITA, Y. FUKUDA, Y. KUTOMI, and N. TAKEUCHI
Emission Processes in Thermally Stimulated Exoelectron Emission and Thermoluminescence of KCl:Ag Single Crystals
61.70 and 78.60; 79.75; S9.11
(b) Vol.158,No.1, 383-390

S.P. KUILA
Photostimulation Studies in Pure and Li-Doped Alkali Halides
78.60; 61.70; 71.55; S9.11
(a) Vol.118,No.1, 311-316

A.D. BELYAEV, V.E. RODIONOV, and N.S. CHERNAYA
Effect of Ultrasound Oscillations on Electroluminescent Properties of MISIM Structures
78.60; 62.80; 73.40; S8.11
(a) Vol.118,No.1, 327-332

C. CHRISTODOULIDES
General-Order Kinetics in Thermoluminescence Resulting from a Distribution of the Frequency Factor
78.60
(a) Vol.118,No.1, 333-342

S.V. MOHARIL and P.D. SAHARE
Langbeinite Based Phosphors for Applications in Thermoluminescence Dosimetry
78.60; S10
(a) Vol.118,No.1, K55-K58

J. MANAM and A.N.C.V. RAO
Influence of Cu^+ Doping on the Thermoluminescence of KCl
78.60; S9.11
(a) Vol.118,No.2, K129-K132

M. STIEBLER, J. REICHARDT, R. HIRRLE, and S. KEMMLER-SACK
Cathodo- and Photoluminescence in the System $Y_{2-x}Tb_xSiO_5:Eu^{3+}$. A Case with an Unusual Effective Energy Transfer in ...
78.55 and 78.60; 71.55; S10.15
(a) Vol.119,No.1, 317-326

J. REICHARDT, M. STIEBLER, R. HIRRLE, and S. KEMMLER-SACK
Cathodo- and Photoluminescence in Oxyorthosilicates of X_1 and X_2 Type: System $Y_{2-x}Gd_xSiO_5:Tb^{3+}$
78.55 and 78.60; 71.55; S10
(a) Vol.119,No.2, 631-642

K. NARASIMHA REDDY, T.S.P.L.N. PRASAD, E. LAXMI NARSAIAH, U.V. SUBBA RAO, and P. PAROLI
Thermoluminescence in $YBa_2Cu_3O_{7-\delta}$ Samples
78.60; S10.15
(a) Vol.119,No.2, 655-660

J. LLOPIS
Luminescence of MgO and CaO Stabilized ZrO_2 Crystals
78.55 and 78.60; S10.1
(a) Vol.119,No.2, 661-668

LI-JIAN MENG and CHANG-HUA LI
Temperature Effect on AC EL Characteristics in $ZnS:ErF_3$ Thin Films
78.60; 78.65; S8.11
(a) Vol.119,No.2, 677-682

S. HILDEBRANDT and W. HERGERT
Unified Theoretical Description of the CL, EBIC, PL, and LBIC Contrast Profile Area of an Individual Surface-Parallel ...
78.55 and 78.60; 72.10
(a) Vol.119,No.2, 689-700

CH. GOPAL REDDY and K. NARASHIMA REDDY
Thermoluminescence Studies in Yttrium Doped KBr Crystals
78.60; S9.11
(a) Vol.120,No.1, 267-272

S. BENCI, F. FERMI, and A.A. NAGORNYI
Annealing of Eu^{2+} Centres in CdF_2 Crystals
78.55 and 78.60; S9.12
(a) Vol.120,No.2, 635-642

R. HIRRLE, J. WIEHL, W. WISCHERT, and S. KEMMLER-SACK
Cathodo- and Photoluminescence in Nd^{3+} and Tb^{3+} Activated Garnets of Type $Cd_{3-x}Ca_xGa_2Ge_3O_{12}$
78.55 and 78.60; 71.55; S11.2
(a) Vol.120,No.2, 643-652

GANGANG CAI, SHOUSHAN WANG, and JIAN ZHANG
Radiation Damage and Recovery of LiF:Mg,Ti-TLD at the Dose of 10^6 Gy Gamma Rays
61.80; 78.60; S9.11
(a) Vol.121,No.1, 57-62

D.P. ERCHAK, R.B. GELFAND, N.M. PENINA, V.F. STELMAKH, V.P. TOLSTYKH, A.G. ULYASHIN, V.S. VARICHENKO, and A.M. ZAITSEV
Point Paramagnetic Defects in Diamond Irradiated by High-Energy Ions
61.70; 76.30; 78.60; S5
(a) Vol.121,No.1, 63-72

H. MENNIGER, M. BEER, R. GLEICHMANN, H. RAIDT, B. ULRICI, and G. VOIGT
Electrical Homogeneity of Semi-Insulating LEC GaAs by Post-Growth Annealing
61.70 and 78.60; 85; S7.12
(a) Vol.121,No.1, 95-104

E.V. KABANOVA, S.N. KULKOV, B.P. GRITZENKO, B.Y. YAKOVLEV, and P.P. PERSTNEV
The Change of Optical and Structural Properties of MgO by the Exposition of Laser Pulse Irradiation
61.80; 78.60; S10.1
(a) Vol.121,No.1, 105-110

B.P. CHANDRA, Y. RAHANGDALE, and M. RAMRAKHIANI
Theory of Mechanical Excitation of Luminescence Centers in Crystals
78.60; S8.11
(a) Vol.121,No.1, 281-288

R. HIRRLE, W. WISCHERT, and S. KEMMLER-SACK
Cathodo- and Photoluminescence in the Mn^{2+} Activated Garnets of Type $Cd_{3-x}Ca_xGa_2Ge_3O_{12}$
78.55 and 78.60; 71.55; S11.2
(a) Vol.121,No.1, 289-298

A.N. GRUZINTSEV, S.V. DEMISHEV, and T.V. NIKIFOROVA
The Influence of Complex Luminescence Centres on the Luminescent Properties of CaS:Eu
71.55 and 78.55; 78.60; S8
(b) Vol.161,No.2, 705-710

SANGEETA, S.C. SABHARWAL, B. GHOSH, and M.K. GUPTA
Thermoluminescence from Europium Doped Calcium Fluoride Crystals
78.60; S9.12
(a) Vol.121,No.2, 657-666

J. BENOIT, P. BENALLOUL, C. BARTHOU, S. CASETTE, and J.C. SORET
De-Excitation Processes and Efficiency in AlE ZnS:Mn Thin Film Electroluminescent Devices
78.60 and 85; S8.11
(a) Vol.122,No.1, 427-438

H. HAEFNER, L. LEHMANN, R. MITDANK, G. OELGART, and E. SCHULZE
Luminescence Characterization of (AlGa)As Single Quantum Wells
73.40 and 78.55; 78.60; S7.15
(a) Vol.122,No.2, 683-694

N. PUHLMANN and G. OELGART
Semiconductor Characterization by Means of EBIC, Cathodo-, and Photoluminescence
78.60; 78.55; S7.12
(a) Vol.122,No.2, 705-714

H.-J. FITTING, E. SCHREIBER, and TH. HINGST
Avalanche Measurement in ZnS by Vacuum Emission
72.20; 78.60; 79.70; 85; S8.11
(a) Vol.122,No.2, K165-K168

78.65. Optical properties of thin films

A. BEL HADJ, B. EL JANI, M. GUITTARD, R. BENNACEUR, and P. GIBART
Photoluminescence of Heavily Doped Sn-GaAs Grown by Metal Organics Vapor Phase Epitaxy
61.70 and 78.65; 73.60; S7.12
(a) Vol.117,No.1, 169-176

V. GRAEFE, D. SCHALCH, and A. SCHARMANN
Hydrogenation of Reactively Sputtered AlN_xO_y Films
61.80 and 78.65; S6
(a) Vol.117,No.1, K23-K28

M.M. EL-OCKER, F. SHARAF, S.M. YOSSEF, F. METAWE, and M. EL-DESOUKI
Optical Behaviour of Lead Selenide Films
68.55 and 78.65; S8
(b) Vol.157,No.2, 609-614

A.A. ZAHAB, M. ABD-LEFDIL, and M. CADENE
Electrical Conductivity, Optical Absorption, and Photoconductivity Measurements of Well Oriented Indium-Telluride Thin Films
73.60 and 78.65; 72.40; S8
(a) Vol.117,No.2, K103-K106

H. BOETTCHER, K. SCHREIBER, and B. VAUPEL
Photosensitivity of Evaporated Mixed AgBr Layers
73.40 and 78.65; S9.15
(a) Vol.117,No.2, K165-K170

V.P. GRIBKOVSKII, L.G. ZIMIN, S.V. GAPONENKO, I.E. MALINOVSKII, P.I. KUZNETSOV, and G.G. YAKUSHCHEVA
Optical Absorption near Excitonic Resonance of the MOCVD-Grown ZnSe Single Crystals
71.35 and 78.65; S8.12
(b) Vol.158,No.1, 359-366

B.I. KHUDIK, V.Z. LOZOVSKII, and A.F. ZHURAVLEV
Nonlinear Electrodynamic Effects Due to Transition Layers in Crystals
78.65
(b) Vol.158,No.1, K89-K94

S.V. IVANOV, P.S. KOPEV, N.N. LEDENTSOV, B.YA. MELTSER, M.YU. NADTOCHY, V.M. USTINOV, and A.M. VASILEV
Ultrathin Layers and Short-Period Superlattices in Semiconductor Structures
73.40; 78.65; S7.12; S7.15
(a) Vol.118,No.1, 169-178

W.A. BADAWY, R.S. MOMTAZ, and E.M. ELGIAR
Solid State Characteristics of Indium-Incorporated TiO_2 Thin Films
68.55 and 78.65; 73.60; S10.1
(a) Vol.118,No.1, 197-202

A.D. BROTHERS, F.J. FEUERBORN, and S.E. BAUER
Excitons in Indium Halide Thin Films
78.65; 71.35; S9
(b) Vol.158,No.2, 503-510

J. PASTRŇÁK, J. OSWALD, M. LÁZNIČKA, A. BOSACCHI, and A. SALOKATVE
The Photoluminescence Spectra of Thin Si-Doped GaAs Layers Grown by MBE
78.55 and 78.65; S7.12
(a) Vol.118,No.2, 567-576

L.G. ZIMIN, S.V. GAPONENKO, V.YU. LEBED, I.E. MALINOVSKII, I.N. GERMANENKO, E.E. PODOROVA, and V.A. TSEKHOMSKII
Nonlinear Optical Absorption in I-VII and II-VI Microcrystallites under Quantum Confinement of Elementary ...
71.35 and 78.65; S8.15; S9
(b) Vol.159,No.1, 267-274

T. ISHIHARA
The Origin of Third-Order Optical Nonlinearity in Exciton Systems
71.35 and 78.20; 78.65; S9
(b) Vol.159,No.1, 371-378

N. PRESSER, G. KUDLEK, J. GUTOWSKI, S. DURBIN, D.R. MENKE, M. KOBAYASHI, and R.L. GUNSHOR
High-Density Spectroscopy of ZnSe/GaAs Epilayers in the Near-Band-Edge Region
71.35 and 78.65; 78.55; S7.12; S8.12
(b) Vol.159,No.1, 443-448

T. TSVETKOVA, B. AMOV, E. VATEVA, and V. AVERYANOV
Ion Implantation Induced Modification of Vitreous Chalcogenides
73.60; 61.80; 78.65; S8.12
(a) Vol.119,No.1, 107-112

B. THOMAS and T.R.N. KUTTY
Formation of Single-Phase Indium Selenide Thin Films by Elemental Evaporation
68.55; 73.60; 78.65; S8.12
(a) Vol.119,No.1, 127-138

P.F. ROBUSTO and R. BRAUNSTEIN
Optical Measurements of the Highly Damped Surface Plasmon of SnTe
71.45 and 78.65; S8
(a) Vol.119,No.1, 145-154

P.F. ROBUSTO and R. BRAUNSTEIN
Optical Measurements of the Surface Plasmon of Indium-Tin Oxide
71.45 and 78.65; S10
(a) Vol.119,No.1, 155-168

G. AMATO, G. BENEDETTO, F. FIZZOTTI, C. MANFREDOTTI, and R. SPAGNOLO
Photothermal and Photoconductive Measurements of Low Pressure Chemically Vapour Deposited Amorphous Silicon
72.40 and 73.60; 78.65; S5.11
(a) Vol.119,No.1, 169-176

V.A. SKRYSHEVSKII and P. KUS
Optical and Electro-Physical Properties of $Gd_1Ba_2Cu_3O_{7-\delta}$ -p-Si Structures at $T>T_c$
73.40 and 74.70; 78.65; S10.15
(a) Vol.119,No.1, 201-208

A. SARKAR, S. CHAUDHURI, and A.K. PAL
Electrical and Optical Properties of Transparent Conducting ZnO Films
73.60 and 78.65; S10.1
(a) Vol.119,No.1, K21-K26

U. TESCHNER and K. HUEBNER
IR-Spectroscopic Data of Thin Insulating Films on Semiconductors. New Methods of Interpretation and Analysis
78.30 and 78.65; S10.1
(b) Vol.159,No.2, 917-926

T. DITTRICH
Correlation between Photoreflectance and CV Measurements of n-GaAs Epitaxial Layers
73.60 and 78.65; S7.12
(a) Vol.119,No.2, 479-486

LI-JIAN MENG and CHANG-HUA LI
Temperature Effect on AC EL Characteristics in $ZnS:ErF_3$ Thin Films
78.60; 78.65; S8.11
(a) Vol.119,No.2, 677-682

A. LEHMANN, W. HEERDEGEN, G. SCHIRMER, H. MUTSCHKE, W. RICHTER, E. HACKER, and R. DOHLE
Structure Correlated Infrared Properties of Fluoride Films
78.30 and 78.65; S9.12
(a) Vol.119,No.2, 683-688

J. GUTOWSKI, N. PRESSER, and G. KUDLEK
Optical Properties of ZnSe Epilayers and Films (Review Article)
78.65; 68.55; 71.35; 73.60; 78.55; S8.12
(a) Vol.120,No.1, 11-60

R.D. GOULD and R.I.R. BLYTH
Conductivity and Absorption Measurement in Nickel Phthalocyanine Thin Films
73.60 and 78.65; S12
(a) Vol.120,No.1, K57-K62

M. STARKE, I. ANDROSCH, and C. HAMANN
A Solid State Electrochromic Cell Using Erbium-Diphthalocyanine
78.65; S12
(a) Vol.120,No.1, K95-K100

N.F. GASHIMZADE
Guided Polariton Waves in Semiconductor Superlattices
71.36; 78.65
(b) Vol.160,No.2, K113-K116

F.M. GASHIMZADE and E.V. TAHIROV
Free-Carrier Absorption in Quantum-Well Structures for Charged Impurity Scattering
78.65
(b) Vol.160,No.2, K177-K182

R. BINDEMANN and O. PAETZOLD
The Influence of Thermal Disorder on the Absorption Edge in the Tauc Region of a-Si:H
78.65; S5.11
(b) Vol.160,No.2, K183-K188

J.C. BUREAU, A. GLACHANT, and B. BALLAND
Thermal Nitridation of SiO_2 Thin Films in NH_3 + H_2O (or O_2) Mixtures
68.55 and 73.60; 78.65; S10.1
(a) Vol.120,No.2, 507-518

T. DUMELOW, A.R. EL GOHARY, K.A. MASLIN, T.J. PARKER, D.R. TILLEY, and S.N. ERSHOV
Analysis of Far Infrared Spectra Showing Bulk and Surface Phonon-Polaritons in CdTe Epilayers on GaAs Substrates
78.30 and 78.65; 71.36; S8.13
(b) Vol.161,No.1, 233-244

E.YU. TURAEV, P.P. SEREGIN, and F.S. NASREDINOV
Tin and Iron Impurity Atom States in Modified As_2Se_3 Films
71.55 and 73.60; 78.65; S8
(b) Vol.161,No.1, 287-292

K.B. SUNDARAM and P. WAHID
Optical Absorption in Cerium Dioxide Thin Films
78.65; S10.1
(b) Vol.161,No.1, K63-K66

J. DUBOWIK, YU.V. KUDTYAVTSEV, and E.E. KRASSOVSKII
Spectroscopic Ellipsometric Investigation of Multilayered and Solid-State Amorphized Co/Zr Films
68.55 and 78.65; 61.40; S1.2
(a) Vol.121,No.1, 173-180

E.P. POKATILOV, S.I. BERIL, V.M. FOMIN, and V.V. KALINOVSKII
The Size-Quantized States of the Wannier-Mott Exciton in Structures with Superthin Films of CdTe
71.35 and 78.65; S8.13
(b) Vol.161,No.2, 603-612

TH. DITTRICH and V.I. TIMOSHENKO
Laser Induced Modification of n-GaAs below the Classical Melting Threshold
61.80 and 78.65; S7.12
(a) Vol.121,No.2, 547-554

U. TESCHNER
Infrared Optical Properties of Amorphous Silicon Oxynitride and Their Theoretical Interpretation
78.30 and 78.65; S6
(a) Vol.121,No.2, 641-656

B.I. KHUDIK, V.Z. LOZOVSKII, and A.F. ZHURAVLEV
Diffuse Scattering of Light on Molecules Adsorbed at a Solid Surface
78.65
(b) Vol.162,No.2, K127-K130

P. DANESH, A. TONEVA, P. PANTCHEV, and V. KUDEYAROVA
Microstructure and Optical Properties in CVD a-Si:H Films
68.55 and 78.65; S5.11
(a) Vol.122,No.2, 599-606

A.M. BARANOV, S.S. GERASIMOVICH, V.M. ELINSON, A.A. KUZIN, V.V. SLEPTSOV, and P.E. KONDRASHOV
Phase Composition Modeling of alpha-C:H Films on the Basis of Their Optical Properties
68.55 and 78.65; S5
(a) Vol.122,No.2, K139-K142

78.70. Other interactions of matter with particles and radiation

V.G. BARYSHEVSKII and A.M. ZAITSEVA
Orientational Effects Due to Traveling of Fast Neutrons and Hard gamma-Quanta through Crystals
61.80; 78.70; S5.12
(b) Vol.157,No.1, 129-136

LING CHEN, GUANGHOU WANG, and MINGKANG TENG
Charge Fluctuation in the High T_c Superconductor YBaCuO in the Transition Temperature Region
74.40; 78.70; S10.15
(b) Vol.157,No.1, 411-416

V. SREERAMALU, V. RAVINDRACHARY, H.R. SREEPAD, A. CHANDRASHEKARA, S. GOPAL, H. SANJEEVAIAH, and B. VISWANATHAN
Structural Relaxation and Crystallization of Metallic Glasses $Fe_{35}Co_{52}V_{13}$ and $Fe_{45}Co_{45}V_6Ni_4$. Positron Annihilation Study
61.40; 78.70; S1.1; S1.2
(a) Vol.117,No.1, 53-60

G. KONTRYM-SZNAJD
Three-Dimensional Image Reconstruction with Application in Positron Annihilation
71.20 and 78.70
(a) Vol.117,No.1, 227-240

K. KRIŠTIAKOVA, G. TUMBEV, and J. KRIŠTIAK
Annihilation of Positrons in $Fe_{80-x}Ni_xB_{20}$ Metallic Glasses
61.40 and 78.70; S1.1
(a) Vol.117,No.1, K5-K8

T. KURIHARA and S. TANIGAWA
The Anomalous Temperature Dependence of Positron Annihilation in Dilute Al-Zn Alloy near the Melting Point
61.70 and 78.70; S4
(a) Vol.117,No.1, K9-K12

B.K. PANDA, D.P. MAHAPATRA, and H.C. PADHI
Compton Profile Anisotropies of GaAs
71.25 and 78.70; S7.12
(b) Vol.158,No.1, 261-266

S. SZUSZKIEWICZ
Interaction of Positrons with the $Co_{66}Fe_4Mo_2Si_{16}B_{12}$ Alloy
78.70; S1.1
(b) Vol.158,No.1, K79-K84

J. DRYZEK and E. ROGOWSKA
Positron Annihilation in Sodium and Copper beta-Vanadium Oxide Bronzes
61.70; 71.25; 78.70; S10.15
(a) Vol.118,No.1, 61-64

G. REITER and P. KINDL
Positron Lifetime Investigations on Linear Polyethylene Compared to Branched Polyethylene
64.60; 78.70; S12
(a) Vol.118,No.1, 161-168

C. MANDE and A.P. DESHPANDE
A Study of Manganese Oxides by EXAFS Spectroscopy
78.70; S10
(b) Vol.158,No.2, 737-742

XIAO-GUANG WANG and HONG ZHANG
A Vacancy Complex in the Dilute Fe-V-N Alloy Identified by Positron Annihilation Spectroscopy
61.70 and 78.70; S1.1
(a) Vol.119,No.1, K15-K20

M.N. GHATIKAR, K.V.S. MAHESH, and S.N. BHATIA
Improved Calculations of Phase Shifts for EXAFS Spectroscopy
61.40; 78.70; S1.1; S1.2
(b) Vol.159,No.2, 551-556

A. CATICHA and S. CATICHA-ELLIS
A Fabry-Perot Interferometer for Hard X-Rays
61.10 and 78.70
(a) Vol.119,No.2, 643-654

H. AOURAG, B. KHELIFA, A. BELAIDI, A. TADJER, M. REZKI, and M. GAMOUDI
Positron Band Structures of GaAs and CdTe
71.25 and 71.60; 78.70; S7.12; S8.13
(b) Vol.160,No.1, 193-200

H. AOURAG, B. KHELIFA, A. BELAIDI, and Z. BELARBI
Positron Distribution in Semiconductors
71.60; 78.70; S5.11; S5.12; S7.1; S8.1
(b) Vol.160,No.1, 201-210

H. AOURAG, B. KHELIFA, A. BELAIDI, and T. KOBAYASHI
Interpretation of the Group-Theoretically Unexpected Shape of the Contour Appearing in the Experimental LCW Data for...
71.25; 71.60; 78.70; S5.11; S5.12
(b) Vol.160,No.2, 561-568

H.R. SREEPAD, V. SREERAMALU, A. CHANDRASHEKARA, V. RAVINDRACHARY, and S. GOPAL
Structural Phase Transition in a Superionic Conductor - $Na_2(La,Al)Zr(PO_4)_3$. Positron Annihilation Study
64.70; 61.40; 78.70; S11
(a) Vol.120,No.2, 411-418

S. CHABIK
Investigation of Interactions of Electrically Charged Hydrogen Atoms with Defects in Nickel by the Positron ...
61.70 and 78.70; S1.2
(a) Vol.120,No.2, K125-K130

W. RUDZIŃSKA, ST. CHABIK, and CZ. SZYMAŃSKI
Investigation of the Solid-Liquid Phase Transition for Sn-5wt%In Alloy by the Positron Annihilation Method
64.70; 78.70; S4
(a) Vol.120,No.2, K145-K148

L.L. BALASHOVA, N.M. KABACHNIK, and V.N. KONDRATEV
Charge Asymmetry in the Dependence of Stopping on Impact Parameter
61.80; 78.70; 79.90; S5.11
(b) Vol.161,No.1, 113-122

V.L. TAVGIN and A.V. STEPANOV
Activation Process Model
66.30 and 78.70
(b) Vol.161,No.1, 123-130

E. DEBȮWSKA and J. RUDZIŃSKA-GIRULSKA
Examination of the Electronic Structure of $Pd_{1-x}Ni_x$ Alloys by Positron Annihilation
71.25; 78.70; S1.2
(b) Vol.161,No.1, 281-286

C.Q. TANG and Y.F. ZHANG
Positron Lifetime Study on the Precipitation Phenomenon in 4Cr3Mo2NiVNb Alloy Steel
61.70 and 78.70; S1.2
(a) Vol.121,No.1, K21-K26

V.A. ARUTYUNOV, N.A. KUDRYASHOV, O.A. MISHIN, V.M. SAMSONOV, M.N. STRIKHANOV, and YU.A. SYTSKO
Synchrotron Radiation, Radiative Losses, and Polarization Effects for Relativistic Beams in Bent Crystals
61.80 and 78.70; S5.11
(b) Vol.161,No.2, 483-500

G. DLUBEK and W. GERBER
Estimation of Fermi Momentum in Al from Fourier Transform of Positron Annihilation Angular Correlation Curve
71.25 and 78.70; S4
(b) Vol.161,No.2, K81-K84

H. MURAKAMI and M. SANO
Positron-Trapping Centers in Potassium-Intercalated Graphite, KC_8 and $KC_8H_{2/3}$
61.70; 78.70; S2; S6
(a) Vol.121,No.2, 461-466

YU.A. NOVIKOV, A.V. RAKOV, and V.P. SHANTAROVICH
Positron Annihilation in Aluminium Films in Electric Contact with Silicon Substrate
73.30; 78.70; S4
(a) Vol.121,No.2, 495-500

J. DRYZEK and M. ROMEK
Positron Annihilation Study of Oxygen Vacancies in the High Temperature Superconductor Y-Ba-Cu-O
61.70 and 78.70; 74.50; S10.15
(a) Vol.121,No.2, K159-K162

N.YU. ARUTYUNOV, V.V. EMTSEV, K. SCHMALZ, and V.YU. TRASHCHAKOV
A Study of Defects in Heat Treated Cz-Silicon by Positron Annihilation
61.70 and 78.70; S5.11
(a) Vol.121,No.2, K163-K166

HONG ZHANG
Positron Trapping in $YBa_2Cu_{3-x}Fe_xO_y$
74.70 and 78.70; S10.15
(a) Vol.121,No.2, K207-K210

H. AOURAG, B. KHELIFA, and A. BELAIDI
Positron Charge Distribution in Aluminium
71.60; 78.70; S4
(b) Vol.162,No.1, K43-K46

V.G. BARYSHEVSKII and S.A. STEPANOV
Effect of Two Critical Angles of X-Ray Surface Diffraction on Secondary Emission Yield
61.10 and 78.70; 68.20
(a) Vol.122,No.1, 69-72

I.E. BONDARENKO and E.B. YAKIMOV
EBIC Investigation of Electrical Activity of Dislocations with Different Impurity Atmospheres in Si
61.70; 78.70; S5.11
(a) Vol.122,No.1, 121-128

MOTOKO KWETE, D. SEGERS, M. DORIKENS, L. DORIKENS-VANPRAET, and P. CLAUWS
Positron Annihilation Study of Defects Created in Silicon Irradiated with Electrons of High Energy
61.80; 78.70; S5.11
(a) Vol.122,No.1, 129-138

V.I. KUSHNIR and E.V. SUVOROV
X-Ray Backscattering on Perfect Crystals
78.70; 61.10; S5.11; S5.12
(a) Vol.122,No.1, 391-404

C.Q. TANG, G.Y. LI, and J.Y. TANG
Positron Lifetime Study on Thermal Fatigue in 5CrNiMo Steel
61.70 and 78.70; 62.20; S1.1
(a) Vol.122,No.1, K11-K14

J. SERNA
Positronium Diffusion in Crystalline Polyethylene
61.40 and 61.70; 78.70; S12
(a) Vol.122,No.2, 489-494

J. WERNISCH, A. SCHOENTHALER, and H.-J. AUGUST
A Method for Determining Detector Efficiencies In-Situ by Variation of the Radiation Incidence Angle
78.70
(a) Vol.122,No.2, 695-704

78.90. Other topics in optical properties of condensed matter and other interactions of matter with particles and radiation

F. ŠROBÁR
Absorptive Optical Bistability. The Case of Fabry-Perot
78.20 and 78.90
(b) Vol.159,No.1, 229-234

A.V. SAVCHUK, E.N. SALKOVA, T.A. SERGAN, and A. HEILMANN
Holographic Recording in Plasma Polymer Germanium Composite Films
78.90; S5.12
(a) Vol.122,No.1, K83-K86

79. Electron and Ion Emission by Liquids and Solids; Impact Phenomena

79.20. Impact phenomena (including electron spectra and sputtering)

A.D. POGREBNYAK
Metastable States and Structural Phase Changes in Metals and Alloys Exposed to High Power Pulsed Ion Beams (Review Article)
64.70 and 79.20; 62.20; 68.45; 68.55; S1.1; S1.2
(a) Vol.117,No.1, 17-52

S. BAUNACK, S. LANDGRAF, and A. ZEHE
Electron Beam Effects Due to Bombardment of CaF_2-Layers Investigated by Auger Electron Spectroscopy
61.80 and 68.60; 79.20; S9.12
(a) Vol.118,No.1, 203-208

E. TAUTE and K. BLUM
On the Charge Distribution of Ions after Auger Neutralisation at Surfaces
79.20; S4
(b) Vol.158,No.2, 551-556

N.A. DOUMANIS and CH.G. TRIKALINOS
On the Transverse Kinetic Energy of Relativistic Channeled Particles
61.80; 79.20
(b) Vol.158,No.2, K101-K104

E.N. IZAKOVICH, N.A. ZHELTOVA, YU.M. SHULGA, I.N. IVLEVA, ZH.D. SOKOLOVSKAYA, A.V. BALUEV, V.V. GANIN, O.S. ROSHUPKINA et al.
Erratum to Study of Ba and Cu(111) Based Inorganic Oxide Compounds (phys. stat. sol. (a) 115, K195 (1989))
74.70; 78.30; 79.20; 79.60
(a) Vol.118,No.2, K139-K140

P. WEISSBRODT and E. NEBAUER
SNMS Study of the Pt Out-Diffusion in the Au/Pt/Ti-n-GaAs System
73.30 and 79.20; S7.12
(a) Vol.119,No.2, K131-K134

R.F. SABIRYANOV, D.L. NOVIKOV, A.L. IVANOVSKII, and V.A. GUBANOV
Quantum-Chemical Calculations of Hydrogen Adsorption on the Polar Surface of Titanium Carbide
68.45 and 73.20; 79.20; S1.61
(b) Vol.160,No.2, 459-472

Z.-J. DING, H. YOSHIKAWA, and R. SHIMIZU
Angular Distribution of Elastically Reflected Electrons from Au
68.20 and 79.20; S1.3
(b) Vol.161,No.1, 257-270

A. REIF, P. STREUBEL, A. MEISEL, and D. ZEISSIG
Characterization of the Diffusion and Reaction Behaviour of Ti/Pt/Au Layer Contacts on GaAs by Means of Auger Electron..
68.48 and 73.40; 79.20; S7.12
(a) Vol.122,No.1, 331-340

79.40. Thermionic emission

K. BIEDRZYCKI
Thermally Stimulated Electron Emission from Triglycine Sulphate Crystals with Well-Defined Initial Macroscopic ...
77.80 and 79.40; S12
(a) Vol.117,No.1, 313-318

79.60. Photoemission and photoelectron spectra

G.I. ROSENMAN, YU.L. CHEPELEV, and E.I. BOIKOVA
Photo-Stimulated Relaxation of Exoemission in $Ba_2NaNb_5O_{15}$
72.40 and 79.60; S11.1
(a) Vol.117,No.1, 259-264

E.N. IZAKOVICH, N.A. ZHELTOVA, YU.M. SHULGA, I.N. IVLEVA, ZH.D. SOKOLOVSKAYA, A.V. BALUEV, V.V. GANIN, O.S. ROSHUPKINA et al.
Erratum to Study of Ba and Cu(111) Based Inorganic Oxide Compounds (phys. stat. sol. (a) 115, K195 (1989))
74.70; 78.30; 79.20; 79.60
(a) Vol.118,No.2, K139-K140

X.Y. ZHOU, H. ZHANG, and Q.R. ZHANG
Substitution Effect of Al and Sn for Cu in the Y-Ba-Cu-O System Studied by X-Ray Photoemission Spectroscopy
74.70 and 79.60; S10.15
(a) Vol.119,No.2, K145-K150

R. FRANKE, T. CHASSÉ, B. AL-ARAJ, P. STREUBEL, and A. MEISEL
Chemical Shifts of Auger Electron and Photoelectron Binding Energies of Phosphorus in Solid Compounds
79.60; S7.1; S11
(b) Vol.160,No.1, 143-152

X.D. LI and L.Z. ZENG
Structure and Reactivity of the System $Si/SiO_2/Fe$
68.48 and 79.60; S1.1; S10.1
(a) Vol.121,No.2, K185-K188

H. SOMMER, A. WEISS, and H. NEUMANN
Valence Band Density of States of $ZnSiP_2$
71.20 and 79.60; S7.16
(b) Vol.162,No.2, K87-K90

H. LU, C.L. BAO, X.N. YU, Q.K. XUE, D.S. TANG, P.J. WU, and Y.Y. XIONG
Crystallization on CuZr Amorphous Alloy Surface. An X-Ray Photoelectron Spectroscopy Study
61.50; 79.60; S1.2
(a) Vol.122,No.2, 481-488

79.70. Field emission and field ionization

H.-J. FITTING, G.O. MUELLER, R. MACH, G.U. REINSPERGER, TH. HINGST, and E. SCHREIBER
Vacuum Emission of Hot Electrons from ZnS
72.20 and 79.70; 85; S8.11
(a) Vol.121,No.1, 305-314

A. DARGYS, N. ŽURAUSKIENĖ, and S. ŽURAUSKAS
Temperature Dependent Field Ionization of Phosphorus Atoms in Silicon
71.55 and 79.70; S5.11
(b) Vol.162,No.1, 183-192

H.-J. FITTING, E. SCHREIBER, and TH. HINGST
Avalanche Measurement in ZnS by Vacuum Emission
72.20; 78.60; 79.70; 85; S8.11
(a) Vol.122,No.2, K165-K168

79.75. Exoelectron emission

B. SORKIN and H. KÄÄMBRE
Exoemission Mechanisms as Discerned via X-Ray Exposure Variations
71.55 and 79.75; S9.11; S10.1
(a) Vol.117,No.1, K33-K36

F. GOLEK, P.W. KLAMUT, B. SUJAK, A.J. ZALESKI, M. CISZEK, and J. OLEJNICZAK
Thermally Stimulated Charge Emission from the High Temperature Superconductor $YBa_2Cu_3O_{7-x}$
74.90 and 79.75; S10.15
(a) Vol.117,No.1, K43-K46

A. TOMITA, Y. FUKUDA, Y. KUTOMI, and N. TAKEUCHI
Emission Processes in Thermally Stimulated Exoelectron Emission and Thermoluminescence of KCl:Ag Single Crystals
61.70 and 78.60; 79.75; S9.11
(b) Vol.158,No.1, 383-390

M.I. MOLOTSKII and S.Z. SHMURAK
Electron Emission at Plastic Deformation of Colored Crystals
61.70; 79.75; S9
(a) Vol.120,No.1, 83-96

I.R. PLUMMER and D.W. TURNER
Imaging and Electron Energy Analysis of Exo-Electron Emission
79.75; 73.20; S9.11; S11.1
(a) Vol.122,No.1, 285-296

79.90. Other topics in emission and impact phenomena of condensed matter

L.L. BALASHOVA, N.M. KABACHNIK, and V.N. KONDRATEV
Charge Asymmetry in the Dependence of Stopping on Impact Parameter
61.80; 78.70; 79.90; S5.11
(b) Vol.161,No.1, 113-122

85. Devices

M.A. DEMYANENKO
The Effect of Back Inversion Channel and Electrically Active Defects on the Nonequilibrium Characteristics of SOS MOS...
73.40 and 85; S5.11
(a) Vol.117,No.2, 601-610

E. NEBAUER
Some Remarks on the Interface Displacement in the AuGeNi-GaAs System
68.48; 85; S7.12
(a) Vol.117,No.2, K111-K114

J.J. LIOU
Modeling the Steady-State Collector Current of AlGaAs/GaAs Heterojunction Bipolar Transistors Using a Generalized ...
73.40 and 85; S7.12; S7.15
(a) Vol.118,No.1, K65-K68

P.S. GLADKOV and K.S. VARBLIANSKA
Ta/GaAs Schottky Barriers Produced by in Situ Sputter Etching, RF Magnetron Sputtering of Ta, and Its Thermal Oxidation
73.30; 85; S1; S7.12
(a) Vol.118,No.2, 479-486

D.A. ARONOV, R. KABULOV, and YU.M. YUABOV
The Photoelectric Characteristics of Solar p-i-n Photocells Based on Hydrogenated Amorphous Silicon with Built-in ...
72.40 and 73.40; 85; S5.11
(a) Vol.118,No.2, 577-590

N.S. NAM, I. ZÓLOMY, and V. SCHILLER
Bipolar Transistor with Tunnel Oxide Emitter Structure (MISET)
73.40 and 85
(a) Vol.118,No.2, 591-596

J.J. LIOU
The Effects of Neutron Irradiation on the Current Gain of AlGaAs/GaAs Heterojunction Bipolar Transistors
73.40 and 85; S7.12; S7.15
(a) Vol.119,No.1, 337-342

I.M. HAFEZ, G. GHIBAUDO, and F. BALESTRA
Numerical and Analytical Modelling of Depletion-Mode MOS Transistors in Ohmic Region at Liquid Helium Temperature
73.40 and 85
(a) Vol.119,No.1, 343-352

P. KUIVALAINEN and K. LINDBERG
Hot Electron Transport in Submicron Semiconductor Devices
72.10 and 73.40; 85; S7.1
(b) Vol.159,No.2, 827-836

I. THURZO and E. PINČÍK
Attempts to Probe the Semi-Insulating GaAs-Plasma Oxide Interface by Small-Signal Charge DLTS
73.40 and 85; S7.12
(a) Vol.119,No.2, 701-710

J. KUZMIK and T. LALINSKY
Correlation between the Backgating Effect and the Gate Length Shortening on GaAs MESFETs
73.40 and 85; S7.12
(a) Vol.119,No.2, K185-K188

H. WENZEL and H.-J. WUENSCHE
A Model for the Calculation of the Threshold Current of SCH-MQW-SAS Lasers
73.40 and 85; S7.12; S7.15
(a) Vol.120,No.2, 661-674

G. KADEN
Zum Einfluss des Oberflaechenpotentials auf die Frequenzdispersion der Admittanz von GaAs-MIS-Teststrukturen ...
73.40 and 85; S7.12
(a) Vol.120,No.2, 675-686

R. SCHLESINGER, S. ROGASCHEWSKI, and P.J. JANIETZ
Electrochemically Deposited Copper Schottky Contacts on n-Type GaAs for Electron-Beam-Induced Current Measurements
73.40 and 85; 73.30; S7.12
(a) Vol.120,No.2, 687-694

D. SUEVA, V. SPASSOV, N. CHIKOV, E.I. VAPIREV, and I. IVANOV
Silicon Avalanche Detectors for X-Ray Analysis
76.80 and 85; S5.11
(a) Vol.120,No.2, K185-K188

H. MENNIGER, M. BEER, R. GLEICHMANN, H. RAIDT, B. ULRICI, and G. VOIGT
Electrical Homogeneity of Semi-Insulating LEC GaAs by Post-Growth Annealing
61.70 and 78.60; 85; S7.12
(a) Vol.121,No.1, 95-104

H.-J. FITTING, G.O. MUELLER, R. MACH, G.U. REINSPERGER, TH. HINGST, and E. SCHREIBER
Vacuum Emission of Hot Electrons from ZnS
72.20 and 79.70; 85; S8.11
(a) Vol.121,No.1, 305-314

G. KADEN
Zum Einfluss des Oberflaechenpotentials auf die Frequenz-dispersion der Admittanz von GaAs-MIT-Teststrukturen ... (II)
73.40 and 85; S7.12
(a) Vol.121,No.1, 315-326

G. EFTEKHARI
Effect of Rapid Thermal Annealing on Anodic Oxides on InP
73.40 and 85; S7.11
(a) Vol.121,No.1, 327-332

A. MZERD, D. SAYAH, P.C. RUSSELL, and B.K. JONES
Capacitance Measurements on Au-$Cd_{1-y}Zn_yS$ Polycrystalline Schottky Diodes
73.40 and 85; S8.15
(a) Vol.121,No.1, 333-340

A. VOHRA and K.K. SRIVASTAVA
Microwave Switching and AC Conductivity of Memory Glass $As_{10}Ge_{15}Te_{75}$
73.40 and 85; 61.40; S8
(a) Vol.121,No.1, 341-348

V.YU. RUD, YU.V. RUD, and M. SERGINOV
Natural Photopleochroism of n-In_2O_3-p-$CdSiAs_2$ Structures
72.40 and 73.40; 85; S7.16; S10.1
(a) Vol.121,No.1, K81-K84

A. POGGI, M. MERLI, E. SUSI, R. ANGELUCCI, and A. SARDO
Effects of Silicon Ion-Beam Mixing on p^+/n Diodes: DLTS Analyses of the Induced Defects
61.80 and 85; S5.11
(a) Vol.121,No.1, K135-K140

G. KISSINGER, K. TITTELBACH-HELMRICH, J. KNOPKE, and U. RETZLAFF
On the Gettering Effiency of Buried Layers in Dielectrically Insulated Structures
64.75 and 85; 71.55; S5.11
(a) Vol.121,No.1, K141-K144

A. MZERD, D. SAYAH, P.C. RUSSELL, and B.K. JONES
Characterization of Electron Traps in Au-$Cd_{1-y}Zn_yS$ Polycrystalline Schottky Diodes by Capacitance Transient ...
71.55 and 73.40; 85; S8.15
(a) Vol.121,No.2, 587-594

N.V. BRILLIANTOV, A.I. RUDENKO, YU.V. SHCHERBAKOV, and V.V. ZVEREV
Anomalous Low Temperature Annealing of Electron Irradiated Lithium-Doped Silicon Solar Cells
61.80; 71.55; 85; S5.11
(a) Vol.121,No.2, K155-K158

P. CHAKRABARTI, M. PURI, A. SINGHAL, and A. MISHRA
Optically Controlled Characteristics of TEGFET
73.40 and 85; S7.12; S7.16
(a) Vol.122,No.1, 405-412

A. SCHENK
Spatially Variable Drift Mobility Model for $Hg_{1-x}Cd_xTe$ Diodes. I. Analytical Base and Fit to Hall Data
73.40 and 85; S8.15
(a) Vol.122,No.1, 413-426

J. BENOIT, P. BENALLOUL, C. BARTHOU, S. CASETTE, and J.C. SORET
De-Excitation Processes and Efficiency in AlE ZnS:Mn Thin Film Electroluminescent Devices
78.60 and 85; S8.11
(a) Vol.122,No.1, 427-438

T. ASAHI, M. YUDASAKA, K. NAKANISHI, S. IWASHIMA, J. AOKI, N. SATO, and H. INOKUCHI
Rectification Properties of In/Tetrabenzo[de,hi,op,st]Pentacene/Au Devices
73.40 and 85; S12
(a) Vol.122,No.1, K97-K100

J. KOLNÍK and M. OŽVOLD
The Influence of Inversion Surface Layers on the Evaluation of the Interface State Energy Distribution from Schottky...
73.30 and 73.40; 85; S5.11; S7.12
(a) Vol.122,No.2, 583-588

J. BARTOŠ, E. PINČÍK, I. THURZO, B. MATÁTKO, and T. LALINSKÝ
Characterization of MAOS Structures on GaAs
73.40 and 85; S7.12
(a) Vol.122,No.2, 715-722

A. SCHENK
Spatially Variable Drift Mobility Model for $Hg_{1-x}Cd_xTe$-Diodes. II. Application to Device Simulation
73.40 and 85; S8.15
(a) Vol.122,No.2, 723-730

F.R. KESSLER, W. KOEPPEN, and G. KRAUSE
External Stabilization of CO_2 Laser Radiation by Means of i-Semiconductor Absorption
78.20 and 85; 78.30; S5.12; S7.13
(a) Vol.122,No.2, 731-740

M. BRAUER
Effect of Surface Charge and Coating on the Field Distribution of Si High-Voltage Thyristor-Like Structures with ...
73.40 and 85; S5.11
(a) Vol.122,No.2, 741-750

G. EFTEKHARI
Characterization of Rapidly Thermally Annealed GaAs and InP Surfaces Using Schottky Barriers
73.30 and 85; S7.11; S7.12
(a) Vol.122,No.2, 751-756

H.-J. FITTING, E. SCHREIBER, and TH. HINGST
Avalanche Measurement in ZnS by Vacuum Emission
72.20; 78.60; 79.70; 85; S8.11
(a) Vol.122,No.2, K165-K168

S. Substances (classification according to physica status solidi)

S1. Transition metals and their alloys and compounds
(for oxides, see S.10)

S. SALEH and M.A. ZAKI EWISS
Studies of the Amorphous-Crystalline Transition in the Metallic Glass $Fe_{32}Ni_{36}Cr_{14}P_{12}B_6$
72.15; S1
(a) Vol.117,No.2, K133-K136

B.A. GREENBERG, M.I. KATSNELSON, V.G. KORESHKOV, YU.N. OSETSKII, G.V. PESCHANSKIKH, A.V. TREFILOV, YU.F. SHAMANAEV et al.
On the Possibility of Describing Lattice Properties of Iridium in Terms of Pseudopotential Theory
63.20 and 71.10; 61.70; 62.20; S1
(b) Vol.158,No.2, 441-456

A.L. IVANOVSKII
Band Structure of Ti, Zr, Hf Cubic Monoborides
71.25; S1
(b) Vol.158,No.2, K129-K132

R. PIRLOT, M. CYAMUKUNGU, L. GRENACS, J. LEHMANN, and K. TOMPA
Relaxation Time Measurements and Hyperfine Fields at ^{12}B in Transition Metals
76.60; S1
(b) Vol.158,No.2, K197-K200

P.S. GLADKOV and K.S. VARBLIANSKA
Ta/GaAs Schottky Barriers Produced by in Situ Sputter Etching, RF Magnetron Sputtering of Ta, and Its Thermal Oxidation
73.30; 85; S1; S7.12
(a) Vol.118,No.2, 479-486

J.C. UPADHYAYA, S. LEHRI, D.K. SHARMA, and S.C. UPADHYAYA
Angular Force Models in the Lattice Dynamics of Cubic Systems
63.20; 62.20; S1
(b) Vol.159,No.2, 659-674

W. WASSERBAECH
Third-Order Elastic Constants of a Cubic Quasi-Isotropic Solid
62.20; S1
(b) Vol.159,No.2, 689-698

P.M. OSSI
Crystal-Glass Phase Transition in Ion Irradiated Binary Systems
61.80 and 64.75; 61.40; S1
(a) Vol.119,No.2, 463-470

I.M. MIKHAILOVSKII and V.B. RABUKHIN
Energy of Boundaries in the Vicinity of a Triple Junction
61.70; S1
(a) Vol.119,No.2, K113-K116

A.M. CHAPLANOV and A.N. SHIBKO
Modification of Metal Films by Pulsed Laser Radiation in Organic Liquids
68.55; S1; S1.61
(a) Vol.120,No.1, K37-K40

A.G . POPOV, YE.V. BELOZEROV, A.G. KUCHIN, A.S. ERMOLENKO, G.M. MAKAROVA, V.S. GAVIKO, and V.I. KHRABROV
Magnetic Properties of $Sm_2(Fe,M)_{17}C_x$ Compounds, M = Co,Al,Ga
75.50; S1
(a) Vol.121,No.1, K111-K116

Y. TAHTAMONI and S. ERKOÇ
Structural Stability and Energetics of X_n(X=V,Cr,Nb,n=3 to 7) Microclusters: Empirical Many-Body Potential Energy ...
61.50; S1; S1.2
(b) Vol.162,No.1, K5-K8

V.G. CHUDINOV, R.M.J. COTTERILL, and V.V. ANDREEV
Kinetics of the Diffuse Processes within a Cascade Region in the Sub-Threshold Stages of F.C.C. and H.C.P. Metals
66.30; S1; S4
(a) Vol.122,No.1, 111-120

G. D ANNA and W. BENOIT
Dislocation Internal Friction Investigation in High Purity Cold Worked Tantalum
61.70 and 62.40; S1
(a) Vol.122,No.1, 153-162

P. ANAGNOSTARAS, I. SIKAKANA, and J.E. LOWTHER
Electronic Structure and Thermodynamic Properties of PdH_x
65.50; 71.20; S1
(b) Vol.162,No.2, 413-420

A.K. SOLANKI, R. AHUJA, and S. AULUCK
Fermi Surface and Mass Enhancement Factor for Niobium
71.25; S1
(b) Vol.162,No.2, 497-508

P. SISODIA and M.P. VERMA
Polycrystalline Elastic Moduli of Some Hexagonal and Tetragonal Materials
62.20; S1; S8; S10
(a) Vol.122,No.2, 525-534

S1.1. Fe

A.D. POGREBNYAK
Metastable States and Structural Phase Changes in Metals and Alloys Exposed to High Power Pulsed Ion Beams (Review Article)
64.70 and 79.20; 62.20; 68.45; 68.55; S1.1; S1.2
(a) Vol.117,No.1, 17-52

V. SREERAMALU, V. RAVINDRACHARY, H.R. SREEPAD, A. CHANDRASHEKARA, S. GOPAL, H. SANJEEVAIAH, and B. VISWANATHAN
Structural Relaxation and Crystallization of Metallic Glasses $Fe_{35}Co_{52}V_{13}$ and $Fe_{45}Co_{45}V_6Ni_4$. Positron Annihilation Study
61.40; 78.70; S1.1; S1.2
(a) Vol.117,No.1, 53-60

B.X. GU, B.G. SHEN, and H.R. ZHAI
Magnetic Properties and Crystallization of Amorphous $Nd_{14}Fe_{81}B_5$ Alloys
75.50; 61.40; S1.1
(a) Vol.117,No.1, 283-290

E.P. ELSUKOV, YU.N. VOROBEV, A.V. TRUBACHEV, and V.A. BARINOV
Structure and Magnetic Properties of Fe-P Electrodeposited Alloys
61.55 and 75.50; 76.80; S1.1
(a) Vol.117,No.1, 291-298

M. JURCZYK and O.D. CHISTYAKOV
Magnetic Properties of Substituted $Nd_{2-x}R_x(Fe,Re,Co)_{14}B$ Compounds (R=Tb or Dy)
75.50; S1.1; S1.2
(a) Vol.117,No.1, 299-304

K. KRIŠTIAKOVA, G. TUMBEV, and J. KRIŠTIAK
Annihilation of Positrons in $Fe_{80-x}Ni_xB_{20}$ Metallic Glasses
61.40 and 78.70; S1.1
(a) Vol.117,No.1, K5-K8

E. TAMURA, R. FEDER, G. WALLER, and U. GRADMANN
Determination of the Surface Magnetization of Fe(110) by Spin-Polarized Low-Energy Electron Diffraction
68.20 and 75.30; 61.14; S1.1
(b) Vol.157,No.2, 627-634

TIESONG ZHAO, ZHONGZHI FAN, and HANMIN JIN
Magnetization Processes of the Rare-Earth and Fe Sublattices in $R_2Fe_{14}B$ (R=Tb,Dy,Ho,Er,and Tm)
75.30; 75.60; S1.1; S1.4
(b) Vol.157,No.2, 677-684

D. PŁUSA, R. PFRANGER, and B. WYSŁOCKI
Magnetic Domain Structure of Sintered $Nd_{15}Fe_{77}B_8$ Permanent Magnet in Rotational Magnetic Field
75.60; S1.1
(a) Vol.117,No.2, 541-548

V.V. SINGER, A.E. VAFIN, L.M. SANDRATSKII, L.P. MOKHRACHEVA, S.V. KONOVALOV, I.Z. RADOVSKII, and P.V. GELD
Magnetic Susceptibility and Valency of Ce and Pr in Their Liquid Binary Alloys with Metals of the Iron Subgroup
75.50; S1.1; S1.2; S1.4
(a) Vol.117,No.2, 577-586

W.A. KACZMAREK, R. BRAMLEY, and A. CALKA
Itinerant Electron Contribution to the Magnetism of Ni-TM Glassy Metallic Magnets
75.50 and 76.30; S1.1; S1.2
(a) Vol.117,No.2, K141-K146

S. SZUSZKIEWICZ
Interaction of Positrons with the $Co_{66}Fe_4Mo_2Si_{16}B_{12}$ Alloy
78.70; S1.1
(b) Vol.158,No.1, K79-K84

K. MAKI and H. KRONMUELLER
Investigation of Demagnetization Processes in $Fe_{14}Nd_2B$-Type Sintered Magnets
75.60; S1.1; S1.4
(a) Vol.118,No.1, 253-270

R. SCHAEFER and A. HUBERT
A New Magnetooptic Effect Related to Non-Uniform Magnetization on the Surface of a Ferromagnet
75.60; 78.20; S1.1; S1.2; S1.4; S11.2
(a) Vol.118,No.1, 271-288

G. SERFŐZŐ, E. KISDI-KOSZÓ, A. SLAWSKA-WANIEWSKA, L. POTOCKY, S. JONELIUNAS, and M. BARAN
Study of Metallic Glasses Rapidly Quenched in Magnetic Field
75.50; 76.50; S1.1
(a) Vol.118,No.1, 307-310

S.KH. KARPENKOV
s-d Spin-Orbit Interaction between Conduction Electrons and d-Spin System
72.15; 71.70; S1.1; S1.2
(b) Vol.158,No.2, K169-K172

H. REUTHER, E. RICHTER, and I.S. SZUECS
CEMS Measurements on Ion Implanted Tool Steels
61.80 and 76.80; S1.1
(a) Vol.118,No.2, 409-414

K.A. AL-SALEH, I.J. JABR, and N.S. SALEH
A Comparison of Ion Beam Induced Atomic Mixing Kinetics of Ti/Si, Fe/Si, and Ni/Si Systems. In Situ Electrical ...
61.80 and 73.40; S1.1; S1.2; S5.11
(a) Vol.118,No.2, 467-472

P.H. DIONISIO, B.A.S. DE BARROS, JR., and I.J.R. BAUMVOL
A Moessbauer Spectroscopic Study of the High-Temperature Oxidation Behaviour of an Austenitic Stainless Steel ...
61.80 and 68.20; 76.80; S1.1
(a) Vol.118,No.2, 473-478

J. ZBROSZCZYK, W. CIURZYŃSKA, J. ŚWIERCZEK, S. SZYMURA, B. WYSŁOCKI, M. BARAN, and A. WIŚNIEWSKI
Effect of Surface Layer Thinning on Approach to Magnetic Saturation in $Co_{64}Fe_{14}Si_{11}B_{11}$ Amorphous Alloy
75.50; S1.1; S1.2
(a) Vol.118,No.2, 525-530

Z.T. SERIKBAEVA, T.M. ZHANTIKIN, and A.K. ZHETBAEV
Moessbauer Study of 30 MeV Proton Dose Dependence of Radiation Damage in Disordered Ni_3Fe
61.80 and 76.80; S1.1
(a) Vol.118,No.2, K99-K102

S.U. JEN, M.C. CHIEN, and C.Y. LEE
Viscous Flow in Amorphous Alloys
61.40 and 64.70; 62.20; S1.1; S1.2
(a) Vol.119,No.1, 35-40

Y.Z. ZHANG, L.H. ZHE, X.Z. DU, R.M. YANG, and R. XU
Effect of Yb on Electrodeposited Amorphous Fe-P Alloys
61.40; S1.1
(a) Vol.119,No.1, K1-K4

XIAO-GUANG WANG and HONG ZHANG
A Vacancy Complex in the Dilute Fe-V-N Alloy Identified by Positron Annihilation Spectroscopy
61.70 and 78.70; S1.1
(a) Vol.119,No.1, K15-K20

BAO-GEN SHEN, LIN-YUAN YANG, JUN-XIAN ZHANG, FENG WO, TAI-SHAN NING, JIAN-GAO ZHAO, HUI-QUN GUO, and WEN-SHAN ZHAN
Hard Magnetic Properties of Heat Treated $Fe_{72.8}Co_{4.7}Nd_4B_{18.5}$ Alloy
75.50; 75.60; S1.1
(a) Vol.119,No.1, K63-K66

A. RAOUFI and S. ARAJS
Anisotropy of Magnetoresistivity in Crystalline and Amorphous Ferromagnets
72.15 and 75.50; S1.1
(a) Vol.119,No.1, K75-K78

M.N. GHATIKAR, K.V.S. MAHESH, and S.N. BHATIA
Improved Calculations of Phase Shifts for EXAFS Spectroscopy
61.40; 78.70; S1.1; S1.2
(b) Vol.159,No.2, 551-556

J. CHOJCAN
The Moessbauer Effect and Positron Annihilation Studies of Iron-Cobalt Alloys
71.20 and 76.80; S1.1
(b) Vol.159,No.2, K59-K62

V.G. CHUDINOV, B.N. GOSHCHITSKII, N.V. MOSEEV, and V.V. ANDREEV
Kinetics of Radiation-Induced Segregation of Impurities in F.C.C. Metals at the Thermal Stage of the Collision Cascade
61.80 and 64.75; S1.1; S4
(a) Vol.119,No.2, 437-442

J. ČERMÁK
Diffusion of ^{51}Cr along High-Diffusivity Paths in Ni-Fe Alloys
66.30; S1.1; S1.2
(a) Vol.119,No.2, 443-454

H.K. LACHOWICZ, E. PULIDO, A. SIEMKO, and A. HERNANDO
Stress-Forced Magnetoresistance in Near-Zero Magnetostrictive Metallic Glasses
72.15 and 75.80; 75.60; S1.1; S1.2
(a) Vol.119,No.2, 595-600

M. HRABCÁK and E. KISDI-KOSZÓ
Modification of Magnetic Properties of Amorphous Fe-Si-B Ribbons by Flash Annealing
75.50; 75.60; S1.1
(a) Vol.119,No.2, 601-606

J. POKLUDA and P. ŠANDERA
On the Relation of Ideal Strength to Other Physical Properties
62.20; S1.1; S1.2
(b) Vol.160,No.1, 89-96

Z.Z. AKSELROD, B.A. KOMISSAROVA, L.N. KRYUKOVA, G.K. RYASNYI, L.G. SHPINKOVA, and A.A. SOROKIN
Electron Quadrupole Interaction of ^{181}Ta in Hf_2Fe Remeasured
71.70; 76.80; S1.1; S4
(b) Vol.160,No.1, 255-262

L.I. KURKINA and O.V. FARBEROVICH
Electron Structure and Magnetic Properties of Small Iron Particles
71.50; 73.20; 75.25; S1.1
(b) Vol.160,No.1, K37-K42

B.X. GU, H. HOMBURG, S. METHFESSEL, and H.R. ZHAI
Crystallization Behavior and Magnetic Properties of Amorphous Nd-Fe-B Thin Films
68.55 and 75.70; 76.80; S1.1
(a) Vol.120,No.1, 159-168

G.D. MAKSIMOVIĆ, Z.S. POPOVIĆ, F.R. VUKAJLOVIĆ, and K. VULETIĆ
Magnetic Properties of Pd_3Fe
75.50; S1.1
(b) Vol.160,No.2, 635-640

A. PAJA
Concentration and Temperature Dependence of the Resistivity of Amorphous $Fe_{1-x}Zr_x$ Alloys
72.15; S1.1
(b) Vol.160,No.2, K123-K128

Z.T. SERIKBAEVA, T.M. ZHANTIKIN, A.K. ZHETBAEV, and K.G. BINNATOV
Moessbauer Investigations of Ordering of the $Ni_3(Fe_{1-x}Mn_x)$ Alloy System
61.55; 76.80; S1.1; S1.2
(a) Vol.120,No.2, 305-314

N.P. KULISH and P.V. PETRENKO
Short-Range Order in the Binary Solid Solutions. Ordering and Its Change on Heating in Fe-Al, Cu-Al, and Ag-Al Alloys
61.55; S1.1; S1.2; S1.3; S4
(a) Vol.120,No.2, 315-326

J. ČERMÁK
Diffusion along Moving Grain Boundaries
66.30; 61.70; S1.1; S1.2
(a) Vol.120,No.2, 351-362

E. MAJKOVA, V. BOHAC, S. LUBY, R. SANDRIK, and M. VESELY
Diffusion of Au Atoms in Ni-Si-B and Fe-Co-B Metallic Glasses
66.30; S1.1; S1.2
(a) Vol.120,No.2, 371-378

N.V. EDNERAL, V.G. KUTILIN, YU.A. LISOVSKII, M.L. SOKOLOVA, A.M. UMPELEV, A.O. TITOV, and L.K. FIONOVA
Thin Film Structures for Magneto-Optical Information Recording
68.55; 75.70; 78.20; S1.1
(a) Vol.120,No.2, 457-466

A.H. WAFIK and A.Z. MOHAMED
Barkhausen Jumps and Microstructure in Molybdenum Eutectoid Steel
75.60; 64.80; S1.1
(a) Vol.120,No.2, 601-608

J. CHEN and H. KRONMUELLER
Properties of Nd-Fe-B Magnets Processed by the Hydrogen Decrepitation Technique
75.60; S1.1
(a) Vol.120,No.2, 617-626

G.F. ZHOU, X.K. SUN, S.Y. FU, Y.C. CHUANG, R. GROESSINGER, and H.R. KIRCHMAYR
Relationship between Microstructure and Coercivity in (Nd,Dy)-(Fe,Co)-B Based Sintered Magnets with Minor Nb and Al ...
75.50; 64.80; S1.1
(a) Vol.120,No.2, 627-634

O. RADULESCU and R. MANAILA
On the Bond Orientational Order in Icosahedral Phases
61.50; 61.55; S1.1; S1.2; S4
(b) Vol.161,No.1, 65-74

SIE CHIN TJONG
Dislocation Microstructures in Fatigued Fe-30Mn-10Al-0.4C Alloy
62.20; 61.70; S1.1
(a) Vol.121,No.1, 119-128

ZHI-DONG ZHANG, X.K. SUN, ZHUONG-WU ZHANG, Y.C. CHUANG, and F.R. DE BOER
Effects of Partial Co Substitution on Magnetic Properties of $(Nd,Gd)_2Fe_{14}B$
75.30; 75.60; S1.1
(a) Vol.121,No.1, 241-248

G.F. ZHOU, S.Y. FU, X.K. SUN, and Y.C. CHUANG
Influence of Annealing on the Magnetic Properties and Microstructure of Nd-Fe-B Based Magnets
75.60; S1.1
(a) Vol.121,No.1, 257-264

B.X. GU, B.G. SHEN, and H.R. ZHAI
Crystallization and Magnetic Properties of Amorphous $Nd_{0.19}Fe_{0.81}$ Alloy
75.50; 61.40; S1.1; S1.4
(a) Vol.121,No.1, K99-K104

R. SHYAM, S.C. UPADHYAYA, and J.C. UPADHYAYA
First-Principles Calculation of the Lattice Dynamics of the $Co_{0.92}Fe_{0.08}$ Alloy
63.20; S1.1; S1.2
(b) Vol.161,No.2, 565-574

R. BRUN DEL RE, J.C. WOOLLEY, M. QUINTERO, and R. TOVAR
T(z) Diagram of $CuIn_{1-z}Fe_zS_2$ Alloys
64.75; 61.55; S1.1
(a) Vol.121,No.2, 483-488

K. WOKULSKA, J. HAERTWIG, and S. GROSSWIG
The Bond Method Used to Determine the Value of the Lattice Parameters a for alpha-Fe
68.70; 61.10; S1.1
(a) Vol.121,No.2, 539-546

ZHI-DONG ZHANG, X.K. SUN, Y.C. CHUANG, F.R. DE BOER, and R.J. RADWAŃSKI
A Model Based on Single-Ion Theory. II. Preferential Substitution and Local-Environment Effects at Different ...
75.50; 75.30; S1.1; S1.2
(a) Vol.121,No.2, 617-626

X.D. LI and L.Z. ZENG
Structure and Reactivity of the System $Si/SiO_2/Fe$
68.48 and 79.60; S1.1; S10.1
(a) Vol.121,No.2, K185-K188

BAO-GEN SHEN and S. METHFESSEL
Crystallization of the Amorphous $Fe_{83}Nd_7B_{10}$ Alloy and the Formation Process of $Nd_2Fe_{23}B_3$ Phase
75.50; S1.1
(a) Vol.121,No.2, K211-K214

N.S. ALY, I. MAHMOUD, M.E. BASSIOUNI, H.A. MOTAWEH, and A.S. ILYUSHIN
Magnetic Phase Diagram of the $Tb_{1-x}Nd_xFe_2$ Quasibinary System
75.30; S1.1
(a) Vol.121,No.2, K219-K222

D. DĂDÂRLAT, M. CHIRTOC, C. NEAMTU, R.M. CÂNDEA, and D. BICANIC
Inverse Photopyroelectric Detection Method
65.90 and 77.70; S1.1; S1.2
(a) Vol.121,No.2, K231-K234

V.N. NOVIKOV
Microcreep of Metallic Glasses at Room Temperature
62.20; 61.40; S1.1; S1.2
(a) Vol.122,No.1, 215-220

K. CHERIFI, P. DONOVAN, C. DUFOUR, PH. MANGIN, and G. MARCHAL
Microstructural Observations of Rare-Earth/Iron Multilayers
68.55; S1.1; S1.4
(a) Vol.122,No.1, 311-322

T. LUCINSKI
The Resistivity of Amorphous $Fe_{1-x}Ti_x$ Thin Films
73.60; 72.15; S1.1
(a) Vol.122,No.1, 323-330

Y. KHAN
Phases Obtained by Continuous Heating of the Amorphous Permanent Magnet Alloy $Nd_{4.4}Fe_{77.8}B_{17.8}$
61.55 and 75.50; S1.1
(a) Vol.122,No.1, K1-K6

V.S. STEPANYUK, A.A. KATSNELSON, O.S. TRUSHIN, and A. SZÁSZ
Structure of the Amorphous Fe-B Alloy Modeled by Molecular Dynamics
61.40; S1.1
(a) Vol.122,No.1, K7-K10

C.Q. TANG, G.Y. LI, and J.Y. TANG
Positron Lifetime Study on Thermal Fatigue in 5CrNiMo Steel
61.70 and 78.70; 62.20; S1.1
(a) Vol.122,No.1, K11-K14

M.A. ZAKI EWISS and S. SALEH
Temperature Dependence of the Resistivity of $Fe_{40}Ni_{40}P_{14}B_6$ Amorphous and Crystalline Alloys
72.15; S1.1
(a) Vol.122,No.1, K63-K68

X.K. SUN, ZHI-GANG ZHAO, LIU WEI, and Y.C. CHUANG
Structure and Magnetic Properties of $Pr_2Fe_{14-x}Co_xB_{1-y}C_y$ Compounds
75.50; S1.1
(a) Vol.122,No.1, K69-K72

I. GOŚCIAŃSKA, H. RATAJCZAK, P. SOVÁK, P. SPIŠÁK, and M. KONČ
The Structure of Fe-Rich Vapor Quenched Fe-Ti Thin Films
68.55; 61.40; 73.60; S1.1; S1.2
(a) Vol.122,No.2, 575-582

JUN-XIAN ZHANG, BAO-GEN SHEN, and WEN-SHAN ZHAN
Magnetic Properties of Amorphous $Nd_4Fe_{96-x}B_x$ Alloys and Influences of Heat Treatment on Its Hard Magnetic Properties
75.30; 75.60; S1.1
(a) Vol.122,No.2, 651-656

W.A. KACZMAREK, R. BRAMLEY, and A. CALKA
ESR Study of Amorphous Metallic Alloys below the Magnetic Multicritical Point
75.50 and 76.30; 75.40; S1.1; S1.2
(a) Vol.122,No.2, 663-676

S1.2. Other 3d metals

J.W. STASIAK, J. KIERUL, and L. KUENNE
Influence of Phonons on the Calculations of the Density of States in Thin Films
71.20 and 73.20; S1.2
(b) Vol.157,No.1, 229-234

R.A. EVARESTOV and V.A. VERYAZOV
The Electronic Structure of Copper Oxide Crystalline Compounds. I. LUC-CNDO Approach to the Electronic Structure of Cu_2O and ...
71.10 and 71.25; 74.10; S1.2; S10.1
(b) Vol.157,No.1, 281-292

E. FLOUDA, N. BOUKOS, and C. PAPASTAIKOUDIS
Hall Coefficient of Dilute CuPd/CuRh/CuZr and CuY Alloys. d-Resonance Scattering
72.15; S1.2
(b) Vol.157,No.1, 351-356

E.A. VASILEV and T.M. TKACHENKO
The ^{119}Sn Atom Dynamics in NiSb-$Ni_{1.5}$ Solid Solutions
76.80; S1.2
(b) Vol.157,No.1, K61-K64

A.D. POGREBNYAK
Metastable States and Structural Phase Changes in Metals and Alloys Exposed to High Power Pulsed Ion Beams (Review Article)
64.70 and 79.20; 62.20; 68.45; 68.55; S1.1; S1.2
(a) Vol.117,No.1, 17-52

V. SREERAMALU, V. RAVINDRACHARY, H.R. SREEPAD, A. CHANDRASHEKARA, S. GOPAL, H. SANJEEVAIAH, and B. VISWANATHAN
Structural Relaxation and Crystallization of Metallic Glasses $Fe_{35}Co_{52}V_{13}$ and $Fe_{45}Co_{45}V_6Ni_4$. Positron Annihilation Study
61.40; 78.70; S1.1; S1.2
(a) Vol.117,No.1, 53-60

YU.N. DEVYATKO, V.N. TRONIN, and V.I. TROYAN
Kinetics of the New Phase Nucleus Growth on a Magnet near the Curie Point
61.50; 75.30; S1.2
(a) Vol.117,No.1, 61-66

J.L. MACQUERON, LL. MAÑOSA, and G. GUÉNIN
High Temperature Martensite Induced by Quenching
64.70; S1.2
(a) Vol.117,No.1, 113-118

P. FELTHAM and N. KAUSER
Solid-Solution Hardening
62.20; S1.2
(a) Vol.117,No.1, 135-140

M. JURCZYK and O.D. CHISTYAKOV
Magnetic Properties of Substituted $Nd_{2-x}R_x(Fe,Re,Co)_{14}B$ Compounds (R=Tb or Dy)
75.50; S1.1; S1.2
(a) Vol.117,No.1, 299-304

S.J. TZENG, Y.D. YAO, and T.H. CHUANG
Electrical Resistivity of Nickel-Rich Nickel-Indium Alloys between 10 and 800 K
72.15; S1.2
(a) Vol.117,No.1, K47-K52

G.E. ABROSIMOVA, N.I. NOVOKHATSKAYA, and A.V. SEREBRYAKOV
The Magnetic Field Effect on the Creep-Induced Structural Anisotropy of Amorphous Co-Fe-Ni-Si-B Alloys
75.30; S1.2
(a) Vol.117,No.1, K67-K70

M. FAEHNLE and J. FURTHMUELLER
On the Change of Magnetostriction by Field Annealing of Amorphous Ferromagnetic Alloys
75.80; S1.2
(a) Vol.117,No.1, K71-K76

A.K. BANDYOPADHYAY and S.K. SEN
Calculation of Self-Interstitial Formation Energy (Both Split and Non-Split) in Noble Metals
61.70; S1.2; S1.3
(b) Vol.157,No.2, 519-530

S. MOCHIZUKI, K. ISHI, and A. JOHGO
Far-Infrared Absorption in Nickel and Copper Microcrystals
78.30; S1.2
(b) Vol.157,No.2, K137-K140

J. ČERMÁK
Grain Boundary Diffusion of Co-60 in Nickel
66.30; S1.2
(a) Vol.117,No.2, 387-394

A.D. SHEIKH-ALI and R.Z. VALIEV
Dislocation Analysis of the Coupling of Grain Boundary Sliding and Migration during the Deformation of Zn Bicrystals
61.70; S1.2
(a) Vol.117,No.2, 429-436

Z.M. FARID, S.A. MAHMOUD, A.R. ALI, and N. EL-NAQUIB
On the Transition in Creep Behaviour during Primary Creep of Cu-Zn Alloys
62.20; S1.2
(a) Vol.117,No.2, 437-446

P. NOZAR, V. SECHOVSKY, M. SOLANKI-MOSER, E. BAUER, and E. GRATZ
Point-Contact Spectroscopy on the $Ce(Cu_xAl_{1-x})_5$ System
71.20; 72.20; 75.30; S1.2; S1.4
(a) Vol.117,No.2, 501-508

R.Z. VALIEV, YA.D. VISHNYAKOV, R.R. MULYUKOV, and G.S. FAINSHTEIN
On the Decrease of Curie Temperature in Submicron-Grained Nickel
75.30; 73.40; S1.2
(a) Vol.117,No.2, 549-554

N. SCHELUDKO and M. MIKHOV
Investigation of Magnetocrystalline Anisotropy of $Mm_2Co_{17-x}Zr_x$ Compounds
75.30; S1.2; S1.4
(a) Vol.117,No.2, 571-576

V.V. SINGER, A.E. VAFIN, L.M. SANDRATSKII, L.P. MOKHRACHEVA, S.V. KONOVALOV, I.Z. RADOVSKII, and P.V. GELD
Magnetic Susceptibility and Valency of Ce and Pr in Their Liquid Binary Alloys with Metals of the Iron Subgroup
75.50; S1.1; S1.2; S1.4
(a) Vol.117,No.2, 577-586

L.M. DI, P.I. LOEFF, and H. BAKKER
Mechanical Amorphisation of Ni_3Sn
64.70; S1.2
(a) Vol.117,No.2, K99-K102

P. CZARNECKI, M. IDZIKOWSKI, and A. WRZECIONO
Magnetoresistance of $Tb_xCo_{70-x}B_{30}$ and $Tm_xCo_{70-x}B_{30}$ Amorphous Ribbons in High Magnetic Fields
72.15; S1.2
(a) Vol.117,No.2, K137-K140

W.A. KACZMAREK, R. BRAMLEY, and A. CALKA
Itinerant Electron Contribution to the Magnetism of Ni-TM Glassy Metallic Magnets
75.50 and 76.30; S1.1; S1.2
(a) Vol.117,No.2, K141-K146

R.A. EVARESTOV and V.A. VERYAZOV
The Electronic Structure of Copper Oxide Crystalline Compounds
II. Chemical Bonding in Copper-Oxygen Crystals
71.20 and 71.25; 74.10; S1.2; S10
(b) Vol.158,No.1, 201-212

A. BUKALUK
AES Depth Profile Studies of Interdiffusion in the Ag-Cu
Bilayer and Multilayer Thin Films
66.30 and 68.48; 61.80; S1.2; S1.3
(a) Vol.118,No.1, 99-108

E.I. SHREDER, I.A. POPOV, E.E. KRASOVSKII, and I.I. SASOVSKAYA
The Optical Properties and Electronic Structure of AgZn Alloy
Undergoing $\beta' \rightleftharpoons \xi$ Structural Transformation
64.70 and 78.20; 71.25; S1.2; S1.3
(a) Vol.118,No.1, 147-152

H. RATAJCZAK, I. GOŚCIAŃSKA, and T. LUCIŃSKI
Influence of Annealing on Saturation Magnetization and Hall
Resistivity in Amorphous $Nd_{12}Co_{82}B_6$ Thin Films
61.40 and 68.55; 73.60; 75.70; S1.2; S1.4
(a) Vol.118,No.1, 189-196

R. SCHAEFER and A. HUBERT
A New Magnetooptic Effect Related to Non-Uniform Magnetization
on the Surface of a Ferromagnet
75.60; 78.20; S1.1; S1.2; S1.4; S11.2
(a) Vol.118,No.1, 271-288

Y. DAI, S.M. LIU, and Q.P. KONG
The Changes of Relaxation Parameters of Grain Boundaries in
the Course of Constant Rate Deformation
62.20; 62.40; S1.2
(a) Vol.118,No.1, K21-K26

R.K. ISLAMGALIEV, N.A. AKHMADEEV, R.R. MULYUKOV, and R.Z. VALIEV
Grain Boundary Influence on the Electrical Resistance of
Submicron Grained Copper
72.15; S1.2
(a) Vol.118,No.1, K27-K30

NINGSHENG LUO, WENLAN XU, and SHUECHU SHEN
Embedded-Atom Method for Surface Lattice Dynamics in Copper
68.30; S1.2
(b) Vol.158,No.2, 493-502

B. REICHERT and CH. JUNG
Electronic Structure of the Charged Copper(001)/Vacuum
Interface
73.20; 73.30; 73.60; S1.2
(b) Vol.158,No.2, K119-K122

S.KH. KARPENKOV
s-d Spin-Orbit Interaction between Conduction Electrons and d-Spin System
72.15; 71.70; S1.1; S1.2
(b) Vol.158,No.2, K169-K172

Y. NAKAMURA, H. KOIKE, and O. NITTONO
Structure of Long Period Superstructure in Cu-Rich gamma-Brass Investigated by Electron Diffraction
61.60 and 61.70; 68.48; S1.2
(a) Vol.118,No.2, 389-400

Q.P. KONG and Y. DAI
The Changes of Relaxation Parameters of Grain Boundaries during Creep in Pure Copper
62.20 and 62.40; S1.2
(a) Vol.118,No.2, 431-440

M. HAMERSKÝ, P. LUKÁČ, and Z. TROJANOVÁ
Stress Relaxation in Cd-Zn Polycrystals
62.20; S1.2
(a) Vol.118,No.2, 455-460

K.A. AL-SALEH, I.J. JABR, and N.S. SALEH
A Comparison of Ion Beam Induced Atomic Mixing Kinetics of Ti/Si, Fe/Si, and Ni/Si Systems. In Situ Electrical ...
61.80 and 73.40; S1.1; S1.2; S5.11
(a) Vol.118,No.2, 467-472

Y. ITOH, H. KADOMATSU, J. SAKURAI, and H. FUJIWARA
Transport Properties of $Yb_xIn_{1-x}Cu_2$
72.15; S1.2; S1.4
(a) Vol.118,No.2, 513-518

T. TYMOSZ, K. ŚWIDERCZAK, and K. BRAŃSKA
Lorentz Microscope Study of Obliquely Evaporated CoCr Thin Films
75.60 and 75.70; S1.2
(a) Vol.118,No.2, 519-524

J. ZBROSZCZYK, W. CIURZYŃSKA, J. ŚWIERCZEK, S. SZYMURA, B. WYSŁOCKI, M. BARAN, and A. WIŚNIEWSKI
Effect of Surface Layer Thinning on Approach to Magnetic Saturation in $Co_{64}Fe_{14}Si_{11}B_{11}$ Amorphous Alloy
75.50; S1.1; S1.2
(a) Vol.118,No.2, 525-530

S.U. JEN, M.C. CHIEN, and C.Y. LEE
Viscous Flow in Amorphous Alloys
61.40 and 64.70; 62.20; S1.1; S1.2
(a) Vol.119,No.1, 35-40

A. CATICHA and S. CATICHA-ELLIS
Dynamical Diffraction of X-Rays by Thin Crystals at Bragg Angles near $\pi/2$
61.10; S1.2
(a) Vol.119,No.1, 47-54

S. PILECKI, V.I. IVANOV, and A.B. LEBEDEV
Amplitude-Dependent Internal Friction in Silicon Bronze at Low Temperatures
62.40; S1.2
(a) Vol.119,No.1, 87-92

H.-A. KUHN and H.G. SOCKEL
Contributions of the Different Phases of Two Nickel-Base Superalloys to the Elastic Behaviour in a Wide Temperature Range
62.20; S1.2; S1.61
(a) Vol.119,No.1, 93-106

F. SACHSLEHNER
Anisotropic Electron Scattering by Phase Boundaries in a Copper-Beryllium Alloy
72.15; S1.2
(a) Vol.119,No.1, 229-236

V.E. ZUBOV, G.S. KRINCHIK, A.S. TABLIN, and A.A. KOSTYURIN
Magnetooptical Techniques in Self-Consistent Determination of Optical and Magnetooptical Constants of Magnetics
78.20; S1.2; S11.2
(a) Vol.119,No.1, 297-306

I. MANNA, S. BADER, W. GUST, and B. PREDEL
Interdiffusion between In Layer and Bulk Cu or Cu-In Alloy
66.30 and 68.48; S1.2
(a) Vol.119,No.1, K9-K14

M.N. GHATIKAR, K.V.S. MAHESH, and S.N. BHATIA
Improved Calculations of Phase Shifts for EXAFS Spectroscopy
61.40; 78.70; S1.1; S1.2
(b) Vol.159,No.2, 551-556

O.M. TOLKACHEV
Spin Fluctuation Mechanism for the Temperature-Induced Magnetic Moment (TIMM) in the Electron Liquid of Y_2Ni_7 Alloy
75.30; S1.2
(b) Vol.159,No.2, K115-K118

J. CERMAK
Diffusion of ^{51}Cr along High-Diffusivity Paths in Ni-Fe Alloys
66.30; S1.1; S1.2
(a) Vol.119,No.2, 443-454

A.S. TAHA and F.H. HAMMAD
The Applications of the Hall-Petch Relation to Microhardness Measurements on Al, Cu, Al-MD 105, and Al-Cu Alloys
62.20; 61.70; 64.80; S1.2; S4
(a) Vol.119,No.2, 455-462

H.K. LACHOWICZ, E. PULIDO, A. SIEMKO, and A. HERNANDO
Stress-Forced Magnetoresistance in Near-Zero Magnetostrictive Metallic Glasses
72.15 and 75.80; 75.60; S1.1; S1.2
(a) Vol.119,No.2, 595-600

W.A. KACZMAREK
Magnetic Structure Rearrangement during Crystallization in $Ni_{73}Mn_5Si_{10}B_{12}$ Ribbons
75.50 and 76.30; S1.2
(a) Vol.119,No.2, K159-K162

A.V. KOROLEV, V.S. GAVIKO, and N.V. MUSHNIKOV
The Magnetic Annealing Effect in Hydrogen Containing Intermetallic $Sm(Fe,Co)_2$ Compounds
75.60 and 75.80; S1.2
(a) Vol.119,No.2, K163-K166

J. POKLUDA and P. ŠANDERA
On the Relation of Ideal Strength to Other Physical Properties
62.20; S1.1; S1.2
(b) Vol.160,No.1, 89-96

V.M. ANISHCHIK and N.N. DOROZHKIN
Electronic Structure of TiB_2 and ZrB_2
71.25; 71.20; S1.2
(b) Vol.160,No.1, 173-178

S. KAHL, R. NEUHAUS, H. PINKVOS, and CH. SCHWINK
The Influence of Tensile Deformation and Atomic Short-Range Order on the Susceptibility of CuMn Spin Glasses
62.20 and 75.40; 61.70; S1.2
(a) Vol.120,No.1, 221-232

T. ZHANG, M.C. KANG, and W.C. LU
Impurity Effect on the Chemisorption of Hydrogen on Supported Catalysts
68.45; S1.2; S10.1
(a) Vol.120,No.1, K41-K44

A.A. REMPEL and A.I. GUSEV
Short-Range Order in Superstructures
61.50; 61.55; S1.2; S1.3
(b) Vol.160,No.2, 389-402

A.K. MEIKAP, A.R. JANA, S.K. DE, and S. CHATTERJEE
Weak Localization and Interaction Effects in Thin Disordered Zinc Films
71.38 and 73.60; S1.2
(b) Vol.160,No.2, 473-482

R. AHUJA, A.K. SOLANKI, and S. AULUCK
Fermi Surface Characteristics of Palladium
71.25; S1.2
(b) Vol.160,No.2, 549-560

V.S. STEPANYUK, A.A. KATSNELSON, O.S. TRUSHIN, P. NAYAK, and A. SZASZ
Electrical Resistivity of Amorphous $Ni_{1-x}P_x$ Alloys Calculated by Molecular Dynamics
72.15; S1.2
(b) Vol.160,No.2, K117-K122

Z.T. SERIKBAEVA, T.M. ZHANTIKIN, A.K. ZHETBAEV, and K.G. BINNATOV
Moessbauer Investigations of Ordering of the $Ni_3(Fe_{1-x}Mn_x)$ Alloy System
61.55; 76.80; S1.1; S1.2
(a) Vol.120,No.2, 305-314

N.P. KULISH and P.V. PETRENKO
Short-Range Order in the Binary Solid Solutions. Ordering and Its Change on Heating in Fe-Al, Cu-Al, and Ag-Al Alloys
61.55; S1.1; S1.2; S1.3; S4
(a) Vol.120,No.2, 315-326

J. ČERMÁK
Diffusion along Moving Grain Boundaries
66.30; 61.70; S1.1; S1.2
(a) Vol.120,No.2, 351-362

E. MAJKOVA, V. BOHAC, S. LUBY, R. SANDRIK, and M. VESELY
Diffusion of Au Atoms in Ni-Si-B and Fe-Co-B Metallic Glasses
66.30; S1.1; S1.2
(a) Vol.120,No.2, 371-378

B.M. LOGINOV and V.T. DEGTYAREV
Regularities Governing the Dislocation Forest Effects on Crystal Resistance to Deformation under Complex Loading ...
61.70; S1.2; S9.11
(a) Vol.120,No.2, 397-410

L.F. CHEN, S. PATU, and C.X. SHI
Effect of Phosphorus Ion Implantation on Fatigue Behaviour of Polycrystalline Nickel
62.20; 61.70; S1.2
(a) Vol.120,No.2, 429-440

P.V. PAVLOV, A.P. PAVLOV, and D.I. TETELBAUM
The Effect of Ion Implantation on the Substructure of Metal Films
68.55; 61.80; S1.2; S4
(a) Vol.120,No.2, 441-446

F. SACHSLEHNER, C. WUTTE, and F. STANGLER
Precipitation Effects on Thermopower of a Commercial Cu-2wt%Be Alloy
72.15; S1.2
(a) Vol.120,No.2, 581-592

S. CHABIK
Investigation of Interactions of Electrically Charged Hydrogen Atoms with Defects in Nickel by the Positron ...
61.70 and 78.70; S1.2
(a) Vol.120,No.2, K125-K130

M.Z. BUTT, Z. RAFI, and M.A. KHAN
Low-Temperature Anomaly in the Strength of Copper-Manganese Alloy Single Crystals
62.20; S1.2
(a) Vol.120,No.2, K149-K154

O. RADULESCU and R. MANAILA
On the Bond Orientational Order in Icosahedral Phases
61.50; 61.55; S1.1; S1.2; S4
(b) Vol.161,No.1, 65-74

J. SCHMELZER
On the Kinetics of Decomposition of Dilute Al-(Zn,Mg)Alloys
64.75; S1.2; S4
(b) Vol.161,No.1, 173-178

S. ERKOC
Structural Stability and Energetics of F.C.C. Metal Microclusters: Empirical Many-Body Potential Energy ...
63.10; 61.55; 62.20; S1.2; S1.3; S4
(b) Vol.161,No.1, 211-216

E. DĘBOWSKA and J. RUDZIŃSKA-GIRULSKA
Examination of the Electronic Structure of $Pd_{1-x}Ni_x$ Alloys by Positron Annihilation
71.25; 78.70; S1.2
(b) Vol.161,No.1, 281-286

G. NEUMANN and D.L. BEKE
Is There a Connection between the Temperature Function of $B\Omega$ and the Curvature of LnD versus 1/T for Self-Diffusion in ...
66.30; S1.2; S2
(b) Vol.161,No.1, K5-K10

J. DUBOWIK, YU.V. KUDTYAVTSEV, and E.E. KRASSOVSKII
Spectroscopic Ellipsometric Investigation of Multilayered and Solid-State Amorphized Co/Zr Films
68.55 and 78.65; 61.40; S1.2
(a) Vol.121,No.1, 173-180

Y.D. YAO, Y.Y. CHEN, S.J. TZENG, and T.H. CHUANG
Electrical Resistivity, Magnetization, and Grain Boundary Precipitates in Ni-Sn Alloys
72.15; 75.60; S1.2
(a) Vol.121,No.1, 213-218

C.Q. TANG and Y.F. ZHANG
Positron Lifetime Study on the Precipitation Phenomenon in 4Cr3Mo2NiVNb Alloy Steel
61.70 and 78.70; S1.2
(a) Vol.121,No.1, K21-K26

J. ČERMÁK
Grain Boundary Diffusion of ^{64}Cu in Cu-Al Alloys
66.30; S1.2
(a) Vol.121,No.1, K31-K36

B. LIPPOLD, J. HERRMANN, W. REICHELT, and H. OPPERMANN
Preparation and Magnetic Investigations on $Mn_2Mo_3O_8$ Single Crystals
75.30; 75.50; S1.2; S10.15
(a) Vol.121,No.1, K91-K94

BAO-GEN SHEN, HUI-QUN GUO, LIN-YUAN YANG, JUN-XIAN ZHANG, and JIAN-GAO ZHAO
Crystallization of Amorphous $Co_{84-x}B_xZr_{16}$ Alloys and Its Influence on Hard Magnetic Properties
75.30; 61.40; 75.60; S1.2
(a) Vol.121,No.1, K105-K110

R. SHYAM, S.C. UPADHYAYA, and J.C. UPADHYAYA
First-Principles Calculation of the Lattice Dynamics of the $Co_{0.92}Fe_{0.08}$ Alloy
63.20; S1.1; S1.2
(b) Vol.161,No.2, 565-574

O.I. DUBROVSKII, S.I. KURGANSKII, L.A. RUBTSOVA, and G.W. WOLF
Non-Relativistic and Scalar-Relativistic LAPW Electronic Energy Bands of (001)Cu Films
71.25 and 73.60; 71.10; S1.2
(b) Vol.161,No.2, 697-704

V.S. STEPANYUK, A.A. KATSNELSON, S. SZASZ, and O.S. TRUSHIN
B-B Direct Contact in Amorphous $Ni_{76}B_{24}$ System Based on Molecular Dynamics
61.40; S1.2
(b) Vol.161,No.2, K77-K80

L.B. ZUEV, V.E. GROMOV, and L.I. GUREVICH
The Effect of Electric Current Pulses on the Dislocation Mobility in Zinc Single Crystals
61.70; S1.2
(a) Vol.121,No.2, 437-444

R. HEIDINGER, H. APPEL, G. THEN, and W.-G. THIES
Observation of Radiation Damage in HfV_2 and Its Hybrides by TDPAC
61.80; 64.80; 74.70; 76.80; S1.2
(a) Vol.121,No.2, 445-454

E.V. TATYANIN and V.G. KURDYUMOV
The Nucleation of the Deformation Induced Amorphous Phase at Twin Boundaries in TiNi Alloy
61.70 and 62.20; 64.70; S1.2
(a) Vol.121,No.2, 455-460

N.H. LUONG, N.P. THUY, L.T. TAI, and N.V. HOANG
Effect of Y Substitution on the Spin Reorientation in $NdFe_{11}Ti$ Compound
75.25; 75.30; S1.2; S1.4
(a) Vol.121,No.2, 607-610

ZHI-DONG ZHANG, X.K. SUN, Y.C. CHUANG, and R.J. RADWAŃSKI
A Model Based on Single-Ion Theory. I. Theoretical Outline
75.50; 75.30; S1.2; S1.4
(a) Vol.121,No.2, 611-616

ZHI-DONG ZHANG, X.K. SUN, Y.C. CHUANG, F.R. DE BOER, and R.J. RADWAŃSKI
A Model Based on Single-Ion Theory. II. Preferential Substitution and Local-Environment Effects at Different ...
75.50; 75.30; S1.1; S1.2
(a) Vol.121,No.2, 617-626

L.I. VERSHININA, S.Z. SKLUEV, V.S. ZHIGALOV, A.G. STEPANOV, G.I. FROLOV, A.S. AVILOV, and S.V. OREKHOV
The Structure of Amorphous Dy-Co Films with Magnetic Anisotropy
61.40 and 75.50; S1.2
(a) Vol.121,No.2, K145-K148

D. DĂDÂRLAT, M. CHIRTOC, C. NEAMTU, R.M. CÂNDEA, and D. BICANIC
Inverse Photopyroelectric Detection Method
65.90 and 77.70; S1.1; S1.2
(a) Vol.121,No.2, K231-K234

A.O. DONGHOL DIALLO, E. MROSAN, and P. ZIESCHE
Model Calculation of the Specific Dislocation Resistivity for Copper
61.70 and 72.10; S1.2
(b) Vol.162,No.1, 237-242

Y. TAHTAMONI and S. ERKOÇ
Structural Stability and Energetics of X_n(X=V,Cr,Nb,n=3 to 7) Microclusters: Empirical Many-Body Potential Energy ...
61.50; S1; S1.2
(b) Vol.162,No.1, K5-K8

M. BAUER, J. MAYER, M. HOHENSTEIN, and M. WILKENS
Quantitative Studies on the Amorphization of Quasicrystals by Means of High Voltage Electron Microscopy
61.40 and 61.80; 64.70; S1.2; S4
(a) Vol.122,No.1, 79-86

V.N. NOVIKOV
Microcreep of Metallic Glasses at Room Temperature
62.20; 61.40; S1.1; S1.2
(a) Vol.122,No.1, 215-220

H. MIZUBAYASHI, H. AGARI, and S. OKUDA
Hydrogen Internal Friction Peak in Amorphous $Cu_{50}Ti_{50}$
61.40 and 62.40; S1.2
(a) Vol.122,No.1, 221-234

N. MATHUR and A.N. NIGAM
Temperature Dependence of Electrolytic Hydrogen Diffusion in Nickel: A Thermoelectric Study
66.30 and 72.15; S1.2
(a) Vol.122,No.1, K59-K62

S. SWAMI, V.K. AGARWAL, S.S. SHARMA, and J. VARMA
Moessbauer and ESR Studies of Polymetallic Nodules
76.30 and 76.80; S1.2
(a) Vol.122,No.1, K77-K82

H. WADA, T. INOUE, M. HADA, M. SHIGA, and Y. NAKAMURA
Effect of Magnetic Phase Transition on the Electronic Specific Heat Coefficient of $Lu(Co_{1-x}Al_x)_2$
65.40; 75.30; S1.2; S1.4
(b) Vol.162,No.2, 407-412

M.K. MISHRA
New Approach to the Lattice Dynamical Study of Cubic Metals
63.20; S1.2
(b) Vol.162,No.2, K73-K76

H. LU, C.L. BAO, X.N. YU, Q.K. XUE, D.S. TANG, P.J. WU, and Y.Y. XIONG
Crystallization on CuZr Amorphous Alloy Surface. An X-Ray Photoelectron Spectroscopy Study
61.50; 79.60; S1.2
(a) Vol.122,No.2, 481-488

FANG YUAN, YENING WANG, and HIUMIN SHEN
Study of the Incommensurate Phase in NiTi(Fe) Alloys
64.70; 62.40; 64.80; S1.2
(a) Vol.122,No.2, 517-524

T. ARAI and S. OKUDA
Relaxation Spectrum Analysis of the Bordoni Peak in Cu by Nonlinear Least-Squares Fitting
62.40; 62.20; S1.2
(a) Vol.122,No.2, 535-544

I. GOŚCIAŃSKA, H. RATAJCZAK, P. SOVÁK, P. SPIŠÁK, and M. KONČ
The Structure of Fe-Rich Vapor Quenched Fe-Ti Thin Films
68.55; 61.40; 73.60; S1.1; S1.2
(a) Vol.122,No.2, 575-582

W.A. KACZMAREK, R. BRAMLEY, and A. CALKA
ESR Study of Amorphous Metallic Alloys below the Magnetic Multicritical Point
75.50 and 76.30; 75.40; S1.1; S1.2
(a) Vol.122,No.2, 663-676

E.M. GOLOLOBOV, E.L. MAGER, and Z.V. MESHEVICH
Ultrasonic Investigations of Structural Instabilities of ZrV_2 under High Hydrostatic Pressures
62.65 and 64.70; S1.2
(a) Vol.122,No.2, K123-K128

A.H. KADIKOVA, N.V. ERSHOV, A.L. AGEEV, YU.V. RUTS, and S.P. SENTEMOV
Investigation of the Local Atomic Structure of the Copper Surface by the EELFS Method
68.20; S1.2
(a) Vol.122,No.2, K143-K146

S1.3. Noble metals

A.K. BANDYOPADHYAY and S.K. SEN
Calculation of Self-Interstitial Formation Energy (Both Split and Non-Split) in Noble Metals
61.70; S1.2; S1.3
(b) Vol.157,No.2, 519-530

A. BUDKOWSKI, V. MARINKOVIC, A. PRODAN, and F.W. BOSWELL
The Two-Dimensional Modulation in Bulk and Thin-Film $Au_{2+x}Cd_{1-x}$
61.55 and 68.55; S1.3
(a) Vol.117,No.2, 351-362

A. BUKALUK
AES Depth Profile Studies of Interdiffusion in the Ag-Cu Bilayer and Multilayer Thin Films
66.30 and 68.48; 61.80; S1.2; S1.3
(a) Vol.118,No.1, 99-108

E.I. SHREDER, I.A. POPOV, E.E. KRASOVSKII, and I.I. SASOVSKAYA
The Optical Properties and Electronic Structure of AgZn Alloy Undergoing $\beta' \rightleftharpoons \zeta$ Structural Transformation
64.70 and 78.20; 71.25; S1.2; S1.3
(a) Vol.118,No.1, 147-152

V.V. BOGDANOV, V.V. ZAKURKIN, D.V. PETROV, and YU.P. PENKOV
Investigation of Proton-Irradiated Platinum by Moessbauer Spectroscopy
61.80 and 76.80; S1.3
(b) Vol.158,No.2, K105-K110

S.N. BISWAS, M.J.P. MURINGER, and C.A. TEN SELDAM
Pressure Dependence of the Shear Modulus of Polycrystalline Silver from 77 to 298 K and up to 253 MPa
62.20; S1.3
(a) Vol.119,No.1, 63-70

C. SCHOBER, V.N. ANTONOV, and P. ZIESCHE
Internal Conversion Coefficient for Valence Electrons of Metals. Theory and Application to ^{110}Ag
71.20 and 71.25; S1.3
(b) Vol.159,No.2, 771-780

G.A. HASSAN and H.H. HASSAN
Strain Hardening Characteristics of 99.70 % Pure Silver
62.20; S1.3
(a) Vol.119,No.2, 471-478

U. WENSCHUH and E. HEINER
Optical Properties of Gold in the Nonequilibrium Thermalized Stage
71.10; 65.40; 71.38; 78.20; S1.3
(b) Vol.160,No.1, 225-236

M.YA. GAMARNIK
On the Physical Nature of the Decrease in Lattice Parameters of Small Metal Particles
61.55; S1.3
(b) Vol.160,No.1, K1-K4

N.S. SALEH, K.A. AL-SALEH, and A.M. AL-SAIE
Kinetics of Ion Beam Induced Mixing in Au/Si System
61.80 and 68.55; 73.60; S1.3; S5.11
(a) Vol.120,No.1, 169-174

A.A. REMPEL and A.I. GUSEV
Short-Range Order in Superstructures
61.50; 61.55; S1.2; S1.3
(b) Vol.160,No.2, 389-402

N.P. KULISH and P.V. PETRENKO
Short-Range Order in the Binary Solid Solutions. Ordering and Its Change on Heating in Fe-Al, Cu-Al, and Ag-Al Alloys
61.55; S1.1; S1.2; S1.3; S4
(a) Vol.120,No.2, 315-326

G. SINGH
Strength and Stability of Silver
62.20; S1.3
(b) Vol.161,No.1, 145-152

S. ERKOÇ
Structural Stability and Energetics of F.C.C. Metal Microclusters: Empirical Many-Body Potential Energy ...
63.10; 61.55; 62.20; S1.2; S1.3; S4
(b) Vol.161,No.1, 211-216

Z.-J. DING, H. YOSHIKAWA, and R. SHIMIZU
Angular Distribution of Elastically Reflected Electrons from Au
68.20 and 79.20; S1.3
(b) Vol.161,No.1, 257-270

D. WANG and P.D. BRISTOWE
Calculation of Diffraction Effects Due to Double Positioning in (111) Gold Bicrystals
61.70; 61.10; S1.3
(b) Vol.161,No.2, 501-512

SIMING HE, MINGZHI LIU, and LINCHANG TANG
Internal Friction Peak Associated with Hydrogen Diffusion in Amorphous $Pd_{77.5}Ag_6Si_{16.5}$ Alloy
62.40; 66.30; S1.3
(a) Vol.121,No.2, K167-K170

U. WENSCHUH and E. HEINER
Nonequilibrium Thermoreflectance Evolution of Laser Irradiated Noble Metals
78.20; 65.40; 71.10; S1.3
(b) Vol.162,No.1, 303-310

P.T. SALO and K. KOKKO
Extraction of Valence Kinetic Energy of Metals from Electron Density
71.10; S1.3; S2
(b) Vol.162,No.1, K29-K32

R. PEREZ
On the Characterization of Crystalline Defects Using CBED Techniques
61.14 and 61.70; S1.3; S5.11
(a) Vol.122,No.1, 51-68

J.Y. KIM and R.E. HUMMEL
Hole and Hillock Formation in Gold Metallizations at Elevated Temperatures Deposited on Titanium, Vanadium and Other ...
61.70 and 68.55; S1.3
(a) Vol.122,No.1, 255-274

E. GRANTSCHAROVA and D. DOBREV
Recrystallization Texture and Microstructure Development in Silver Films
68.55; S1.3
(a) Vol.122,No.1, K31-K34

T. KAMINO, M. TOMITA, I. OHTSUKA, and H. SAKA
High Resolution Transmission Electron Microscopy Observation of the Motion of a Surface Dislocation on a Gold Particle
61.16; 61.70; S1.3
(a) Vol.122,No.2, K105-K108

S1.4. Lanthanides

TIESONG ZHAO, ZHONGZHI FAN, and HANMIN JIN
Magnetization Processes of the Rare-Earth and Fe Sublattices in $R_2Fe_{14}B$ (R=Tb,Dy,Ho,Er,and Tm)
75.30; 75.60; S1.1; S1.4
(b) Vol.157,No.2, 677-684

A. TARI and J.S. HWANG
Magnetisation and ESR Studies of $Gd(Ni_{1-x}Rh_x)_5$ Compounds
75.60 and 76.30; S1.4
(b) Vol.157,No.2, 701-710

P. NOZAR, V. SECHOVSKY, M. SOLANKI-MOSER, E. BAUER, and E. GRATZ
Point-Contact Spectroscopy on the $Ce(Cu_xAl_{1-x})_5$ System
71.20; 72.20; 75.30; S1.2; S1.4
(a) Vol.117,No.2, 501-508

N. SCHELUDKO and M. MIKHOV
Investigation of Magnetocrystalline Anisotropy of $Mm_2Co_{17-x}Zr_x$ Compounds
75.30; S1.2; S1.4
(a) Vol.117,No.2, 571-576

V.V. SINGER, A.E. VAFIN, L.M. SANDRATSKII, L.P. MOKHRACHEVA, S.V. KONOVALOV, I.Z. RADOVSKII, and P.V. GELD
Magnetic Susceptibility and Valency of Ce and Pr in Their Liquid Binary Alloys with Metals of the Iron Subgroup
75.50; S1.1; S1.2; S1.4
(a) Vol.117,No.2, 577-586

N.K. GHOSH, S.C. GHOSH, and R.L. SARKAR
Electron-Phonon Interaction Induced Hybridisation in Mixed-Valence Systems
71.38; S1.4
(b) Vol.158,No.1, 223-228

H. RATAJCZAK, I. GOŚCIAŃSKA, and T. LUCIŃSKI
Influence of Annealing on Saturation Magnetization and Hall Resistivity in Amorphous $Nd_{12}Co_{82}B_6$ Thin Films
61.40 and 68.55; 73.60; 75.70; S1.2; S1.4
(a) Vol.118,No.1, 189-196

K. MAKI and H. KRONMUELLER
Investigation of Demagnetization Processes in $Fe_{14}Nd_2B$-Type Sintered Magnets
75.60; S1.1; S1.4
(a) Vol.118,No.1, 253-270

R. SCHAEFER and A. HUBERT
A New Magnetooptic Effect Related to Non-Uniform Magnetization on the Surface of a Ferromagnet
75.60; 78.20; S1.1; S1.2; S1.4; S11.2
(a) Vol.118,No.1, 271-288

Y. ITOH, H. KADOMATSU, J. SAKURAI, and H. FUJIWARA
Transport Properties of $Yb_xIn_{1-x}Cu_2$
72.15; S1.2; S1.4
(a) Vol.118,No.2, 513-518

G. FISCHER and A. HERR
Mean Magnetic Moments of Polycrystalline Ce Compounds in a Tetragonal Crystal Field
75.10; S1.4
(b) Vol.159,No.1, K23-K26

P. SVOBODA, M. DIVIŠ, J. BISCHOF, Z. SMETANA, R. ČERNÝ, and J. BURIÁNEK
The Magnetic Properties of $SmCu_5$
75.50; S1.4
(a) Vol.119,No.1, K67-K70

K.A. KIKOIN, G.P. NIZHNIKOVA, and O.V. FARBEROVICH
Optical Properties of the Mixed-Valent TmSe
71.25; 71.50; 78.20; S1.4; S8
(b) Vol.159,No.2, 743-754

R.B. PRABHU and A.F. BARBOSA
Magnetoresistivity of a Crystal Field Split Kondo System
72.15; S1.4
(b) Vol.161,No.1, 309-318

B.X. GU, B.G. SHEN, and H.R. ZHAI
Crystallization and Magnetic Properties of Amorphous $Nd_{0.19}Fe_{0.81}$ Alloy
75.50; 61.40; S1.1; S1.4
(a) Vol.121,No.1, K99-K104

E.M. LEVIN, B.S. KUZHEL, O.I. BODAK, B.D. BELAN, and I.N. STETS
Concentrated Kondo Systems in Solid Solutions on the Base of Europium Ternary Compounds
72.10; 64.75; 71.25; S1.4
(b) Vol.161,No.2, 783-796

N.H. LUONG, N.P. THUY, L.T. TAI, and N.V. HOANG
Effect of Y Substitution on the Spin Reorientation in $NdFe_{11}Ti$ Compound
75.25; 75.30; S1.2; S1.4
(a) Vol.121,No.2, 607-610

ZHI-DONG ZHANG, X.K. SUN, Y.C. CHUANG, and R.J. RADWAŃSKI
A Model Based on Single-Ion Theory. I. Theoretical Outline
75.50; 75.30; S1.2; S1.4
(a) Vol.121,No.2, 611-616

K. CHERIFI, P. DONOVAN, C. DUFOUR, PH. MANGIN, and G. MARCHAL
Microstructural Observations of Rare-Earth/Iron Multilayers
68.55; S1.1; S1.4
(a) Vol.122,No.1, 311-322

H. WADA, T. INOUE, M. HADA, M. SHIGA, and Y. NAKAMURA
Effect of Magnetic Phase Transition on the Electronic Specific Heat Coefficient of $Lu(Co_{1-x}Al_x)_2$
65.40; 75.30; S1.2; S1.4
(b) Vol.162,No.2, 407-412

G.V. IONOVA and A.V. NIKOLAEV
Isostructural gamma-alpha Transition in Metallic Cerium as a Mott Transition
64.60 and 71.20; S1.4
(b) Vol.162,No.2, 451-460

I. MAHMOUD, A.M. HAFEZ, H.A. MOTAWEH, and N.S. ALY
Effect of Nickel on the C_{14} Structure Stability in the $Er(Mn_{1-x}Ni_x)_2$ Quasibinary System
61.55; S1.4
(a) Vol.122,No.2, K113-K116

S1.5. Actinides

T. PALEWSKI
Metamagnetism in UAs-YSe Solid Solution
75.30; S1.5
(a) Vol.117,No.2, K151-K154

S1.61. Carbides, silicides, germanides

A.P. DOSTANKO, V.V. SHATALOV, and G.M. SHAKHLEVICH
Investigation of the Phase Formation Process in Ni-Based Binary Alloy-Silicon Systems
68.48 and 68.55; S1.61; S5.11
(a) Vol.117,No.1, 213-220

V.YU. PETUKHOV, I.B. KHAIBULLIN, M.M. ZARIPOV, E. WIESER, R. GROETZSCHEL, and H. BARTSCH
Formation of Epitaxial $NiSi_2$ by High Dose Implantation and Rapid Thermal Annealing
61.80 and 68.55; S1.61
(a) Vol.117,No.2, 477-484

A.Z. MENSHIKOV, A.P. VOKHMYANIN, and YU.A. DOROFEEV
Magnetic Structure and Phase Transformations in Mn_5Si_3
75.30; 61.55; S1.61
(b) Vol.158,No.1, 319-328

H.-A. KUHN and H.G. SOCKEL
Contributions of the Different Phases of Two Nickel-Base Superalloys to the Elastic Behaviour in a Wide Temperature Range
62.20; S1.2; S1.61
(a) Vol.119,No.1, 93-106

K.-H. BAETHER and G. ZIES
Film Thickness Effects and Their Influence on the Stability of the Cr-Si-O/V-O/Al Film System
68.55 and 73.40; S1.61; S4
(a) Vol.119,No.2, 503-510

A.M. CHAPLANOV and A.N. SHIBKO
Modification of Metal Films by Pulsed Laser Radiation in Organic Liquids
68.55; S1; S1.61
(a) Vol.120,No.1, K37-K40

R.F. SABIRYANOV, D.L. NOVIKOV, A.L. IVANOVSKII, and V.A. GUBANOV
Quantum-Chemical Calculations of Hydrogen Adsorption on the Polar Surface of Titanium Carbide
68.45 and 73.20; 79.20; S1.61
(b) Vol.160,No.2, 459-472

R.L. SOMMER and F.P. LIVI
Barkhausen Noise Measurements in Small Samples of (110)[001] Silicon-Iron
75.60; S1.61
(a) Vol.120,No.2, 609-616

S1.62. Nitrides, phosphides, arsenides

M. MANZEL, E. STEINBEISS, and W. SCHUEPPEL
Magnetic Properties of Sputter-Synthesized CoN_x-Films
75.70; S1.62
(a) Vol.119,No.1, 279-284

V.S. STEPANYUK, A. SZÁSZ, A.A. KATSNELSON, A.A. GRIGORENKO, and O.V. FARBEROVICH
On the Electronic Structure of Small Ni-P Clusters
71.25; S1.62
(b) Vol.160,No.1, 219-224

S.K. KULSHRESHTHA
On Moessbauer Spectrum Analysis for Spin Orientation in Fe_2As
76.80; S1.62
(b) Vol.160,No.1, 313-320

Z. WOKULSKI
Mechanical Properties of TiN Whiskers
62.20 and 68.70; S1.62
(a) Vol.120,No.1, 175-184

W. LUBCZYŃSKI, J. CISOWSKI, and J.C. PORTAL
Shubnikov-de Haas Effect in $Cd_{3-x}Zn_xAs_2$ Alloys
71.25; 72.20; S1.62
(a) Vol.120,No.2, 525-530

S1.63. Sulfides, selenides, tellurides

J. MOLENDA, A. STOKLOSA, DO THAN, and T. BAK
Electrical Properties of $Ti_{1+y}S_2$ at Temperatures 77 to 300K
72.15; S1.63
(a) Vol.119,No.1, 221-228

J. MOLENDA, A. STOKŁOSA, S. MROWEC, and DO THAN
On the Defect Structure and Electronic Properties of Titanium Disulfide
72.15; 61.70; S1.63
(a) Vol.119,No.2, 571-582

W. MAYR and A. PHILIPP
Charge Density Wave and Contact Effects in $NbSe_3$ Observed with Microwave Hamonic Mixing
72.15; S1.63
(b) Vol.160,No.2, 613-624

R.A. YAKSHIBAEV, A.F. NADEZHDINA, and R.F. ALMUKHAMETOV
Structural Features of Niobium Diselenide Intercalated by Copper at High Temperatures
61.60; 65.70; S1.63
(a) Vol.121,No.1, K5-K8

M. STORDEUR
The Thermoelectric Figure of Merit in the Mixed Crystal System p-$(Bi_{1-x}Sb_x)_2Te_3$
72.20; 72.90; S1.63; S8.15
(b) Vol.161,No.2, 831-836

S2. Alkali metals and their alloys and compounds
(for alkali halides, see S9.11)

A. DEROUICHE and J.L. BRETONNET
The Long-Wavelength Limit of the Structure Factor of Liquid Alkali Metals Using a One-Component Plasma System
61.25; 65.50; S2
(b) Vol.157,No.1, 61-64

A.K. MISHRA, B.B. SAHAY, and K.K. MUKHERJEE
Structure and Electrical Resistivity of Alkali-Alkali and Lithium-Based Liquid Binary Alloys
61.25; 72.15; S2
(b) Vol.157,No.1, 85-92

T. DAS, T. NAMMALVAR, L.B. BHUIYAN, and R.N. JOARDER
Computation of Vacancy Formation Energy of Liquid Metals
61.25; 61.70; S2; S4
(b) Vol.157,No.1, 93-100

S. PATRA and A. KUMAR
Effect of Lattice Vibrations on the Change in Work Function
68.30 and 73.30; S2
(b) Vol.157,No.1, 209-220

Z.A. GURSKII and YA.G. CHUSHAK
Lattice Dynamics of Binary Alloys
63.20; S2
(b) Vol.157,No.2, 557-566

S.K. MISHRA and T.N. SINGH
Lattice Dynamics of Alkali Metals
63.20; 62.20; S2
(b) Vol.158,No.1, 153-164

E. HAQUE and A.K.M.A. ISLAM
Hydrides and Deuterides of Lithium and Sodium. I. Model Potentials and Their Use in Perfect Lattice
63.20; S2
(b) Vol.158,No.2, 457-468

M. AHMAD and S.J. GLADYSZ
The Dynamic Spin Susceptibility and l=0.1 Spin Waves in a Charged Fermi Liquid of Alkali Metals in a DC Magnetic Field
75.30; 67.50; S2
(b) Vol.159,No.2, K119-K124

D. BHANDARI, A. PRATAP, N.S. SAXENA, and M.P. SAKSENA
Collective Excitations in a Liquid Alloy
63.50; S2
(b) Vol.160,No.1, 83-88

YU.N. ZHURAVLEV and A.S. POPLAVNOI
Band Structure of Sodium and Lithium Azides
71.25; S2
(b) Vol.160,No.1, 153-164

R.V. GOPALA RAO and U. BANDYOPADHYAY
An Evaluation of Structural and Transport Properties of Molten Potassium-Rubidium Alloy
61.40; 62.20; 66.30; S2
(b) Vol.160,No.2, 403-414

K.C. JAIN, N. GUPTA, A. PRATAP, and N.S. SAXENA
Dynamical Correlations in Liquid Sodium
63.50; S2
(b) Vol.160,No.2, 433-442

G.J. VÁZQUEZ
Specific Heat and Elastic Constants for Martensitic Forms of Lithium
62.20 and 65.40; S2
(b) Vol.160,No.2, K89-K92

O. AKINLADE
A Study of the Structure and Thermodynamics of Liquid Metals Based on Some Reference Systems
61.25 and 65.50; S2; S3; S4
(b) Vol.161,No.1, 75-84

S.V. FOMICHEV, M.I. KATSNELSON, V.G. KORESHKOV, and A.V. TREFILOV
On the Effect of Three-Body Interactions and Proximity of the Fermi Level to the Brillouin Zone Faces on Elastic ...
62.20 and 71.25; S2; S3; S4
(b) Vol.161,No.1, 153-164

K.C. JAIN, NEELAM GUPTA, and N.S. SAXENA
Dynamical Correlations in Liquid Metals: K and Cs
63.50; S2
(b) Vol.161,No.1, 201-210

G. NEUMANN and D.L. BEKE
Is There a Connection between the Temperature Function of B and the Curvature of LnD versus 1/T for Self-Diffusion in ...
66.30; S1.2; S2
(b) Vol.161,No.1, K5-K10

S.D. ALTEKAR and A.K. RAY
A Correlation Study of Small Potassium Cationic Clusters
61.50; 71.45; S2
(b) Vol.161,No.2, 463-472

T. DAS, L.B. BHUIYAN, and R.N. JOARDER
Ashcroft-Langreth Prescription for Effective Core Diameter in Liquids: A Variational Analysis
61.20; S2; S5.2
(b) Vol.161,No.2, 473-482

H. MURAKAMI and M. SANO
Positron-Trapping Centers in Potassium-Intercalated Graphite, KC_8 and $KC_8H_{2/3}$
61.70; 78.70; S2; S6
(a) Vol.121,No.2, 461-466

U. WALZER
Systematization of Binary Intermetallic Compounds
61.50; 71.10; 71.45; S2
(b) Vol.162,No.1, 75-88

R. KASCHNER and TH. VOJTA
Accurate Numerical Calculation of the Electron Density at a Bijellic Interface within the Gradient Method
68.48 and 73.40; S2
(b) Vol.162,No.1, 141-146

T. SOMA, M. HASEBE, and H.-MATSUO KAGAYA
Lattice Dynamical Contribution to the Bulk and Thermal Properties of Alkali Metals
65.50; 65.70; S2
(b) Vol.162,No.1, K21-K24

P.T. SALO and K. KOKKO
Extraction of Valence Kinetic Energy of Metals from Electron Density
71.10; S1.3; S2
(b) Vol.162,No.1, K29-K32

K.C. JAIN, N. GUPTA, and N.S. SAXENA
Dynamics of Liquid $Na_{0.5}K_{0.5}$ Alloy
63.50; S2
(b) Vol.162,No.2, 395-406

S3. Alkaline earth metals and their alloys and compounds
(for alkaline earth halides, see S9.12)

N. SINGH and S.P. SINGH
Lattice Dynamics of Ca and Sr in B.C.C. and F.C.C. Phases. A Dynamical Treatment
63.20; S3
(b) Vol.158,No.2, 433-440

M.L. NÓ, A. OLEAGA, C. ESNOUF, and J. SAN JUAN
Internal Friction at Medium Temperatures in High Purity Magnesium
62.40; S3
(a) Vol.120,No.2, 419-428

O. AKINLADE
A Study of the Structure and Thermodynamics of Liquid Metals Based on Some Reference Systems
61.25 and 65.50; S2; S3; S4
(b) Vol.161,No.1, 75-84

S.V. FOMICHEV, M.I. KATSNELSON, V.G. KORESHKOV, and A.V. TREFILOV
On the Effect of Three-Body Interactions and Proximity of the Fermi Level to the Brillouin Zone Faces on Elastic ...
62.20 and 71.25; S2; S3; S4
(b) Vol.161,No.1, 153-164

S4. Other metals and their alloys

T. DAS, T. NAMMALVAR, L.B. BHUIYAN, and R.N. JOARDER
Computation of Vacancy Formation Energy of Liquid Metals
61.25; 61.70; S2; S4
(b) Vol.157,No.1, 93-100

E. GERLACH
A Contribution to DC Transport in Thin Bi Films
73.60; S4
(b) Vol.157,No.1, 189-198

D. BANERJEE/CHAKRAVORTY and R. BHATTACHARYA
Magnetic Properties of Single Crystals of Bismuth Doped with Lead and Tin
75.20; S4
(b) Vol.157,No.1, 443-450

XIAN-FANG ZHU, LI-DE ZHANG, and ZI-YING JIANG
Effect of Yttrium on Grain Boundary Internal Friction in Commericial Aluminium
62.40; 61.70; S4
(a) Vol.117,No.1, 105-112

V.M. KOLESHKO and I.V. KIRYUSHIN
Electromigration, Stress, and Pulse Effects in Thin Film Conductors. Model
68.60; S4
(a) Vol.117,No.1, 177-190

A. HU, S.S. JIANG, H. CHEN, W. LIU, Y. QIU, and D. FENG
Anomalous Annealing Behaviour of Ta/Al Periodic Multilayer Films
68.55; S4
(a) Vol.117,No.1, 221-226

T. KURIHARA and S. TANIGAWA
The Anomalous Temperature Dependence of Positron Annihilation in Dilute Al-Zn Alloy near the Melting Point
61.70 and 78.70; S4
(a) Vol.117,No.1, K9-K12

A. SAKAKIBARA, T. KANADANI, M. YAMADA, and M. OHTA
Effect of Surface on the Reversion of Age-Hardened Al-Zn Alloy and Role of Grain Boundaries
62.20; S4
(a) Vol.117,No.1, K19-K22

T. KANADANI and A. SAKAKIBARA
Effect of Soft Surface Layer on the Fatigue Strength of Age-Hardened Al-2mass%Cu Alloy
62.20; S4
(a) Vol.117,No.2, K97-K98

F.M. MUNTYANU, M.I. ONU, and V.G. KISTOL
Quantum Oscillations of Galvanomagnetic Effects in Bismuth Bicrystals
72.15; S4
(b) Vol.158,No.1, 301-306

F. VOELKLEIN and E. KESSLER
Thermal Conductivity and Thermoelectric Figure of Merit of Thin Antimony Films
72.15 and 73.60; S4
(b) Vol.158,No.2, 521-530

E. TAUTE and K. BLUM
On the Charge Distribution of Ions after Auger Neutralisation at Surfaces
79.20; S4
(b) Vol.158,No.2, 551-556

V.G. CHUDINOV and V.V. ANDREEV
Atomic Mechanism of Solid-Liquid Phase Transition in F.C.C. Metals with a Large Stacking-Fault Energy
64.70; 61.40; 61.70; S4
(a) Vol.118,No.2, 415-424

A.G. BUDREVICH, A.S. LYUTOVICH, and B.L. OKSENGENDLER
Microselectivity Energy Dependence of Growing Film Orientation during Ion-Molecular Flow Condensation
68.55; S4
(a) Vol.119,No.1, K27-K30

R.V. GOPALA RAO and A. SATPATHY
Computation of the Structure Factors of the Peculiar Metals Sb, Bi, Ga, and Sn Using Lebowitzs Solution of Hard-Sphere ...
61.25; S4
(b) Vol.159,No.2, 545-550

V.F. LEMBERG and S.G. PSAKHIE
Analytical Representation of the Thermodynamic Functions of Simple Metals in the Model of Self-Consistent Einstein...
65.50; 63.20; 64.10; S4
(b) Vol.159,No.2, 645-658

S.A. OSTANIN, N.A. SHILKOVA, and V.P. SHIROKOVSKII
Optical Absorption in Heavy Transition B.C.C. Metals
78.20; 71.10; S4
(b) Vol.159,No.2, 893-902

G.A. MALYGIN
Dislocation Density Evolution Equation and Strain Hardening of F.C.C. Crystals
61.70; 62.20; S4
(a) Vol.119,No.2, 423-436

V.G. CHUDINOV, B.N. GOSHCHITSKII, N.V. MOSEEV, and V.V. ANDREEV
Kinetics of Radiation-Induced Segregation of Impurities in F.C.C. Metals at the Thermal Stage of the Collision Cascade
61.80 and 64.75; S1.1; S4
(a) Vol.119,No.2, 437-442

A.S. TAHA and F.H. HAMMAD
The Applications of the Hall-Petch Relation to Microhardness Measurements on Al, Cu, Al-MD 105, and Al-Cu Alloys
62.20; 61.70; 64.80; S1.2; S4
(a) Vol.119,No.2, 455-462

K.-H. BAETHER and G. ZIES
Film Thickness Effects and Their Influence on the Stability of the Cr-Si-O/V-O/Al Film System
68.55 and 73.40; S1.61; S4
(a) Vol.119,No.2, 503-510

S.P. ZAKHAROV and V.V. USTINOV
The High Frequency Quantum Size Effect in the Impedance of Metal Films
73.60; 72.15; S4
(b) Vol.160,No.1, 97-104

Z.Z. AKSELROD, B.A. KOMISSAROVA, L.N. KRYUKOVA, G.K. RYASNYI, L.G. SHPINKOVA, and A.A. SOROKIN
Electron Quadrupole Interaction of ^{181}Ta in Hf_2Fe Remeasured
71.70; 76.80; S1.1; S4
(b) Vol.160,No.1, 255-262

E.V. BOGDANOV, D.A. KOMARKOV, and W. KRAAK
Pressure Dependence of the Direct Energy Gap in $Bi_{1-x}Sb_x$ Semiconductor Alloys
71.25 and 72.20; S4
(b) Vol.160,No.1, K33-K36

S. TAKAMURA and M. KOBIYAMA
New Peak in Internal Friction after Deformation of Irradiated Al Alloys
61.80 and 62.40; S4
(a) Vol.120,No.1, 119-124

Y.B. XIA, Z.G. WANG, and R.H. WANG
Secondary Hardening and Dislocation Evolution in Low Cycle Fatigue of Polycrystalline Aluminium
62.20; 61.70; S4
(a) Vol.120,No.1, 125-132

M.A. KENAWY, M.S. SAKR, E.M. SAKR, H.A. ZAYED, and N.O. MOURAD
Effect of Dissolution of the beta-Phase on the Work Hardening Characteristics of Al-4.5wt%Mg Alloy
62.20 and 64.75; S4
(a) Vol.120,No.1, K31-K36

C.A. MOYER, S. ARAJS, Y.D. YAO, and Y.Y. CHEN
Electrical Resistivity of Metal Matrix Composites
72.15; S4
(a) Vol.120,No.1, K71-K72

M. ZAŁUZNY
The Influence of the Size Quantization on the Cyclotron Effective Mass and the Density of States in Bi Films
71.25; S4
(b) Vol.160,No.2, 541-548

V.I. GOSTISHEV, S.E. DEMYANOV, V.R. SOBOL, and D.V. PASHIK
On the Anisotropy of Electron-Dislocation Scattering in Aluminum in Charge Transport
61.70 and 72.15; S4
(b) Vol.160,No.2, K133-K136

N.P. KULISH and P.V. PETRENKO
Short-Range Order in the Binary Solid Solutions. Ordering and Its Change on Heating in Fe-Al, Cu-Al, and Ag-Al Alloys
61.55; S1.1; S1.2; S1.3; S4
(a) Vol.120,No.2, 315-326

P.V. PAVLOV, A.P. PAVLOV, and D.I. TETELBAUM
The Effect of Ion Implantation on the Substructure of Metal Films
68.55; 61.80; S1.2; S4
(a) Vol.120,No.2, 441-446

D.B. TRAN THOAI and E. ZEITLER
Inelastic Scattering of Fast Electrons by Thin Metal Slabs
71.45 and 73.20; 61.80; S4
(a) Vol.120,No.2, 467-474

W. RUDZIŃSKA, ST. CHABIK, and CZ. SZYMAŃSKI
Investigation of the Solid-Liquid Phase Transition for Sn-5wt%In Alloy by the Positron Annihilation Method
64.70; 78.70; S4
(a) Vol.120,No.2, K145-K148

V.G. SHEPELEVICH
Structure and Electrical Properties of Quickly-Quenched Foils of Antimony and Its Alloys
61.55 and 72.15; S4
(a) Vol.120,No.2, K173-K176

O. RADULESCU and R. MANAILA
On the Bond Orientational Order in Icosahedral Phases
61.50; 61.55; S1.1; S1.2; S4
(b) Vol.161,No.1, 65-74

O. AKINLADE
A Study of the Structure and Thermodynamics of Liquid Metals Based on Some Reference Systems
61.25 and 65.50; S2; S3; S4
(b) Vol.161,No.1, 75-84

S.V. FOMICHEV, M.I. KATSNELSON, V.G. KORESHKOV, and A.V. TREFILOV
On the Effect of Three-Body Interactions and Proximity of the Fermi Level to the Brillouin Zone Faces on Elastic ...
62.20 and 71.25; S2; S3; S4
(b) Vol.161,No.1, 153-164

J. SCHMELZER
On the Kinetics of Decomposition of Dilute Al-(Zn,Mg)Alloys
64.75; S1.2; S4
(b) Vol.161,No.1, 173-178

S. ERKOÇ
Structural Stability and Energetics of F.C.C. Metal Microclusters: Empirical Many-Body Potential Energy ...
63.10; 61.55; 62.20; S1.2; S1.3; S4
(b) Vol.161,No.1, 211-216

T.S. KÊ and Q.F. FANG
Evolution of the Anomalously Amplitude Dependent Internal Friction Peaks in Al-Mg Solid Solutions and the Discovery ...
62.40; S4
(a) Vol.121,No.1, 139-148

S.P. WU, L.D. ZHANG, G.Y. TIAN, and J.Z. LU
High-Temperature Internal Friction Peaks in Dilute Al-Sn Polycrystals
62.40; S4
(a) Vol.121,No.1, 149-156

F. MUELLER, H.-J. KOEHLER, P. WEBER, K. BLUETHNER, and H.-G. MEYER
A Microwave Circuit for a 1 V Josephson Voltage Standard Operated at 35 GHz
74.50; S4
(a) Vol.121,No.1, K73-K76

G. DLUBEK and W. GERBER
Estimation of Fermi Momentum in Al from Fourier Transform of Positron Annihilation Angular Correlation Curve
71.25 and 78.70; S4
(b) Vol.161,No.2, K81-K84

M.A. KENAWY, M.S. SAKR, E.M. SAKR, H.A. ZAYED, and N.O. MOURAD
Transient and Steady-State Creep of Al-4.5wt%Mg during Phase Transformation
62.20; S4
(a) Vol.121,No.2, 467-474

G.Y. TIAN, J.Z. LU, S.P. WU, and L.D. ZHANG
High-Temperature Internal Friction Peak in Al-0.008wt%Sn Single Crystals
62.40; S4
(a) Vol.121,No.2, 475-482

YU.A. NOVIKOV, A.V. RAKOV, and V.P. SHANTAROVICH
Positron Annihilation in Aluminium Films in Electric Contact with Silicon Substrate
73.30; 78.70; S4
(a) Vol.121,No.2, 495-500

F. VOELKLEIN and U. DILLNER
Mobilities and Concentrations of Charge Carriers in Polycrystalline $Bi_{0.87}Sb_{0.13}$ Films
73.60; S4
(b) Vol.162,No.1, 147-154

H. AOURAG, B. KHELIFA, and A. BELAIDI
Positron Charge Distribution in Aluminium
71.60; 78.70; S4
(b) Vol.162,No.1, K43-K46

M. BAUER, J. MAYER, M. HOHENSTEIN, and M. WILKENS
Quantitative Studies on the Amorphization of Quasicrystals by Means of High Voltage Electron Microscopy
61.40 and 61.80; 64.70; S1.2; S4
(a) Vol.122,No.1, 79-86

V.G. CHUDINOV, R.M.J. COTTERILL, and V.V. ANDREEV
Kinetics of the Diffuse Processes within a Cascade Region in the Sub-Threshold Stages of F.C.C. and H.C.P. Metals
66.30; S1; S4
(a) Vol.122,No.1, 111-120

J. YAN, Z.G. WANG, T.Y. ZHANG, Y.Y. GUO, and C.H. SHIH
Latent Hardening in Aluminum Single Crystals under Cyclic Loading
62.20; S4
(a) Vol.122,No.1, 195-206

Q. TAN and T.S. KÊ (GE TINGSUI)
Activation Energy Associated with the Nonlinear Internal Friction Peaks (P1′ and P1′′ Peaks) in Cold-Worked and ...
62.40; S4
(a) Vol.122,No.1, K25-K30

Š. VARGA and M. KRAJČÍ
Calculation of the Electronic Structure of the (Al,Zn)-Mg Quasicrystal System
71.20; S4
(b) Vol.162,No.2, K77-K82

S5. Nonmetallic elements

J.G. WEN and R. WANG
Two-Beam Dynamical Theory of the Displacement and Splitting of Diffraction Fringes Caused by a Stacking Fault
61.14 and 61.70; S5
(a) Vol.117,No.1, 73-84

V.I. MIKLA, D.G. SEMAK, A.V. MATELESHKO, and A.A. BAGANICH
Photosensitized Deep Gap Center Relaxation in Amorphous Selenium Layers Containing As Additions
71.55; 72.40; 73.60; S5
(a) Vol.117,No.1, 241-250

F. BECHSTEDT and D. REICHARDT
Calculation of Surface-Induced Core-Level Shifts for Covalent Semiconductors C, Si, Ge, and alpha-Sn. II. (100)2x1 Surfaces
68.20 and 73.20; S5; S5.11; S5.12
(b) Vol.157,No.2, 567-576

S.A. LEE and S.M. LINDSAY
Surface Brillouin Scattering from Graphite
68.30 and 78.35; S5
(b) Vol.157,No.2, K83-K86

A.F. AKKERMAN and V.E. FAIN
Calculation of Energy Loss Characteristics of Accelerated Ions Passing through Thin Targets
61.80; S5
(b) Vol.158,No.1, 67-72

P. KLEINERT
A Field-Theoretical Treatment of Third-Sound Localization
68.15; 68.30; S5
(b) Vol.158,No.2, 539-550

P. MIKULSKI, K. FABISIAK, S. ORZESZKO, and F. ROZPŁOCH
The Effect of Starting Material on the Properties of Soft a-C:H Films Obtained by the Glow Discharge Technique
68.55; 78.30; S5
(a) Vol.119,No.2, K127-K130

D.P. ERCHAK, R.B. GELFAND, N.M. PENINA, V.F. STELMAKH, V.P. TOLSTYKH, A.G. ULYASHIN, V.S. VARICHENKO, and A.M. ZAITSEV
Point Paramagnetic Defects in Diamond Irradiated by High-Energy Ions
61.70; 76.30; 78.60; S5
(a) Vol.121,No.1, 63-72

M.YA. GAMARNIK
Size Changes of Lattice Parameters in Ultradisperse Diamond and Silicon
61.55; 61.10; S5; S5.11
(b) Vol.161,No.2, 457-462

P.N. GORLEI, V.V. KOSHMAN, A.I. SKITSKO, L.S. SOLONCHUK, V.V. KHOMYAK, and V.A. SHENDEROVSKII
Longitudinal Hall Effect in p-Te
72.80; S5
(b) Vol.161,No.2, K91-K94

O.A. GOLIKOVA, N. AMANDZHANOV, M.M. KAZANIN, G.M. KLIMASHIN, and V.V. KUTASOV
Electrical Activity of Impurities in beta-Rhombohedral Boron
71.55; 72.80; 76.80; S5
(a) Vol.121,No.2, 579-586

A. LEHMANN, E. BROESE, and B. SCHROETER
Vibrational Modes of Ge_xSe_{1-x} Films
63.20 and 78.30; S5
(b) Vol.162,No.1, K53-K58

YU.I. MAZUR, YU.G. RUBO, O.V. SNITKO, and M.V. STRIKHA
Auger Recombination via Defects in Tellurium
72.10; 71.25; 72.40; S5
(b) Vol.162,No.2, 531-538

W. GRUENEWALD and J. ULLMANN
Cross-Sectional Transmission Electron Microscopy Investigations of RF-Sputtered a-C Films
68.55; 61.16; 68.60; S5
(a) Vol.122,No.2, K129-K132

A.M. BARANOV, S.S. GERASIMOVICH, V.M. ELINSON, A.A. KUZIN, V.V. SLEPTSOV, and P.E. KONDRASHOV
Phase Composition Modeling of alpha-C:H Films on the Basis of Their Optical Properties
68.55 and 78.65; S5
(a) Vol.122,No.2, K139-K142

S5.11. Silicon

Y.F. CHEN and C.K. WONG
Microscopic Model of Metastable Changes in Hydrogenated Amorphous Silicon
61.40; S5.11
(b) Vol.157,No.1, 101-106

K.L. SOROKINA
The Meyer-Neldel Rule in a-Si:H and the Correlation Energy
71.20 and 72.20; S5.11
(b) Vol.157,No.1, K43-K46

J. SAID, H. JAOUEN, G. GHIBAUDO, and I. STOEMENOS
Electrical and Physical Investigation of Defect Annihilation in Arsenic Implanted Silicon
61.70; 61.80; S5.11
(a) Vol.117,No.1, 99-104

R.K. BHAN, S.K. LOMASH, and K.C. CHHABRA
Determination and Study of Field Dependent Defect Density in Thermal SiO_2 Using 1,1,1 Trichloroethane (TCA)
61.70 and 68.48; S5.11; S12
(a) Vol.117,No.1, 203-212

A.P. DOSTANKO, V.V. SHATALOV, and G.M. SHAKHLEVICH
Investigation of the Phase Formation Process in Ni-Based Binary Alloy-Silicon Systems
68.48 and 68.55; S1.61; S5.11
(a) Vol.117,No.1, 213-220

R.J. TARENTO
Injection Effects on Cathodoluminescence Experiments. Dember-Electric Field
78.60; S5.11
(a) Vol.117,No.1, 305-312

YU.A. MANZHOSOV and G.A. KACHURIN
Large Area SOI Films Crystallized by Nanosecond Laser Pulses
68.55; S5.11
(a) Vol.117,No.1, K29-K32

F. BECHSTEDT and D. REICHARDT
Calculation of Surface-Induced Core-Level Shifts for Covalent Semiconductors C, Si, Ge, and alpha-Sn. II. (100)2x1 Surfaces
68.20 and 73.20; S5; S5.11; S5.12
(b) Vol.157,No.2, 567-576

U. PIETSCH
Lattice Compression from Conduction Band Electrons in As-Implanted Silicon
61.50; 62.30; S5.11
(b) Vol.157,No.2, K73-K76

H. GDANITZ and K. SCHMALZ
On the Backside Gettering of Au in Silicon
66.30; S5.11
(a) Vol.117,No.2, 395-402

M.D. DE COTEAU, P.R. WILSHAW, and R. FALSTER
Gettering of Copper to Oxidation Induced Stacking Faults in Silicon
61.70; S5.11
(a) Vol.117,No.2, 403-408

M.A. DEMYANENKO
The Effect of Back Inversion Channel and Electrically Active Defects on the Nonequilibrium Characteristics of SOS MOS...
73.40 and 85; S5.11
(a) Vol.117,No.2, 601-610

J.L. LINDSTROEM, B.G. SVENSSON, W.-X. NI, and M. WILLANDER
Infrared Absorption of Oxygen-Doped Silicon Grown by Molecular Beam Epitaxy
78.30; S5.11
(a) Vol.117,No.2, K171-K176

V.M. BABICH, M.YA. VALAKH, V.B. KOVALCHUK, G.YU. RUDKO, and N.I. SHAKHRAYCHUK
Photoluminescence and Electrical Properties of Germanium-Doped and Thermally Annealed Silicon
78.55; S5.11
(a) Vol.117,No.2, K185-K188

S.I. KHLUPIN and A.F. AKKERMAN
Radial Energy Transfer Density Distribution around the Fast Ion Tracks in Silicon and Germanium
61.80; S5.11; S5.12
(b) Vol.158,No.1, 63-66

A. ANAGNOSTOPOULOS, G.L. BLERIS, and I.Y. YANCHEV
Percolation Conductivity in Polycrystalline Silicon at Moderate Electrical Fields
72.20; S5.11
(b) Vol.158,No.1, K43-K48

M. WATTENBACH, C. KISIELOWSKI-KEMMERICH, H. ALEXANDER, V.V. KVEDER, T.R. MCHEDLIDZE, and YU.A. OSIPYAN
Electric-Dipole Spin Resonance of Dislocations in Plastically Deformed p-Type Silicon
71.55 and 76.30; S5.11
(b) Vol.158,No.1, K49-K54

S. FUJITA and S. WATANABE
On the Orientation Dependence of the Cyclotron Resonance Peaks for Holes in Ge and Si
71.25 and 76.30; S5.11; S5.12
(b) Vol.158,No.1, K69-K74

I.M. FILIPPOV, L.A. KAZAKEVICH, and P.F. LUGAKOV
The Effect of Irradiation and Heat Treatment on the Dislocation Recombination Activity in n-Type Silicon
61.70 and 71.55; S5.11
(a) Vol.118,No.1, 53-60

I.I. KOLKOVSKII, P.F. LUGAKOV, V.V. LUKYANITSA, and V.V. SHUSHA
Formation of Thermal Donors during Annealing of Radiation Defects in Float-Zone Dislocations-Free n-Si
61.70 and 71.55; S5.11
(a) Vol.118,No.1, 65-70

J. REICHEL, M. REICHE, and E. ZUBER
Formation of New Donors in Cz-Grown Silicon at Temperatures below 500 °C
61.70; S5.11
(a) Vol.118,No.1, 85-98

Y.Y. LOGINOV, V.M. LENCHENKO, P.D. BROWN, and J. WOODS
TEM Study of the Formation of Structural Defects in Ion Implanted Silicon at Depths of $h \lesssim R_p$ and $h \gtrsim 2R_p$
61.70; 61.80; S5.11
(a) Vol.118,No.1, 117-130

TENG-GE MA, XIAN-JUN YANG, JI-SHENG ZHANG, and PEI-HSIN TSIEN
Localization of the Boundaries in Laser Recrystallization of SOI with Antireflective Stripe
61.80 and 68.55; 73.60; S5.11
(a) Vol.118,No.1, 179-184

L. SZARO and J. MISIEWICZ
The AC Surface Photovoltage on Silicon under Subbandgap Illumination
73.20 and 73.30; 72.40; S5.11
(a) Vol.118,No.1, 185-188

C.A. LONDOS, K. EFTAXIAS, and V. HADJICONTIS
Correlation of Solubilities of Various Elements in Silicon
66.30; S5.11
(a) Vol.118,No.1, K13-K16

FANGQING ZHANG, DEYAN HE, ZHIZHONG SONG, and GUANGHUA CHEN
Boron Diffusion in Device-Like B-Doped a-SiC:H/a-Si:H Heterojunction
66.30; S5.11
(a) Vol.118,No.1, K17-K20

DEYAN HE, FANGQING ZHANG, and GUANGHUA CHEN
A Direct Measurement of Thermal Equilibrium Defects in Hydrogenated Amorphous Silicon-Nitrogen Alloy Films
61.70 and 76.30; S5.11
(a) Vol.118,No.1, K39-K42

GUANG-XU CHENG, HUA XIA, KUN-JI CHEN, WEI ZHANG, and XING-KUI ZHANG
Raman Measurement of the Grain Size for Silicon Crystallites
78.30; S5.11
(a) Vol.118,No.1, K51-K54

EN-GE WANG, LI-YUAN ZHANG, HUAI-YU WANG, and WEI-MIN JIN
Antisite Defects in Si-AlAs/GaAs Modulation-Doped Superlattices
71.55; 71.20; S5.11; S7.15
(b) Vol.158,No.2, 589-594

H.-J. UNGER
Covalent Bonds in Amorphous Semiconductors and Their Influence on the Structure of the Energy Spectrum
71.50; 71.10; 71.25; S5.11; S5.12
(b) Vol.158,No.2, 609-628

J. D ALBUQUERQUE E CASTRO, R. BECHARA MUNIZ, and L.E. OLIVEIRA
A Model Calculation of the Dielectric Band Structure of Diamond and Silicon
71.25 and 77.90; S5.1; S5.11
(b) Vol.158,No.2, 743-752

D. WINDISCH and P. BECKER
Silicon Lattice Parameters as an Absolute Scale of Length for High Precision Measurements of Fundamental Constants
61.60 and 61.70; S5.11
(a) Vol.118,No.2, 379-388

K.A. AL-SALEH, I.J. JABR, and N.S. SALEH
A Comparison of Ion Beam Induced Atomic Mixing Kinetics of Ti/Si, Fe/Si, and Ni/Si Systems. In Situ Electrical ...
61.80 and 73.40; S1.1; S1.2; S5.11
(a) Vol.118,No.2, 467-472

G.A. ADEGBOYEGA, A. POGGI, and E. SUSI
Some Electrical Characteristics of Fe and Its Annealing Behaviour in FZ Silicon
71.55; 72.20; S5.11
(a) Vol.118,No.2, 491-496

D.A. ARONOV, R. KABULOV, and YU.M. YUABOV
The Photoelectric Characteristics of Solar p-i-n Photocells Based on Hydrogenated Amorphous Silicon with Built-in ...
72.40 and 73.40; 85; S5.11
(a) Vol.118,No.2, 577-590

G. DIMOPOULOS
A New Model of Transconductance for SOI Gated Double Injection Structures
73.40; S5.11
(a) Vol.118,No.2, K73-K78

G. AMATO, G. BENEDETTO, F. FIZZOTTI, C. MANFREDOTTI, and R. SPAGNOLO
Photothermal and Photoconductive Measurements of Low Pressure Chemically Vapour Deposited Amorphous Silicon
72.40 and 73.60; 78.65; S5.11
(a) Vol.119,No.1, 169-176

M.R. MURTI and K.V. REDDY
Grain Boundary Effects on the Carrier Mobility of Polysilicon
72.20; 61.70; S5.11
(a) Vol.119,No.1, 237-240

WANLU WANG and KEJUN LIAO
Effect of the Surface Condition on the Internal Stress of a-Si:H, a-SiN_x:H, and a-Si:H/a-SiN_x:H Heterojunction Films
68.25; S5.11
(a) Vol.119,No.1, K31-K34

M.KH. KHOKONOV and H. NITTA
Channeling of Antiprotons and Electrons
61.80; S5.11
(b) Vol.159,No.2, 589-596

M. ABBAS, H. KERKOW, and R. WEDELL
Heavy Ion Channeling Implantation Processes in Single Crystals
61.80; S5.11
(b) Vol.159,No.2, 597-608

C. ENGLER
Adsorption of H_2O (Molecular and Dissociative) on Si(100) and Si(111) Surfaces Considered by an Empirical Pair Potential Method
68.20; S5.11
(b) Vol.159,No.2, 721-730

M. WAUTELET
On the Transitions between the Crystalline, Amorphous, and Liquid Phases of Silicon and Germanium, when Their Size ...
64.70; 68.60; S5.11; S5.12
(b) Vol.159,No.2, K43-K46

X.J. WU, F.H. LI, H.Q. HOU, J.M. ZHOU, and H. HASHIMOTO
Direct Observation of Interfaces and Microstructures in GaAs/Si and GaAs/$Ga_{1-x}Al_xAs$/Si by Transmission Electron Microscopy
68.55; 68.48; S5.11; S7.13; S7.15
(a) Vol.119,No.2, 511-522

A. FEDOTOV, B. EVTODYI, L. FIONOVA, YU. ILYASHUK, E. KATZ, and L. POLYAK
Electrical Activity of Grain Boundaries in Shaped Grown Silicon
61.70 and 68.55; 73.20; 73.40; S5.11
(a) Vol.119,No.2, 523-534

J. REICHEL and M. REICHE
The Importance of Self-Interstitials to the Defect Formation in Silicon during Quenching
61.70; S5.11
(a) Vol.119,No.2, K103-K106

H. AOURAG, B. KHELIFA, A. BELAIDI, and Z. BELARBI
Positron Distribution in Semiconductors
71.60; 78.70; S5.11; S5.12; S7.1; S8.1
(b) Vol.160,No.1, 201-210

A.G. KAZANSKII, W. FUHS, and H. MELL
On the Fermi-Level Position in a-Si:H(P)
73.60; S5.11
(b) Vol.160,No.1, K25-K28

P. ZAUMSEIL and U. WINTER
Characterization of Boron Implanted Silicon by X-Ray Triple-Crystal Diffractometry
61.10 and 61.70; S5.11
(a) Vol.120,No.1, 67-76

N.S. SALEH, K.A. AL-SALEH, and A.M. AL-SAIE
Kinetics of Ion Beam Induced Mixing in Au/Si System
61.80 and 68.55; 73.60; S1.3; S5.11
(a) Vol.120,No.1, 169-174

V.V. ARISTOV, V.N. MORDKOVICH, A.YU. NIKULIN, A.A. SNIGIREV, U. WINTER, YU.N. EROKHIN, and P. ZAUMSEIL
Possibilities of X-Ray Interference Diffractometry for the Investigation of Ion-Doped Layers
61.10 and 61.70; S5.11
(a) Vol.120,No.1, K1-K6

Y. KAMIURA, M. YONETA, and F. HASHIMOTO
Hole Traps Produced by High-Temperature Heat Treatments of Silicon above 1300 °C
61.70; 71.55; 66.30; S5.11
(a) Vol.120,No.1, K11-K14

V.V. EMTSEV, V.A. KHRAMTSOV, and K. SCHMALZ
On the A-Centre Formation in Heat-Treated Cz-Silicon
61.70; S5.11
(a) Vol.120,No.1, K15-K18

DEYAN HE, FANGQING ZHANG, and GUANGHUA CHEN
Thermal Equilibration in Boron-Doped Hydrogenated Amorphous Silicon-Carbon Alloy Films
68.60; S5.11
(a) Vol.120,No.1, K49-K52

H. KITAGAWA and S. TANAKA
Electrically Active Nickel in Silicon Studied by DLTS in Several Kinds of Silicon Diodes
71.55; S5.11
(a) Vol.120,No.1, K67-K70

G. SCHRAMM
Determination of the Free Exciton Capture Cross Section of As Doped Silicon by Photoluminescence
78.55; 71.35; S5.11
(a) Vol.120,No.1, K101-K104

Š. PICK, M. TOMÁŠEK, G.V. GADIYAK, and YU.N. MOROKOV
Interactions on Silicon Surfaces and Theory of Moments
68.25 and 68.40; 73.20; S5.11
(b) Vol.160,No.2, 483-490

H. AOURAG, B. KHELIFA, A. BELAIDI, and T. KOBAYASHI
Interpretation of the Group-Theoretically Unexpected Shape of the Contour Appearing in the Experimental LCW Data for...
71.25; 71.60; 78.70; S5.11; S5.12
(b) Vol.160,No.2, 561-568

YANG-FANG CHEN
Persistent Photoconductivity in Doping Modulated Amorphous Silicon Multilayers
73.60; 72.40; S5.11
(b) Vol.160,No.2, K103-K108

R. BINDEMANN and O. PAETZOLD
The Influence of Thermal Disorder on the Absorption Edge in the Tauc Region of a-Si:H
78.65; S5.11
(b) Vol.160,No.2, K183-K188

A.M. MAZZONE
Monte-Carlo Simulation of the Backscattering Effects of Loop Dislocations in a Silicon Crystal
61.70; 61.80; S5.11
(a) Vol.120,No.2, 379-386

S.V. KOVESHNIKOV, S.V. NOSENKO, and E.B. YAKIMOV
The Peculiarities of Deep Level Defect Passivation in Si by Atomic Hydrogen
61.70 and 71.55; 66.30; S5.11
(a) Vol.120,No.2, 391-396

A.N. NAZAROV, V.S. LYSENKO, S.A. VALIEV, M.M. LOKSHIN,
A.S. TKACHENKO, and I.A. KUNITSKII
Flash Lamp Annealing and RF Plasma Annealing of Al-SiO_2-Si Structures
73.40; 73.20; S5.11; S10.1
(a) Vol.120,No.2, 447-456

V.V. ARTAMONOV, V.S. LYSENKO, A.N. NAZAROV, B.D. NECHIPORUK,
V.V. STRELCHUK, and M.YA. VALAKH
Radio-Frequency Plasma Treatment and Thermal Annealing in Implanted Si. Raman Study
61.80 and 68.55; 78.30; S5.11
(a) Vol.120,No.2, 475-484

J. ATAEV, O.I. KONKOV, E.I. TERUKOV, V.A. VASSILEV,
and A.S. VOLKOV
Influence of Preparation Conditions on the Radiative Recombination in a-Si:H Films
68.55 and 78.55; S5.11
(a) Vol.120,No.2, 519-524

J. KANIEWSKI, M. KANIEWSKA, W. JUNG, and T. PIOTROWSKI
Iron-Related Deep States in Chemical Vapour Deposited Si
71.55 and 73.40; S5.11
(a) Vol.120,No.2, 531-538

O.O. AWADELKARIM, A. HENRY, B. MONEMAR, and J.L. LINDSTROEM
The Role of Group-V Impurity in Defect Formation in Irradiated Silicon
71.55; 61.70; S5.11
(a) Vol.120,No.2, 539-546

A. WOLKENBERG
Exact Resistivity Calculation of Heavily Doped Silicon
72.20; S5.11
(a) Vol.120,No.2, 567-574

T.S. SHI, A. DE VEIRMAN, J. VAN LANDUYT, G.Q. YANG, and C.L. LIN
TEM Study of the Molecular Effect in Phosphorus Implanted Silicon
61.80; S5.11
(a) Vol.120,No.2, K131-K132

XIAOTONG LIN, ZHIPU YOU, and HUIFEN GUO
An Enhancement Phenomenon of the Minority Carrier Lifetime in Annealed Silicon
72.40; 61.70; S5.11
(a) Vol.120,No.2, K177-K180

G. SCHRAMM
Quantitative Analysis of As Doped Silicon by Photoluminescence
78.55; S5.11
(a) Vol.120,No.2, K181-K184

D. SUEVA, V. SPASSOV, N. CHIKOV, E.I. VAPIREV, and I. IVANOV
Silicon Avalanche Detectors for X-Ray Analysis
76.80 and 85; S5.11
(a) Vol.120,No.2, K185-K188

C. KISIELOWSKI-KEMMERICH
Vacancies and Their Complexes in the Core of Screw Dislocations. Models which Account for ESR-Investigation ... (Review Article)
61.70; 76.30; S5.11
(b) Vol.161,No.1, 11-42

G.S. MYAKENKAYA, G.L. GUTSEV, N.P. AFANASEVA, V.A. EVSEEV, and R.F. KONOPLEVA
Study of Lithium Interaction with Lattice Defects in Silicon
61.70 and 61.80; 72.20; S5.11
(b) Vol.161,No.1, 91-104

L.L. BALASHOVA, N.M. KABACHNIK, and V.N. KONDRATEV
Charge Asymmetry in the Dependence of Stopping on Impact Parameter
61.80; 78.70; 79.90; S5.11
(b) Vol.161,No.1, 113-122

V.I. KHRUPA, D.O. GRIGOREV, V.V. NIKOLAEV, L.I. DATSENKO, and V.B. MOLODKIN
Influence of Defects on the Thickness Dependences of Integral Reflectivity of Dislocation-Free Silicon Crystals
61.10 and 61.70; S5.11
(a) Vol.121,No.1, 11-20

G.A. ADEGBOYEGA and A. POGGI
Intrinsic Gettering of Cr Impurities in p-Type Cz Silicon
71.55; 72.80; 78.30; S5.11
(a) Vol.121,No.1, 181-186

V.I. BLYNSKII, A.YU. KULIKOV, S.YU. RAKHLEY, and V.I. OSINSKII
Photoelectric Crosstalk between Silicon Photoelements
72.40; 78.40; S5.11
(a) Vol.121,No.1, 227-234

R.K. KARAKHANYAN, P.L. ALEKSANYAN, and Y.K. MANUCHAROVA
On Electron Double Diffraction in the Formation of Kikuchi Patterns
61.14; S5.11
(a) Vol.121,No.1, K1-K4

A.I. KURBATOV, M. MAEDER, N.N. SOKOLOV, A. TEMPEL, J. TOBISCH, and A. ZEHE
gamma-Ray Diffraction Study of Thin CaF_2/Si(111) Heterostructures Grown by MBE
68.55; S5.11; S9.12
(a) Vol.121,No.1, K65-K68

A. POGGI, M. MERLI, E. SUSI, R. ANGELUCCI, and A. SARDO
Effects of Silicon Ion-Beam Mixing on p^+/n Diodes: DLTS Analyses of the Induced Defects
61.80 and 85; S5.11
(a) Vol.121,No.1, K135-K140

G. KISSINGER, K. TITTELBACH-HELMRICH, J. KNOPKE, and U. RETZLAFF
On the Gettering Effiency of Buried Layers in Dielectrically Insulated Structures
64.75 and 85; 71.55; S5.11
(a) Vol.121,No.1, K141-K144

M.YA. GAMARNIK
Size Changes of Lattice Parameters in Ultradisperse Diamond and Silicon
61.55; 61.10; S5; S5.11
(b) Vol.161,No.2, 457-462

V.A. ARUTYUNOV, N.A. KUDRYASHOV, O.A. MISHIN, V.M. SAMSONOV, M.N. STRIKHANOV, and YU.A. SYTSKO
Synchrotron Radiation, Radiative Losses, and Polarization Effects for Relativistic Beams in Bent Crystals
61.80 and 78.70; S5.11
(b) Vol.161,No.2, 483-500

K. LENKEIT, CH. TRIKALINOS, L.L. BALASHOVA, N.M. KABACHNIK, and V.I. SHULGA
Monte Carlo Simulation Analysis of Proton-Energy Spectra for Axial Channeling in Silicon in the Intermediate Energy Region
61.80; S5.11
(b) Vol.161,No.2, 513-524

GUANGHUA CHEN, JINZHANG XU, and FANGQING ZHANG
The ESR Studies of B-Doped a-Si:H Thin Films Prepared by Reactive-Sputtering
61.40 and 76.30; 68.55; S5.11
(a) Vol.121,No.2, 501-506

K. LENKEIT, A. PIRRWITZ, A.I. NIKIFOROV, B.Z. KANTER, S.I. STENIN, and V.P. POPOV
Investigations of MBE Grown Superlattices with Sb Delta-Layers in Silicon by Rutherford Backscattering Spectrometry
68.55; 61.80; S5.11
(a) Vol.121,No.2, 523-532

P. ZAUMSEIL, U. WINTER, I.A. VARTANYANTS, and I.YU. KHARITONOV
Comparison between Two X-Ray Methods for Deformation Profile Determination
61.10; S5.11
(a) Vol.121,No.2, K149-K154

N.V. BRILLIANTOV, A.I. RUDENKO, YU.V. SHCHERBAKOV, and V.V. ZVEREV
Anomalous Low Temperature Annealing of Electron Irradiated Lithium-Doped Silicon Solar Cells
61.80; 71.55; 85; S5.11
(a) Vol.121,No.2, K155-K158

N.YU. ARUTYUNOV, V.V. EMTSEV, K. SCHMALZ, and V.YU. TRASHCHAKOV
A Study of Defects in Heat Treated Cz-Silicon by Positron Annihilation
61.70 and 78.70; S5.11
(a) Vol.121,No.2, K163-K166

YINYUE WANG, FANGQING ZHANG, and GUANGHUA CHEN
Behavior of the Thermal Equilibrium in Reactively Sputtered a-$Si_{1-x}Ge_x$:H Alloy Films Studied by Conductivity and...
73.60; 68.55; S5.11; S5.12
(a) Vol.121,No.2, K177-K180

WANLU WANG
Experimental Studies of Light-Induced Stress in a-Si:H/a-SiN_x:H Multilayer Films
68.25; S5.11; S6
(a) Vol.121,No.2, K189-K192

V.A. GURTOV and P.A. RAIKERUS
Radiation Induced Current in Silicon Dioxide Considering Diffusion Currents
73.40; S5.11
(a) Vol.121,No.2, K199-K200

P. MALY, A. SVOBODA, and P. BUSA
Picosecond Transient Concentration Grating in Silicon
78.20; S5.11
(a) Vol.121,No.2, K223-K226

A. DARGYS, N. ŽURAUSKIENĖ, and S. ŽURAUSKAS
Temperature Dependent Field Ionization of Phosphorus Atoms in Silicon
71.55 and 79.70; S5.11
(b) Vol.162,No.1, 183-192

R. PEREZ
On the Characterization of Crystalline Defects Using CBED Techniques
61.14 and 61.70; S1.3; S5.11
(a) Vol.122,No.1, 51-68

I.E. BONDARENKO and E.B. YAKIMOV
EBIC Investigation of Electrical Activity of Dislocations with Different Impurity Atmospheres in Si
61.70; 78.70; S5.11
(a) Vol.122,No.1, 121-128

MOTOKO KWETE, D. SEGERS, M. DORIKENS, L. DORIKENS-VANPRAET, and P. CLAUWS
Positron Annihilation Study of Defects Created in Silicon Irradiated with Electrons of High Energy
61.80; 78.70; S5.11
(a) Vol.122,No.1, 129-138

Y. NISHINO, T. NISHIKAWA, and S. ASANO
Strain Aging in Czochralski-Grown Silicon Crystals
61.70; 62.20; S5.11
(a) Vol.122,No.1, 163-170

J.D. STEPHENSON
Step-Scanning WSR Section Topography for Indirect (Point Defect) Characterization of Dislocation-Free Si Wafers
61.70; S5.11
(a) Vol.122,No.1, 171-186

Z. CHEN and L.C. BURTON
Electrical Properties and Their Depth Variation in Poly-Si under AM1 Illumination
72.20; 72.40; S5.11
(a) Vol.122,No.1, 361-370

V.I. KUSHNIR and E.V. SUVOROV
X-Ray Backscattering on Perfect Crystals
78.70; 61.10; S5.11; S5.12
(a) Vol.122,No.1, 391-404

È. SOBESLAVSKY and W. SKORUPA
Peculiarities in the Modeling of High Dose Implantation of Nitrogen on Silicon Targets
61.80; S5.11
(a) Vol.122,No.1, K19-K24

D. HOEHL, V. HEERA, H. BARTSCH, K. WOLLSCHLAEGER, W. SKORUPA, and V. VOELSKOW
Temperature Dependence of Electron Beam Induced Epitaxial Crystallization of Silicon
68.48 and 68.55; S5.11
(a) Vol.122,No.1, K35-K38

V.I. PUNEGOV, A.P. PETRAKOV, and N.A. TIKHONOV
X-Ray Diffraction on Laser Disturbed Near-Surface Crystal Layers
61.10; S5.11
(a) Vol.122,No.2, 449-458

J. KOLNÍK and M. OŽVOLD
The Influence of Inversion Surface Layers on the Evaluation of the Interface State Energy Distribution from Schottky...
73.30 and 73.40; 85; S5.11; S7.12
(a) Vol.122,No.2, 583-588

P. DANESH, A. TONEVA, P. PANTCHEV, and V. KUDEYAROVA
Microstructure and Optical Properties in CVD a-Si:H Films
68.55 and 78.65; S5.11
(a) Vol.122,No.2, 599-606

J. BARTOŠ and L. TESAŘ
Some Electrical and Optical Properties of Nickel-Related Deep Levels in Silicon
61.70 and 71.55; 73.40; S5.11
(a) Vol.122,No.2, 607-616

H. LEMKE
Properties of Silicon Crystals Doped with Zirconium or Hafnium
71.55; S5.11
(a) Vol.122,No.2, 617-630

M. BRAUER
Effect of Surface Charge and Coating on the Field Distribution of Si High-Voltage Thyristor-Like Structures with ...
73.40 and 85; S5.11
(a) Vol.122,No.2, 741-750

J.D. STEPHENSON
Quasi-Realtime Characterization of Microchip Wafers Using Scanning WSR-Section Topography
61.70; S5.11
(a) Vol.122,No.2, K121-K122

M. MAEDER, J. RICHTER, and A. ZEHE
Surface Reordering of (111)Si 7x7 to 1x1 Reconstruction during In-Situ Annealing
68.20; 61.14; S5.11
(a) Vol.122,No.2, K133-K134

S.S. SIMEONOV, E.I. KAFEDJIISKA, and A.L. GUERASSIMOV
Deep Levels in n-Type Si Substrates Subjected to Ion Cleaning
71.55; 61.80; S5.11
(a) Vol.122,No.2, K147-K150

S5.12. Germanium

V.G. BARYSHEVSKII and A.M. ZAITSEVA
Orientational Effects Due to Traveling of Fast Neutrons and Hard gamma-Quanta through Crystals
61.80; 78.70; S5.12
(b) Vol.157,No.1, 129-136

F. BECHSTEDT and D. REICHARDT
Calculation of Surface-Induced Core-Level Shifts for Covalent Semiconductors C, Si, Ge, and alpha-Sn. II. (100)2x1 Surfaces
68.20 and 73.20; S5; S5.11; S5.12
(b) Vol.157,No.2, 567-576

S.I. KHLUPIN and A.F. AKKERMAN
Radial Energy Transfer Density Distribution around the Fast Ion Tracks in Silicon and Germanium
61.80; S5.11; S5.12
(b) Vol.158,No.1, 63-66

S. FUJITA and S. WATANABE
On the Orientation Dependence of the Cyclotron Resonance Peaks for Holes in Ge and Si
71.25 and 76.30; S5.11; S5.12
(b) Vol.158,No.1, K69-K74

H.-J. UNGER
Covalent Bonds in Amorphous Semiconductors and Their Influence on the Structure of the Energy Spectrum
71.50; 71.10; 71.25; S5.11; S5.12
(b) Vol.158,No.2, 609-628

A. GHOSHAL, M. MONDAL, and K.P. GHATAK
On the Burstein-Moss Shift in 3D Quantum Dots of Ge
73.40; S5.12
(b) Vol.158,No.2, K123-K128

M. WAUTELET
On the Transitions between the Crystalline, Amorphous, and Liquid Phases of Silicon and Germanium, when Their Size ...
64.70; 68.60; S5.11; S5.12
(b) Vol.159,No.2, K43-K46

J. MUSILOVÁ
Method of the Exact Determination of Transition Energies from Modulated Piezoreflectance Spectra of Cubic Crystals
78.20; 71.20; S5.12
(a) Vol.119,No.2, 669-676

H. AOURAG, B. KHELIFA, A. BELAIDI, and Z. BELARBI
Positron Distribution in Semiconductors
71.60; 78.70; S5.11; S5.12; S7.1; S8.1
(b) Vol.160,No.1, 201-210

E.P. SINYAVSKII and E.YU. SAFRONOV
The Optical Properties of Impurity Center Short-Range Potentials in the Magnetic Field
78.50 and 78.55; 71.55; S5.12
(b) Vol.160,No.1, 357-364

K. EFTAXIAS and V. HADJICONTIS
Comments on the Diffusion of Ga in Ge
66.30; S5.12
(b) Vol.160,No.1, K9-K12

T. HATTORI, T. TAKADA, K. KAGAWA, A. MITSUISHI, and Y. KAMIURA
Copper Related Acceptors in Quenched Germanium
71.55 and 78.50; S5.12
(a) Vol.120,No.1, K63-K66

H. AOURAG, B. KHELIFA, A. BELAIDI, and T. KOBAYASHI
Interpretation of the Group-Theoretically Unexpected Shape of the Contour Appearing in the Experimental LCW Data for...
71.25; 71.60; 78.70; S5.11; S5.12
(b) Vol.160,No.2, 561-568

O. PACHEROVÁ
The Calculation of the Coplanar Four-Beam X-Ray Dynamical Diffraction
61.10; S5.12
(a) Vol.120,No.2, 339-350

G.A. MATVEEV, I.M. TSIDILKOVSKI, and A.T. LONCHAKOV
On the Anisotropy of Scattering in Doped n-Ge at Low-Temperatures
71.30 and 72.20; S5.12
(b) Vol.161,No.1, 303-308

E.A. KONDRASHKINA, D.V. NOVIKOV, and S.A. STEPANOV
Study of X-Ray Surface Back Diffraction. Recording of a Diffracted Beam
61.10; 68.20; S5.12
(a) Vol.121,No.1, K9-K14

YINYUE WANG, FANGQING ZHANG, and GUANGHUA CHEN
Behavior of the Thermal Equilibrium in Reactively Sputtered a-$Si_{1-x}Ge_x$:H Alloy Films Studied by Conductivity and...
73.60; 68.55; S5.11; S5.12
(a) Vol.121,No.2, K177-K180

V.I. KUSHNIR and E.V. SUVOROV
X-Ray Backscattering on Perfect Crystals
78.70; 61.10; S5.11; S5.12
(a) Vol.122,No.1, 391-404

A.V. SAVCHUK, E.N. SALKOVA, T.A. SERGAN, and A. HEILMANN
Holographic Recording in Plasma Polymer Germanium Composite Films
78.90; S5.12
(a) Vol.122,No.1, K83-K86

F.R. KESSLER, W. KOEPPEN, and G. KRAUSE
External Stabilization of CO_2 Laser Radiation by Means of i-Semiconductor Absorption
78.20 and 85; 78.30; S5.12; S7.13
(a) Vol.122,No.2, 731-740

S5.2. Solidified Gases

N.C. BACALIS and N.I. PAPANICOLAOU
Electron Momentum Distribution and Compton Profiles of Crystalline Rare Gases by the Augmented-Plane-Wave Method
71.25; 71.10; S5.2
(b) Vol.157,No.1, 293-304

W. LAASCH, H. HAGEDORN, T. KLOIBER, and G. ZIMMERER
Fine Structure of the Luminescence of Solid Neon and Its Relation to Exciton Trapping and Desorption Induced by ...
78.55; 71.35; S5.2
(b) Vol.158,No.2, 753-768

J.S. ARELLANO and G.J. VÁZQUEZ
A First Principles Calculation of the Critical Temperature of Superconducting Metallic Hydrogen
74.10; S5.2
(b) Vol.158,No.2, K161-K164

I.YA. FUGOL, A.M. RATNER, and E.M. YURTAEVA
On Conditions of Spectroscopic Manifestation of Dynamical Jahn-Teller Effect
71.70; 78.50; S5.2
(b) Vol.160,No.1, 245-254

NGUYEN TANG and VU VAN HUNG
Investigation of the Thermodynamic Properties of Anharmonic Crystals by the Momentum Method. II. Comparison of ...
65.50; S5.2
(b) Vol.161,No.1, 165-172

T. DAS, L.B. BHUIYAN, and R.N. JOARDER
Ashcroft-Langreth Prescription for Effective Core Diameter in Liquids: A Variational Analysis
61.20; S2; S5.2
(b) Vol.161,No.2, 473-482

W.Y. ZENG and A. ANDERSON
Lattice Dynamics of Ammonia
63.20; S5.2
(b) Vol.162,No.1, 111-118

NGUYEN TANG and VU VAN HUNG
Investigation of the Thermodynamic Properties of Anharmonic Crystals by the Momentum Method. III. Thermodynamic ...
65.50; 65.40; 65.70; S5.2
(b) Vol.162,No.2, 371-378

NGUYEN TANG and VU VAN HUNG
Investigation of the Thermodynamic Properties of Anharmonic Crystals by the Momentum Method. IV. The Limiting of ...
65.50; 63.90; 64.60; S5.2
(b) Vol.162,No.2, 379-386

O.S. BINBREK and A. ANDERSON
Lattice Dynamics of Sulphur Dioxide
63.20; S5.2
(b) Vol.162,No.2, 421-428

S6. Group IV compounds
(for transition metal compounds, see S1.61)

V. GRAEFE, D. SCHALCH, and A. SCHARMANN
Hydrogenation of Reactively Sputtered AlN_xO_y Films
61.80 and 78.65; S6
(a) Vol.117,No.1, K23-K28

GUANGHUA CHEN, WEI JIA, and FANGQING ZHANG
ESR Study on Annealing Behavior of Reactively Sputtered $a\text{-}GeN_x$:H Films
68.60 and 76.30; S6
(a) Vol.118,No.2, K115-K118

E.A. REPNIKOVA, V.A. GURTOV, and Z.V. PANOVA
The Effect of Synthesis Conditions on the Short-Range Order Characteristics in Silicon Nitride Layers
68.55; S6
(a) Vol.119,No.1, 113-126

T. SOMA, N. MORI, and H.-MATSUO KAGAYA
Pressure-Volume Relation and Pressure-Induced Phase Transition of the Ge-Sn System
64.30; 64.70; S6
(b) Vol.159,No.2, K51-K54

H. AOURAG, B. SOUDINI, B. KHELIFA, and A. BELAIDI
Electron and Positron Distributions in Grey Tin
71.20 and 71.60; S6
(b) Vol.161,No.2, 685-696

K.K. BOURDELLE, N.G. CHECHENIN, A.S. AKHMANOV, A.YU. POROIKOV, and A.V. SUVOROV
Melting and Damage Production in Silicon Carbide under Pulsed Laser Irradiation
61.80; 64.75; S6
(a) Vol.121,No.2, 399-406

H. MURAKAMI and M. SANO
Positron-Trapping Centers in Potassium-Intercalated Graphite, KC_8 and $KC_8H_{2/3}$
61.70; 78.70; S2; S6
(a) Vol.121,No.2, 461-466

U. TESCHNER
Infrared Optical Properties of Amorphous Silicon Oxynitride and Their Theoretical Interpretation
78.30 and 78.65; S6
(a) Vol.121,No.2, 641-656

V.P. PARKHUTIK, YU.E. MAKUSHOK, S.YU. BORISOV, D.V. YAKOVLEV, J.M. MARTINEZ-DUART, J.M. ALBELLA, and A. CLIMENT
Plasma Anodization of Silicon Nitride
68.55; S6
(a) Vol.121,No.2, K181-K184

WANLU WANG
Experimental Studies of Light-Induced Stress in a-Si:H/a-SiN_x:H Multilayer Films
68.25; S5.11; S6
(a) Vol.121,No.2, K189-K192

L.I. BEREZHINSKII, S.I. VLASKINA, G.I. LYASHENKO, and V.E. RODIONOV
On the Possibility of Obtaining the Crystalline Phase in Amorphous Films of Silicon Carbide on Silicon at Low Temperature
68.55; 73.60; 78.55; S6
(a) Vol.121,No.2, K193-K198

L. SKALA, V.M. KENKRE, M.W. WEISER, and J.D. KATZ
Remarks on the Thermodynamics of Sintered Ceramics
65.50; 64.60; S6; S10.1
(b) Vol.162,No.2, 429-434

V.I. GAVRILENKO, A.V. POSTNIKOV, N.I. KLYUI, and V.G. LITOVCHENKO
Energy Band Structure and Optical Properties of Wurtzite-Structure of Silicon Carbide Crystals
71.25 and 78.20; S6
(b) Vol.162,No.2, 477-488

S7. Group V compounds
(for transition metal compounds, see S1.62)

J.M. DUSSEAU, J.L. ROBERT, and S. ABDALLA
Quantitative Interpretation of Shubnikov-de Haas and Hall Effects in Amorphous Thin Films of Cd_3As_2
73.60; S7
(a) Vol.120,No.1, 151-158

S7.1. III-V compounds

DOAN NHAT QUANG
Auger Recombination in Disordered Semiconductors. Effect of the Band-Gap Fluctuation
71.25 and 72.20; S7.1
(b) Vol.157,No.1, 235-246

L. PAVESI and F.-K. REINHART
Transfer Matrix Method to Compute Energy Levels of Superlattices
71.25 and 73.40; S7.1
(b) Vol.157,No.2, 615-626

K. KRÁL, B. HEJDA, and Z. KHÁS
Hot LO-Phonon Quasi-Steady State Distribution Function in Polar Semiconductors
63.20 and 72.10; S7.1; S8.1
(b) Vol.157,No.2, 667-676

D. FROEHLICH, CH. NEUMANN, B. UEBBING, and R. WILLE
Experimental Investigation of Three-Level Optical Stark Effect in Semiconductors
71.35 and 78.20; 73.40; S7.1; S10.1
(b) Vol.159,No.1, 297-308

Y.M. GU, L. FRITSCHE, and N.E. CHRISTENSEN
Chemical Trends in the Formation and Binding Energy of Native Defects in III-V Semiconductors
61.70 and 71.55; S7.1
(b) Vol.159,No.2, 617-628

P. KUIVALAINEN and K. LINDBERG
Hot Electron Transport in Submicron Semiconductor Devices
72.10 and 73.40; 85; S7.1
(b) Vol.159,No.2, 827-836

V.I. MITROKHIN, S.I. REMBEZA, V.V. SVIRIDOV, and N.P. YAROSLAVTSEV
Acousting Probing of Deep Centres in III-V Semiconductors
71.55; 62.40; S7.1
(a) Vol.119,No.2, 535-544

R. FRANKE, T. CHASSÉ, B. AL-ARAJ, P. STREUBEL, and A. MEISEL
Chemical Shifts of Auger Electron and Photoelectron Binding Energies of Phosphorus in Solid Compounds
79.60; S7.1; S11
(b) Vol.160,No.1, 143-152

H. AOURAG, B. KHELIFA, A. BELAIDI, and Z. BELARBI
Positron Distribution in Semiconductors
71.60; 78.70; S5.11; S5.12; S7.1; S8.1
(b) Vol.160,No.1, 201-210

N.A. SEMIKOLENOVA, V.A. BOGDANOVA, and A.S. SEMIKOLENOV
The Ordering Phenomenon in the Highly Doped III-V Semiconductor Materials
61.14; S7.1
(a) Vol.120,No.2, K121-K124

M. ZAŁUŻNY
Coupling between Intersubband Modes in Wide Quantum Wells
71.25 and 73.90; S7.1
(b) Vol.161,No.1, K19-K24

B.P. SINGH
Analysis of Ionic Radii, Second Neighbour Interactions, and Electronegativity Difference for Ionic Crystals
61.50; 71.45; S7.1; S8.1; S9.11
(b) Vol.162,No.2, 329-338

S.YU. KASANDROV
Tight-Binding Description of Surface States in Semiconductors
73.20; S7.1
(b) Vol.162,No.2, 435-442

S7.11. Phosphides

V. GEIST, D. STEPHAN, and D. FINK
The Effect of the Lateral and Longitudinal Spread of Fast Protons on Strain Production in GaP and GaAs
61.10 and 61.80; S7.11; S7.12
(a) Vol.117,No.1, 67-72

L.F. ZAKHARENKOV, V.V. KOZLOVSKII, and B.A. SHUSTROV
Transmutation Doping of Indium Phosphide and Gallium Arsenide Due to Protons and alpha-Particles
61.80; 71.55; S7.11; S7.12
(a) Vol.117,No.1, 85-90

A.A. KOLB, I.G. MEGELA, A.P. BUTURLAKIN, and D.B. GOYER
Low-Temperature Annealing of Radiation Defects in Electron-Irradiated Gallium Phosphide
61.80 and 78.30; S7.11
(a) Vol.118,No.1, K9-K12

R. PAESSLER
Novel Method for Detecting Thermal Trap Depths from Phonon-Broadened Photoneutralization Cross Section Bands
71.55; S7.11
(b) Vol.158,No.2, K143-K148

C. ELL and H. HAUG
Absorption and Optical Gain Spectra and Band Gap Renormalization of Highly Excited Quantum Well Systems
71.25 and 78.20; 73.40; S7.11; S7.15
(b) Vol.159,No.1, 117-124

E. KLIER
Direct Piezoelectric Effect in Semi-Insulating InP:Fe
77.60; S7.11
(a) Vol.119,No.1, K79-K82

A. BECKMANN, K. DETTMER, and F.R. KESSLER
Determination of the Indirect Exciton Energy Gap of GaP Using Thermomodulation in the Temperature Region 80 to 350 K
71.35; 78.20; S7.11
(b) Vol.159,No.2, 765-770

P. DEMKIN, V. LAZAUSKAS, and Z. JANUŠKEVIČIUS
Investigation of Isovalent Impurity States in GaP Using the Extended Hueckel and the Mulliken-Wolfsberg-Helmholtz Method
71.55; S7.11
(b) Vol.159,No.2, K63-K66

G. EFTEKHARI
Electrical Degradation of InP Schottky Barriers Due to Mercury Contamination
73.30; S7.11
(a) Vol.120,No.2, K155-K158

ZHANG-LONG XU, KA-DI ZHU, and SHI-WEI GU
A Quantum Mechanical Treatment of Surface Polaritons in Semi-Infinite Dielectric Systems
71.36 and 73.90; S7.11
(b) Vol.161,No.1, K25-K28

G. EFTEKHARI
Effect of Rapid Thermal Annealing on Anodic Oxides on InP
73.40 and 85; S7.11
(a) Vol.121,No.1, 327-332

TH. DITTRICH
Laser Induced Generation of a Micro-Crack Structure on (111) GaP
61.80; S7.11
(a) Vol.121,No.1, K15-K16

G. EFTEKHARI
Characterization of Rapidly Thermally Annealed GaAs and InP Surfaces Using Schottky Barriers
73.30 and 85; S7.11; S7.12
(a) Vol.122,No.2, 751-756

S7.12. Arsenides

A. ZEHE and D. DE LA CRUZ
Effect of Fluctuations in delta-Doping Superlattices on Donor-Acceptor Pair Luminescence
71.55; 78.55; S7.12
(b) Vol.157,No.1, 305-310

M. SAID and M.A. KANEHISA
Excited Acceptor States in II-VI and III-V Semiconductors
71.55; S7.12; S8.13
(b) Vol.157,No.1, 311-322

T. TAKAI, D.K. CHOI, Y.T. THATHACHARI, T. HALICIOGLU, and W.A. TILLER
A Potential Energy Function for GaAs Systems
63.10; 68.20; S7.12
(b) Vol.157,No.1, K13-K18

S. ERKOC, T. HALICIOGLU, and W.A. TILLER
Simulation Calculations on Energetics and Reconstruction Patterns for the GaAs(100) Surface
68.20; S7.12
(b) Vol.157,No.1, K23-K28

V. GEIST, D. STEPHAN, and D. FINK
The Effect of the Lateral and Longitudinal Spread of Fast Protons on Strain Production in GaP and GaAs
61.10 and 61.80; S7.11; S7.12
(a) Vol.117,No.1, 67-72

L.F. ZAKHARENKOV, V.V. KOZLOVSKII, and B.A. SHUSTROV
Transmutation Doping of Indium Phosphide and Gallium Arsenide Due to Protons and alpha-Particles
61.80; 71.55; S7.11; S7.12
(a) Vol.117,No.1, 85-90

A. BEL HADJ, B. EL JANI, M. GUITTARD, R. BENNACEUR, and P. GIBART
Photoluminescence of Heavily Doped Sn-GaAs Grown by Metal Organics Vapor Phase Epitaxy
61.70 and 78.65; 73.60; S7.12
(a) Vol.117,No.1, 169-176

T. SOMA, S. TAHATA, and H.-MATSUO KAGAYA
Interstitial Atom in the Zincblende-Type Lattice
61.70; S7.12
(b) Vol.157,No.2, 509-518

L. KONCZEWICZ, E. LITWIN-STASZEWSKA, A. MASLOWSKA, R. PIOTRZKOWSKI, J.L. ROBERT, and J.P. ANDRE
Investigation of Electron Mobility in a Two-Dimensional Electron Gas. GaAs-$Al_xGa_{1-x}As$ Heterostructures. Application of ...
73.40; S7.12; S7.15
(b) Vol.157,No.2, 593-600

N.L. DMITRUK, N.A. FIDRYA, O.V. SNITKO, and E.F. VENGER
Surface Relief and Dissipation Influence on Plasmon-Phonon Polariton Dispersion of Polar Semiconductors
68.20 and 71.36; 71.45; S7.12
(b) Vol.157,No.2, 601-608

H. ALEXANDER, S. DIETRICH, M. HUEHNE, M. KOLBE, and G. WEBER
EBIC Microscopy Applied to Glide Dislocations
61.70; 71.55; 72.20; S7.12
(a) Vol.117,No.2, 417-428

E.G. SALMAN, A.N. KORSHUNOV, and V.N. VERTOPRAKHOV
A DLTS Study of InAs MIS Structures
71.55 and 73.40; S7.12
(a) Vol.117,No.2, 509-514

H. WITTE, W. SIEGEL, G. KUEHNEL, T. FLADE, and H.A. SCHNEIDER
Medium-Deep Acceptors in LEC GaAs
71.55; 72.20; S7.12
(a) Vol.117,No.2, 527-534

C.S. WU, C.S. PAI, S.J. PEARTON, F. REN, E. LANE, and D.M. SCHLEICH
Schottky Barrier Enhancement on n-Type GaAs by As Implantation
73.60; S7.12
(a) Vol.117,No.2, K107-K110

E. NEBAUER
Some Remarks on the Interface Displacement in the AuGeNi-GaAs System
68.48; 85; S7.12
(a) Vol.117,No.2, K111-K114

L.P. WANG, L.Z. Zhang, W.X. ZHU, X.T. LU, and G.G. QIN
The Behaviour of the Hydrogen-Localized Vibrational Mode in Proton-Implanted GaAs
63.20 and 78.50; S7.12
(b) Vol.158,No.1, 113-118

B. SUKUMAR and K. NAVANEETHAKRISHNAN
Diamagnetic Susceptibility of a Donor in a GaAs/$Ga_{1-x}Al_xAs$ Quantum Well Heterostructure and Its Pressure Dependence
73.40 and 75.20; 71.55; S7.12; S7.15
(b) Vol.158,No.1, 193-200

B.K. PANDA, D.P. MAHAPATRA, and H.C. PADHI
Compton Profile Anisotropies of GaAs
71.25 and 78.70; S7.12
(b) Vol.158,No.1, 261-266

S.V. IVANOV, P.S. KOPEV, N.N. LEDENTSOV, B.YA. MELTSER, M.YU. NADTOCHY, V.M. USTINOV, and A.M. VASILEV
Ultrathin Layers and Short-Period Superlattices in Semiconductor Structures
73.40; 78.65; S7.12; S7.15
(a) Vol.118,No.1, 169-178

V.N. BRUDNYI and V.V. PESHEV
Electric Field Effect on the Temperature Dependence of the (E_C-0.33eV) Center Introduction Rate under gamma-Irradiation ...
71.55; 61.80; S7.12
(a) Vol.118,No.1, 219-224

J. PASTRŇÁK, J. OSWALD, O. PETŘÍČEK, and V.N. OVSYUK
The Degradation of Band Edge Luminescence of GaAs under Laser Light Excitation
78.55; 68.20; S7.12
(a) Vol.118,No.1, 317-326

J.J. LIOU
Modeling the Steady-State Collector Current of AlGaAs/GaAs Heterojunction Bipolar Transistors Using a Generalized ...
73.40 and 85; S7.12; S7.15
(a) Vol.118,No.1, K65-K68

DOAN NHAT QUANG
Auger Recombination in Heavily Doped Semiconductors
71.25 and 72.20; S7.12
(b) Vol.158,No.2, 669-684

J. BĄK-MISIUK, G. KUEHNEL, W. SIEGEL, and U. PIETSCH
The Possible Influence of the Free Carrier Redistribution within the Conduction Band on the Thermal Expansion ...
65.70; 72.20; S7.12
(b) Vol.158,No.2, K111-K114

P.S. GLADKOV and K.S. VARBLIANSKA
Ta/GaAs Schottky Barriers Produced by in Situ Sputter Etching, RF Magnetron Sputtering of Ta, and Its Thermal Oxidation
73.30; 85; S1; S7.12
(a) Vol.118,No.2, 479-486

M. KANIEWSKA and J. KANIEWSKI
Influence of DX Centers on Photoresponse of GaAs/AlGaAs Double Heterostructures
73.40; 71.55; 72.40; S7.12; S7.15
(a) Vol.118,No.2, 487-490

J. PASTRŇÁK, J. OSWALD, M. LÁZNIČKA, A. BOSACCHI, and A. SALOKATVE
The Photoluminescence Spectra of Thin Si-Doped GaAs Layers Grown by MBE
78.55 and 78.65; S7.12
(a) Vol.118,No.2, 567-576

D.S. CHEMLA, S. SCHMITT-RINK, W.H. KNOX, D.A.B. MILLER, K.W. GOOSSEN, and G. HASNAIN
Ultrafast Nonlinear Optical Effects in Biased Semiconductor Quantum Wells
71.35 and 78.20; 73.40; S7.12
(b) Vol.159,No.1, 11-28

I. ABRAM
Photon Echoes and the Structure of Wannier Excitons
71.35 and 78.20; S7.12; S7.15
(b) Vol.159,No.1, 29-34

W. SCHAEFER and K. HENNEBERGER
Pulse Propagation and Carrier Kinetics in Laser Excited Semiconductors
72.10 and 78.20; S7.12
(b) Vol.159,No.1, 59-70

E. LACH, V.D. KULAKOVSKII, A. FORCHEL, T.L. REINECKE, J. STRAKA, D. GRUETZMACHER, and G. WEIMANN
Single and Many Particle Effects in the Emission Spectra of Laterally Homogeneous 2D Plasmas
71.35 and 78.55; 73.40; S7.12; S7.15
(b) Vol.159,No.1, 125-132

Y. MASUMOTO, S. IDESHITA, and T. WAMURA
Nonlinear Optical Study of Semiconductor Microstructures, CuCl Quantum Dots, and GaAs Quantum Wells
71.35 and 78.20; 73.40; S7.12; S9
(b) Vol.159,No.1, 133-142

L. WENDLER, F. BECHSTEDT, and M. FIEDLER
Quasi-Two-Dimensional Screening of the Electron-Hole Interaction in Modulation-Doped Quantum Wells
71.35 and 73.40; S7.12; S7.15
(b) Vol.159,No.1, 143-154

K.-H. SCHLAAD, CH. WEBER, D.S. CHEMLA, J. CUNNINGHAM, D.A.B. MILLER, C.V. HOOF, G. BORGHS, G. WEIMANN, W. SCHLAPP et al.
Nonlinear Optical Properties of GaAs/(AlGa)As Multiple Quantum Well Structures under Quasistationary High Excitation ...
71.35 and 78.20; 73.40; S7.12; S7.15
(b) Vol.159,No.1, 173-180

J.L. OUDAR, B. SFEZ, R. KUSZELEWICZ, J.C. MICHEL, and R. AZOULAY
Optical Nonlinearities of GaAs-Based Epitaxial Structures for All-Optical Switching
78.20; 68.55; S7.12; S7.15
(b) Vol.159,No.1, 181-190

I.L. ALEINER, L.M. CHISTYAKOVA, S.I. KOKHANOVSKII, R.P. SEISYAN, A.A. TOROPOV, and T.V. SHUBINA
Passive and Active Q-Switching in an AlGaAs/GaAs SQW Laser. Giant Stark Shift in the Laser p-n SQW Heterostructure
71.35 and 78.20; S7.12; S7.15
(b) Vol.159,No.1, 197-204

H. HAUG, L. BANYAI, J. LIEBLER, and T. WICHT
The Pulsed Nonresonant Optical Stark Effect and the Urbach Tail in Semiconductors
71.35 and 78.20; S7.12; S8.11; S9
(b) Vol.159,No.1, 309-316

N. PRESSER, G. KUDLEK, J. GUTOWSKI, S. DURBIN, D.R. MENKE, M. KOBAYASHI, and R.L. GUNSHOR
High-Density Spectroscopy of ZnSe/GaAs Epilayers in the Near-Band-Edge Region
71.35 and 78.65; 78.55; S7.12; S8.12
(b) Vol.159,No.1, 443-448

A. FORCHEL, B.E. MAILE, H. LEIER, G. MAYER, and R. GERMANN
Optical Spectroscopy on One-Dimensional Compound Semiconductor Structures
71.35 and 78.20; S7.12; S7.15
(b) Vol.159,No.1, 457-470

N.S. AVERKIEV, V.M. ASNIN, A.A. ROGACHEV, A.YU. SILOV, and V.I. STEPANOV
Photoluminescence Studies of Anomalous Diamagnetic Properties of the 2D e-h System on GaAs Surface
75.20 and 78.55; 73.40; S7.12
(b) Vol.159,No.1, 471-476

YU.B. BOLKHOVITYANOV and A.E. DOLBAK
Surface Composition of Epitaxial Films of Gallium Arsenide and Related Compounds after Termination of Their Growth ...
68.20 and 68.55; S7.12
(a) Vol.119,No.1, 139-144

A.E. TSURKAN, T.D. SHEMYAKOVA, S.P. MEDVETSKII, and L.A. NAZARENKO
Spectral Response and Detectivity of InAs Hybrid Structures
68.48; 73.60; S7.12; S10.1
(a) Vol.119,No.1, 191-200

J.J. LIOU
The Effects of Neutron Irradiation on the Current Gain of AlGaAs/GaAs Heterojunction Bipolar Transistors
73.40 and 85; S7.12; S7.15
(a) Vol.119,No.1, 337-342

J.M. SALLESE, D. LAVIELLE, J. SINGLETON, A. LEVCURAS, J.-C. GRENET, P. GIBART, and J.-C. PORTAL
Non Emergence of DX States in GaAs:Te under Hydrostatic Pressure up to 1.5 GPa
71.55; S7.12
(a) Vol.119,No.1, K41-K46

J.TH. ZETTLER, TH. DITTRICH, and L. SCHROTTKE
Modulation Ellipsometry - a New Technique for the Characterization of Semiconductor Materials and Complex ...
78.20; S7.12
(a) Vol.119,No.1, K91-K96

H. LEON, F. COMAS, and M. SUHRKE
Influence of Level Broadening in Mobility Calculations for Low-Dimensional Semiconductor Heterostructures
72.10 and 73.40; S7.12
(b) Vol.159,No.2, 731-742

T. DITTRICH
Correlation between Photoreflectance and CV Measurements of n-GaAs Epitaxial Layers
73.60 and 78.65; S7.12
(a) Vol.119,No.2, 479-486

M. LESZCZYNSKI, A. KULIK, and P. CIEPIELEWSKI
Lattice Parameters of GaInP/GaAs Epi-Layers at Temperatures 77 to 752 K
68.48 and 68.55; S7.12; S7.15
(a) Vol.119,No.2, 495-502

J.E. BISBEE and N.C. HALDER
Activation Energy and Distribution Function of the EL2 Defect Level in Si-Implanted GaAs
61.70 and 71.55; S7.12
(a) Vol.119,No.2, 545-554

I. THURZO and E. PINČÍK
Attempts to Probe the Semi-Insulating GaAs-Plasma Oxide Interface by Small-Signal Charge DLTS
73.40 and 85; S7.12
(a) Vol.119,No.2, 701-710

P. WEISSBRODT and E. NEBAUER
SNMS Study of the Pt Out-Diffusion in the Au/Pt/Ti-n-GaAs System
73.30 and 79.20; S7.12
(a) Vol.119,No.2, K131-K134

G. BACQUET, J. FRANDON, J. BANDET, F. FABRE, R. TAOUINT, C. FONTAINE, and A. MUNOZ-YAGUE
Optical Orientation in GaAs Grown on Si
78.20; S7.12
(a) Vol.119,No.2, K181-K184

J. KUZMIK and T. LALINSKY
Correlation between the Backgating Effect and the Gate Length Shortening on GaAs MESFETs
73.40 and 85; S7.12
(a) Vol.119,No.2, K185-K188

H. LEON and F. COMAS
Confined-LO-Phonon-Limited Mobility in a Semiconductor Quantum Well
72.10 and 73.40; S7.12; S7.15
(b) Vol.160,No.1, 105-116

H. RUECKER, M. MORA-RAMOS, and F. COMAS
Magnetopolaron in a Quantum Well. LO-Phonon Confinement Effects
71.38 and 73.40; 63.20; S7.12
(b) Vol.160,No.1, 117-126

H. AOURAG, B. KHELIFA, A. BELAIDI, A. TADJER, M. REZKI, and M. GAMOUDI
Positron Band Structures of GaAs and CdTe
71.25 and 71.60; 78.70; S7.12; S8.13
(b) Vol.160,No.1, 193-200

S.V. DUBONOS and S.V. KOVESHNIKOV
Formation of High Mobile Defects in GaAs under Ar-Plasma Etching
61.70; 71.55; S7.12
(a) Vol.120,No.1, 77-82

B. THOMAS, M.W. AHKTAR, D.V. MORGAN, M.A. MOHAMMED, and D.E. DAVIES
A Low-Cost Ion Assisted Evaporation Technique for AuGe/n-GaAs Contact Fabrication
73.40; S7.12
(a) Vol.120,No.1, K45-K48

L.G. GERCHIKOV and A.V. SUBASHIEV
Interface States in Subband Structure of Semiconductor Quantum Wells
71.25 and 73.40; S7.12; S7.15
(b) Vol.160,No.2, 443-458

A. GOLTZENÉ and C. SCHWAB
Qualitative and Quantitative Differentiation of Paramagnetic Anion-Antisite-Related Spectra in GaAs
71.70 and 76.30; S7.12
(b) Vol.160,No.2, 649-660

N.N. FALEEV, L.I. FLAKS, S.G. KONNIKOV, I.K. SOLOMIN, and S.V. BATASHOVA
The Influence of the Directed Displacement of Atomic Planes on the X-Ray Diffraction Rocking Curves
61.10 and 68.55; S7.12; S7.15
(a) Vol.120,No.2, 327-338

H. WENZEL and H.-J. WUENSCHE
A Model for the Calculation of the Threshold Current of SCH-MQW-SAS Lasers
73.40 and 85; S7.12; S7.15
(a) Vol.120,No.2, 661-674

G. KADEN
Zum Einfluss des Oberflaechenpotentials auf die Frequenz-dispersion der Admittanz von GaAs-MIS-Teststrukturen ...
73.40 and 85; S7.12
(a) Vol.120,No.2, 675-686

R. SCHLESINGER, S. ROGASCHEWSKI, and P.J. JANIETZ
Electrochemically Deposited Copper Schottky Contacts on n-Type GaAs for Electron-Beam-Induced Current Measurements
73.40 and 85; 73.30; S7.12
(a) Vol.120,No.2, 687-694

E.L. IVCHENKO, V.P. KOCHERESHKO, I.N. URALTSEV, and D.R. YAKOVLEV
Reflectivity and Photoreflectivity in Superlattices and Quantum Wells
73.40 and 78.20; 71.35; S7.12; S7.15
(b) Vol.161,No.1, 217-222

E.L. GARTSTEIN
Multibeam Calculation of the Intensity Distribution for an Imperfect Crystal in Laue Geometry
61.10; 61.70; S7.12
(a) Vol.121,No.1, 37-44

H. MENNIGER, M. BEER, R. GLEICHMANN, H. RAIDT, B. ULRICI, and G. VOIGT
Electrical Homogeneity of Semi-Insulating LEC GaAs by Post-Growth Annealing
61.70 and 78.60; 85; S7.12
(a) Vol.121,No.1, 95-104

G. KADEN
Zum Einfluss des Oberflaechenpotentials auf die Frequenz-dispersion der Admittanz von GaAs-MIT-Teststrukturen ... (II)
73.40 and 85; S7.12
(a) Vol.121,No.1, 315-326

T. IKARI, H. YOKOYAMA, K. MAEDA, and K. FUTAGAMI
Near Band Edge Photoacoustic Spectra of GaAs below Room Temperature
78.20; 71.50; S7.12
(a) Vol.121,No.1, K125-K128

TH. DITTRICH and V.I. TIMOSHENKO
Laser Induced Modification of n-GaAs below the Classical Melting Threshold
61.80 and 78.65; S7.12
(a) Vol.121,No.2, 547-554

R. PAESSLER
Zero-Phonon versus Classical Absorption Thresholds Due to Photoexcitation Cross-Section Bands of the A Centre in GaAs
71.55; S7.12
(b) Vol.162,No.1, K47-K52

F. SAJOVEC, R. WOLF, A. FATTAH, K. BICKMANN, H. WENZL, G. NAGEL, H. RUEFER, E. TOMZIG, and P. DE BIÈVRE
Defect Analysis on GaAs Crystals by Precision Measurements of Density and Lattice Parameter
61.70; 61.60; S7.12
(a) Vol.122,No.1, 139-152

I. THURZO, V. NÁDAŽDY, and E. PINČÍK
Electric-Field-Enhanced Emission from a Discrete Energy Level at the GaAs-Oxide Interface
73.40; 71.55; S7.12
(a) Vol.122,No.1, 275-284

A. REIF, P. STREUBEL, A. MEISEL, and D. ZEISSIG
Characterization of the Diffusion and Reaction Behaviour of Ti/Pt/Au Layer Contacts on GaAs by Means of Auger Electron..
68.48 and 73.40; 79.20; S7.12
(a) Vol.122,No.1, 331-340

P. CHAKRABARTI, M. PURI, A. SINGHAL, and A. MISHRA
Optically Controlled Characteristics of TEGFET
73.40 and 85; S7.12; S7.16
(a) Vol.122,No.1, 405-412

V. HOLÝ, J. KUBĚNA, and K. PLOOG
X-Ray Analysis of Structural Defects in a Semiconductor Superlattice
61.10 and 61.70; S7.12
(b) Vol.162,No.2, 347-362

J. KOLNÍK and M. OŽVOLD
The Influence of Inversion Surface Layers on the Evaluation of the Interface State Energy Distribution from Schottky...
73.30 and 73.40; 85; S5.11; S7.12
(a) Vol.122,No.2, 583-588

N. PUHLMANN and G. OELGART
Semiconductor Characterization by Means of EBIC, Cathodo-, and Photoluminescence
78.60; 78.55; S7.12
(a) Vol.122,No.2, 705-714

J. BARTOŠ, E. PINČÍK, I. THURZO, B. MATÁTKO, and T. LALINSKÝ
Characterization of MAOS Structures on GaAs
73.40 and 85; S7.12
(a) Vol.122,No.2, 715-722

G. EFTEKHARI
Characterization of Rapidly Thermally Annealed GaAs and InP Surfaces Using Schottky Barriers
73.30 and 85; S7.11; S7.12
(a) Vol.122,No.2, 751-756

G. GRUMMT, R. PICKENHAIN, and G. OELGART
Photocurrent Investigation on a Graded Heterojunction
73.40; 72.40; S7.12
(a) Vol.122,No.2, 765-772

S7.13. Antimonides

B.A. ARONZON and I.M. TSIDILKOVSKII
Magnetic-Field-Induced Localization of Electrons in Fluctuation-Potential Wells of Impurities (Review Article)
71.30; 72.20; 75.40; S7.13; S8.15
(b) Vol.157,No.1, 17-60

P. AGHAMKAR and P.K. SEN
Parametric Excitation of Laser in Magnetoactive Doped Piezoelectric Semiconductors
78.20; S7.13
(b) Vol.157,No.2, 735-744

H. GERSTENBERG and W. GLAESER
Transmutation Doping and Lattice Defects in Degenerate InSb
61.70 and 72.20; 71.55; S7.13
(a) Vol.118,No.1, 241-252

M. DEY (BOSE) and S. GHOSH
Amplification of Acoustohelicon Waves in Piezoelectric Semiconductors. Effect of Nonlinearity in Electron ...
72.10 and 72.50; 71.25; S7.13
(b) Vol.159,No.2, 817-826

X.J. WU, F.H. LI, H.Q. HOU, J.M. ZHOU, and H. HASHIMOTO
Direct Observation of Interfaces and Microstructures in GaAs/Si and GaAs/$Ga_{1-x}Al_xAs$/Si by Transmission Electron Microscopy
68.55; 68.48; S5.11; S7.13; S7.15
(a) Vol.119,No.2, 511-522

S.A. STUDENIKIN
Dispersion Properties of Indium Antimonide under a Strong Magnetic Field
78.20; S7.13
(b) Vol.161,No.1, K67-K70

V.I. BARBASHOV and T.V. RODZINA
The Reverse Motion of Screw Dislocations in Indium Antimonide Crystals
61.70; S7.13
(a) Vol.121,No.1, K27-K30

N.S. AVERKIEV, B.N. KALININ, A.V. LOSEV, A.A. ROGACHEV, and A.S. FILIPCHENKO
Recombination Radiation from Heavily-Doped n-Type Indium Antimonide Crystals
78.55; S7.13
(a) Vol.121,No.1, K129-K134

M.R. SUROWIEC
Possibility of a New Type Lomer-Cottrell Lock Formation
61.70; S7.13
(a) Vol.122,No.1, K15-K18

F.R. KESSLER, W. KOEPPEN, and G. KRAUSE
External Stabilization of CO_2 Laser Radiation by Means of i-Semiconductor Absorption
78.20 and 85; 78.30; S5.12; S7.13
(a) Vol.122,No.2, 731-740

S7.15. Mixed crystals

L. KONCZEWICZ, E. LITWIN-STASZEWSKA, A. MASLOWSKA, R. PIOTRZKOWSKI, J.L. ROBERT, and J.P. ANDRE
Investigation of Electron Mobility in a Two-Dimensional Electron Gas. GaAs-$Al_xGa_{1-x}As$ Heterostructures. Application of ...
73.40; S7.12; S7.15
(b) Vol.157,No.2, 593-600

T. TAKAI, T. HALICIOGLU, and W.A. TILLER
A Model Potential Energy Function for Systems Containing Al, Ga, and As Atoms
63.10; 68.20; S7.15
(b) Vol.157,No.2, K77-K82

R. MITDANK, A. STURM, and B. JACOBS
Investigation of Excess Carrier Diffusion in $Ga_{0.35}Al_{0.65}As$ by Cathodoluminescence
73.60; 66.30; S7.15
(a) Vol.117,No.2, 485-494

N.A. GUSHCHINA and V.K. NIKULIN
A Tight-Binding Calculation of Band Discontinuities in the $Al_xGa_{1-x}As$/GaAs(AlAs) Heterostructures
71.25 and 73.40; S7.15
(b) Vol.158,No.1, 175-180

B. SUKUMAR and K. NAVANEETHAKRISHNAN
Diamagnetic Susceptibility of a Donor in a GaAs/$Ga_{1-x}Al_xAs$ Quantum Well Heterostructure and Its Pressure Dependence
73.40 and 75.20; 71.55; S7.12; S7.15
(b) Vol.158,No.1, 193-200

S.V. IVANOV, P.S. KOPEV, N.N. LEDENTSOV, B.YA. MELTSER, M.YU. NADTOCHY, V.M. USTINOV, and A.M. VASILEV
Ultrathin Layers and Short-Period Superlattices in Semiconductor Structures
73.40; 78.65; S7.12; S7.15
(a) Vol.118,No.1, 169-178

J. BĄK-MISIUK, J. WOLF, and U. PIETSCH
The Thermal Expansion Coefficient of $A^{III}B^{V}$ Multilayer Structures
65.70 and 68.60; S7.15
(a) Vol.118,No.1, 209-218

J.J. LIOU
Modeling the Steady-State Collector Current of AlGaAs/GaAs Heterojunction Bipolar Transistors Using a Generalized ...
73.40 and 85; S7.12; S7.15
(a) Vol.118,No.1, K65-K68

H. RODRÍGUEZ-COPPOLA, J.L. MARTÍNEZ-CUELLAR, and R. PÉREZ-ÁLVAREZ
Two-Band Envelope Function Model for Graded Interfaces
71.25 and 73.40; S7.15
(b) Vol.158,No.2, 481-492

EN-GE WANG, LI-YUAN ZHANG, HUAI-YU WANG, and WEI-MIN JIN
Antisite Defects in Si-AlAs/GaAs Modulation-Doped Superlattices
71.55; 71.20; S5.11; S7.15
(b) Vol.158,No.2, 589-594

M. KANIEWSKA and J. KANIEWSKI
Influence of DX Centers on Photoresponse of GaAs/AlGaAs Double Heterostructures
73.40; 71.55; 72.40; S7.12; S7.15
(a) Vol.118,No.2, 487-490

I. ABRAM
Photon Echoes and the Structure of Wannier Excitons
71.35 and 78.20; S7.12; S7.15
(b) Vol.159,No.1, 29-34

C. ELL and H. HAUG
Absorption and Optical Gain Spectra and Band Gap Renormalization of Highly Excited Quantum Well Systems
71.25 and 78.20; 73.40; S7.11; S7.15
(b) Vol.159,No.1, 117-124

E. LACH, V.D. KULAKOVSKII, A. FORCHEL, T.L. REINECKE, J. STRAKA, D. GRUETZMACHER, and G. WEIMANN
Single and Many Particle Effects in the Emission Spectra of Laterally Homogeneous 2D Plasmas
71.35 and 78.55; 73.40; S7.12; S7.15
(b) Vol.159,No.1, 125-132

L. WENDLER, F. BECHSTEDT, and M. FIEDLER
Quasi-Two-Dimensional Screening of the Electron-Hole Interaction in Modulation-Doped Quantum Wells
71.35 and 73.40; S7.12; S7.15
(b) Vol.159,No.1, 143-154

L.V. BUTOV, V.D. KULAKOVSKII, and T.G. ANDERSSON
Photoluminescence of a Dense Quasi-2D Electron-Hole System in Modulation Doped Heterostructure with Strained InGaAs QW
71.35 and 78.55; 73.40; S7.15
(b) Vol.159,No.1, 165-172

K.-H. SCHLAAD, CH. WEBER, D.S. CHEMLA, J. CUNNINGHAM, D.A.B. MILLER, C.V. HOOF, G. BORGHS, G. WEIMANN, W. SCHLAPP et al.
Nonlinear Optical Properties of GaAs/(AlGa)As Multiple Quantum Well Structures under Quasistationary High Excitation ...
71.35 and 78.20; 73.40; S7.12; S7.15
(b) Vol.159,No.1, 173-180

J.L. OUDAR, B. SFEZ, R. KUSZELEWICZ, J.C. MICHEL, and R. AZOULAY
Optical Nonlinearities of GaAs-Based Epitaxial Structures for All-Optical Switching
78.20; 68.55; S7.12; S7.15
(b) Vol.159,No.1, 181-190

I.L. ALEINER, L.M. CHISTYAKOVA, S.I. KOKHANOVSKII, R.P. SEISYAN, A.A. TOROPOV, and T.V. SHUBINA
Passive and Active Q-Switching in an AlGaAs/GaAs SQW Laser. Giant Stark Shift in the Laser p-n SQW Heterostructure
71.35 and 78.20; S7.12; S7.15
(b) Vol.159,No.1, 197-204

H. FIESELER, R. SCHWABE, and J.L. STAEHLI
Electron-Hole Plasma in Cross-Over-Near Ga(As,P)
71.35 and 78.55; S7.15
(b) Vol.159,No.1, 411-416

A. FORCHEL, B.E. MAILE, H. LEIER, G. MAYER, and R. GERMANN
Optical Spectroscopy on One-Dimensional Compound Semiconductor Structures
71.35 and 78.20; S7.12; S7.15
(b) Vol.159,No.1, 457-470

R. GOLDHAHN, H. ROEPPISCHER, M. GERICKE, and C.-E. RICHTER
Characterization of $Al_xGa_{1-x}As$/GaAs Heterostructures by Electroreflectance Depth Profiling
73.40 and 78.20; S7.15
(a) Vol.119,No.1, 183-190

J.J. LIOU
The Effects of Neutron Irradiation on the Current Gain of AlGaAs/GaAs Heterojunction Bipolar Transistors
73.40 and 85; S7.12; S7.15
(a) Vol.119,No.1, 337-342

M. LESZCZYNSKI, A. KULIK, and P. CIEPIELEWSKI
Lattice Parameters of GaInP/GaAs Epi-Layers at Temperatures 77 to 752 K
68.48 and 68.55; S7.12; S7.15
(a) Vol.119,No.2, 495-502

X.J. WU, F.H. LI, H.Q. HOU, J.M. ZHOU, and H. HASHIMOTO
Direct Observation of Interfaces and Microstructures in GaAs/Si and GaAs/$Ga_{1-x}Al_xAs$/Si by Transmission Electron Microscopy
68.55; 68.48; S5.11; S7.13; S7.15
(a) Vol.119,No.2, 511-522

B. PODOR
On the Concentration and Composition Dependence of the Acceptor Energy in $Al_xGa_{1-x}As$:Ge (x<0.40)
71.55; S7.15
(a) Vol.119,No.2, K135-K140

H. LEON and F. COMAS
Confined-LO-Phonon-Limited Mobility in a Semiconductor Quantum Well
72.10 and 73.40; S7.12; S7.15
(b) Vol.160,No.1, 105-116

L.G. GERCHIKOV and A.V. SUBASHIEV
Interface States in Subband Structure of Semiconductor Quantum Wells
71.25 and 73.40; S7.12; S7.15
(b) Vol.160,No.2, 443-458

N.N. FALEEV, L.I. FLAKS, S.G. KONNIKOV, I.K. SOLOMIN, and S.V. BATASHOVA
The Influence of the Directed Displacement of Atomic Planes on the X-Ray Diffraction Rocking Curves
61.10 and 68.55; S7.12; S7.15
(a) Vol.120,No.2, 327-338

H. WENZEL and H.-J. WUENSCHE
A Model for the Calculation of the Threshold Current of SCH-MQW-SAS Lasers
73.40 and 85; S7.12; S7.15
(a) Vol.120,No.2, 661-674

E.L. IVCHENKO, V.P. KOCHERESHKO, I.N. URALTSEV, and D.R. YAKOVLEV
Reflectivity and Photoreflectivity in Superlattices and Quantum Wells
73.40 and 78.20; 71.35; S7.12; S7.15
(b) Vol.161,No.1, 217-222

H.H. WEBER
A Shorter Wavelength Mode Shift of Ridge-Waveguided 1.3 μm Lasers
78.45; 78.20; S7.15
(a) Vol.121,No.2, K239-K242

H. HAEFNER, L. LEHMANN, R. MITDANK, G. OELGART, and E. SCHULZE
Luminescence Characterization of (AlGa)As Single Quantum Wells
73.40 and 78.55; 78.60; S7.15
(a) Vol.122,No.2, 683-694

S7.16. Higher compounds

V.N. BRUDNYI, S.I. BORISENKO, and A.I. POTAPOV
Electrical and Optical Properties and Fermi Level Pinning in Electron Irradiated $ZnSnAs_2$
72.20 and 78.20; 71.55; S7.16
(a) Vol.118,No.2, 505-512

J.W. TOMM, K.H. HERRMANN, H. BOETTNER, M. TACKE, and A. LAMBRECHT
A Luminescence Study in the $Pb_{1-x}Eu_xSe$ System
73.40 and 78.55; 72.40; S7.16
(a) Vol.119,No.2, 711-720

K.P. GHATAK and M. MONDAL
On a Simplified Analysis of the Magnetic Susceptibilities of Electrons in Ultrathin Films of Non-Parabolic ...
75.20 and 75.70; 71.25; S7.16
(b) Vol.160,No.2, 673-682

V.YU. RUD, YU.V. RUD, and M. SERGINOV
Natural Photopleochroism of $n-In_2O_3-p-CdSiAs_2$ Structures
72.40 and 73.40; 85; S7.16; S10.1
(a) Vol.121,No.1, K81-K84

V.YU. RUD, YU.V. RUD, and M. SERGINOV
Photoelectric Properties of $H_2O-p-CdSiAs_2$:In Structures
72.40 and 73.40; S7.16
(a) Vol.121,No.2, K171-K176

P. CHAKRABARTI, M. PURI, A. SINGHAL, and A. MISHRA
Optically Controlled Characteristics of TEGFET
73.40 and 85; S7.12; S7.16
(a) Vol.122,No.1, 405-412

H. SOMMER, A. WEISS, and H. NEUMANN
Valence Band Density of States of $ZnSiP_2$
71.20 and 79.60; S7.16
(b) Vol.162,No.2, K87-K90

A.S. POPOV and A.Y. BAHNEV
Photoconductivity Spectra of Fe-Doped GaInP Epitaxial Layers
72.40 and 73.40; S7.16
(a) Vol.122,No.2, 569-574

S8. Chalcogenides
(for transition metal compounds, see S1.63)

K.D. TOVSTYUK, I.O. POLYAKOV, E.I. SLINKO, N.K. TOVSTYUK, and A.G. KHANDOZHKO
Lithium Impurity States in PbTe
64.60; 61.70; 76.60; S8
(b) Vol.157,No.1, 151-158

S.K. PLACHKOVA, L.E. SHELIMOVA, and O.G. KARPINSKII
X-Ray Study of the Structural Phase Transitions in $Ge_{1-x}Ag_{x/2}Bi_{x/2}Te$ Solid Solutions
64.70; 61.10; S8
(a) Vol.117,No.1, 155-162

M.M. EL-OCKER, F. SHARAF, S.M. YOSSEF, F. METAWE, and M. EL-DESOUKI
Optical Behaviour of Lead Selenide Films
68.55 and 78.65; S8
(b) Vol.157,No.2, 609-614

A.I. BOBRYSHEVA, G.P. GOLUBEV, I.I. DOBÎNDĂ, E.A. ZHUKOV, D.G. LUCHINSKII, and YU.G. SHEKUN
Picosecond Kinetics of Photoluminescence in GaSe
78.55; S8
(a) Vol.117,No.2, 587-592

A.A. ZAHAB, M. ABD-LEFDIL, and M. CADENE
Electrical Conductivity, Optical Absorption, and Photoconductivity Measurements of Well Oriented Indium-Telluride Thin Films
73.60 and 78.65; 72.40; S8
(a) Vol.117,No.2, K103-K106

A.P. MIKHAILOVSKII, A.M. POLUBOTKO, V.D. PROCHUKHAN, and YU.V. RUD
Gapless State in $CuFeS_2$
71.25 and 75.10; S8
(b) Vol.158,No.1, 229-238

G. KLUGE, M. GRAUROCK, R. STROBEL, M. LENZNER, H. BERGNER, V. BRUECKNER, and P. SUEPTITZ
Picosecond Photoconductivity in Amorphous Germanium Chalcogenides
72.40; S8
(b) Vol.158,No.1, K37-K42

O. FOLTIN
Simulation of the Oblique Incidence Specular Reflectivity of Uniaxial Semiconductors at Plasma Resonance
78.30; S8
(a) Vol.118,No.1, K43-K46

C. JULIEN, A. CHEVY, and D. SIAPKAS
Optical Properties of In_2Se_3 Phases
78.50; 78.30; S8
(a) Vol.118,No.2, 553-560

T. TSVETKOVA
Modification Effect of Ni Ion Implantation on the Electrical Properties of Vitreous As_2Se_3
61.80 and 72.80; S8
(a) Vol.118,No.2, K83-K86

P.F. ROBUSTO and R. BRAUNSTEIN
Optical Measurements of the Highly Damped Surface Plasmon of SnTe
71.45 and 78.65; S8
(a) Vol.119,No.1, 145-154

D.M. ZAYACHUK, D.D. IVANCHUK, R.D. IVANCHUK, S.S. MASLYANCHUK, and V.I. MIKITYUK
The Effect of Gadolinium Doping on the Physical Properties of Lead Telluride
71.55; 72.80; 75.50; S8
(a) Vol.119,No.1, 215-220

A.A. ZAHAB, M. ABD-LEFDIL, and M. CADENE
Rectifying and Photovoltaic Parameters of Indium Telluride(p)-Cadmium Sulfide(n) Thin Film Heterojunctions
72.40 and 73.40; S8
(a) Vol.119,No.1, K35-K40

A.T. NAGAT, G.A. GAMAL, Y.H. GAMEEL, and N.M. MOHAMED
Phototransport Studies of Thallium Monosulfide Single Crystals
72.40; S8
(a) Vol.119,No.1, K47-K52

K.A. KIKOIN, G.P. NIZHNIKOVA, and O.V. FARBEROVICH
Optical Properties of the Mixed-Valent TmSe
71.25; 71.50; 78.20; S1.4; S8
(b) Vol.159,No.2, 743-754

S. BISWAS and R. BHATTACHARYA
Magnetic Properties of Single Crystals of Bismuth Telluride Doped with 0.2 at% Lead and Its Thermoelectric Power
75.20; 71.25; 72.20; S8
(b) Vol.159,No.2, 851-860

A.M. ELKORASHY
Optical Constants of Tin Sulphide Single Crystals Measured by Interference Method
78.20; S8
(b) Vol.159,No.2, 903-916

N.T. MAMEDOV and N.K. MOROZ
Tl-Me Bond and Semiconductor-to-Metal Transition in $TlMeX_2$ Low-Dimensional Crystals
72.60; S8
(b) Vol.159,No.2, K83-K86

M.A. ALDZHANOV, N.G. GUSEINOV, G.D. SULTANOV, and M.D. NADZHAFZADE
Magnetic Heat Capacity and Susceptibility of the Pseudo-One-Dimensional Magnetic Systems $TlFeS_2$ and $TlFeSe_2$
75.40; 75.50; S8
(b) Vol.159,No.2, K107-K110

A. KUMAR, P. AGARWAL, and J.S.P. RAI
Crystallization Kinetics in Glassy $Se_{80}Te_{20-x}Sb_x$
61.40; S8
(a) Vol.119,No.2, K97-K102

A.K. ABASS, Z.A. AHMED, and R.M. SAMUEL
Optical Properties of Chemically Deposited FeS_2 Coatings
78.20; S8
(a) Vol.120,No.1, 247-252

M. STORDEUR
Consideration to the Nernst-Ettingshausen (NE) Effect in Thermoelectric p-$(Bi_{1-x}Sb_x)_2Te_3$ Single Crystals
72.20; S8
(b) Vol.160,No.2, K145-K148

V.T. MASLYUK
Stables States and Ordering in Amorphous Polymer Semiconductors
61.40; 65.50; 71.45; S8
(b) Vol.161,No.1, 43-54

E.YU. TURAEV, P.P. SEREGIN, and F.S. NASREDINOV
Tin and Iron Impurity Atom States in Modified As_2Se_3 Films
71.55 and 73.60; 78.65; S8
(b) Vol.161,No.1, 287-292

YU.P. GNATENKO, YU.I. ZHIRKO, and Z.D. KOVALYUK
Optical Properties of GaSe Crystals Containing Mn Impurity Atoms. I. Exciton-Phonon Interaction
71.35 and 78.30; S8
(b) Vol.161,No.1, 419-426

YU.P. GNATENKO, YU.I. ZHIRKO, and Z.D. KOVALYUK
Optical Properties of GaSe Crystals Containing Mn Impurity Atoms. II. Excitonic and Impurity Emission
78.55; 71.35; 71.55; S8
(b) Vol.161,No.1, 427-434

M. GUYMONT, A. TOMAS, M. JULIEN-POUZOL, S. JAULMES, and M. GUITTARD
Structure and Vacancy Ordering in $Cu_2Er_{2/3}S_2$. Electron and X-Ray Diffraction and High Resolution Electron Microscopy ...
61.60; 61.10; S8
(a) Vol.121,No.1, 21-28

A. FOITZIK, W. SKROTZKI, and P. HAASEN
Dislocation Microstructure on {100} and {110} Slip Planes in Lead Sulphide
61.70; S8
(a) Vol.121,No.1, 81-94

A. VOHRA and K.K. SRIVASTAVA
Microwave Switching and AC Conductivity of Memory Glass $As_{10}Ge_{15}Te_{75}$
73.40 and 85; 61.40; S8
(a) Vol.121,No.1, 341-348

A.N. GRUZINTSEV, S.V. DEMISHEV, and T.V. NIKIFOROVA
The Influence of Complex Luminescence Centres on the Luminescent Properties of CaS:Eu
71.55 and 78.55; 78.60; S8
(b) Vol.161,No.2, 705-710

W. KULLMANN, G. EICHHORN, H. RAUH, R. GEICK, G. ECKOLD, and U. STEIGENBERGER
Lattice Dynamics and Phonon Dispersion in the Narrow Gap Semiconductor Bi_2Te_3 with Sandwich Structure
63.20; S8
(b) Vol.162,No.1, 125-140

A.I. DMITRIEV, Z.D. KOVALYUK, V.I. LAZORENKO, and G.V. LASHKAREV
Two-Dimensional Character of Electron Gas in Layered InSe Crystals
72.20; 71.25; S8
(b) Vol.162,No.1, 213-226

H. ATMANI and C. VAUTIER
Crystallization of Amorphous BiSe Alloys: Advantages and Limits of a Proposed Theoretical Model
61.50 and 68.55; S8
(a) Vol.122,No.1, 73-78

M. QUINTERO, R. TOVAR, C. BELLABARBA, and J.C. WOOLLEY
Temperature Variation of Optical Energy Gap and Deformation Potentials in $AgInTe_2$
71.25; 78.20; S8
(b) Vol.162,No.2, 517-522

E. LOPEZ-CRUZ
A Comment on the Temperature Dependence of the Band Gap of FeS_2
78.20; S8
(b) Vol.162,No.2, K131-K134

P. SISODIA and M.P. VERMA
Polycrystalline Elastic Moduli of Some Hexagonal and Tetragonal Materials
62.20; S1; S8; S10
(a) Vol.122,No.2, 525-534

S8.1. II-VI compounds

K. KRÁL, B. HEJDA, and Z. KHÁS
Hot LO-Phonon Quasi-Steady State Distribution Function in Polar Semiconductors
63.20 and 72.10; S7.1; S8.1
(b) Vol.157,No.2, 667-676

L.G. ZIMIN, S.V. GAPONENKO, I.E. MALINOVSKII, V.YU. LEBED, P.I. KUZNETSOV, G.G. YAKUSHCHEVA, and A.V. KUZNETSOV
Excitonic Optical Nonlinearity in II-VI MOCVD-Grown Crystals
71.35 and 78.20; S8.1
(b) Vol.159,No.1, 449-456

H. AOURAG, B. KHELIFA, A. BELAIDI, and Z. BELARBI
Positron Distribution in Semiconductors
71.60; 78.70; S5.11; S5.12; S7.1; S8.1
(b) Vol.160,No.1, 201-210

D. AGASSI and T.K. CHU
Strain-Induced Superconductivity in Lead Salt Superlattices
74.70; 73.40; S8.1
(b) Vol.160,No.2, 601-612

A.T. NAGAT, S.A. HUSSEIN, Y.H. GAMEEL, and G.A. GAMAL
Switching Effect with Memory in Thallium Sulfide Single Crystals
72.20; S8.1
(a) Vol.121,No.2, K201-K206

J.W. TOMM, K.H. HERRMANN, and A.E. YUNOVICH
Infrared Photoluminescence in Narrow-Gap Semiconductors (Review Article)
78.55; 78.45; S8.1; S8.15
(a) Vol.122,No.1, 11-42

B.P. SINGH
Analysis of Ionic Radii, Second Neighbour Interactions, and Electronegativity Difference for Ionic Crystals
61.50; 71.45; S7.1; S8.1; S9.11
(b) Vol.162,No.2, 329-338

S8.11. Sulfides

M. DEY (BOSE) and S. GHOSH
Amplification of Acoustic Waves in Magnetised High Resistivity Piezoelectric Semiconductors. Effect of Nonuniform ...
62.65 and 72.50; S8.11
(b) Vol.157,No.1, 159-166

D. BOULANGER and R. PARROT
Molecular Perturbation Model for the Spin-Lattice Coupling Coefficients of $^6A_1$ States to E_θ Strains
71.70; S8.11
(b) Vol.157,No.1, 323-330

V.V ZUBRITSKII and R.P. PURLYS
The Relation between the Orientation of Streamer Discharges and Anisotropy of Thermal Vibrations of Atoms in II-VI ...
63.20; 71.55; 72.20; 77.50; S8.11
(b) Vol.158,No.1, 91-102

WAN-LUN YU
Crystal Field Effect on the g-Factors of ^6S-State Ions
71.70; 76.30; S8.11; S10.1; S11.1
(b) Vol.158,No.1, K13-K16

A.D. BELYAEV, V.E. RODIONOV, and N.S. CHERNAYA
Effect of Ultrasound Oscillations on Electroluminescent Properties of MISIM Structures
78.60; 62.80; 73.40; S8.11
(a) Vol.118,No.1, 327-332

K.-H. PANTKE, H. OVER, and I. BROSER
Excitation of the Impurity Emission Due to the Mixed Mode Exciton Absorption in CdS
78.55; 71.35; S8.11
(a) Vol.118,No.1, 343-348

WEN-CHEN ZHENG
An Estimate of the Spin-Lattice Coupling Coefficients for Trigonal $CdS:Mn^{2+}$ Crystals
71.70; 76.30; S8.11
(b) Vol.158,No.2, K137-K142

W. RUDOLPH, J. PULS, F. HENNEBERGER, and D. LAP
Femtosecond Studies of Transient Nonlinearities in Wide-Gap II-VI Semiconductor Compounds
71.35 and 78.20; S8.11; S8.12; S8.15
(b) Vol.159,No.1, 49-58

N.A. GIPPIUS, A.L. IVANOV, S.G. TIKHODEEV, and S.G. VYGOVSKII
Phonoriton Transient Phenomena
78.20; 71.35; 71.36; S8.11
(b) Vol.159,No.1, 71-80

F. JAHNKE and K. HENNEBERGER
Nonequilibrium Description of the Intense Resonant Laser Excitation of a Semiconductor
71.35 and 72.30; 78.20; S8.11
(b) Vol.159,No.1, 81-90

J. HOLLANDT, J. GUTOWSKI, and I. BROSER
A Detailed Study Concerning the Main Aspects Influencing Thermally Induced Optical Bistability of CdS
78.20; S8.11
(b) Vol.159,No.1, 205-212

V.A. STADNIK
Multistability of Absorption Domains in Bulk CdS Due to Band Edge Nonlinearities
71.35 and 78.20; S8.11
(b) Vol.159,No.1, 241-248

H. ROSSMANN, A. SCHUELZGEN, F. HENNEBERGER, and M. MUELLER
Quantum Confined DC Stark Effect in Microcrystallites
71.35 and 78.20; S8.11
(b) Vol.159,No.1, 287-290

H. HAUG, L. BANYAI, J. LIEBLER, and T. WICHT
The Pulsed Nonresonant Optical Stark Effect and the Urbach Tail in Semiconductors
71.35 and 78.20; S7.12; S8.11; S9
(b) Vol.159,No.1, 309-316

R. ZIMMERMANN
Transverse Relaxation and Polarization Specifics in the Dynamical Stark Effect
71.35 and 78.20; S8.11
(b) Vol.159,No.1, 317-326

CH. SPIEGELBERG, J. PULS, and F. HENNEBERGER
Nonlinear Optical Properties of CdS at Room Temperature
71.35 and 78.20; S8.11
(b) Vol.159,No.1, 353-362

T. HOENIG, J. GUTOWSKI, and I. BROSER
Carrier-Induced Pulse Narrowing in Indium-Doped CdS
71.55 and 78.20; 71.36; S8.11
(b) Vol.159,No.1, 363-370

V.P. GRIBKOVSKII, V.A. ZYULKOV, A.E. KAZACHENKO, and V.A. SHAKIN
Second Harmonic Generation on the Grating of Optical Inhomogeneities in ZnSe and ZnS
78.20; S8.11; S8.12
(b) Vol.159,No.1, 379-386

V.D. EGOROV, HOANG XUAN NGUYEN, R. ZIMMERMANN, V.S. DNEPROVSKII, M. KASCHKE, and D.S. KHECHINASHVILI
Mott Transition in CdS Revealed by Picosecond Nonlinear Absorption Measurements
71.35 and 78.20; S8.11
(b) Vol.159,No.1, 403-410

A.L. IVANOV and V.V. PANASHCHENKO
The Exciton-Biexciton Spectrum Renormalization in the Presence of a Polariton Pumping Wave. The New Mechanism, Exciton ...
71.35 and 78.20; 71.36; S8.11
(b) Vol.159,No.1, 417-424

V.A. KHARCHENKO, G.V. MIKHAILOV, D.K. NELSON, and B.S. RAZBIRIN
Impact Excitation and Auger Decay of Bound Exciton Complexes in a Dense Exciton Gas
71.35 and 78.55; S8.11
(b) Vol.159,No.1, 425-430

U. NEUKIRCH, I. BROSER, and R. RASS
Nonlinear Luminescence of Heavily Doped CdS:In Crystals
78.55; 71.45; S8.11
(b) Vol.159,No.1, 431-436

K.-H. PANTKE, H. OVER, and I. BROSER
Dissipation of Excitonic Polaritons in CdS
71.36 and 78.20; S8.11
(b) Vol.159,No.1, 437-442

LI-JIAN MENG and CHANG-HUA LI
Temperature Effect on AC EL Characteristics in ZnS:ErF_3 Thin Films
78.60; 78.65; S8.11
(a) Vol.119,No.2, 677-682

M. DEY (BOSE) and S. GHOSH
Acoustic Second Harmonic Generation in Magnetized High Resistivity Piezoelectric Semiconductors. Effect of ...
72.20 and 72.50; 77.60; S8.11
(b) Vol.161,No.1, 319-328

M.A. JAKOBSON, V.D. KAGAN, R. KATILIUS, and G.O. MUELLER
Impact Ionization of Free and Bound Excitons by Free Carriers in an Electric Field in CdS: Investigation by Optical ...
78.55; 71.35; S8.11
(b) Vol.161,No.1, 395-404

B.P. CHANDRA, Y. RAHANGDALE, and M. RAMRAKHIANI
Theory of Mechanical Excitation of Luminescence Centers in Crystals
78.60; S8.11
(a) Vol.121,No.1, 281-288

H.-J. FITTING, G.O. MUELLER, R. MACH, G.U. REINSPERGER, TH. HINGST, and E. SCHREIBER
Vacuum Emission of Hot Electrons from ZnS
72.20 and 79.70; 85; S8.11
(a) Vol.121,No.1, 305-314

L.L. GOLIK, M.M. GUTMAN, and V.E. PAKSEEV
Nonlinear Oscillations and Chaos in CdS Single Crystals under Temperature-Electric Instability Conditions
72.40; 72.70; S8.11
(b) Vol.162,No.1, 199-212

J. BENOIT, P. BENALLOUL, C. BARTHOU, S. CASETTE, and J.C. SORET
De-Excitation Processes and Efficiency in AlE ZnS:Mn Thin Film Electroluminescent Devices
78.60 and 85; S8.11
(a) Vol.122,No.1, 427-438

A. KARA, J.J. COUDERC, C. LEVADE, and G. VANDERSCHAEVE
Memory Photoelastic Effects in ZnS Sphalerite
62.20 and 78.20; S8.11
(a) Vol.122,No.2, 545-554

H.-J. FITTING, E. SCHREIBER, and TH. HINGST
Avalanche Measurement in ZnS by Vacuum Emission
72.20; 78.60; 79.70; 85; S8.11
(a) Vol.122,No.2, K165-K168

S8.12. Selenides

V.A. KASIYAN, D.D. NEDEOGLO, A.V. SIMASHKEVICH, and I.N. TIMCHENKO
Metal-Dielectric Transition in n-ZnSe Obtained by Doping with Shallow Donor Impurity
72.20; 71.30; S8.12
(b) Vol.157,No.1, 341-350

M. KARAI, K. KIDO, H. NAITO, K. KUROSAWA, M. OKUDA, T. FUJINO, and M. KITAGAWA
Deep Trap Levels in Zn-Annealed ZnSe Single Crystals
71.55; 73.30; S8.12
(a) Vol.117,No.2, 515-526

V.A. KASIYAN, D.D. NEDEOGLO, A.V. SIMASHKEVICH, and I.N. TIMCHENKO
Coulomb Gap in n-ZnSe and Its Critical Behaviour in the Region of the Metal-Dielectric Transition
71.30 and 72.60; S8.12
(b) Vol.158,No.1, 293-300

V.P. GRIBKOVSKII, L.G. ZIMIN, S.V. GAPONENKO, I.E. MALINOVSKII, P.I. KUZNETSOV, and G.G. YAKUSHCHEVA
Optical Absorption near Excitonic Resonance of the MOCVD-Grown ZnSe Single Crystals
71.35 and 78.65; S8.12
(b) Vol.158,No.1, 359-366

W. RUDOLPH, J. PULS, F. HENNEBERGER, and D. LAP
Femtosecond Studies of Transient Nonlinearities in Wide-Gap II-VI Semiconductor Compounds
71.35 and 78.20; S8.11; S8.12; S8.15
(b) Vol.159,No.1, 49-58

A.I. BOBRYSHEVA and S.S. RUSSU
Optical Properties of Biexcitons in ZnSe/$Zn_{1-x}Mn_xSe$ Quantum Wells
71.35 and 78.20; 73.40; S8.12; S8.15
(b) Vol.159,No.1, 155-164

Y.Z. HU, S.W. KOCH, M. LINDBERG, and N. PEYGHAMBARIAN
Theoretical and Experimental Results on Coulomb Effects in Semiconductor Quantum Dots
71.35 and 78.20; S8.12
(b) Vol.159,No.1, 249-258

I. BALSLEV
Induced Absorption and the Dynamical Stark Shift Associated with Biexcitons
71.35 and 78.20; S8.12; S9
(b) Vol.159,No.1, 339-346

V.P. GRIBKOVSKII, V.A. ZYULKOV, A.E. KAZACHENKO, and V.A. SHAKIN
Second Harmonic Generation on the Grating of Optical Inhomogeneities in ZnSe and ZnS
78.20; S8.11; S8.12
(b) Vol.159,No.1, 379-386

N. PRESSER, G. KUDLEK, J. GUTOWSKI, S. DURBIN, D.R. MENKE,
M. KOBAYASHI, and R.L. GUNSHOR
High-Density Spectroscopy of ZnSe/GaAs Epilayers in the Near-Band-Edge Region
71.35 and 78.65; 78.55; S7.12; S8.12
(b) Vol.159,No.1, 443-448

T. TSVETKOVA, B. AMOV, E. VATEVA, and V. AVERYANOV
Ion Implantation Induced Modification of Vitreous Chalcogenides
73.60; 61.80; 78.65; S8.12
(a) Vol.119,No.1, 107-112

B. THOMAS and T.R.N. KUTTY
Formation of Single-Phase Indium Selenide Thin Films by Elemental Evaporation
68.55; 73.60; 78.65; S8.12
(a) Vol.119,No.1, 127-138

K. KUMAZAKI, K. IMAI, and A. ODAJIMA
Estimation of Strains in MBE Grown ZnSe Films by Raman Scattering
63.20 and 68.30; 68.60; S8.12
(a) Vol.119,No.1, 177-182

I.CH. AKOPYAN, B.V. NOVIKOV, R.V. GRIGOREV, S.R. GRIGOREV,
H. ROEPPISCHER, and N. STEIN
Photoluminescence Study of Na and Li Acceptors in ZnSe
78.55; 61.80; S8.12
(a) Vol.119,No.2, K167-K172

U. WOGGON, M. MUELLER, I. RUECKMANN, J. KOLENDA, and M. PETRAUSKAS
Influence of Hydrogenation of Defects at the Microcrystallite-Glass Interface on the Nonlinear Optics ...
61.70 and 78.20; S8.12
(b) Vol.160,No.1, K79-K84

J. GUTOWSKI, N. PRESSER, and G. KUDLEK
Optical Properties of ZnSe Epilayers and Films (Review Article)
78.65; 68.55; 71.35; 73.60; 78.55; S8.12
(a) Vol.120,No.1, 11-60

S.A. MARSHALL, D.R. YODER-SHORT, C. YU, Y.N. ZHANG,
and J.K. FURDYNA
Super Hyperfine Structure in the EPR Spectrum of Divalent Manganese in Single Crystal Zinc Selenide
71.70 and 76.30; S8.12
(b) Vol.160,No.2, 591-600

A.O. ALEKSANYAN, V.A. GANSHIN, YU.N. KORKISHKO, and V.A. FEDOROV
Equilibrium and Kinetics of Ion Exchange in Zinc Selenide Single Crystals
66.30 and 68.45; 68.40; S8.12
(b) Vol.161,No.2, 629-636

S8.13. Tellurides

M. SAID and M.A. KANEHISA
Excited Acceptor States in II-VI and III-V Semiconductors
71.55; S7.12; S8.13
(b) Vol.157,No.1, 311-322

I.N. YURIICHUK, S.V. MELNICHUK, and V.A. SHENDEROVSKII
Local Modes in CdTe Doped with Transition Element Impurities
63.20; S8.13
(b) Vol.157,No.1, K19-K22

H. ZIMMERMANN, R. BOYN, C. MICHEL, and P. RUDOLPH
Absorption-Calibrated Determination of Impurity Concentrations in CdTe from Excitonic Photoluminescence
71.55; 78.55; S8.13
(a) Vol.118,No.1, 225-234

I. RUECKMANN, J. KORNACK, J. KOLENDA, and M. PETRAUSKAS
Transient Energy Transfer in ps Two-Photon Absorption Experiments in CdTe at Room Temperature
78.20; S8.13
(b) Vol.158,No.2, 769-780

F. EL-AKKAD and A. AL-SHAHRANY
Electron Diffusion Length in ZnTe
72.80; S8.13
(a) Vol.118,No.2, K79-K82

A.V. NURMIKKO and R.L. GUNSHOR
Excitons and Phonons in CdTe/MnTe and ZnTe/MnTe Single Quantum Wells. Aspects of Confinement and Optical Phonon Coupling
71.35 and 78.20; 73.40; S8.13
(b) Vol.159,No.1, 107-116

F. RODRÍGUEZ, A. CAMACHO, L. QUIROGA, and R. BAQUERO
Electronic Structure of (001) Semiconducting MTe Surfaces (M=Zn,Cd,Hg)
71.25 and 73.20; S8.13
(b) Vol.160,No.1, 127-136

H. AOURAG, B. KHELIFA, A. BELAIDI, A. TADJER, M. REZKI, and M. GAMOUDI
Positron Band Structures of GaAs and CdTe
71.25 and 71.60; 78.70; S7.12; S8.13
(b) Vol.160,No.1, 193-200

K. SADAIYANDI and K. RAMACHANDRAN
Chlorine Diffusion in CdTe
66.30; S8.13
(b) Vol.160,No.2, K85-K88

U. BECKER, P. RUDOLPH, R. BOYN, M. WIENECKE, and I. UTKE
Characterization of p-Type CdTe Bridgman Crystals by Infrared Extinction Spectra
64.75 and 78.30; S8.13
(a) Vol.120,No.2, 653-660

T. DUMELOW, A.R. EL GOHARY, K.A. MASLIN, T.J. PARKER, D.R. TILLEY, and S.N. ERSHOV
Analysis of Far Infrared Spectra Showing Bulk and Surface Phonon-Polaritons in CdTe Epilayers on GaAs Substrates
78.30 and 78.65; 71.36; S8.13
(b) Vol.161,No.1, 233-244

P. JANDL, M. RICK, and J. ROSENZWEIG
Defect Analysis of Bipolar CdTe Using Photc-Hall and Photoconductivity Measurements in the Temperature Range ...
72.20; S8.13
(a) Vol.121,No.1, 219-226

T. BERGUNDE, M. WIENECKE, and B. THOMAS
High Mobility HgTe and High Resistivity CdTe Grown by MOVPE on GaAs(001)
68.48 and 73.60; S8.13
(a) Vol.121,No.1, K55-K60

E.P. POKATILOV, S.I. BERIL, V.M. FOMIN, and V.V. KALINOVSKII
The Size-Quantized States of the Wannier-Mott Exciton in Structures with Superthin Films of CdTe
71.35 and 78.65; S8.13
(b) Vol.161,No.2, 603-612

V. SLUSARENKO, Y. BURKI, W. CZAJA, and H. BERGER
Temperature Dependence of the Oxygen Absorption Band in ZnTe:O
78.50; S8.13
(b) Vol.161,No.2, 897-906

F. TAKEDA, K. MATSUURA, and S. SETO
Admittance Spectroscopy and Photoluminescence Characterization of High-Quality p-Type Cadmium Telluride
71.55 and 78.55; 72.80; S8.13
(a) Vol.121,No.2, 595-606

A. KANOUN, S. ALAYA, and H. MAAREF
Spectroscopic Properties of Pr^{3+} and Nd^{3+} Ions in Zinc Tellurite Glasses
71.55; 78.55; S8.13
(b) Vol.162,No.2, 523-530

J. TOUŠEK
Photovoltaic Effect in Schottky Diodes. Contribution to Theory and Experiment
72.40 and 73.30; S8.13
(a) Vol.122,No.2, 757-764

K.S. CHOI, B.J. KOO, M. JUNG, H.J. SHIN, J.H. LEE, S.B. KIM, S.K. CHANG, and H.L. PARK
(100) CdTe Thin Film Growth by Temperature Gradient Vapor Transport (TGVT)
68.55; 78.40; 78.55; S8.13
(a) Vol.122,No.2, K135-K138

S8.15. Mixed crystals

B.A. ARONZON and I.M. TSIDILKOVSKII
Magnetic-Field-Induced Localization of Electrons in Fluctuation-Potential Wells of Impurities (Review Article)
71.30; 72.20; 75.40; S7.13; S8.15
(b) Vol.157,No.1, 17-60

I.S. BAKSHEE, L.A. KARACHEVTSEVA, A.V. LYUBCHENKO, V.A. PETRYAKOV, E.A. SALKOV, and B.I. KHIZHNYAK
Superposition of Burst Noises as the Origin of 1/f Noise Due to Grain Boundaries in $Cd_xHg_{1-x}Te$
72.70; 61.70; S8.15
(a) Vol.117,No.1, K37-K42

T. PIOTROWSKI, J.W. TOMM, and N. PUHLMANN
Photoluminescence Studies of $Hg_{1-x}Mn_xTe$ Single Crystals
78.55; S8.15
(a) Vol.117,No.2, K181-K184

Y. MADHAVAN, K. RAMACHANDRAN, and T.M. HARIDASAN
Tellurium Diffusion in the Mixed System $Cd_xHg_{1-x}Te$
66.30; S8.15
(b) Vol.158,No.2, 427-432

K. ZINK, A. KROST, H. NELKOWSKI, J. SAHM, and H. STUTENBECKER
Energy Transfer Mechanisms in $Zn_{1-x}Mn_xSe$ $(0.005<x<0.017)$ Detected by ODMR
71.70; 76.30; 78.55; S8.15
(b) Vol.158,No.2, 603-608

A. GRAF, R. OBERMAIER, and W. GEBHARDT
Spin Glass Properties of $Zn_{1-x}Mn_xSe$
75.50; S8.15
(b) Vol.158,No.2, 695-706

Z. KUČERA, P. HLIDEK, P. HOESCHL, V. KOUBELE, V. PROSSER, and M. ZVARÁ
Magneto-Optical Spectroscopy of p-Type $Hg_{0.8}Cd_{0.2}Te$
72.40; 71.55; S8.15
(b) Vol.158,No.2, K173-K178

W. RUDOLPH, J. PULS, F. HENNEBERGER, and D. LAP
Femtosecond Studies of Transient Nonlinearities in Wide-Gap II-VI Semiconductor Compounds
71.35 and 78.20; S8.11; S8.12; S8.15
(b) Vol.159,No.1, 49-58

A.I. BOBRYSHEVA and S.S. RUSSU
Optical Properties of Biexcitons in $ZnSe/Zn_{1-x}Mn_xSe$ Quantum Wells
71.35 and 78.20; 73.40; S8.12; S8.15
(b) Vol.159,No.1, 155-164

L.G. ZIMIN, S.V. GAPONENKO, V.YU. LEBED, I.E. MALINOVSKII, I.N. GERMANENKO, E.E. PODOROVA, and V.A. TSEKHOMSKII
Nonlinear Optical Absorption in I-VII and II-VI Microcrystallites under Quantum Confinement of Elementary ...
71.35 and 78.65; S8.15; S9
(b) Vol.159,No.1, 267-274

D. RICARD, P. ROUSSIGNOL, F. HACHE, and CH. FLYTZANIS
Nonlinear Optical Properties of Quantum Confined Semiconductor Microcrystallites
78.20; S8.15
(b) Vol.159,No.1, 275-286

A. MZERD, D. SAYAH, I.J. SAUNDERS, and B.K. JONES
Electrical Properties of $Cd_{1-y}Zn_yS$ and CdS(In) Thin Films Prepared by Pyrolytic Spray Technique
73.60; 71.55; S8.15
(a) Vol.119,No.2, 487-494

W. HOERSTEL, A. KLIMAKOV, R. KRAMER, and A. SCHMIEDE
Defect-Induced Conduction Anomalies in n-Type $Hg_{1-x}Cd_xTe$
72.20; S8.15
(a) Vol.119,No.2, 589-594

P.I. BARANSKII, A.E. BELYAEV, and O.P. GORODNICHII
Specific Features of Kinetic Effects Measured in (Cd,Hg)Te in Quantizing Magnetic Fields
72.20; S8.15
(b) Vol.160,No.1, 291-296

K. KUMAZAKI
Dielectric Constant in $Zn_xHg_{1-x}Se$ Determined by Raman Scattering
77.20; S8.15
(b) Vol.160,No.2, K173-K176

H. HEUKENKAMP and W. HOERSTEL
Trap Tunneling in HgCdTe n^+-p Junctions Fabricated by Ion Implantation
73.40; 71.55; S8.15
(a) Vol.120,No.2, 485-496

S. BHOWMIK, C.K. SARKAR, and K. SANTRA
The Effect of the Quantum Screening on the Energy Loss Rates of Hot Electrons in n-$Hg_{0.8}Cd_{0.2}Te$ in the Extreme Quantum .
71.25 and 72.20; S8.15
(b) Vol.161,No.1, 329-334

S. JIANG, S.C. SHEN, and W. GIRIAT
Absorption Edge and New Absorption Peaks of $Cd_{1-x}Mn_xS$ under Hydrostatic Pressure
78.40; S8.15
(b) Vol.161,No.1, K71-K76

CHANGHE HUANG, ZHENZHONG YU, and DINGYUAN TANG
Defect Related Absorption in p-Type $Hg_{1-x}Cd_xTe$ Alloys
78.50; 61.70; S8.15
(a) Vol.121,No.1, 273-280

A. MZERD, D. SAYAH, P.C. RUSSELL, and B.K. JONES
Capacitance Measurements on Au-$Cd_{1-y}Zn_yS$ Polycrystalline Schottky Diodes
73.40 and 85; S8.15
(a) Vol.121,No.1, 333-340

W. KRAAK, J. KALDASCH, P. GILLE, TH. SCHURIG, and R. HERRMANN
Magnetotransport Properties and Subband Structure of the Two-Dimensional Electron Gas in the Inversion Layer of ...
73.40; S8.15
(b) Vol.161,No.2, 613-628

M. STORDEUR
The Thermoelectric Figure of Merit in the Mixed Crystal System p-$(Bi_{1-x}Sb_x)_2Te_3$
72.20; 72.90; S1.63; S8.15
(b) Vol.161,No.2, 831-836

F.M. GASHIMZADE, T.G. ISMAILOV, and R.S. NADIRZADE
Interband Resonance Raman Scattering in p-Type Semimagnetic Semiconductors in Quantizing Magnetic Field
78.30; S8.15
(b) Vol.161,No.2, 861-868

F.S. NASREDINOV, E.YU. TURAEV, P.P. SEREGIN, H.B. RAKHMATULLAEV, and M.K. BAKHADIRKHANOV
Mechanism of the Two-Electron Exchange between Neutral and Ionized Centres of Tin in the $PbS_{1-x}Se_x$ Solid Solutions
61.70 and 71.55; 76.80; S8.15
(a) Vol.121,No.2, 571-578

A. MZERD, D. SAYAH, P.C. RUSSELL, and B.K. JONES
Characterization of Electron Traps in Au-$Cd_{1-y}Zn_yS$ Polycrystalline Schottky Diodes by Capacitance Transient ...
71.55 and 73.40; 85; S8.15
(a) Vol.121,No.2, 587-594

K.H. HERRMANN, K.-P. MOELLMANN, and J.W. TOMM
On the Broadening Mechanisms near the E_0 Transition in Narrow-Gap (Hg,Cd)Te
78.20; 71.25; 78.55; S8.15
(a) Vol.121,No.2, 635-640

Y. MADHAVAN, K. PONNAMMAL, K. RAMACHANDRAN, and T.M. HARIDASAN
Mean Square Displacement (MSD) of Neighbours around a Vacancy in Some Zinc Based Tellurides
63.20; 61.70; S8.15
(b) Vol.162,No.1, 119-124

J.W. TOMM, K.H. HERRMANN, and A.E. YUNOVICH
Infrared Photoluminescence in Narrow-Gap Semiconductors (Review Article)
78.55; 78.45; S8.1; S8.15
(a) Vol.122,No.1, 11-42

A. BAIDULLAEVA, B.K. DAULETMURATOV, V.A. GNATYUK, and P.E. MOZOL
Surface States of Laser-Irradiated Cadmium and Tellurium-Containing Solid Solutions of II-VI Compounds
73.20 and 73.40; 72.40; S8.15
(a) Vol.122,No.1, 243-248

A. SCHENK
Spatially Variable Drift Mobility Model for $Hg_{1-x}Cd_xTe$ Diodes. I. Analytical Base and Fit to Hall Data
73.40 and 85; S8.15
(a) Vol.122,No.1, 413-426

A. ROGALSKI and K. JÓŹWIKOWSKI
Intrinsic Carrier Concentration and Effective Masses in $Hg_{1-x}Mn_xTe$
72.20; S8.15
(a) Vol.122,No.1, K39-K44

A. SCHENK
Spatially Variable Drift Mobility Model for $Hg_{1-x}Cd_xTe$-Diodes. II. Application to Device Simulation
73.40 and 85; S8.15
(a) Vol.122,No.2, 723-730

S8.16. Higher compounds

L.I. KOROLEVA and T.V. VIROVETS
Critical Behaviour of Semiconductor Spin Glasses $(CuCr_2Se_4)_x(Co_{0.5}Me_{0.5}Cr_2Se_4)_{1-x}$ (Me=In,Ga, $0<x<0.1$)
75.30; S8.16
(b) Vol.157,No.1, 431-442

N.T. MAMEDOV and A.M. PANICH
Phase Transition in $TlGaTe_2$
64.70 and 76.60; S8.16
(a) Vol.117,No.1, K15-K18

E.A. EIVAZOV, A.F. SAFAROV, S.M. ATAKISHIEV, and YA.M. ABASOV
The EPR Spectrum of $Co_{0.7}Cu_{0.3}Cr_2S_{4-x}Se_x$ System
76.30; S8.16
(a) Vol.117,No.2, K147-K150

R.M. SARDARLY, R.A. ALIEV, E.M. KERIMOVA, N.A. BAKHYSHOV, and I.SH. SADYKHOV
Rearrangement Features of the Vibrational Spectra of $TlInS_{2x}Se_{2(1-x)}$ and $TlGa_xIn_{1-x}Se_2$ Solid Solutions
63.20 and 78.30; 64.75; S8.16
(b) Vol.158,No.1, 73-80

N.M. GASANLY, S.A. EL-HAKIM, L.G. GASANOVA, and A.Z. MAGOMEDOV
Lattice Vibrations in $CuIn_5S_8$ Crystals
63.20 and 78.30; S8.16
(b) Vol.158,No.1, K1-K4

N.M. GASANLY, A.G. GUSEINOV, E.A. ASLANOV, and S.A. EL-HAMID
Infrared Reflection Spectra of $Cu_3B_5^{III}C_9^{VI}$ Single Crystals
78.30; 63.20; S8.16
(b) Vol.158,No.1, K85-K88

M. GUYMONT, A. TOMAS, and M. PALAZZI
Short-Range Order in $EuLiS_2$ by Electron Microscopy
61.60; S8.16
(a) Vol.118,No.1, 29-40

N.S. ORLOVA, G.A. TURTSEVICH, O.E. KOCHKARIK, and I.V. BODNAR
X-Ray Study of Debye-Waller Factors and Debye Characteristic Temperatures in $AgInC_2$ Compounds
63.20; S8.16
(a) Vol.118,No.1, 141-146

I.V. BODNAR, G.F. SMIRNOVA, A.G. KAROZA, and A.P. CHERNYAKOVA
Vibrational Spectra of $CuGaS_2$ and $CuGaSe_2$ Compounds and $CuGaS_{2x}Se_{2(1-x)}$ Solid Solutions
63.20; 64.75; 78.30; S8.16
(b) Vol.158,No.2, 469-474

S.K. CHANG, H.L. PARK, H.K. KIM, J.S. HWANG, C.H. CHUNG, and W.T. KIM
Co^{2+} in $CuGa_{1-x}Al_xSe_2$ and Valence Band Offset in $CuGa_{1-x}Al_xSe_2/CuGaSe_2$
73.40; S8.16
(b) Vol.158,No.2, K115-K118

D. SCHIKORA, J.W. TOMM, K. RAKKENUS, and M. PESSA
Growth and Optical Characterization of HgTe-CdTe Superlattices
78.55; S8.16
(a) Vol.118,No.2, K123-K128

D. BALTRUNAS, R. MIKAITIS, V.YU. SLIVKA, and YU.M. VYSOCHANSKII
Manifestation of the Ferroelectric Phase Transition in the Moessbauer Parameters of the System $(Sn_xPb_{1-x})_2P_2S_6$
64.70 and 76.80; S8.16
(a) Vol.119,No.1, 71-76

S. VES
High Pressure Raman Study of the Chain Chalcogenide $TlInTe_2$
63.20 and 78.30; 64.60; S8.16
(b) Vol.159,No.2, 699-706

H.C. GUPTA, A. PARASHAR, V.B. GUPTA, and B.B. TRIPATHI
Lattice Vibrations in Spinel-Type Compounds $CdCr_2S_4$ and $CdCr_2Se_4$
63.20; S8.16
(b) Vol.160,No.1, K19-K24

I.M. TSIDILKOVSKII, YU.G. ARAPOV, N.G. GLUZMAN, N.K. LERINMAN, L.D. SABIRSYANOVA, and M.L. ZVEREVA
Magnetic-Field-Induced Metal-Insulator Transition in $Hg_{1-x}Cd_xTe$ and $Hg_{1-x-y}Cd_xMn_yTe$
71.30; S8.16
(b) Vol.160,No.2, K141-K144

T. KANOMATA, T. KAWASHIMA, T. KANEKO, H. TAKAHASHI, and T. MORI
Pressure Effect on the Lattice Parameters of RMn_2X_2 (R=Y,La,Nd and X=Ge,Si)
61.60 and 62.20; S8.16
(a) Vol.120,No.2, K117-K120

A.T. NAGAT, G.A. GAMAL, and S.A. HUSSEIN
Growth and Characterization of Ternary Compound $TlGaTe_2$ Single Crystals
72.80; S8.16
(a) Vol.120,No.2, K163-K168

R.A. YAKSHIBAEV, N.N. MUKHAMADEEVA, and R.F. KADRGULOV
Phase Relations, Ionic Transport, and Diffusion in the Alloys of Ag_2Te-Cu_2Te Mixed Conductors
61.60 and 64.80; 66.30; S8.16
(a) Vol.121,No.1, 111-118

V.M. BURLAKOV, A.A. ANIKIEV, and M.R. YAKHEEV
Evidence for Order Parameter Spatial Fluctuations in $TlGa(Se_{1-x}S_x)_2$ Crystals by Raman Spectroscopy
78.30; 77.80; S8.16
(b) Vol.161,No.2, 883-890

I.T. BODNAR and I.V. BODNAR
Dispersion and Temperature Dependence of Refractive Indicies of Some $A^IB^{III}C_2^{VI}$ Compounds
78.20; S8.16
(a) Vol.121,No.2, K247-K250

H.L. PARK, S.Y. GEE, S.K. CHANG, S. PARK, S.K. LEE, J.Y. LEE, C.S. YOON, J.E. KIM, H.Y. PARK, W.T. KIM, and H. LIM
Solid Solubility of Selenium in $AgGaSe_2$
64.75; 78.30; S8.16
(b) Vol.162,No.1, K9-K14

S.I. RADAUTSAN, I.M. TIGINYANU, V.V. URSAKII, V.M. FOMIN, and E.P. POKATILOV
The Peculiarities of the Temperature Broadening of Raman Light Scattering Lines in $Zn(Cd)Ga_2Se_4$ Single Crystals
78.30; S8.16
(b) Vol.162,No.1, K63-K66

B. FERNÁNDEZ and S.M. WASIM
Sound Velocities and Elastic Moduli in $CuInTe_2$ and $CuInSe_2$
62.20 and 62.65; S8.16
(a) Vol.122,No.1, 235-242

P. DUBČEK, B. ETLINGER, K. FURIĆ, and M. KRANJČEC
Raman Spectra of $(Ga_xIn_{1-x})_2Se_3$
78.30; S8.16
(a) Vol.122,No.1, K87-K90

R. GRILL
Self-Consistent Calculation of Density of States in Impure Semiconductors
71.20; S8.16
(b) Vol.162,No.2, 509-516

A.N. VEIS, D.D. KODITSA, and D.V. GITSU
Optical Properties of p-$TlSbTe_2$ Single Crystals
78.30; S8.16
(a) Vol.122,No.2, K177-K182

S9. Halides

YA.O. DOVGYI and I.V. KITYK
Antiresonances in the Fundamental Spectra of CdI_2-4H and $CdBr_2$-6R Crystals
71.35 and 78.40; S9
(b) Vol.157,No.1, 473-480

I.R. SHEIN and A.L. IVANOVSKII
Electronic Structure of Fluorite-Like TiF_2
71.25; S9
(b) Vol.157,No.1, K29-K32

A. SVOBODA, J. VEČEŘ, C. BÁRTA, and I. PELANT
Kinetics of Two-Photon Excited Luminescence of Hg_2Cl_2 Crystals
78.55; S9
(b) Vol.157,No.1, K69-K72

K. SOMAIAH, M. VENKATA NARAYANA, and L.H. BRIXNER
Thermally Stimulated Luminescence of LaOBr:Tm
78.60; S9
(a) Vol.117,No.1, K81-K84

M. NIKL and K. POLAK
Luminescence Kinetics of PbF_2 Single Crystals
78.55; S9
(a) Vol.117,No.1, K89-K92

D. SKRZYPEK
Critical Behaviour of the Magnetic Resonance in $KNi_xMn_{1-x}F_3$
75.30 and 76.30; S9
(b) Vol.157,No.2, 695-700

R. SATHYAMOORTHY, SA.K. NARAYANDASS, C. BALASUBRAMANIAN, and D. MANGALARAJ
Structure, Dielectric, and AC Conduction Studies on Yttrium Fluoride Thin Films
68.55 and 73.60; 77.55; S9
(a) Vol.117,No.2, 495-500

T. TAKEUCHI
Electronic Structure of F Centers in LiH by SCF $X\alpha$ Method
71.55; 61.70; S9
(b) Vol.158,No.1, 277-282

R. LÉVY, M.J.M. GOMES, B. KIPPELEN, and B. HOENERLAGE
Coherent Signal Generation in CuCl by Light-Induced Grating and Induced Biexciton Decay
78.40; 71.35; S9
(b) Vol.158,No.1, 391-396

A.D. BROTHERS, F.J. FEUERBORN, and S.E. BAUER
Excitons in Indium Halide Thin Films
78.65; 71.35; S9
(b) Vol.158,No.2, 503-510

M.J.M. GOMES, B. KIPPELEN, R. LEVY, J.B. GRUN, and B. HOENERLAGE
Time-Resolved Four-Wave Mixing Experiments in CuCl
71.35 and 78.20; S9
(b) Vol.159,No.1, 101-106

Y. MASUMOTO, S. IDESHITA, and T. WAMURA
Nonlinear Optical Study of Semiconductor Microstructures, CuCl Quantum Dots, and GaAs Quantum Wells
71.35 and 78.20; 73.40; S7.12; S9
(b) Vol.159,No.1, 133-142

P. GILLIOT, B. HOENERLAGE, R. LEVY, and J.B. GRUN
Nonlinearities of CuCl Microcrystals
71.35 and 78.20; S9
(b) Vol.159,No.1, 259-266

L.G. ZIMIN, S.V. GAPONENKO, V.YU. LEBED, I.E. MALINOVSKII, I.N. GERMANENKO, E.E. PODOROVA, and V.A. TSEKHOMSKII
Nonlinear Optical Absorption in I-VII and II-VI Microcrystallites under Quantum Confinement of Elementary ...
71.35 and 78.65; S8.15; S9
(b) Vol.159,No.1, 267-274

H. HAUG, L. BANYAI, J. LIEBLER, and T. WICHT
The Pulsed Nonresonant Optical Stark Effect and the Urbach Tail in Semiconductors
71.35 and 78.20; S7.12; S8.11; S9
(b) Vol.159,No.1, 309-316

I. BALSLEV
Induced Absorption and the Dynamical Stark Shift Associated with Biexcitons
71.35 and 78.20; S8.12; S9
(b) Vol.159,No.1, 339-346

M. KUWATA-GONOKAMI, R. SHIMANO, J. IWAMATSU, H. AKIYAMA, T. KUGA, and M. MATSUOKA
Optical Nonlinearity and Dynamics of Exciton-Biexciton System in CuCl in the Weak Excitation Regime
71.35 and 78.20; S9
(b) Vol.159,No.1, 347-352

T. ISHIHARA
The Origin of Third-Order Optical Nonlinearity in Exciton Systems
71.35 and 78.20; 78.65; S9
(b) Vol.159,No.1, 371-378

ZHENGJU YANG, J. ORTIZ-LOPEZ, and F. LUTY
Theoretical Investigation of Ferroelastic Phase Transition of Pure and Mixed Alkali Cyanide Systems by ... III. The ...
64.60; S9
(b) Vol.159,No.2, 629-644

I.KH. AKOPYAN, A.A. KLOCHIKHIN, B.V. NOVIKOV, M.YA. VALAKH, A.P. LITVINCHUK, and I. KOSAZKII
Optical Spectroscopy of the Superionic Crystals (Review Article)
78.30 and 78.40; 64.70; 71.25; 71.35; 78.55; S9
(a) Vol.119,No.2, 363-416

M.I. MOLOTSKII and S.Z. SHMURAK
Electron Emission at Plastic Deformation of Colored Crystals
61.70; 79.75; S9
(a) Vol.120,No.1, 83-96

P. SAINT-GRÉGOIRE, R. ALMAIRAC, J. LAPASSET, and J. MORET
Structural Phase Transitions in $KFeF_4$ and Related Materials
64.70; S9
(a) Vol.120,No.1, K25-K30

S.K. SHARMA, N. SHUKLA, and G.C. TRIGUNAYAT
TEM and X-Ray Studies on Melt-Grown CdI_2 and $CdBr_2$ Crystals
61.60; 61.70; S9
(a) Vol.121,No.1, 29-36

S9.11. Alkali halides

R.V. GOPALA RAO and R. VENKATESH
Theoretical Method of Evaluation of Heats of Vaporisation of Fused Salts
61.20; 65.50; S9.11
(b) Vol.157,No.1, 65-70

R. CYWIŃSKI, E. MUGEŃSKI, and I. SOKÓLSKA
Eu^{2+}-Mn^{2+} Energy Transfer in NaCl and KCl Crystals
61.70; 78.50; S9.11
(b) Vol.157,No.1, 137-144

B. SORKIN and H. KÄÄMBRE
Exoemission Mechanisms as Discerned via X-Ray Exposure Variations
71.55 and 79.75; S9.11; S10.1
(a) Vol.117,No.1, K33-K36

K.R. MURALI and Y.V.G.S. MURTI
Color Centers in Irradiated $NaF:Mn^{++}$ Crystals
61.70; 78.55; 78.60; S9.11
(a) Vol.117,No.2, 377-386

B.G. DICK and M. GUTOWSKI
Quantum Chemical Calculation of Possible Forms of Electronic-Vibrational Coupling for $F_H(CN^-)$ Defects
71.55; S9.11
(a) Vol.117,No.2, K115-K118

J. SHANKER and M. KUMAR
Studies on Melting of Alkali Metals (Review Article)
64.70; 63.75; S9.11
(b) Vol.158,No.1, 11-50

V.A. LOBATCH, B.E. KULYABIN, V.P. ZHUKOV, and N.I. MEDVEDEVA
The Electronic Structure of the Alkaline-Earth Oxides. I.The Differences between MgO and LiF, NaF Electronic Structures
71.20 and 71.25; S9.11; S10.1
(b) Vol.158,No.1, 239-248

D. FROEHLICH, ST. KIRCHHOFF, P. KOEHLER, and W. NIESWAND
Magnetic Field Splitting and Dispersion of the Lower 1S-Exciton Polariton in KI and RbI
71.35 and 71.70; 71.36; 78.20; S9.11
(b) Vol.158,No.1, 267-276

P.S. GUIMARÃES and M.L. DE SIQUEIRA
The Crystalline Cluster Method for Ionic Solids: The F Center in NaCl
61.70 and 71.55; S9.11
(b) Vol.158,No.1, 283-292

A. TOMITA, Y. FUKUDA, Y. KUTOMI, and N. TAKEUCHI
Emission Processes in Thermally Stimulated Exoelectron Emission and Thermoluminescence of KCl:Ag Single Crystals
61.70 and 78.60; 79.75; S9.11
(b) Vol.158,No.1, 383-390

T. OHGAKU and N. TAKEUCHI
Study on Dislocation-Impurity Interaction by the Blaha Effect
62.20; 61.70; S9.11
(a) Vol.118,No.1, 153-160

S.P. KUILA
Photostimulation Studies in Pure and Li-Doped Alkali Halides
78.60; 61.70; 71.55; S9.11
(a) Vol.118,No.1, 311-316

V. HIZHNYAKOV
Stimulated Echo and Vibrational Dynamics
78.40; 78.45; 78.50; S9.11
(b) Vol.158,No.2, 725-736

J. MANAM and A.N.C.V. RAO
Influence of Cu^+ Doping on the Thermoluminescence of KCl
78.60; S9.11
(a) Vol.118,No.2, K129-K132

N.M. OLEKHNOVICH and A.V. PUSHKAREV
Birefringence and Depolarization Effects of X-Ray Scattering in Laue Geometry for Highly Distorted Crystals
61.10; 61.70; S9.11
(a) Vol.119,No.1, 27-34

T. KAMIKAWA
A New Hole Centre in Neutron Irradiated LiF Crystals
61.70; 76.30; S9.11
(b) Vol.159,No.2, 571-576

P.G. BARANOV, V.V. DYAKONOV, N.G. ROMANOV, and V.A. VETROV
Frenkel Defect Formation by Intense UV Light Irradiation in the Absorption Bands of Tl^+-Like Ions in Alkali Halides
61.70; S9.11
(b) Vol.159,No.2, K33-K36

G. LIFANTE, P. SILFSTEN, F. CUSSÓ, and F. JAQUE
$(F^{2+})_H$-Centre Production in Molecule-Doped KCl
61.70; S9.11
(b) Vol.159,No.2, K37-K42

G.R. AUGST and YU.I. VLADISLAVSKII
A Theoretical Study of Diffuse Scattering by Clusters in Alkali Halides
61.12 and 61.70; S9.11
(b) Vol.160,No.1, 71-82

A.V. EGRANOV, V.V. OTROSHOK, and B.P. CHERNYAGO
Infrared Absorption Due to U-Centres in Highly Doped $LiF:H^-$ Crystals
61.70 and 78.30; S9.11
(b) Vol.160,No.1, K75-K78

CH. GOPAL REDDY and K. NARASHIMA REDDY
Thermoluminescence Studies in Yttrium Doped KBr Crystals
78.60; S9.11
(a) Vol.120,No.1, 267-272

M. TANIBAYASHI
A Method to Determine the Internal Stress and the Stress Exponent m of the Dislocation Velocity
62.20; S9.11
(a) Vol.120,No.1, K19-K24

S.V. NISTOR, GH. STOICESCU, and C.D. MATEESCU
Incorporation and Aggregation Properties of Bismuth Ions in KCl Crystals
61.70; 66.30; 78.50; S9.11
(b) Vol.160,No.2, 423-432

E. STEFANOVICH, E. SHIDLOVSKAYA, A. SHLUGER, and M. ZAKHAROV
The Modification of the INDO Calculation Scheme and Parametrization for Ionic Crystals
71.10; 71.25; S9.11; S10; S11.1
(b) Vol.160,No.2, 529-540

B.M. LOGINOV and V.T. DEGTYAREV
Regularities Governing the Dislocation Forest Effects on Crystal Resistance to Deformation under Complex Loading ...
61.70; S1.2; S9.11
(a) Vol.120,No.2, 397-410

Y.V.G.S. MURTI, G.A. POPSON, and A.L. LASKAR
Correlation of Melting Temperature and Formation Enthalpy of Schottky Defects in Ionic Solids
61.70; S9.11; S10.1
(a) Vol.120,No.2, K133-K138

A. KALNIŅŠ, I. PLAVIŅA, A.I. POPOV, and A. TALE
Determination of the Effective Absorption Cross-Section of F-Centres in KBr-In by Photostimulated Luminescence
61.70; 78.55; S9.11
(b) Vol.161,No.1, 85-90

O. SILD
The Model of Rotation of NO_2^- Impurity Molecule in KCl Crystal
61.70; 78.30; 78.55; S9.11
(b) Vol.161,No.1, 105-112

V.P. GUPTA and S.K. SIPANI
The Analysis of Rydberg Potential and van der Waals Interactions in Alkali Halides
63.90; 62.20; 65.70; S9.11
(b) Vol.161,No.1, 131-144

H.-B. ABEL
Low Frequency SERS from Ag Colloids in Alkali Halide Crystals
78.30; 61.70; S9.11
(b) Vol.161,No.1, 435-446

F. MAES, F. CALLENS, P. MATTHYS, and E. BOESMAN
EPR of S_2^- in NaBr
61.70 and 76.30; S9.11
(b) Vol.161,No.1, K1-K4

W. KUCHARCZYK
Correlation between Electronic Polarizabilities of Ions and Ratio of Their Radii in Alkali Halides
77.30; S9.11
(b) Vol.161,No.1, K59-K62

GANGANG CAI, SHOUSHAN WANG, and JIAN ZHANG
Radiation Damage and Recovery of LiF:Mg,Ti-TLD at the Dose of 10^6 Gy Gamma Rays
61.80; 78.60; S9.11
(a) Vol.121,No.1, 57-62

A.CH. LUSHCHIK and A.G. FRORIP
Thermalized and Hot Interstitial Halogen Ions in Alkali Halides
61.70 and 78.55; 61.80; 76.30; S9.11
(b) Vol.161,No.2, 525-536

S. DIXIT and R. NIWAS
Cohesive Energies, Equation of State, and Thermoelastic Properties of Mixed Alkali Halide Crystals
62.20 and 64.30; 65.70; S9.11
(b) Vol.161,No.2, 583-596

V.A. LOBATCH, I.R. RUBIN, and P.V. LUSHNIKOV
Electronic Structure of Alkaline-Earth Oxides.
V. Peculiarities of the Hole-Trapping in NaF and MgO
71.20 and 71.25; S9.11; S10.1
(b) Vol.161,No.2, 647-660

H. OPYRCHAŁ and M. MANFREDI
Spectroscopic Evidences of Eu-V Dipole Precipitation in KCl:Eu Crystals
61.70; 71.55; 78.55; S9.11
(a) Vol.121,No.2, 407-414

S. DATTA ROY (PAUL) and S. SENGUPTA
A Greens Function Estimation of Lattice Relaxation for Isovalent Substitutional Defects in Alkali Halide Crystals ...
61.70; S9.11
(b) Vol.162,No.1, 89-98

A.A. O'CONNEL-BRONIN and A.B. TRESHCHALOV
Exciton Reflection Spectra of NaI Crystals. Effect of Surface Irradiation by Photons and Incidence Angle Dependence
71.35 and 78.40; 73.20; S9.11
(b) Vol.162,No.1, 289-296

I.R. PLUMMER and D.W. TURNER
Imaging and Electron Energy Analysis of Exo-Electron Emission
79.75; 73.20; S9.11; S11.1
(a) Vol.122,No.1, 285-296

B.P. SINGH
Analysis of Ionic Radii, Second Neighbour Interactions, and Electronegativity Difference for Ionic Crystals
61.50; 71.45; S7.1; S8.1; S9.11
(b) Vol.162,No.2, 329-338

T. YANAGIHARA and K. YOMOGITA
Potential Energy of an Argon Atom Adsorbed on the Stepped and Kinked Surfaces of a Sodium Chloride Crystal
68.30; 68.45; S9.11
(b) Vol.162,No.2, 443-450

V.I. TYUTYUNNIK and O.I. TYUTYUNNIK
Phonon Structure of Raman Scattering Spectra of LiH Crystals
63.20 and 78.30; S9.11
(b) Vol.162,No.2, 597-604

V. HIZHNYAKOV, M. KARKLINYA, V. NAGIRNYI, L. NAGLI, and S. ZAZUBOVICH
On the Reasons of the Off-Centre Position of Excited Ga^+ and In^+ Ions in Alkali Halides
71.55 and 78.55; S9.11
(b) Vol.162,No.2, K91-K94

A.G. LIPSON, V.A. KUZNETSOV, D.M. SAKOV, and YU.P. TOPOROV
Electrical and Magnetic Properties of the Polymer-Ceramics Produced on the Basis of Lithium Fluoride
73.40 and 75.50; S9.11; S12
(a) Vol.122,No.2, 563-568

S9.12. Alkaline earth halides

A.K. SHARMA and S.C. GOYAL
Second-Order Pressure Derivatives of SOE Constants of Fluorite Structure Solids
62.20; S9.12
(b) Vol.157,No.1, 145-150

G.P. TRIBERIS and N. GUSKOS
Electronic Conduction in $CdF_2:Gd^{3+}$ Single Crystals before and after Electrical Conversion
72.80; S9.12
(b) Vol.158,No.1, K17-K20

S. BAUNACK, S. LANDGRAF, and A. ZEHE
Electron Beam Effects Due to Bombardment of CaF_2-Layers Investigated by Auger Electron Spectroscopy
61.80 and 68.60; 79.20; S9.12
(a) Vol.118,No.1, 203-208

R.V. GOPALA RAO and U. BANDYOPADHYAY
Application of the One-Component Plasma Approximation to Structural Studies of Like Charges in Molten Salts
61.25; S9.12
(a) Vol.118,No.2, 365-370

D. DRAHEIM, A. TEMPEL, A. ZEHE, and D. BAITHER
TEM Investigations of MBE Grown CaF_2 Strained Layers on (111) Silicon
68.55; 61.70; S9.12
(a) Vol.119,No.1, 209-214

A. LEHMANN, W. HEERDEGEN, G. SCHIRMER, H. MUTSCHKE, W. RICHTER, E. HACKER, and R. DOHLE
Structure Correlated Infrared Properties of Fluoride Films
78.30 and 78.65; S9.12
(a) Vol.119,No.2, 683-688

S. BENCI, F. FERMI, and A.A. NAGORNYI
Annealing of Eu^{2+} Centres in CdF_2 Crystals
78.55 and 78.60; S9.12
(a) Vol.120,No.2, 635-642

A.I. KURBATOV, M. MAEDER, N.N. SOKOLOV, A. TEMPEL, J. TOBISCH, and A. ZEHE
gamma-Ray Diffraction Study of Thin CaF_2/Si(111) Heterostructures Grown by MBE
68.55; S5.11; S9.12
(a) Vol.121,No.1, K65-K68

SANGEETA, S.C. SABHARWAL, B. GHOSH, and M.K. GUPTA
Thermoluminescence from Europium Doped Calcium Fluoride Crystals
78.60; S9.12
(a) Vol.121,No.2, 657-666

E. LAREDO, A. BELLO, N. SUAREZ, and M. DÍAZ
Comparison of Analyzing Procedures for Determining the Relaxation Parameters from ITC Curves
77.30; S9.12
(a) Vol.121,No.2, K235-K238

S9.15. Mixed crystals

J.M. RÉAU, J. SÉNÉGAS, M. ELOMARI, J.P. LAVAL, and B. FRIT
Ordre à courte distance et conductivité ionique au sein de la solution $Pb_{1-x}In_xF_{2+x}$
61.60 and 66.30; 64.75; S9.15
(a) Vol.117,No.2, 409-416

H. BOETTCHER, K. SCHREIBER, and B. VAUPEL
Photosensitivity of Evaporated Mixed AgBr Layers
73.40 and 78.65; S9.15
(a) Vol.117,No.2, K165-K170

V.K. VORONKOVA, L.V. MOSINA, A.E. USACHEV, and YU.V. YABLOKOV
Weak Pair Exchange Interactions between Cobalt Ions in the Quasi-One-Dimensional Antiferromagnet $CsCoCl_3$
76.30; 71.70; S9.15
(b) Vol.158,No.1, 337-346

M. KOLESÍK, D. TUNEGA, and B.P. SOBOLEV
A Study of the Disorder in Heavily Doped $Ba_{1-x}La_xF_{2+x}$ by Raman Scattering
78.30; S9.15
(b) Vol.160,No.1, 375-380

S.K. GUPTA, V.V. GLADKII, and K. LAL
Relaxation Effects in the Vicinity of the Incommensurate-Commensurate Phase Transition in Rb_2ZnCl_4 ...
64.70; S9.15
(a) Vol.121,No.1, K47-K50

M. EL OMARI, J.M. RÉAU, and J. SÉNÉGAS
Correlations entre propriétés électriques et ordre à courte distance au sein de quelques solutions solides basées sur...
66.30; 64.75; S9.15
(a) Vol.121,No.2, 415-428

S9.16. Higher compounds

J. CZAPLICKI, N. WEIDEN, and A. WEISS
Dynamics of Protons in $Cd(NH_3)_6X_2$ Type Compounds
76.60; S9.16
(a) Vol.117,No.2, 555-562

M. HIDAKA, H. FUJII, M. NISHI, and B.M. WANKLYN
Magnetic Phase Transition in the Layer Compound $CsVF_4$
75.30; S9.16
(a) Vol.117,No.2, 563-570

M. SIESKIND, D. AYACHOUR, J.-C. MERLE, and J.-C. BOULOU
Crystal Growth, Raman and Infrared Spectra of BaFI Single Crystals
63.20 and 78.30; S9.16
(b) Vol.158,No.1, 103-112

L.K. AMINOV and YU.YU. KOSTETSKII
Temperature Dependence of EPR Linewidth of Impurities in Van-Vleck Paramagnets
71.70 and 76.30; S9.16
(b) Vol.158,No.2, 595-602

Z. TYLCZYŃSKI and P. PISKUNOWICZ
Dilatometric Studies of $(NH_4)_2ZnCl_4$ in the Temperature Range from 100 to 450 K
64.70 and 65.70; S9.16
(a) Vol.118,No.2, 441-446

V. GOMATHY, C. BASU, and U.S. GHOSH
Configuration Interaction and Crystal Field Splitting of Cu^{2+} Ion in D_{2d} and D_{3h} Symmetry
71.70; S9.16; S11
(b) Vol.160,No.2, 583-590

S.M. AKHMIN, V.P. MEIKLYAR, A.E. USACHEV, and YU.V. YABLOKOV
$^{59}Co^{2+}$ ENDOR in the Tetragonal Phase of $CsCaCl_3$
76.30 and 76.70; 64.60; S9.16
(b) Vol.160,No.2, K149-K152

V.M. BURLAKOV, S.A. KISELEV, V.N. PETUKHOV, and A.A. ANIKIEV
The Dynamics of the Domain Boundaries and Soft Lattice Mode in SbSI Crystals
77.80; 65.40; S9.16
(b) Vol.161,No.1, 387-394

A.A. BOGUSLAVSKII, YU.N. IVANOV, A.I. KRIEGER, A.K. MOSKALEV, V.I. PAKHOMOV, and R.SH. LOTFULLIN
NMR Investigation of Quadrupole Interaction and Chemical Shift ^{133}Cs in Cs_2HgBr_4 Crystal
76.60 and 77.80; S9.16
(b) Vol.161,No.1, K49-K54

H. KAHNT and J.M. REAU
Mixed Mobile Ion Effect of Anionic Conduction in Lead Halogenosilicate Glasses
61.40 and 66.30; S9.16
(a) Vol.121,No.1, 73-80

WEN-CHEN ZHENG
Spin-Lattice Coupling Coefficients G_{ij} for Trigonal $Al_2O_3:Fe^{3+}$ and $ZnSiF_6 * 6\ H_2O:Ni^{2+}$ Crystals
71.70; 76.30; S9.16; S10
(b) Vol.161,No.2, 711-716

T. TSUBOI and P. SILFSTEN
Detection of Structural Phase Transition in $BaMnF_4$ by Luminiscence Lifetime
64.70 and 78.55; S9.16
(b) Vol.162,No.1, K15-K20

S.M. AKHMIN, V.P. MEIKLYAR, A.E. USACHEV, YU.V. YABLOKOV, and S.V. YURTAEVA
Electron-Nuclear Interactions of the Ce^{3+} Ion in $Cs_2NaInCl_6$
76.70; S9.16
(b) Vol.162,No.2, K107-K112

S10. Oxides

J. KRUPSKI
Interfacial Capacitance
73.40; S10
(b) Vol.157,No.1, 199-208

A.A. KORNIENKO, A.A. KAMINSKII, and E.B. DUNINA
Dependence of the Line Strength of f-f Transitions on the Manifold Energy. II. Analysis of Pr^{3+} in $KPrP_4O_{12}$
71.10 and 78.20; 71.70; S10
(b) Vol.157,No.1, 267-274

E. MATSUBARA, Y. WASEDA, S. HASHIMOTO, and K. INOMATA
Structure of Ferromagnetic $0.5La_{1-x}Sr_xMnO_3 * 0.5B_2O_3$ (x=0.3,0.5,and 1.0) Oxide Glasses by X-Ray Diffraction
61.40; S10
(a) Vol.117,No.1, K1-K4

A.H. WAFIK and A.A. SATTAR
Dependence of the Irreversible Magnetization in $Co_{1-a}Cd_aFe_2O_4$ on the Magnetizing Field
75.60; S10
(a) Vol.117,No.1, K61-K66

R. SALAM
Trapping Parameters of Electronic Defect States in Indium Tin Oxide from AC Conductivity
73.60; 71.55; S10
(a) Vol.117,No.2, 535-540

R.A. EVARESTOV and V.A. VERYAZOV
The Electronic Structure of Copper Oxide Crystalline Compounds II. Chemical Bonding in Copper-Oxygen Crystals
71.20 and 71.25; 74.10; S1.2; S10
(b) Vol.158,No.1, 201-212

J.J. BARA, B.F. BOGACZ, A. POLACZEK, M. PĘKAŁA, and A. GRODZIŃSKI
The Temperature Induced Transition in the theta-Type Vanadium Oxide Bronzes
61.60; 64.70; 76.80; S10
(a) Vol.118,No.1, 41-52

S.V. MOHARIL and P.D. SAHARE
Langbeinite Based Phosphors for Applications in Thermoluminescence Dosimetry
78.60; S10
(a) Vol.118,No.1, K55-K58

A.A. KAMINSKII, B.V. MILL, and A.V. BUTASHIN
Stimulated Emission from Nd^{3+} Ions in Acentric $LaBGeO_5$ Crystals
78.20; 78.30; 78.45; 78.55; S10
(a) Vol.118,No.1, K59-K64

C. MANDE and A.P. DESHPANDE
A Study of Manganese Oxides by EXAFS Spectroscopy
78.70; S10
(b) Vol.158,No.2, 737-742

N.N. KHUSNATDINOV, V.F. PETRENKO, and A.N. TURANOV
Intrinsic Photoconductivity of Hexagonal Ice
61.70 and 72.40; S10
(a) Vol.118,No.2, 401-408

F. MICHARD and A. ZAREMBOWITCH
Changes in the Propagation of Ultrasonic Waves in Silica Aerogels
62.65; 78.35; S10
(a) Vol.118,No.2, 461-466

J.A. BAQUEDANO and J.M. CABRERA
Absorption Bands Associated to Photorefractive Erasure in $Bi_{12}SiO_{20}$
71.55; 72.40; 78.50; S10
(a) Vol.118,No.2, 497-504

U. HEIBER
On the Domain Configuration of Finite Uniaxial Crystals
75.60; S10
(a) Vol.118,No.2, 531-538

P.F. ROBUSTO and R. BRAUNSTEIN
Optical Measurements of the Surface Plasmon of Indium-Tin Oxide
71.45 and 78.65; S10
(a) Vol.119,No.1, 155-168

K. IWAUCHI and Y. IKEDA
Magnetic Properties of Sintered Fe_2TiO_5
75.50; S10
(a) Vol.119,No.1, K71-K74

N.P. KOLMAKOVA, I.B. KRYNETSKII, M.M. LUKINA, and A.A. MUKHIN
Crystal Field and Metamagnetic Behavior of Rare Earth Orthoaluminates: $DyAlO_3$
75.30; 71.70; S10
(b) Vol.159,No.2, 845-850

J. REICHARDT, M. STIEBLER, R. HIRRLE, and S. KEMMLER-SACK
Cathodo- and Photoluminescence in Oxyorthosilicates of X_1 and X_2 Type: System $Y_{2-x}Gd_xSiO_5:Tb^{3+}$
78.55 and 78.60; 71.55; S10
(a) Vol.119,No.2, 631-642

V.A. LOBATCH, A.B. SOBOLEV, and I.R. RUBIN
The Electronic Structure of Alkaline-Earth Oxides. II. Crystalline Cluster Approach Formalism
71.20 and 71.25; S10
(b) Vol.160,No.1, 165-172

W. SCHILLER and K. WESCHKE
Calculation of the Normal-State Resistivity for High-T_c Superconductors
72.15 and 74.70; S10
(b) Vol.160,No.1, 297-304

A.A. KAMINSKII, B.V. MILL, A.V. BUTASHIN, K. KURBANOV, and L.A. POLYAKOVA
Spectroscopy of Disordered $La_3Ga_5SiO_{14}$ Single Crystals Doped with Pr^{3+}, Ho^{3+}, and Er^{3+} Ions
78.55; 71.55; 78.50; S10
(a) Vol.120,No.1, 253-266

V.A. LOBATCH and I.R. RUBIN
Electronic Structure of Alkaline-Earth Oxides. III. Resonant Behaviour of MgO Valence Band Top
71.20 and 71.25; S10
(b) Vol.160,No.2, 505-516

E. STEFANOVICH, E. SHIDLOVSKAYA, A. SHLUGER, and M. ZAKHAROV
The Modification of the INDO Calculation Scheme and Parametrization for Ionic Crystals
71.10; 71.25; S9.11; S10; S11.1
(b) Vol.160,No.2, 529-540

S.I. KLOKISHNER, A.V. KORYACHENKO, and B.S. TSUKERBLAT
Charge Ordering of Multielectronic Clusters in Mixed-Valence Molecular Crystals
71.70 and 75.30; 64.60; S10
(b) Vol.160,No.2, 641-648

W. WALZ and H. KRONMUELLER
Point Defects in Electron-Irradiated Stoichiometric Magnetite
61.70 and 75.50; S10
(b) Vol.160,No.2, 661-672

E.V. SVIRIDOV, V.A. ALYOSHIN, YU.I. GOLOVKO, I.N. ZAKHARCHENKO, V.M. MUKHORTOV, and V.P. DUDKEVICH
Roentgenographic and Electron-Microscopic Data on Heteroepitaxial $PbTiO_3$/(001)MgO Films
68.55; 77.80; S10
(a) Vol.121,No.1, 157-162

J. LI and J. BEYNON
Fine-Structure in the I-U Characteristics of Cu-SiO Thin-Film Sandwich Devices
73.40 and 73.60; S10
(a) Vol.121,No.1, 163-172

QIANG WANG, BINGLIN GU, and XIAOWEN ZHANG
Preliminary Studies on the Stabilities of Three Types of Ordered Structures in the $A(B^1_{0.5}B^2_{0.5})O_3$ System
64.70; 61.60; S10
(b) Vol.161,No.2, 537-542

WEN-CHEN ZHENG
Spin-Lattice Coupling Coefficients G_{ij} for Trigonal $Al_2O_3:Fe^{3+}$ and $ZnSiF_6$ * 6 $H_2O:Ni^{2+}$ Crystals
71.70; 76.30; S9.16; S10
(b) Vol.161,No.2, 711-716

V.V. BRYKSIN, A.V. GOLTSEV, S.D. KHANIN, A.V. NOVOTELNOVA, and A.N. VASILEV
Nonlinear Current-Voltage Characteristics of Ta_2O_5 and Nb_2O_5 Amorphous Oxides
72.20; S10
(b) Vol.161,No.2, 777-782

I. ŚLEDZIŃSKA, A. MURASIK, and A. WIŚNIEWSKI
Linear-Chain Antiferromagnetism in alpha-NiC_2O_4 * 2 D_2O
61.60 and 75.40; S10
(b) Vol.161,No.2, K99-K102

M.D. KANNAN, SA.K. NARAYANDASS, C. BALASUBRAMANIAN, and D. MANGALARAJ
Aging, Annealing, and Dielectric Properties of Neodymium Oxide Thin Films
73.60; 77.55; S10
(a) Vol.121,No.2, 515-522

H.A. KASSIM
Dipole Polarizability of O^{2-} Ions near the (011) and the (111) Surface of Alkaline Earth Oxides
73.20 and 77.30; S10
(b) Vol.162,No.1, 297-302

M. SASAKI, G.X. TAI, and M. INOUE
Transient Thermoelectric Effect in Quasi-Two-Dimensional eta-Mo_4O_{11} Crystal
72.20; 71.30; S10
(b) Vol.162,No.2, 553-558

K. IWAUCHI
Ionic Conductivity and Dielectric Relaxation of beta-Al_2O_3 Fiber Superionic Conductor
66.30; 77.40; S10
(a) Vol.122,No.2, 503-510

P. SISODIA and M.P. VERMA
Polycrystalline Elastic Moduli of Some Hexagonal and Tetragonal Materials
62.20; S1; S8; S10
(a) Vol.122,No.2, 525-534

A.H. WAFIK
Dependence of the Irreversible Magnetization in $Co_{1-a}Cu_aFe_2O_4$ on the Magnetizing Field
75.60; S10
(a) Vol.122,No.2, 657-662

S10.1. Simple oxides

R.A. EVARESTOV and V.A. VERYAZOV
The Electronic Structure of Copper Oxide Crystalline Compounds. I. LUC-CNDO Approach to the Electronic Structure of Cu_2O and ...
71.10 and 71.25; 74.10; S1.2; S10.1
(b) Vol.157,No.1, 281-292

A.K. ZHETBAEV, K.M. DONBAEV, and M.B. ABLANOV
Spin Reorientation in Irradiated Hematite at 7 to 60 K
61.70 and 76.80; S10.1
(b) Vol.157,No.1, K55-K60

B. SORKIN and H. KÄÄMBRE
Exoemission Mechanisms as Discerned via X-Ray Exposure Variations
71.55 and 79.75; S9.11; S10.1
(a) Vol.117,No.1, K33-K36

KH.B. GEZALOV, A.M. GASANOV, A.A. GARIBOV, M.M. ALIEV, and N.I. MUSAEV
The Nature of Paramagnetic Centers in gamma-Irradiated Boron Oxides
61.80 and 76.30; S10.1
(a) Vol.117,No.1, K57-K60

XI CHEN, CHAO DING, YU WANG, XIAN-TING ZEN, and BAO-SHENG LUO
The Structures of $YBa_2Cu_3O_{6+\varepsilon}$ Superconducting and Nonsuperconducting Phases
61.60 and 74.70; S10.1
(a) Vol.117,No.2, 335-340

T. KUROBORI, E. YOSHIDA, S. NAKAMURA, N. TAKEUCHI, K. KAWABATA, and Y. FUKUDA
Optical Properties of a Modified F^+ Center in CaO Crystal Doped with Impurities
78.55; 61.70; S10.1
(a) Vol.117,No.2, K177-K180

E. PIGORSCH and W.E. STEGER
Identification of LiO Bands in the Infrared Spectra of the Insertion Compound delta-LiV_2O_5
61.70 and 78.30; S10.1
(a) Vol.117,No.2, K189-K192

V.A. LOBATCH, B.E. KULYABIN, V.P. ZHUKOV, and N.I. MEDVEDEVA
The Electronic Structure of the Alkaline-Earth Oxides. I.The Differences between MgO and LiF, NaF Electronic Structures
71.20 and 71.25; S9.11; S10.1
(b) Vol.158,No.1, 239-248

T. PISARKIEWICZ and A. KOLODZIEJ
Nonparabolicity of the Conduction Band Structure in Degenerate Tin Dioxide
71.25; S10.1
(b) Vol.158,No.1, K5-K8

WAN-LUN YU
Crystal Field Effect on the g-Factors of ^6S-State Ions
71.70; 76.30; S8.11; S10.1; S11.1
(b) Vol.158,No.1, K13-K16

W.A. BADAWY, R.S. MOMTAZ, and E.M. ELGIAR
Solid State Characteristics of Indium-Incorporated TiO_2 Thin Films
68.55 and 78.65; 73.60; S10.1
(a) Vol.118,No.1, 197-202

G. BEHR, G. KRABBES, J. WERNER, P. DORDOR, and J.-P. DOUMERC
Crystal Growth and Electrical Properties of Sb-Doped SnO_2 Single Crystals
72.80; 61.50; S10.1
(a) Vol.118,No.2, K91-K94

D. FROEHLICH, CH. NEUMANN, B. UEBBING, and R. WILLE
Experimental Investigation of Three-Level Optical Stark Effect in Semiconductors
71.35 and 78.20; 73.40; S7.1; S10.1
(b) Vol.159,No.1, 297-308

A. MYSYROWICZ, D.W. SNOKE, and J.P. WOLFE
Progress on Bose-Einstein Condensation in Excitons
71.35 and 78.55; S10.1
(b) Vol.159,No.1, 387-402

JIZHONG ZHANG
Fractal Growth of MoO_3 Crystals during Vapor-Phase Deposition
61.50 and 68.45; S10.1
(a) Vol.119,No.1, 41-46

A.E. TSURKAN, T.D. SHEMYAKOVA, S.P. MEDVETSKII, and L.A. NAZARENKO
Spectral Response and Detectivity of InAs Hybrid Structures
68.48; 73.60; S7.12; S10.1
(a) Vol.119,No.1, 191-200

H. HIKITA, K. TAKEDA, and Y. KIMURA
EPR of New Cr^{3+} Centers in SnO_2 Crystals
76.30; S10.1
(a) Vol.119,No.1, 251-258

A. SARKAR, S. CHAUDHURI, and A.K. PAL
Electrical and Optical Properties of Transparent Conducting ZnO Films
73.60 and 78.65; S10.1
(a) Vol.119,No.1, K21-K26

F. PIO, M. GUZZI, G. SPINOLO, and M. MARTINI
Intrinsinc and Impurity-Related Point Defects in Amorphous Silica: A Spectroscopic Study
61.70; 78.50; 78.55; S10.1
(b) Vol.159,No.2, 577-588

U. TESCHNER and K. HUEBNER
IR-Spectroscopic Data of Thin Insulating Films on Semiconductors. New Methods of Interpretation and Analysis
78.30 and 78.65; S10.1
(b) Vol.159,No.2, 917-926

WEN-CHEN ZHENG
Spin-Lattice Coupling Coefficients F_{ij} for d^8 Ions in Cubic Symmetry
71.70; 76.30; S10.1
(b) Vol.159,No.2, K67-K72

V.N. BOGDANOV, I.M. BROVCHENKO, L.V. MAKSIMOV, A.R. SILIN, and O.V. YANUSH
Spectroscopic Optical and Acoustic Investigations of Radiation Modified Silica Glasses
61.40 and 78.20; 61.80; 62.65; S10.1
(a) Vol.119,No.2, 621-630

J. LLOPIS
Luminescence of MgO and CaO Stabilized ZrO_2 Crystals
78.55 and 78.60; S10.1
(a) Vol.119,No.2, 661-668

E.M. DIANOV, V.O. SOKOLOV, and V.B. SULIMOV
Theory of Silicon Vacancy in Vitreous Silica
61.70 and 71.55; S10.1
(b) Vol.160,No.1, 263-274

WEN-CHEN ZHENG
Spin-Lattice Coupling Coefficients for $CaO:Ni^{2+}$ Crystals
71.70; 76.30; S10.1
(b) Vol.160,No.1, K43-K48

J. KASPRZAK
The Influence of Electron Hopping on the Quadrupole Relaxation in Cu_2O
76.60; 72.20; S10.1
(b) Vol.160,No.1, K53-K56

V.V. HARUTUNIAN, V.A. GEVORKIAN, N.E. GRIGORIAN, and G.N. ERITSIAN
Study of Optical Properties of alpha-Al_2O_3 in the Vacuum Ultraviolet Region of Synchrotron Radiation
78.40; S10.1
(b) Vol.160,No.1, K71-K74

F.G. REQUEJO, A.G. BIBILONI, H. SAITOVITCH, and P.R.J. SILVA
Study of Point Defects in Irradiated and Reduced MoO_3 through Hyperfine Interactions on ^{99}Mo
61.70; 61.80; S10.1
(a) Vol.120,No.1, 105-118

T. ZHANG, M.C. KANG, and W.C. LU
Impurity Effect on the Chemisorption of Hydrogen on Supported Catalysts
68.45; S1.2; S10.1
(a) Vol.120,No.1, K41-K44

A. CORNET, A. MIRALLES, O. RUIZ, and J.R. MORANTE
Near Infrared Photoluminescence of ZnO:Co Varistors
78.55; S10.1
(a) Vol.120,No.1, K105-K110

N.I. MEDVEDEVA, V.P. ZHUKOV, M.YA. KHODOS, and V.A. GUBANOV
The Electronic Structure and Cohesive Energy of HfO_2, ZrO_2, TiO_2, and SnO_2 Crystals
71.25; S10.1
(b) Vol.160,No.2, 517-528

A.N. NAZAROV, V.S. LYSENKO, S.A. VALIEV, M.M. LOKSHIN, A.S. TKACHENKO, and I.A. KUNITSKII
Flash Lamp Annealing and RF Plasma Annealing of Al-SiO_2-Si Structures
73.40; 73.20; S5.11; S10.1
(a) Vol.120,No.2, 447-456

J.C. BUREAU, A. GLACHANT, and B. BALLAND
Thermal Nitridation of SiO_2 Thin Films in NH_3 + H_2O (or O_2) Mixtures
68.55 and 73.60; 78.65; S10.1
(a) Vol.120,No.2, 507-518

Y.V.G.S. MURTI, G.A. POPSON, and A.L. LASKAR
Correlation of Melting Temperature and Formation Enthalpy of Schottky Defects in Ionic Solids
61.70; S9.11; S10.1
(a) Vol.120,No.2, K133-K138

E.M. DIANOV, A.V. EGIBYAN, S.A. AKOPYAN, A.E. RUSTAMYAN, V.O. SOKOLOV, and V.B. SULIMOV
Computer Simulation on the Threefold Coordinated Oxygen Atom in Vitreous Silicon Dioxide
61.40; 61.70; 71.50; S10.1
(b) Vol.161,No.1, 55-64

V.A. LOBATCH, I.R. RUBIN, and A.B. SOBOLEV
Electronic Structure of Alkaline-Earth Oxides. IV. CaO and SrO Ground State in the Crystalline Cluster Approach
71.20 and 71.25; S10.1
(b) Vol.161,No.1, 271-280

K.B. SUNDARAM and P. WAHID
Optical Absorption in Cerium Dioxide Thin Films
78.65; S10.1
(b) Vol.161,No.1, K63-K66

E.V. KABANOVA, S.N. KULKOV, B.P. GRITZENKO, B.Y. YAKOVLEV, and P.P. PERSTNEV
The Change of Optical and Structural Properties of MgO by the Exposition of Laser Pulse Irradiation
61.80; 78.60; S10.1
(a) Vol.121,No.1, 105-110

A. POEPPL and G. VOELKEL
ESR and Photo-ESR Investigations of the V_0^+ Centre in ZnO Raw Material and Li-Doped ZnO Ceramic Powder
71.55; 76.30; S10.1
(a) Vol.121,No.1, 195-204

Y.S. TANG, J.E. CASTLE, H. LIU, J.F. WATTS, and N.K. HUANG
Rutherford Backscattering and X-Ray Photoelectron Spectroscopy Studies of Dual Ion Beam Deposited Zirconium Oxide Films
68.55; S10.1
(a) Vol.121,No.1, K61-K64

A. POEPPL and G. VOELKEL
ESR and Photo-ESR Investigations of the Pb^{3+} Centre in Undoped ZnO Ceramics
71.55; 76.30; S10.1
(a) Vol.121,No.1, K69-K72

V.YU. RUD, YU.V. RUD, and M. SERGINOV
Natural Photopleochroism of n-In_2O_3-p-$CdSiAs_2$ Structures
72.40 and 73.40; 85; S7.16; S10.1
(a) Vol.121,No.1, K81-K84

V.A. LOBATCH, I.R. RUBIN, and P.V. LUSHNIKOV
Electronic Structure of Alkaline-Earth Oxides.
V. Peculiarities of the Hole-Trapping in NaF and MgO
71.20 and 71.25; S9.11; S10.1
(b) Vol.161,No.2, 647-660

T.L. CHOY and Y.Y. YEUNG
Local Distortion and Crystal Field Parameters for Cr^{3+} at Orthorhombic Sites in MgO
76.30; S10.1
(b) Vol.161,No.2, K107-K110

R. ASHOKAN, V. GOPAL, and K.C. CHHABRA
IR Absorption Spectra of SiO_2 Films Grown by Photo-CVD
68.55 and 78.30; S10.1
(a) Vol.121,No.2, 533-538

X.D. LI and L.Z. ZENG
Structure and Reactivity of the System Si/SiO_2/Fe
68.48 and 79.60; S1.1; S10.1
(a) Vol.121,No.2, K185-K188

A.B. SOBOLEV, A.N. VARAKSIN, O.A. KEDA, and A.P. KHAIMENOV
Electronic Structure and Charge State of Oxygen Vacancies in ZrO_2 Perfect Crystals
71.55; 61.70; S10.1
(b) Vol.162,No.1, 165-172

E.M. DIANOV, V.O. SOKOLOV, and V.B. SULIMOV
Theory of Germanium-Related Defects in Vitreous Silicon Dioxide. I. Germanium Atom in Silicon Site
71.55; S10.1
(b) Vol.162,No.1, 173-182

H.-J. FITTING, P. MAGDANZ, W. MEHNERT, D. HECHT, and TH. HINGST
Charge Trap Spectroscopy in Single and Multiple Layer Dielectrics
73.40; S10.1
(a) Vol.122,No.1, 297-310

V.K. ADAMCHUK, V.V. AFANASEV, and A.V. AKULOV
The Electron Trap Activation in Thermal SiO_2
71.55; 73.40; S10.1
(a) Vol.122,No.1, 347-354

T. KUROBORI and N. TAKEUCHI
On the Lifetime in Magnesium-Doped Calcium Oxide Crystals
78.55; 61.70; 71.55; S10.1
(a) Vol.122,No.1, K91-K96

L. SKALA, V.M. KENKRE, M.W. WEISER, and J.D. KATZ
Remarks on the Thermodynamics of Sintered Ceramics
65.50; 64.60; S6; S10.1
(b) Vol.162,No.2, 429-434

C.B. STEELE and J. BEYNON
Non-Activated Conduction in Au/SiO_x Thin Films
72.80; S10.1
(a) Vol.122,No.2, K169-K170

S10.15. Mixed crystals

B.E. VUGMEISTER and T.V. ANTIMIROVA
Simple Model for Ferroelectric Phase Transition in $KTa_{1-f}Nb_fO_3$
64.60 and 77.80; S10.15
(b) Vol.157,No.1, 183-188

C. YU, C. LAI, S.A. MARSHALL, D.R. YODER-SHORT, and Y.N. ZHANG
Electron Paramagnetic Resonance Spectra Observed in Single Crystals of $YBa_2Cu_3O_{7-\delta}$
74.60; 76.30; S10.15
(b) Vol.157,No.1, 379-388

LING CHEN, GUANGHOU WANG, and MINGKANG TENG
Charge Fluctuation in the High T_c Superconductor YBaCuO in the Transition Temperature Region
74.40; 78.70; S10.15
(b) Vol.157,No.1, 411-416

H. MIZUBAYASHI, H. AKINAGA, and K. TAKITA
Low-Temperature Elastic Properties of $Nd_{1+x}Ba_{2-x}Cu_3O_{7-\delta}$
62.40; 74.70; S10.15
(a) Vol.117,No.1, 125-134

V.D. KUSHKOV, A.V. ZVERLIN, A.M. ZASLAVSKII, and A.V. MELNIKOV
Cubic-Tetragonal Phase Transformation in Films in the Zirconia-Rare Earth Oxide Binary Systems
64.70; S10.15
(a) Vol.117,No.1, K13-K14

F. GOLEK, P.W. KLAMUT, B. SUJAK, A.J. ZALESKI, M. CISZEK, and J. OLEJNICZAK
Thermally Stimulated Charge Emission from the High Temperature Superconductor $YBa_2Cu_3O_{7-x}$
74.90 and 79.75; S10.15
(a) Vol.117,No.1, K43-K46

N. KRISTOFFEL and P. RUBIN
Localized One-Electron States of Impurities on Cu Sites in the CuO_2 Planes of Oxide Superconductors
71.55 and 74.70; S10.15
(b) Vol.157,No.2, K93-K98

C. LUCCHINI, C. GIOVANNELLA, R. MESSI, B. LECUYER, L. FRUCHTER, and M. IANNUZZI
AC Susceptibility and Non-Linear Dynamics in Granular Superconductors
74.30; S10.15
(b) Vol.157,No.2, K123-K126

S. WAPLAK and N.N. KOLPAKOVA
Phase Transitions in $Cd_2Nb_2O_7$ Studied by EPR of Cr^{3+} Ion
64.70; 76.30; S10.15
(a) Vol.117,No.2, 461-466

M.N. KHAN
A Study of the Effects of CuO on the Structure and Optical Absorption of Tellurite Glasses
78.30; 61.40; S10.15
(a) Vol.117,No.2, 593-600

M. MANZEL, L. ILLGEN, and R. HERGT
Superconducting Bi-Sr-Ca-Cu-O Thin Films on YAG
74.70; S10.15
(a) Vol.117,No.2, K119-K120

E.M. GOLOLOBOV, N.A. PRYTKOVA, ZH.M. TOMILO, D.M. TURTSEVICH, M.S. TSELUEVSKII, and N.M. SHIMANSKAYA
Properties of High T_c Superconducting Metal Oxide $Bi_2Sr_2CaCu_2O_y$ with Boron Nitride Additions
61.50 and 74.70; S10.15
(a) Vol.117,No.2, K121-K126

XIN JIN, XIAOHUI WANG, YI TANG, MU LU, YITONG ZANG, WUMING CHEN, JIRONG JIN, HUIMIN SHAO, and XIXIAN YAO
The Irreversibility of the Critical Currents of Polycrystalline Bulk $EuBa_2Cu_3O_{7-x}$ with Cycles in an Applied Magnetic Field
74.70; S10.15
(a) Vol.117,No.2, K127-K132

T. KRYSZTOŃ
On the Flux Penetration in the Anisotropic Uniaxial Superconductor
74.60; S10.15
(b) Vol.158,No.1, K21-K26

XIAO-GUANG LI, XIANGLEI MAO, LIN CHEN, and YUHENG ZHANG
Study on the Formation Kinetics of Flux Lines in High-T_c Superconductors
74.60; 74.70; S10.15
(b) Vol.158,No.1, K27-K32

C. SUŁKOWSKI, K. ROGACKI, Z. BUKOWSKI, R. HORYŃ, and E. TROJNAR
Correlation between Superconducting Transition Temperature, Thermopower, and Resistivity for $YBa_2Cu_3O_{7-x}$
74.70; S10.15
(b) Vol.158,No.1, K33-K36

J. DRYZEK and E. ROGOWSKA
Positron Annihilation in Sodium and Copper beta-Vanadium Oxide Bronzes
61.70; 71.25; 78.70; S10.15
(a) Vol.118,No.1, 61-64

B. GILLOT and A. ROUSSET
On the Limit of Aluminium Substitution in Fe_3O_4 and gamma-Fe_2O_3
61.60; 78.30; S10.15
(a) Vol.118,No.1, K5-K8

S. GRANDE, B. LIPPOLD, G. MAYER, and H. SCHMIEDEL
NQR Measurements of ^{63}Cu in $YBa_2Cu_3O_{7-x}$
74.70 and 76.60; S10.15
(b) Vol.158,No.2, 707-714

YIFENG WANG, WENHUI SU, HONGJIAN LIU, and LIPING LI
Orthorhombic Phase and Oxygen Deficiencies in the Electron Type Superconductor $Ln_{2-x}Ce_xCuO_{4-y}$
64.70 and 74.70; S10.15
(a) Vol.119,No.1, 55-62

V.A. SKRYSHEVSKII and P. KUS
Optical and Electro-Physical Properties of $Gd_1Ba_2Cu_3O_{7-\delta}$ -p-Si Structures at $T>T_c$
73.40 and 74.70; 78.65; S10.15
(a) Vol.119,No.1, 201-208

R. HERGT, W. ANDRAE, K. FISCHER, N.M. TCHEBOTAEV, and S.L. TOWN
Pinning in Twinned Single Crystals of Superconducting $REBa_2Cu_3O_{7-x}$ Investigated by Torque Magnetometry
74.60 and 74.70; 61.70; S10.15
(a) Vol.119,No.1, 241-250

J. PIECHOTA and A. PAJĄCZKOWSKA
Magnetic Properties of $YBa_2(Cu_{1-x}Ni_x)_3O_{7-\delta}$
75.30; 74.70; S10.15
(a) Vol.119,No.1, 271-278

M. STIEBLER, J. REICHARDT, R. HIRRLE, and S. KEMMLER-SACK
Cathodo- and Photoluminescence in the System $Y_{2-x}Tb_xSiO_5:Eu^{3+}$. A Case with an Unusual Effective Energy Transfer in ...
78.55 and 78.60; 71.55; S10.15
(a) Vol.119,No.1, 317-326

XICHUNNG JIN, LINJIANG SHEN, ZUGANG LI, HUIMIN SHAO, HUAQIN WANG, YITONG ZHANG, LONG HAI, and MEISHENG HU
Microdomain Structure and Temperature Behaviour of $YBa_2(Cu_{1-x}Fe_x)_3O_{7-y}$
61.50; 74.50; S10.15
(a) Vol.119,No.1, K5-K8

J. MUCHA and A. JEŻOWSKI
Evidence of Phase Transition in $La_{1.85}CuO_{4+\delta}$ by Thermal Conductivity
74.30; S10.15
(a) Vol.119,No.1, K53-K58

SHA JIAN, LIU WEI, C.F. ZHU, S.X. WANG, N.L. WANG, X.A. CHENG, L.Z. CAO, Z.J. CHENG, Z.Y. CHENG, and Q.R. ZHANG
Superconductivity Affected by Zr Doping in $Bi_2Sr_2Ca_1Cu_2O_y$ Ceramic
74.70; S10.15
(a) Vol.119,No.1, K59-K62

N. FLOROS, L.J. DUNNE, and J.N. MURRELL
The Relative Energies of Ionic Structures for the High-T_c Superconductor $YBa_2Cu_3O_7$ and Implications for the ...
74.30; S10.15
(b) Vol.159,No.2, 781-786

K.P. SINHA and M. SINGH
Correlated Pair-Transfer Mechanism for High Temperature Superconductivity. The Case of $YBa_2Cu_3O_{6+x}$
74.60; S10.15
(b) Vol.159,No.2, 787-796

G. FUCHS, A. GLADUN, S.-L. DRECHSLER, R. MUELLER, M. RITSCHEL,
G. KRABBES, N. KOPPETZKI, and A. GILEWSKI
Effect of Oxygen Concentration on the Resistive
Superconducting Transition of $YBa_2Cu_3O_{7-x}$ in High Magnetic Field
74.10 and 74.60; 74.40; S10.15
(b) Vol.159,No.2, 797-804

P.P. SEREGIN, F.S. NASREDINOV, V.F. MASTEROV, and G.T. DARIBAEVA
Electron Structure and Crystalline Field Parameters Determined
by the Emission Moessbauer Spectroscopy for the Copper...
76.80; 74.70; S10.15
(b) Vol.159,No.2, K97-K102

YUPING SUN, JIANYI JIANG, FANCHUN ZHENG, HUAQING YIN, and JIAJU DU
Synthetic Condition and Properties of the Bi-Pb-Sr-Ca-Cu-O
Superconductor with Single High-T_c Phase
74.70; 64.80; S10.15
(a) Vol.119,No.2, 555-560

K. NARASIMHA REDDY, T.S.P.L.N. PRASAD, E. LAXMI NARSAIAH,
U.V. SUBBA RAO, and P. PAROLI
Thermoluminescence in $YBa_2Cu_3O_{7-\delta}$ Samples
78.60; S10.15
(a) Vol.119,No.2, 655-660

XIAO-GUANG LI, XIANDONG WANG, XIANGLEI MAO, LIN CHEN, XU WANG,
YU WANG, and YUHENG ZHANG
The Effect of La Doping on the Superconductivity of
$Bi_{1.7}Pb_{0.3}Sr_{2-x}La_xCa_2Cu_3O_y$
74.70; S10.15
(a) Vol.119,No.2, K141-K144

X.Y. ZHOU, H. ZHANG, and Q.R. ZHANG
Substitution Effect of Al and Sn for Cu in the Y-Ba-Cu-O
System Studied by X-Ray Photoemission Spectroscopy
74.70 and 79.60; S10.15
(a) Vol.119,No.2, K145-K150

S.N. BARILO, V.I. GATALSKAYA, G.V. GATALSKII, A.P. GES,
S.A. GURETSKII, F.P. KORSHUNOV, L.A. KUROCHKIN et al.
Electron Irradiation of Single-Crystalline $Bi_2Sr_2CaCu_2O_{8+\delta}$
61.80 and 74.70; S10.15
(a) Vol.119,No.2, K151-K154

Š. CHROMIK, V. ŠTRBIK, Š. BEŇAČKA, R. ADAM, and M. JERGEL
Thin $YBa_2Cu_3O_{7-x}$ Films Prepared by Low Temperature
Coevaporation on Si Substrates without Buffer Layer
74.70; 68.55; S10.15
(a) Vol.119,No.2, K155-K158

L. KOVÁCS, K. POLGÁR, R. CAPELLETTI, and C. MORA
Diffusion of Hydrogen Isotopes in Pure and Mg-Doped $LiNbO_3$ Crystals
66.30; 78.30; S10.15
(a) Vol.120,No.1, 97-104

W. KRAAK, A. KRAPF, G. NACHTWEI, H.-U. MUELLER, H. SCHOETZ, H. DWELK, R. HERRMANN, P. VASEK, and P. SVOBODA
Preparation, Structure, and Physical Properties of Bi-Sr-Ca-Cu-O Superconducting Ceramics
74.70; S10.15
(a) Vol.120,No.1, 185-198

N. KAWAHARA, S. KAWABATA, H. ENAMI, T. SHINOHARA, H. HOSHIZAKI, M. HASEGAWA, SH. ASAI, and T. IMURA
Preparation of Highly Oriented Bulky Bi(Pb)-Sr-Ca-Cu-O Superconductor by Magnetized Twin-Roll
74.70; S10.15
(a) Vol.120,No.1, 199-208

J. VON CIEMINSKI, H.TH. LANGHAMMER, and H.-P. ABICHT
Peculiar Electromechanical Properties of Some $Ba(Ti,Sn)O_3$ Ceramics
77.60; 77.20; 77.30; S10.15
(a) Vol.120,No.1, 285-294

P.W. KLAMUT, I. BENZAR, A.J. ZALESKI, and M. DYBIEC
Resistivity of $Nd_{1.85}Ce_{0.15}CuO_{4-x}$ in High Magnetic Fields
74.70; S10.15
(a) Vol.120,No.1, K73-K78

N. KRISTOFFEL
On the Vibronic Nature of the Tetragonal-to-Orthorhombic Structural Transition in $La_{2-x}M_xCuO_4$
64.70; 63.20; 74.70; S10.15
(b) Vol.160,No.2, K93-K96

S.A. SERGEENKOV
On Logarithmic CVC of High-T_c Superconductivity Glass Model
74.20; 61.40; S10.15
(b) Vol.160,No.2, K137-K140

T. KOPTE and E. HEGENBARTH
Thermal Conductivity of High-T_c YBCO-Ceramics in Magnetic Fields up to 10 T
66.70 and 74.70; S10.15
(a) Vol.120,No.2, 547-556

U.-C. BOEHNKE, H. BOERNER, J. HERRMANN, B. LIPPOLD, CH. SEMMELHACK, L. WALDOESTL, and M. WURLITZER
Some Investigations on Praseodymium-Substituted Yttrium-Barium-Copper-Oxide
74.70; 61.60; 75.30; S10.15
(a) Vol.120,No.2, 557-566

M. CHYBICKI and J. LIWO
keV-Proton Induced Surface Conductivity of Lead-Silicate Glass
73.25; S10.15
(a) Vol.120,No.2, K159-K162

E. ALTSHULER, S. GARCÍA, and A. AGUILAR
Hysteretic Critical Currents in Y-Ba-Cu-O Superconductors. A Microstructural Approach
74.70; S10.15
(a) Vol.120,No.2, K169-K172

A.M. GABOVICH, D.P. MOISEEV, A.S. SHPIGEL, and A.I. VOITENKO
Josephson Tunneling Critical Current between Superconductors with Charge- or Spin-Density Waves
74.50; 74.20; S10.15
(b) Vol.161,No.1, 293-302

U.C. JOHRI, R.M. SINGRU, and D. BAHADUR
Magnetic and Moessbauer Studies of the System $LaCo_{1-x}Zr_xO_3$ (x=0.1,0.3,0.5)
76.80; 75.50; S10.15
(b) Vol.161,No.1, 357-366

A.M. SHAMAH, N.K. MINA, and M.K. EL-MOUSLY
Structural Investigation of the High T_c Superconducting Oxide System $La_{1-x}Pr_xCaBaCu_3O_7$
61.55 and 74.70; S10.15
(a) Vol.121,No.1, 235-240

T. KOPTE and E. HEGENBARTH
Long-Time Effects in the Thermal Conductivity of YBCO
66.70 and 74.70; S10.15
(a) Vol.121,No.1, K17-K20

F.P. KORSHUNOV, G.V. GATALSKII, V.I. GATALSKAYA, E.K. STRIBUK, and V.T. TROSHCHINSKII
Low-Temperature Electron Irradiation of Thallium and Bismuth High-T_c Superconducting Ceramics
74.70; S10.15
(a) Vol.121,No.1, K85-K90

B. LIPPOLD, J. HERRMANN, W. REICHELT, and H. OPPERMANN
Preparation and Magnetic Investigations on $Mn_2Mo_3O_8$ Single Crystals
75.30; 75.50; S1.2; S10.15
(a) Vol.121,No.1, K91-K94

E. ANACHKOVA, M. GOSPODINOV, A. NIKOLOV, P. SVESTAROV, N. PETKOV, YU. MARKOV, M. LIMONOV, and G. BRUCHMAN
Fundamental Vibrations in $Pb_5(GeO_4)(VO_4)_2$ Crystals
63.20; 78.30; S10.15
(b) Vol.161,No.2, 575-582

W. SCHILLER and K. WESCHKE
Calculation of the Normal-State Thermal Conductivity of the Lattice for High-T_c Superconductors
74.30 and 74.70; S10.15
(b) Vol.161,No.2, 755-768

C. LANDRON, D. RUFFIER, B. DUBOIS, P. ODIER, D. BONNIN, and H. DEXPERT
Etude par EXAFS du rôle des precurseurs dans la synthèse de ZrO_2-Y_2O_3
61.10 and 61.55; S10.15
(a) Vol.121,No.2, 359-368

E.S. TSIKHOTSKII, I.N. ZAKHARCHENKO, and P.V. DUDKEVICH
Linear Defects and Ferroelectric Phase Transition in $BaTiO_3$ Crystals
61.70; 77.80; S10.15
(a) Vol.121,No.2, 391-398

J. DRYZEK and M. ROMEK
Positron Annihilation Study of Oxygen Vacancies in the High Temperature Superconductor Y-Ba-Cu-O
61.70 and 78.70; 74.50; S10.15
(a) Vol.121,No.2, K159-K162

HONG ZHANG
Positron Trapping in $YBa_2Cu_{3-x}Fe_xO_y$
74.70 and 78.70; S10.15
(a) Vol.121,No.2, K207-K210

N. GUSKOS, G.P. TRIBERIS, M. CALAMIOTOU, CH. TRIKALINOS, A. KOUFOUDAKIS, C. MITROS, H. GAMARI-SEALE, and D. NIARCHOS
Temperature Dependence of the EPR Spectra of $GdBa_2Cu_3O_{7-\delta}$ Compounds in Orthorombic and Tetragonal Phase
72.20 and 74.70; 76.30; S10.15
(b) Vol.162,No.1, 243-250

J. ECHIGOYA and K. NARITA
A Study of $La(Mn_{1-x}Ni_x)O_3$ and Its Bonding with Cubic ZrO_2
61.60; 64.75; S10.15
(a) Vol.122,No.1, 87-96

M.D. GLINCHUK, I.P. BYKOV, V.M. KURLIAND, M. BOUDYŠ, and K. NEJEZCHLEB
Valency States and Distribution of Manganese Ions in PZT Ceramics Simultaneously Doped with Mn and Nb
71.55; 76.30; S10.15
(a) Vol.122,No.1, 341-346

G. SRINIVASAN, GUO-MEI WU, and T.T. SRINIVASAN
Surface Resistance at Microwave Frequencies in Transition Metal Ion Substituted Yttrium Barium Copper Oxide ...
73.25 and 74.10; S10.15
(a) Vol.122,No.1, 355-360

O.YU. POLYAK, R.KH. TUKHVATULIN, KIM GEN CHAN, E.M. GASANOV, and E.M. IBRAGIMOVA
Effect of gamma-Irradiation on $YBa_2Cu_3O_{7-x}$ Ceramics and Monocrystals in the Superconducting State
74.70; S10.15
(a) Vol.122,No.1, K45-K50

R. HERGT, W. ANDRAE, and W. SCHUEPPEL
Pinning of Vortices at Twin Boundaries in $YBa_2Cu_3O_{7-x}$ Single Crystals
74.60 and 74.70; 61.70; S10.15
(a) Vol.122,No.1, K51-K54

J. SHA, C.F. ZHU, Z.H. HE, X.-G. LI, Z.P. SU, L.Z. CAO, Q.R. ZHANG, H.R. YI, and H.C. LI
Nonlinear I-V Characteristics of $GdBa_2Cu_3O_y$ Epitaxial Films Related with the Flux Creep
74.60 and 74.70; S10.15
(a) Vol.122,No.1, K55-K58

F.J. OWENS
Non-Resonant Microwave Absorption in N=3 Phase of Bi-Pb-Sr-Ca-Cu-O High-Temperature Superconductor
74.30; 76.30; S10.15
(b) Vol.162,No.2, 565-574

N. GUSKOS, M. CALAMIOTOU, S.M. PARASKEVAS, A. KOUFOUDAKIS, C. MITROS, H. GAMARI-SEALE, J. KURIATA, L. SADLOWSKI, and M. WABIA
On the Influence of Oxygen Deficiency on EPR Spectra of $NdBa_2Cu_3O_{7-\delta}$
76.30; 74.70; S10.15
(b) Vol.162,No.2, K101-K106

I.A. GOHAR, W. SCHMITZ, P. PAUFLER, and R.M. ELSHAZLY
Comments on the Structure of Mixed Alkali-Borosilicate Glasses Containing Chromium and/or Manganese
61.40 and 61.60; S10.15
(a) Vol.122,No.2, 469-474

E.I. TOCHITSKII, O.G. SVIRIDOVICH, E.M. GOLOLOBOV, and D.M. TURTSEVICH
Phase Transformations in $Y_1Ba_2Cu_3O_{7-x}$ Films Prepared by Pulsed Laser Deposition
64.70 and 68.55; 74.70; S10.15
(a) Vol.122,No.2, 555-562

O. EIBL, J.M. TRISCONE, and O. FISCHER
Imaging of $YBa_2Cu_3O_7/PrBa_2Cu_3O_7$ Superlattices by TEM
68.55 and 73.40; 64.80; S10.15
(a) Vol.122,No.2, 589-598

M. POPESCU, G. ALDICA, C. SÂRBU, P. PĂUŞESCU, E. CRUCEANU, F. VASILIU, and M. APOSTOL
Crystalline Phases in Superconducting $Bi:Sr:Ca:Cu_2:O_x$ Ceramic
61.60 and 74.70; S10.15
(a) Vol.122,No.2, 637-644

M. FUKUHARA, A. BHALLA, and R.E. NEWNHAM
Morphotropic Phase Boundary in the $Pb(Zr_xTi_{1-x})O_3$ System
77.30 and 77.90; S10.15
(a) Vol.122,No.2, 677-682

C. BOULESTEIX, Z.C. KANG, P. AUDIBERT, S. NITSCHE, and O. MONNEREAU
Evolution of BiSrCaCuO (2212) Samples by Air Exposure, Showing a Drastic Change of Modulation Very Likely Due to Water..
61.10; S10.15
(a) Vol.122,No.2, K101-K104

X.A. CHENG, Z.H. HE, H. ZHANG, X.Y. ZHOU, S.X. WANG, M.H. FANG, J.S. XIA, J. SHA, Z.P. SU, and Q.R. ZHANG
Crystal Structure and Superconductivity of $Y_{1-x}Pr_xBa_2Cu_3O_{7-y}$ Compounds
74.70; S10.15
(a) Vol.122,No.2, K151-K156

S. MOHAN, T. RADJA KOUMAR, and M. GAL
High T_c Superconductivity of $Ba_2ErCu_3O_7$
74.70; S10.15
(a) Vol.122,No.2, K157-K160

W. ANDRAE, J. BETZ, B. BRUNNER, R. HERGT, H. LENGFELLNER,
K.F. RENK, W. SCHUEPPEL, and K. STEENBECK
In-Plane Rotational Magnetic Losses in a Laser Deposited $YBa_2Cu_3O_{7-x}$ Film
74.60; 74.70; S10.15
(a) Vol.122,No.2, K161-K164

S11. Other higher compounds

M.E. KASSEM, N.G. GOMAA, A.M. EL-KHATIB, and S.H. KANDIL
Fast Neutron Irradiation Effect and Thermal Properties of Doped Nonstoichiometric Lithium Potassium Sulphate Crystals
65.40; 61.80; S11
(a) Vol.117,No.1, 119-124

S. AHMED, A.M. SHAMAH, H.A. KHALIK, A. IBRAHIM, and F. HANNA
Electrical and Thermal Studies of $[(NH_4)_{1-x}K_x]_2SO_4$ Crystals at High Temperature
64.70; 72.20; S11
(a) Vol.117,No.1, 163-168

P.S. GOYAL, R. CHAKRAVARTHY, B.A. DASANNACHARYA, S.K. KULSHRESHTHA, M.S. SASTRY, and J. TOMKINSON
Neutron Incoherent Inelastic Scattering from $(NH_4)_xRb_{1-x}H_2PO_4$
64.60; 61.12; S11
(b) Vol.157,No.2, 547-556

W. SALEJDA and N.A. DZHAVADOV
Phase Transition in $Rb_3H(SeO_4)_2$-Type Crystals. I. The Symmetry Analysis of Proton Ordering
64.60; 61.50; 65.50; S11
(b) Vol.158,No.1, 119-126

S.G. KOZLOVA, YU.G. KRIGER, and S.P. GABUDA
Anisotropic Reorientation of Octahedral Ions According to NMR Relaxation in Solid State
76.60; 71.70; S11
(b) Vol.158,No.1, 329-336

B. SALEJDA and N.A. DZHAVADOV
Phase Transition in $Rb_3H(SeO_4)_2$-Type Crystals.
II. The Molecular Field Approximation
64.60; 65.50; S11
(b) Vol.158,No.2, 475-480

GUOYIN SHEN, CHANGQING XU, and GUIRU BAI
Cu^{2+}-Cu^{2+} Pair Effect on EPR Parameters in C_{2v} Symmetry Crystals
71.70 and 76.30; S11
(b) Vol.158,No.2, K185-K190

V.K. JAIN and G. LEHMANN
Electron Paramagnetic Resonance of Mn^{2+} in Orthorhombic and Higher Symmetry Crystals (Review Article)
76.30; 64.60; S11
(b) Vol.159,No.2, 495-544

H.L. PARK, G.C. KIM, HYUN KUK KIM, and S.W . SHIN
Study on the Temperature Dependent Dielectric Constant of the $BaTiO_3$-$Cd_5(PO_4)_3Cl$ System
77.80; S11
(a) Vol.119,No.2, K177-K180

R. FRANKE, T. CHASSÉ, B. AL-ARAJ, P. STREUBEL, and A. MEISEL
Chemical Shifts of Auger Electron and Photoelectron Binding Energies of Phosphorus in Solid Compounds
79.60; S7.1; S11
(b) Vol.160,No.1, 143-152

H.A. BUCKMASTER and K. KUMAR
The C_{3h} Symmetry EPR Spectral Parameters for $(0.005Gd,0.995La)(C_2H_5SO_4)_3$ * 9 H_2O and 9 D_2O
76.30; 71.70; S11
(b) Vol.160,No.1, 349-356

V. GOMATHY, C. BASU, and U.S. GHOSH
Configuration Interaction and Crystal Field Splitting of Cu^{2+} Ion in D_{2d} and D_{3h} Symmetry
71.70; S9.16; S11
(b) Vol.160,No.2, 583-590

J.E. GARBARCZYK
Study of Protonic Diffusion in $H_{4.5}UO_2(IO_6)_{1.3}$ * m H_2O Using PFG NMR Method
66.30; 76.60; S11
(a) Vol.120,No.2, 387-390

H.R. SREEPAD, V. SREERAMALU, A. CHANDRASHEKARA, V. RAVINDRACHARY, and S. GOPAL
Structural Phase Transition in a Superionic Conductor - $Na_2(La,Al)Zr(PO_4)_3$. Positron Annihilation Study
64.70; 61.40; 78.70; S11
(a) Vol.120,No.2, 411-418

R. CACH and Z. CZAPLA
Dielectric State Equation Parameters of Monoclinic Telluric Acid Ammonium Phosphate
77.80; 77.20; S11
(a) Vol.121,No.1, 299-304

A.P. PECHONII, T.V. ANTIMIROVA, M.D. GLINCHUK, and I.M. SMOLYANINOV
An ESR Study of Li^+ Impurity Ion Deformational Fields in $KTaO_3$
76.30; 61.70; S11
(b) Vol.161,No.2, 843-852

T. FUKAMI
X-Ray Study of Crystal Structure of $NH_4H_2AsO_4$ in the Paraelectric and Antiferroelectric Phases
61.60; S11
(a) Vol.121,No.2, 383-390

S.K. MISRA and JIANSHENG SUN
EPR of a Gd^{3+}-Doped $NH_4Pr(SO_4)_2*4H_2O$ Single Crystal. Study of Phase Transition and Superposition-Model Calculation of ..
76.30; S11
(b) Vol.162,No.1, 265-274

R. HRABAŃSKI and J. LECH
Optimization of Spin Hamiltonian Parameters by the Method of Non-Linear Least Squares Fitting
76.30; S11
(b) Vol.162,No.1, 275-280

SH.M. EFENDIEV, V. GAVRYUSHIN, G. RACIUKAITIS, V.E. BAGIEV, V. KAZLAUSKAS, G. PUZONAS, R.A. ALIEV, and E.R. MUSTAFAEV
Induced Absorption Spectroscopy by Lattice Defects in Sillenite-Type Crystals
78.50; 71.55; S11
(b) Vol.162,No.1, 281-288

S.K. MISRA, JIANSHENG SUN, and U. ORHUN
EPR of a VO^{2+}-Doped $Fe(NH_4)_2(SO_4)_2*6H_2O$ Single Crystal: VO^{2+}-Fe^{2+} Exchange Interaction and Spin-Lattice Relaxation ...
76.30; S11
(b) Vol.162,No.2, 585-596

S11.1. Salts of oxy-acids

M.H. BROOKER
A Search for a Phase Transition in $Sr(NO_3)_2$ by Raman Spectroscopy. Evidence for Partial Disorder
64.60 and 78.30; 63.20; 63.50; S11.1
(b) Vol.157,No.1, 167-182

WEN-CHEN ZHENG
Spin-Lattice Coupling Coefficients F_{ij} for $SrTiO_3:Cr^{3+}$ Crystal
71.70; 76.30; S11.1
(b) Vol.157,No.1, K33-K38

G.I. ROSENMAN, YU.L. CHEPELEV, and E.I. BOIKOVA
Photo-Stimulated Relaxation of Exoemission in $Ba_2NaNb_5O_{15}$
72.40 and 79.60; S11.1
(a) Vol.117,No.1, 259-264

J. MANGIN, G. JEANDEL, and G. MARNIER
Temperature Dependence of Polarization in $KTiOPO_4$ Single Crystals
77.30; 77.70; S11.1
(a) Vol.117,No.1, 319-324

P. MORETTI, P. THEVENARD, G. GODEFROY, R. SOMMERFELDT, P. HERTEL, and E. KRAETZIG
Waveguides in Barium Titanate by Helium Implantation
78.20; 61.80; S11.1
(a) Vol.117,No.1, K85-K88

R. DEBNATH, J. CHAUDHURY, and S. CHANDRA BERA
Optical Properties and Nature of Coordination of Cu^+ Ions in Calcium-Metaphosphate Glass
78.50 and 78.55; 71.55; S11.1
(b) Vol.157,No.2, 723-734

XIQI FENG, DONGSHENG WANG, and JIZHOU ZHANG
NMR Spectra of Mg Nuclei in Mg-Doped $LiNbO_3$ Crystals
76.60; S11.1
(b) Vol.157,No.2, K127-K130

T. FUKAMI
Refinement of the Crystal Structure of KH_2PO_4 in the Ferroelectric Phase
61.50 and 77.80; S11.1
(a) Vol.117,No.2, K93-K96

V. ZUBKUS, P. KUNDROTAS, S. LAPINSKAS, and V. KASHCHEEV
Diffusion in Crystals with Competing Interactions of Migrating Ions
66.30; S11.1
(b) Vol.158,No.1, 51-62

G. VIELSACK and R. VON BALTZ
Dielectric Function of La_2CuO_4
71.25; 78.20; S11.1
(b) Vol.158,No.1, 249-260

WAN-LUN YU
Crystal Field Effect on the g-Factors of ^6S-State Ions
71.70; 76.30; S8.11; S10.1; S11.1
(b) Vol.158,No.1, K13-K16

I. LICEA, A. IOANID, and A. DAFINEI
High-Temperature Phase Transition in KH_2PO_4
64.70; 65.40; S11.1
(a) Vol.118,No.1, 131-140

R. SANGHVI and S. GHOSH
Amplitude Modulation and Demodulation of Electromagnetic Waves in Magnetised Material with Strain-Dependent Dielectric..
72.30; 62.65; S11.1
(a) Vol.118,No.1, 235-240

HONG WANG, JINKE WEN, BIN LI, and HUAFU WANG
Infrared Absorption Study of OH^- in $LiNbO_3$:Mg and $LiNbO_3$:Mg:Fe Crystals
61.70 and 78.30; S11.1
(a) Vol.118,No.1, K47-K50

E.K. SADYKOV and A.I. SKVORTSOV
Theory of R.F. Moessbauer Spectra of the Magnetics with Magnetic Anisotropy
76.60 and 76.80; S11.1
(b) Vol.158,No.2, 685-694

D.L. DZHAPARIDZE, S.V. ALCHANGYAN, D.M. DARASELIA, and T.I. SANADZE
EPR and RFDS Study of Gd^{3+} in alpha-$LiIO_3$ Single Crystals
76.30; S11.1
(b) Vol.158,No.2, K195-K196

J. ECHIGOYA, S. HIRATSUKA, and H. SUTO
Structure and Low-Temperature Phase of $LaCoO_3$
61.60; S11.1
(a) Vol.118,No.2, 371-378

P.S. GOYAL, R. CHAKRAVARTHY, B.A. DASANNACHARYA, and C.J. CARLILE
Reorientations of NH^{4+} Ions in the Ferroelectric Phase of $(NH_4)_2SO_4$
63.90; 64.70; 77.80; S11.1
(a) Vol.118,No.2, 425-430

W. ZAPART
Possibility of Simultaneously Incommensurate and Ferroelastic QQPhase in $RbIn(MoO_4)_2$ by EPR of Cr^{3+} Ion
64.70 and 76.30; S11.1
(a) Vol.118,No.2, 447-454

P. SIVAPRASAD, K. RAMESH, and Y.P. REDDY
Optical and EPR Studies of Cu^{2+} in $MgNH_4PO_4 * 6 H_2O$
76.30 and 78.50; 71.55; S11.1
(a) Vol.118,No.2, K103-K106

I.O. TROYANCHUK, S.N. PASTUSHONOK, A.K. BOGUSH, and V.I. PAVLOV
Magnetic Properties of Nonstoichiometric Orthomanganites
75.50; S11.1
(a) Vol.118,No.2, K111-K114

K. NITSCH, M. NIKL, Č. BARTA, D. SCHULTZE, A. TŘÍSKA, and R. UECKER
Preparation and Emission Properties of $NaBi(WO_4)_2$ and $NaBi(WO_4)_2$:Ce Single Crystals
78.40 and 78.55; S11.1
(a) Vol.118,No.2, K133-K138

V.A. GANSHIN and YU.N. KORKISHKO
Proton Exchange in Lithium Niobate and Lithium Tantalate Single Crystals: Regularities and Specific Features (Review Article)
61.40 and 68.48; 78.20; S11.1
(a) Vol.119,No.1, 11-26

R. KRAIENHEMKE, P. SEMIG, and F. FISCHER
Travelling Solvent Zone Melting Growth and Optical Properties of Lead Activated Calcite Crystals
78.50 and 78.55; 61.50; S11.1
(a) Vol.119,No.1, 327-336

XIQI FENG, LIANAN TANG, LIANGYING XU, and XIAOKE JIANG
Photoconductivity of Nd,Mg-Doped $LiNbO_3$ Crystals
72.40; S11.1
(a) Vol.119,No.2, 561-570

G.C. BHAR, S. DAS, and P.K. DATTA
Efficient Frequency Doubling of Nd Laser Radiation
78.20; S11.1
(a) Vol.119,No.2, K173-K176

A.A. ALYBAKOV, O.M. ARBOTOEV, V.A. GUBANOVA, and I.K. TURDALIEV
Paramagnetic Centers in X-Irradiated LiH_2PO_4 Single Crystals
76.30; S11.1
(a) Vol.120,No.1, K85-K88

J. MANGIN, A. KHODJAOUI, and G. MARNIER
Optical Absorption of KTP Single Crystals
78.20; 78.40; S11.1
(a) Vol.120,No.1, K111-K116

E. STEFANOVICH, E. SHIDLOVSKAYA, A. SHLUGER, and M. ZAKHAROV
The Modification of the INDO Calculation Scheme and Parametrization for Ionic Crystals
71.10; 71.25; S9.11; S10; S11.1
(b) Vol.160,No.2, 529-540

I.V. STASYUK, S.S. KOTSUR, and V.P. TUPYCHAK
On the Theory of the Faraday Effect in Dielectric Ionic Crystals and Crystals with Structural Phase Transition of ...
78.20; S11.1
(b) Vol.160,No.2, 683-696

J.M. WESSELINOWA
The Dielectric Function and Reflectivity in KDP-Type Ferroelectrics
77.80; S11.1
(b) Vol.160,No.2, 697-704

FU-ZHEN LI and ZHAO-MIN LI
The Eight-Order Perturbation Formula of Zero-Field Splitting for $3d^5$ Ions in D_{3d} Site and Its Application to $CaCO_3:Mn^{2+}$
71.70; 76.30; S11.1
(b) Vol.161,No.1, K29-K34

E. POSSENRIEDE, O.F. SCHIRMER, and G. GODEFROY
ESR of Nd^{3+} in $BaTiO_3$
76.30; S11.1
(b) Vol.161,No.1, K55-K58

W. ZAPART and M.B. ZAPART
Incommensurate Phase in $KSc(MoO_4)_2$ by EPR of Cr^{3+}
64.70 and 76.30; S11.1
(a) Vol.121,No.1, K43-K46

L. SADLOWSKI, J. KURIATA, B. BOJANOWSKI, J. WALCZAK, M. KURZAWA, M.L. FALIN, and V.V. IZOTOV
EPR Study of Antiferromagnetic Phase Transition in $Fe_4V_2Mo_3O_{20}$
76.30; 75.50; S11.1
(a) Vol.121,No.1, K95-K98

U. VAN OLFEN
Light-Induced Transparency in Absorbing Powders
78.35; 72.40; S11.1
(a) Vol.121,No.1, K121-K124

A.D. VISHNOI, V.B. SAPRE, and C. MANDE
A Study of the White Lines Associated with the Rare Earth L_{II} and L_{III} Absorption Discontinouities in Some Oxides ...
78.50; 71.55; S11.1
(b) Vol.161,No.2, 891-896

QIREN ZHANG and XIQI FENG
Defect Structures and Densities of Mg-Doped Lithium Niobate
61.70; S11.1
(a) Vol.121,No.2, 429-436

R. STYRKOWIEC, W.A. SYSŁO, A. WIECZOREK, and Z. CZAPLA
Splitting of the Incommensurate-Commensurate Phase Transition of NH_4HSeO_4 Crystals under Uniaxial Mechanical Stress
77.80; S11.1
(a) Vol.121,No.2, K227-K230

P. GÓRSKI and W. KUCHARCZYK
The Fourth-Order Electrooptic Effect in KDP and ADP Crystals
78.20; S11.1
(a) Vol.121,No.2, K243-K246

KANG-WEI ZHOU, SANG-BO ZHAO, PING-FENG WU, and JUN-KAI XIE
EPR Parameters and Spectral Fine Structure of Ni^{2+} in $LiNbO_3$
71.70 and 76.30; 78.50; S11.1
(b) Vol.162,No.1, 193-198

S. DEFREGGER, G.F. ENGEL, and P.W. KREMPL
The Temperature Dependence of the Optical Second Harmonic Generation Coefficient d_{11} of Quartz Homeotypic $AlPO_4$
78.20; 64.60; S11.1
(b) Vol.162,No.1, 311-320

YIFAN HU and RONGCHANG YIN
The Gas-Sensing Characteristics of Bismuth Chromate Material
68.45 and 73.60; S11.1
(a) Vol.122,No.1, 249-254

I.R. PLUMMER and D.W. TURNER
Imaging and Electron Energy Analysis of Exo-Electron Emission
79.75; 73.20; S9.11; S11.1
(a) Vol.122,No.1, 285-296

T. FUKAMI
Refinement of the Crystal Structure of RbH_2PO_4 in the Ferroelectric Phase
61.60 and 77.80; S11.1
(a) Vol.122,No.2, K117-K120

S11.2. Spinels, garnets, ferrites

M.L. MARÍN, M. ORTUÑO, A. HERNÁNDEZ, and J. ABELLÁN
Percolative Treatment of the Verwey Transition in Cobalt-Iron and Nickel-Iron Ferrites
71.30 and 75.40; S11.2
(b) Vol.157,No.1, 275-280

A.G. SHISHKOV, E.N. ILICHEVA, E.V. KOCHETKOVA, and YU.N. FEDYUNIN
On the Choice of the Pulse Waveform for Reading without Distortion of the Vertical Bloch Line Information
75.70; 75.60; S11.2
(a) Vol.117,No.1, K53-K56

A.G. MAKHANEK
On the Mechanism of Two-Photon Absorption by Nd^{3+} Ions in Yttrium Aluminium Garnet and Yittrium Lithium Fluoride in ...
71.70; 78.50; S11.2
(b) Vol.157,No.2, 641-648

YOU XU, GUI-LING YANG, DA-PING CHU, and HONG-RU ZHAI
Theory of the Single Ion Magnetocrystalline Anisotropy of 3d Ions
75.30; S11.2
(b) Vol.157,No.2, 685-694

M. KUČERA, V.N. KOLOBANOV, V.V. MIKHAILIN, P.A. OREKHANOV, and V.N. MAKHOV
Reflection Spectra of Some Garnet and Orthoferrite Single Crystals in Vacuum Ultraviolet
78.20; 78.40; S11.2
(b) Vol.157,No.2, 745-752

V.A. KOSTYLEV, B.A. GIZHEVSKII, A.A. SAMOKHVALOV, M.I. AUSLENDER, and N.G. BEBENIN
Anisotropy of Magnetoresistance of the p-Type Ferromagnetic Semiconductor $HgCr_2Se_4$
72.20; S11.2
(b) Vol.158,No.1, 307-318

K.G. NIKIFOROV, M. BARAN, V.K. BELYAEV, L.YA. PASENKO, S.I. RADAUTSAN, and A. WISNIEWSKI
Magnetization and Chromium Ions with Minority Valence in $CdCr_2S_4$ and $HgCr_2Se_4$ Magnetic Semiconductors
75.30; 75.60; S11.2
(b) Vol.158,No.1, K63-K68

T. ABRITTA, F.H. BLAK, and R.J.M. DA FONSECA
Photoacoustic Measurements of $LiGa_5O_8:Fe^{3+}$
78.50 and 78.55; S11.2
(b) Vol.158,No.1, K75-K78

R. SCHAEFER and A. HUBERT
A New Magnetooptic Effect Related to Non-Uniform Magnetization on the Surface of a Ferromagnet
75.60; 78.20; S1.1; S1.2; S1.4; S11.2
(a) Vol.118,No.1, 271-288

D. MARTÍNEZ, C. DE FRANCISCO, J. IÑIGUEZ, J. RIVAS, and F. WALZ
Influence of Ba Substitutions on Magnetic After-Effects in Magnetite Containing Vacancies
75.60; S11.2
(a) Vol.118,No.1, K31-K34

K. MELZER, J. SUWALSKI, and M. ŁUKASIAK
Moessbauer Study on the Spin Arrangement in $BaFe_{12}O_{19}$ and $BaCo_{0.8}Ti_{0.8}Fe_{10.4}O_{19}$
75.25 and 76.80; S11.2
(a) Vol.118,No.1, K35-K38

V.N. BERZHANSKII, I.I. EVSTAFEV, and V.I. IVANOV
Charge Carrier Transfer in Ferrite Spinels Affected by Magnetic Ordering
72.20; 75.40; S11.2
(b) Vol.158,No.2, 643-652

V.P. GOGIN, V.A. MATVEEV, V.G. TASHIROV, YU.G. RAYZANOV, and A.A. ROMANYUKHA
Compensation Mechanism of the Temperature Stabilization of the Bubble Collapse Field in Eu-Tm Garnet
75.70; 75.60; S11.2
(a) Vol.118,No.2, K107-K110

H. PFEIFFER and W. SCHUEPPEL
Investigation of Magnetic Properties of Barium Ferrite Powders by Remanence Curves
75.60; S11.2
(a) Vol.119,No.1, 259-270

S.E. SARKISOV, A.A. KAMINSKII, and A.V. BUTASHIN
Manifestation of Structural Disorder Peculiarities of $Ca_3Ga_2Ge_3O_{12}$-$Ca_3(Nb/Ga)_2Ga_3O_{12}$ and BaF_2-YF_3 Crystalline ...
78.30; S11.2
(a) Vol.119,No.1, 285-296

V.E. ZUBOV, G.S. KRINCHIK, A.S. TABLIN, and A.A. KOSTYURIN
Magnetooptical Techniques in Self-Consistent Determination of Optical and Magnetooptical Constants of Magnetics
78.20; S1.2; S11.2
(a) Vol.119,No.1, 297-306

K.M. MUKIMOV, B.YU. SOKOLOV, and U.V. VALIEV
The Faraday Effect of Rare-Earth Ions in Garnets
78.20; 75.30; S11.2
(a) Vol.119,No.1, 307-316

S.R. ROTMAN
Energy Transfer in Iron-Doped Yttrium Aluminum Garnet
78.55; S11.2
(a) Vol.119,No.1, K87-K90

A.M. PRITULOV, A.P. SURZHIKOV, V.A. KOZHEMYAKIN, YU.N. AFANASIEV, A.P. VORONIN, O.S. GRIBKOV, and G.R. KARAGEDOV
Radiation-Thermal Packing of Lithium Ferrite Compacts
61.40 and 61.80; S11.2
(a) Vol.119,No.2, 417-422

S.E. SARKISOV, A.A. KAMINSKII, A.V. BUTASHIN, YU.E. PERLIN, V.N. ENAKII, and M.G. BLAZHA
Selective Spectroscopy, Luminescence Kinetics of Pr^{3+} Ions, and Phonon Density Calculation in Multicentred and ...
78.55; 63.20; 71.55; 78.30; S11.2
(a) Vol.119,No.2, 607-620

E.P. NAIDEN, V.I. MALTSEV, and G.I. RYABTSEV
Magnetic Structure and Spin-Orientational Transitions of Hexaferrites of the $BaCo_{2-x}Zn_xFe_{16}O_{27}$ System
75.50; S11.2
(a) Vol.120,No.1, 209-220

W.A. KACZMAREK
Magnetic Hyperfine Interactions in $BaIn_{1.5}Fe_{10.5}O_{19}$ Ferrite
76.80; 75.50; S11.2
(a) Vol.120,No.1, K79-K84

H. PFEIFFER, W. SCHUEPPEL, and TH. KLUPSCH
On the Magnetic Properties of Planar Hexaferrite Powders
75.60; S11.2
(a) Vol.120,No.1, K89-K92

A.YA. FISHMAN, M.A. IVANOV, V.YA. MITROFANOV, and A.A. SHEMYAKOV
Hyperfine Interactions of Jahn-Teller Ion Mn^{3+} in Cubic Magnets with Spinel and Garnet Structures
75.50; 76.20; S11.2
(b) Vol.160,No.2, K153-K158

M. BHAGAVANTHA REDDY and P. VENUGOPAL REDDY
Electrical Conductivity of Some Mixed Li-Ti Ferrites
72.20; S11.2
(a) Vol.120,No.2, 575-580

R. HIRRLE, J. WIEHL, W. WISCHERT, and S. KEMMLER-SACK
Cathodo- and Photoluminescence in Nd^{3+} and Tb^{3+} Activated Garnets of Type $Cd_{3-x}Ca_xGa_2Ge_3O_{12}$
78.55 and 78.60; 71.55; S11.2
(a) Vol.120,No.2, 643-652

D. VLADIKOVA, L. ILKOV, and S. KARBANOV
Optimal Prefiring Conditions of Substituted Nickel Ferrites for Microwave Applications
64.75 and 75.50; S11.2
(a) Vol.121,No.1, 249-256

S. UNNIKRISHNAN and D.K. CHAKRABARTY
Magnetic Properties and Domain State of the Magnesium Ferrite-Magnesium Titanate Solid Solutions
75.50; 76.80; S11.2
(a) Vol.121,No.1, 265-272

R. HIRRLE, W. WISCHERT, and S. KEMMLER-SACK
Cathodo- and Photoluminescence in the Mn^{2+} Activated Garnets of Type $Cd_{3-x}Ca_xGa_2Ge_3O_{12}$
78.55 and 78.60; 71.55; S11.2
(a) Vol.121,No.1, 289-298

D.A. GUSEINOV and M.A. ALYANOV
Heat Capacity and Thermal Expansion of $TiFe_2Se_4$
65.40; 75.40; S11.2
(a) Vol.121,No.1, K51-K54

A.A. SOROKIN, D.D. PERLOV, R.Z. LEVITIN, N.P. KOLMAKOVA, and A.P. DODOKIN
Determination of Ground Quasidoublet Parameters for Ho^{3+} Ion in $(Ho_xY_{1-x})_3Al_5O_{12}$ from Low-Temperature Magnetic ...
71.70 and 75.50; S11.2
(b) Vol.161,No.2, K103-K106

A. LARGETEAU, J.M. RÉAU, and J. RAVEZ
Dispersions diélectrique et d'impédance et énergies d'activation de céramiques à couches d'arrêt de type ...
77.20; 64.80; 77.40; S11.2
(a) Vol.121,No.2, 627-634

P. NOVOTNÝ and R. GEMPERLE
Wall Hysteresis Loops Exhibiting Asymmetry
75.60; S11.2
(a) Vol.121,No.2, K215-K218

V.M. TALANOV
Structural Modelling of Low-Symmetry Phases of Spinels. I. Phases with 1:1 Octahedral Order
61.50; 64.60; S11.2
(b) Vol.162,No.1, 61-74

R.J.M. DA FONSECA and T. ABRITTA
Photoluminescence of Octahedrally Coordinated Mn^{2+} in Lithium Gallate
78.55; S11.2
(b) Vol.162,No.1, K59-K62

A. LARGETEAU, J.P. BONNET, P. DORDOR, and J. RAVEZ
Propriétés électriques des joints de grains d'une céramique ferrite Ni-Zn à couches d'arrêt
72.20; S11.2
(a) Vol.122,No.1, 371-376

V.M. TALANOV
Structural Modelling of Low-Symmetry Phases of Spinels. II. Phases with 1:3 Octahedral Order
61.60; S11.2
(b) Vol.162,No.2, 339-346

V.V. IVANOV and V.M. TALANOV
Modeling of the Structure of the Ordered Spinel-Like Phases (of Type 2:1)
61.60; S11.2
(a) Vol.122,No.2, K109-K112

A. GRUSKOVÁ, J. LIPKA, J. SLÁMA, I. TÓTH, M. SEBERÍNI, and P. KABOŠ
Moessbauer Effect and Its Application in the Investigation of the Substituted Li Ferrites Obtained from the Liquid Phase
75.30 and 75.50; 76.80; S11.2
(a) Vol.122,No.2, K171-K176

S12. Organic compounds

R.K. BHAN, S.K. LOMASH, and K.C. CHHABRA
Determination and Study of Field Dependent Defect Density in Thermal SiO_2 Using 1,1,1 Trichloroethane (TCA)
61.70 and 68.48; S5.11; S12
(a) Vol.117,No.1, 203-212

K. BIEDRZYCKI
Thermally Stimulated Electron Emission from Triglycine Sulphate Crystals with Well-Defined Initial Macroscopic ...
77.80 and 79.40; S12
(a) Vol.117,No.1, 313-318

J.P. SRIVASTAVA, ASITA KULSHRESHTHA, and H. RAUH
FIR Reflectivity and Vibrational Spectra of $(CH_3NH_3)_2ZnCl_4$ Crystals
78.30; 63.20; S12
(a) Vol.117,No.1, K77-K80

L.J. DUNNE and A. TANG KAI
Excited-State Molecular Superconductivity
74.20; S12
(b) Vol.157,No.2, 657-666

P. WUENSCHE and A. MUELLER
Interpretation of Thermally Stimulated Processes by an Experimentally Determined Distribution of the Activation Energy
65.90; S12
(a) Vol.117,No.2, 447-460

A.M. MANSANARES, F.C.G. GANDRA, E.C. DA SILVA, H. VARGAS, M. DO CARMO, R.S. VARELLA, and F. GALEMBECK
Photoacoustic Detection of Magnetite-Crystal Formation from Iron(III) Hydroxide Acetate
75.50 and 76.50; S12
(a) Vol.117,No.2, K155-K160

S.R. STOYANOV
Heat Pretreatment Effect on the Pyroelectric Current of Pure and Sb^{3+} Doped Triglycine Sulphate Monocrystals
77.70; S12
(a) Vol.117,No.2, K161-K164

G. REITER and P. KINDL
Positron Lifetime Investigations on Linear Polyethylene Compared to Branched Polyethylene
64.60; 78.70; S12
(a) Vol.118,No.1, 161-168

M.P. PETROV and P.D. SIMOVA
Textural and Electrooptical Investigations of the HOBA and OOBA Cholesteric Phases Induced by Chiral Dopants
61.40; 78.20; S12
(a) Vol.118,No.2, 357-364

S.S. BAWA, K. SAXENA, A.M. BIRADAR, K.K. SAINI, and S. CHANDRA
Bistable Switching and Molecular Reorientation of Ferroelectric Liquid Crystals by Current Analysis
77.30 and 77.80; S12
(a) Vol.118,No.2, 561-566

M. ZĄBKOWSKA-WACŁAWEK and W. WOJCIECHOWSKI
Electrical and Magnetic Properties of Copper Phthalocyanine-Carbon Black Mixtures
72.80; 76.30; S12
(a) Vol.118,No.2, K87-K90

H. KRESSE and F. KREMER
Dielectric Relaxation of a Swallow-Tailed Compound at the Phase Transition from the Nematic into the Isotropic Phase
77.40; 64.70; S12
(a) Vol.118,No.2, K119-K122

M. MAĆKOWIAK, N. WEIDEN, and A. WEISS
Phase Transitions and Molecular Dynamics in Tris(Methylammonium) Nonabromodiantimonate(III) and Its ...
63.20 and 64.70; 76.60; S12
(a) Vol.119,No.1, 77-86

H. KRESSE, K. JAHN, and D. DEMUS
Investigation of the Temperature Dependence of Ferroelectricity in Smectic C Liquid Crystals
77.80; S12
(a) Vol.119,No.1, K83-K86

R.A. MORENO, C.X. CARDOSO, and M. YOSHIDA
Hole Injection in Teflon FEP Films
72.20 and 73.60; S12
(a) Vol.119,No.2, 583-588

S.K. GHOSHAL
Photoinduced Crystallization of Organic Molecular Solids from Vapour and Solution Phases
61.50; 64.60; S12
(a) Vol.120,No.1, 61-66

P. WUENSCHE
Influence of Kinetics of Thermally Stimulated Processes on Experimentally Determined Distribution of the Activation ...
65.90; S12
(a) Vol.120,No.1, 139-150

V. LHOTSKÁ, J. FOUSEK, and N. NEUMANN
Possible Origin of Dielectric Loss Behaviour in Samples of TGS-Type Crystals
77.40; 77.80; S12
(a) Vol.120,No.1, 273-284

R.D. GOULD and R.I.R. BLYTH
Conductivity and Absorption Measurement in Nickel Phthalocyanine Thin Films
73.60 and 78.65; S12
(a) Vol.120,No.1, K57-K62

P. KOZIOŁ and R. JAKUBAS
Effect of Hydrostatic Pressure on the Phase Transitions in $(CH_3NH_3)_3Sb_2Br_9$
77.70; S12
(a) Vol.120,No.1, K93-K94

M. STARKE, I. ANDROSCH, and C. HAMANN
A Solid State Electrochromic Cell Using Erbium-Diphthalocyanine
78.65; S12
(a) Vol.120,No.1, K95-K100

D.M. PETTY and M.R. WILLIS
The Effect of Current Injection on Photoconduction in Organic Semiconductors and Insulators
72.40; S12
(a) Vol.120,No.2, 593-600

M. DANZIGER and G. ANDRASSY
Experimental and Theoretical Study of Heavy Ion Irradiation Processes in Polymer Materials
61.80 and 68.60; 73.60; S12
(b) Vol.161,No.1, 223-232

A.P. TYUTNEV, S.G. BOEV, V.N. ABRAMOV, E.A. KUZMENKOV, and A.V. VANNIKOV
Track Effects in Radiation Induced Conductivity of Polymers
61.80 and 72.20; S12
(a) Vol.121,No.1, 205-212

M. BUIVYDAS, O. ADOMENIENE, and A. BEGANSKIENE
Model of Polarization Reversal Kinetics after an Excitation by an Asymmetric Rectangular Shaped Voltage Pulse
77.30; S12
(b) Vol.161,No.2, K113-K118

M. ZĄBKOWSKA-WACŁAWEK, P. TALIK, and W. WACLAWEK
On the Compensation Behaviour of Copper Phthalocyanine Films in NO_x Ambient
73.60; S12
(a) Vol.121,No.2, 489-494

T.G. ABDEL-MALIK, A.A. AHMAD, and A.S. RIAD
Dark and Photoelectric Investigation on Thin Films of beta-Cobalt Phthalocyanine Dispersed in Polymers
73.60; 72.40; S12
(a) Vol.121,No.2, 507-514

H. BOETTCHER, U. WEIH, J. OPFERMANN, and R. ACKERMANN
Electrical Properties of Pure and Mixed Evaporated Dye Layers
72.40 and 73.60; S12
(a) Vol.121,No.2, 563-570

C. BENOIT, M. ROLLAND, and M. PALPACUER
Infrared Properties of Polyacetylene. Experiments and Theory (Review Article)
78.30; S12
(b) Vol.162,No.1, 9-60

T. ASAHI, M. YUDASAKA, K. NAKANISHI, S. IWASHIMA, J. AOKI, N. SATO, and H. INOKUCHI
Rectification Properties of In/Tetrabenzo[de,hi,op,st]Pentacene/Au Devices
73.40 and 85; S12
(a) Vol.122,No.1, K97-K100

HAOLIANG LI, S. RENTSCH, JIPING YANG, M. LENZNER, and H. BERGNER
Picosecond Photoconductivity Studies on Phenyl-Substituted Poly(Phenylenevinylene)s Derivatives
72.40; S12
(b) Vol.162,No.2, 545-552

J. SERNA
Positronium Diffusion in Crystalline Polyethylene
61.40 and 61.70; 78.70; S12
(a) Vol.122,No.2, 489-494

A.G. LIPSON, V.A. KUZNETSOV, D.M. SAKOV, and YU.P. TOPOROV
Electrical and Magnetic Properties of the Polymer-Ceramics Produced on the Basis of Lithium Fluoride
73.40 and 75.50; S9.11; S12
(a) Vol.122,No.2, 563-568

S12.1. Quasi-one-dimensional systems

L.S. BRIZHIK, A.S. DAVYDOV, and V.N. ERMAKOV
The Magnetic Field Influence on a Superconducting Bisoliton Condensate
74.20; 71.45; S12.1
(b) Vol.157,No.1, 417-424

L.S. BRIZHIK
Bisoliton in Constant Magnetic Field
74.20; 71.45; S12.1
(b) Vol.157,No.2, 649-656

M.T. PRIMATAROWA and D.I. PUSHKAROV
On the Theory of Solitary Excitons in One-Dimensional Systems
71.35; S12.1
(b) Vol.159,No.1, K17-K22

R. TURCU
AC Conductivity of ClO_4^- Doped Polythiophene Films
73.60; S12.1
(a) Vol.119,No.2, K121-K126

A.S. DAVYDOV
Foundations of the Quasi-One-Dimensional Bisoliton Model of Superconductivity of Ceramic Oxides
71.45 and 74.20; S12.1
(b) Vol.160,No.1, 305-312

G. BOCHNACKA, V.A. OSIPOV, and V.K. FEDYANIN
Polaron States of the Linear Diatomic Polymer
71.38; S12.1
(b) Vol.160,No.1, K29-K32

S. SOEDERHOLM, J. HELLBERG, J. KRZYSTEK, G. OLOVSSON, I. OLOVSSON, and J.U. VON SCHUETZ
Electronic Properties and Crystal Structure of a Methylthio Substituted Naphthalene Cation Radical Salt: $(DMbTN)_3(ClO_4)_2$
61.65 and 72.20; 71.55; 75.30; 76.30; S12.1
(b) Vol.161,No.1, 245-256

A.A. KISELEV and A.V. LIAPTSEV
A Model Theory of Structural and Superconducting Phase Transitions in Chain-Like Systems
74.20; 64.60; S12.1
(b) Vol.161,No.2, 717-730

V.M. LOKTEV
On the Carrier Spectrum of CuO_2 Planes in High-T_c Superconductors and Its Dependence on Stoichiometry
74.20; 71.45; S12.1
(b) Vol.161,No.2, 731-744

V.N. KADANTSEV and L.N. LUPICHEV
The Capture of an Extra Electron by Supersonic Acoustic Soliton in a Molecular Chain in Presence of Thermal Vibrations
71.45 and 72.10; S12.1
(b) Vol.161,No.2, 769-776

L.N. GUMEN and O.V. USATENKO
Phonon Spectrum of the One-Dimensional Two-Component Quasiperiodic Crystal
63.20; S12.1
(b) Vol.162,No.2, 387-394

H. BOETTGER and G. KASNER
Multifractal, Localized, and Extended Wave Functions in Generalized Fibonacci Chains
71.10; 71.20; S12.1
(b) Vol.162,No.2, 489-496

Author Index

Ya.M. Abasov	(a) 117,K147	76.30; S8.16
A.K. Abass	(a) 120, 247	78.20; S8
M. Abbas	(b) 159, 597	61.80; S5.11
S. Abdalla	(a) 120, 151	73.60; S7
T.G. Abdel-Malik	(a) 121, 507	73.60; 72.40; S12
F.Kh. Abdullaev	(b) 158,K201	77.80
M. Abd-Lefdil	(a) 117,K103	73.60 and 78.65; 72.40; S8
	(a) 119, K35	72.40 and 73.40; S8
H.-B. Abel	(b) 161, 435	78.30; 61.70; S9.11
J. Abellán	(b) 157, 275	71.30 and 75.40; S11.2
T.Sh. Abesadze	(b) 161,K111	76.60
H.-P. Abicht	(a) 120, 285	77.60; 77.20; 77.30; S10.15
M.B. Ablanov	(b) 157, K55	61.70 and 76.80; S10.1
A.O. Aboyan	(a) 118, 11	61.10 and 61.70
	(a) 118, 21	61.10
	(a) 122, 43	61.10
I. Abram	(b) 159, 29	71.35 and 78.20; S7.12; S7.15
V.N. Abramov	(a) 121, 205	61.80 and 72.20; S12
T. Abritta	(b) 158, K75	78.50 and 78.55; S11.2
	(b) 162, K59	78.55; S11.2
C. Abromeit	(b) 157, 117	61.70
G.E. Abrosimova	(a) 117, K67	75.30; S1.2
R. Ackermann	(a) 121, 563	72.40 and 73.60; S12
R. Adam	(a) 119,K155	74.70; 68.55; S10.15
V.K. Adamchuk	(b) 159, K77	72.10
	(a) 122, 347	71.55; 73.40; S10.1
G.A. Adegboyega	(a) 118, 491	71.55; 72.20; S5.11
	(a) 121, 181	71.55; 72.80; 78.30; S5.11
O. Adomeniene	(b) 161,K113	77.30; S12
A.M. Afanasev	(a) 117, 341	61.10; 68.20
	(a) 122, 459	61.10
V.V. Afanasev	(a) 122, 347	71.55; 73.40; S10.1
N.P. Afanaseva	(b) 161, 91	61.70 and 61.80; 72.20; S5.11
Yu.N. Afanasiev	(a) 119, 417	61.40 and 61.80; S11.2
H. Agari	(a) 122, 221	61.40 and 62.40; S1.2
P. Agarwal	(a) 119, K97	61.40; S8
V.K. Agarwal	(a) 122, K77	76.30 and 76.80; S1.2
D. Agassi	(b) 160, 601	74.70; 73.40; S8.1
A.L. Ageev	(b) 160, 625	76.80
	(a) 122,K143	68.20; S1.2
P. Aghamkar	(b) 157, 735	78.20; S7.13
A. Aguilar	(a) 120,K169	74.70; S10.15
M.W. Ahktar	(a) 120, K45	73.40; S7.12
A.A. Ahmad	(a) 121, 507	73.60; 72.40; S12
M. Ahmad	(b) 159,K119	75.30; 67.50; S2
	(b) 160, K65	67.50 and 75.30

S. Ahmed	(a) 117, 163	64.70; 72.20; S11
Z.A. Ahmed	(a) 120, 247	78.20; S8
R. Ahuja	(b) 160, 549	71.25; S1.2
	(b) 162, 497	71.25; S1
E.C. Aifantis	(b) 157, 117	61.70
N.A. Akhmadeev	(a) 118, K27	72.15; S1.2
A.S. Akhmanov	(a) 121, 399	61.80; 64.75; S6
S.M. Akhmin	(b) 160,K149	76.30 and 76.70; 64.60; S9.16
	(b) 162,K107	76.70; S9.16
H. Akinaga	(a) 117, 125	62.40; 74.70; S10.15
O. Akinlade	(b) 161, 75	61.25 and 65.50; S2; S3; S4
H. Akiyama	(b) 159, 347	71.35 and 78.20; S9
A.F. Akkerman	(b) 158, 63	61.80; S5.11; S5.12
	(b) 158, 67	61.80; S5
I.Ch. Akopyan	(a) 119,K167	78.55; 61.80; S8.12
I.Kh. Akopyan	(a) 119, 363	78.30 and 78.40; 64.70; 71.25; 71.35; 78.55; S9
S.A. Akopyan	(b) 161, 55	61.40; 61.70; 71.50; S10.1
Z.Z. Akselrod	(b) 160, 255	71.70; 76.80; S1.1; S4
A.V. Akulov	(a) 122, 347	71.55; 73.40; S10.1
S. Alaya	(b) 162, 523	71.55; 78.55; S8.13
J.M. Albella	(a) 121,K181	68.55; S6
J. d'Albuquerque e Castro	(b) 158, 743	71.25 and 77.90; S5.1; S5.11
S.V. Alchangyan	(b) 158,K195	76.30; S11.1
G. Aldica	(a) 122, 637	61.60 and 74.70; S10.15
M.A. Aldzhanov	(b) 159,K107	75.40; 75.50; S8
I.L. Aleiner	(b) 159, 197	71.35 and 78.20; S7.12; S7.15
A.O. Aleksanyan	(b) 161, 629	66.30 and 68.45; 68.40; S8.12
P.L. Aleksanyan	(a) 121, K1	61.14; S5.11
H. Alexander	(a) 117, 417	61.70; 71.55; 72.20; S7.12
	(b) 158, K49	71.55 and 76.30; S5.11
A.R. Ali	(a) 117, 437	62.20; S1.2
M.M. Aliev	(a) 117, K57	61.80 and 76.30; S10.1
R.A. Aliev	(b) 158, 73	63.20 and 78.30; 64.75; S8.16
	(b) 162, 281	78.50; 71.55; S11
R. Almairac	(a) 120, K25	64.70; S9
R.F. Almukhametov	(a) 121, K5	61.60; 65.70; S1.63
S.D. Altekar	(b) 161, 463	61.50; 71.45; S2
E. Altshuler	(a) 120,K169	74.70; S10.15
N.S. Aly	(a) 121,K219	75.30; S1.1
	(a) 122,K113	61.55; S1.4
M.A. Alyanov	(a) 121, K51	65.40; 75.40; S11.2
A.A. Alybakov	(a) 120, K85	76.30; S11.1
V.A. Alyoshin	(a) 121, 157	68.55; 77.80; S10
B. Al-Araj	(b) 160, 143	79.60; S7.1; S11
A.M. Al-Saie	(a) 120, 169	61.80 and 68.55; 73.60; S1.3; S5.11

K.A. Al-Saleh	(a) 118, 467	61.80 and 73.40; S1.1; S1.2; S5.11
	(a) 120, 169	61.80 and 68.55; 73.60; S1.3; S5.11
A. Al-Shahrany	(a) 118, K79	72.80; S8.13
N. Amandzhanov	(a) 121, 579	71.55; 72.80; 76.80; S5
G. Amato	(a) 119, 169	72.40 and 73.60; 78.65; S5.11
L.K. Aminov	(b) 158, 595	71.70 and 76.30; S9.16
B. Amov	(a) 119, 107	73.60; 61.80; 78.65; S8.12
E. Anachkova	(b) 161, 575	63.20; 78.30; S10.15
P. Anagnostaras	(b) 162, 413	65.50; 71.20; S1
A. Anagnostopoulos	(b) 158, K43	72.20; S5.11
A. Anderson	(b) 162, 111	63.20; S5.2
	(b) 162, 421	63.20; S5.2
T.G. Andersson	(b) 159, 165	71.35 and 78.55; 73.40; S7.15
W. Andrae	(a) 119, 241	74.60 and 74.70; 61.70; S10.15
	(a) 122, K51	74.60 and 74.70; 61.70; S10.15
	(a) 122,K161	74.60; 74.70; S10.15
G. Andrassy	(b) 161, 223	61.80 and 68.60; 73.60; S12
J.P. André	(b) 157, 593	73.40; S7.12; S7.15
V.V. Andreev	(a) 118, 415	64.70; 61.40; 61.70; S4
	(a) 119, 437	61.80 and 64.75; S1.1; S4
	(a) 122, 111	66.30; S1; S4
	(a) 122, 187	64.70; 61.70
I. Androsch	(a) 120, K95	78.65; S12
R. Angelucci	(a) 121,K135	61.80 and 85; S5.11
A.A. Anikiev	(b) 161, 387	77.80; 65.40; S9.16
	(b) 161, 883	78.30; 77.80; S8.16
V.M. Anishchik	(b) 160, 173	71.25; 71.20; S1.2
H.J. Anklam	(b) 160, 491	68.55
T.V. Antimirova	(b) 157, 183	64.60 and 77.80; S10.15
	(b) 161, 843	76.30; 61.70; S11
V.N. Antonov	(b) 159, 771	71.20 and 71.25; S1.3
J. Aoki	(a) 122, K97	73.40 and 85; S12
H. Aourag	(b) 160, 193	71.25 and 71.60; 78.70; S7.12; S8.13
	(b) 160, 201	71.60; 78.70; S5.11; S5.12; S7.1; S8.1
	(b) 160, 561	71.25; 71.60; 78.70; S5.11; S5.12
	(b) 161, 685	71.20 and 71.60; S6
	(b) 162, K43	71.60; 78.70; S4
P.A. Apanasevich	(b) 161, 869	71.55 and 78.50; 71.38
M. Apostol	(a) 122, 637	61.60 and 74.70; S10.15
A. Apostolov	(a) 121, 45	61.50; 61.70; 66.30
H. Appel	(a) 121, 445	61.80; 64.80; 74.70; 76.80; S1.2

T. Arai	(a) 122, 535	62.40; 62.20; S1.2
S. Arajs	(a) 119, K75	72.15 and 75.50; S1.1
	(a) 120, K71	72.15; S4
Yu.G. Arapov	(b) 160,K141	71.30; S8.16
O.M. Arbotoev	(a) 120, K85	76.30; S11.1
J.S. Arellano	(b) 158,K161	74.10; S5.2
V.V. Aristov	(a) 120, K1	61.10 and 61.70; S5.11
V.E. Arkhincheev	(b) 161, 815	72.20
D.A. Aronov	(a) 118, 577	72.40 and 73.40; 85; S5.11
B.A. Aronzon	(b) 157, 17	71.30; 72.20; 75.40; S7.13; S8.15
E.Z. Arshakyan	(a) 122, 43	61.10
V.V. Artamonov	(a) 120, 475	61.80 and 68.55; 78.30; S5.11
N.Yu. Arutyunov	(a) 121,K163	61.70 and 78.70; S5.11
V.A. Arutyunov	(b) 161, 483	61.80 and 78.70; S5.11
T. Asahi	(a) 122, K97	73.40 and 85; S12
Sh. Asai	(a) 120, 199	74.70; S10.15
S. Asano	(a) 122, 163	61.70; 62.20; S5.11
R. Ashokan	(a) 121, 533	68.55 and 78.30; S10.1
B.M. Askerov	(b) 158,K179	72.10; 73.60
E.A. Aslanov	(b) 158, K85	78.30; 63.20; S8.16
V.M. Asnin	(b) 159, 471	75.20 and 78.55; 73.40; S7.12
J. Ataev	(a) 120, 519	68.55 and 78.55; S5.11
S.M. Atakishiev	(a) 117,K147	76.30; S8.16
H. Atmani	(a) 122, 73	61.50 and 68.55; S8
K.Z. Atsagortsyan	(b) 158, 213	61.80 and 71.10
P. Audibert	(a) 122,K101	61.10; S10.15
S. Auerbach	(b) 158, 407	61.50; 61.10; 61.12
G.R. Augst	(b) 160, 71	61.12 and 61.70; S9.11
H.-J. August	(a) 122, 695	78.70
S. Auluck	(b) 160, 549	71.25; S1.2
	(b) 162, 497	71.25; S1
M.I. Auslender	(b) 158, 307	72.20; S11.2
N.S. Averkiev	(b) 159, 471	75.20 and 78.55; 73.40; S7.12
	(a) 121,K129	78.55; S7.13
V. Averyanov	(a) 119, 107	73.60; 61.80; 78.65; S8.12
A.S. Avilov	(a) 121,K145	61.40 and 75.50; S1.2
I. Avramov	(a) 120, 133	62.10; 64.60
O.O. Awadelkarim	(a) 120, 539	71.55; 61.70; S5.11
D. Ayachour	(b) 158, 103	63.20 and 78.30; S9.16
R. Azoulay	(b) 159, 181	78.20; 68.55; S7.12; S7.15
Yu.A. Babanov	(b) 160, 625	76.80
V.M. Babich	(a) 117,K185	78.55; S5.11
N.C. Bacalis	(b) 157, 293	71.25; 71.10; S5.2
G. Bacquet	(a) 119,K181	78.20; S7.12

W.A. Badawy	(a) 118, 197	68.55 and 78.65; 73.60; S10.1
S. Bader	(a) 119, K9	66.30 and 68.48; S1.2
K.-H. Baether	(a) 119, 503	68.55 and 73.40; S1.61; S4
A.A. Baganich	(a) 117, 241	71.55; 72.40; 73.60; S5
V.E. Bagiev	(b) 162, 281	78.50; 71.55; S11
D. Bahadur	(b) 161, 357	76.80; 75.50; S10.15
A.Y. Bahnev	(a) 122, 569	72.40 and 73.40; S7.16
Guiru Bai	(b) 158,K185	71.70 and 76.30; S11
A. Baidullaeva	(a) 122, 243	73.20 and 73.40; 72.40; S8.15
V. Baier	(a) 118, K69	66.70
D. Baither	(a) 119, 209	68.55; 61.70; S9.12
T. Bąk	(a) 119, 221	72.15; S1.63
A.S. Bakai	(b) 158, 81	64.60
M.K. Bakhadirkhanov	(a) 121, 571	61.70 and 71.55; 76.80; S8.15
N.A. Bakhyshov	(b) 158, 73	63.20 and 78.30; 64.75; S8.16
H. Bakker	(a) 117, K99	64.70; S1.2
I.S. Bakshee	(a) 117, K37	72.70; 61.70; S8.15
J. Bąk-Misiuk	(a) 118, 209	65.70 and 68.60; S7.15
	(b) 158,K111	65.70; 72.20; S7.12
L.L. Balashova	(b) 161, 113	61.80; 78.70; 79.90; S5.11
	(b) 161, 513	61.80; S5.11
C. Balasubramanian	(a) 117, 495	68.55 and 73.60; 77.55; S9
	(a) 121, 515	73.60; 77.55; S10
T. Balcerzak	(b) 157,K103	72.10 and 75.10
	(b) 160, K61	75.10; 75.30
	(b) 160,K129	72.10 and 75.10
	(b) 161, 335	75.10
F. Balestra	(a) 119, 343	73.40 and 85
B. Balland	(a) 120, 507	68.55 and 73.60; 78.65; S10.1
I. Balslev	(b) 159, 339	71.35 and 78.20; S8.12; S9
D. Baltrūnas	(a) 119, 71	64.70 and 76.80; S8.16
R. von Baltz	(b) 158, 249	71.25; 78.20; S11.1
A.V. Baluev	(a) 118,K139	74.70; 78.30; 79.20; 79.60
J. Bandet	(a) 119,K181	78.20; S7.12
A.K. Bandyopadhyay	(b) 157, 519	61.70; S1.2; S1.3
U. Bandyopadhyay	(a) 118, 365	61.25; S9.12
	(b) 160, 403	61.40; 62.20; 66.30; S2
D. Banerjee/Chakravorty	(b) 157, 443	75.20; S4
J. Bánkuti	(a) 117, 265	74.50; 74.70
	(a) 117, 273	74.20 and 74.50
L. Banyai	(b) 159, 309	71.35 and 78.20; S7.12; S8.11; S9
C.L. Bao	(a) 122, 481	61.50; 79.60; S1.2
J.A. Baquedano	(a) 118, 497	71.55; 72.40; 78.50; S10
R. Baquero	(b) 160, 127	71.25 and 73.20; S8.13
J.J. Bara	(a) 118, 41	61.60; 64.70; 76.80; S10
M. Baran	(b) 158, K63	75.30; 75.60; S11.2
	(a) 118, 307	75.50; 76.50; S1.1
	(a) 118, 525	75.50; S1.1; S1.2

A.M. Baranov	(a) 122,K139	68.55 and 78.65; S5
P.G. Baranov	(b) 159, K33	61.70; S9.11
P.I. Baranskii	(b) 160, 291	72.20; S8.15
V.I. Barbashov	(a) 121, K27	61.70; S7.13
A.F. Barbosa	(b) 161, 309	72.15; S1.4
P.I. Bardetskii	(b) 159, 755	71.35 and 71.36
S.N. Barilo	(a) 119,K151	61.80 and 74.70; S10.15
V.A. Barinov	(a) 117, 291	61.55 and 75.50; 76.80; S1.1
B.A.S. de Barros, Jr.	(a) 118, 473	61.80 and 68.20; 76.80; S1.1
C. Bárta	(b) 157, K69	78.55; S9
	(a) 118,K133	78.40 and 78.55; S11.1
C. Barthou	(a) 122, 427	78.60 and 85; S8.11
J. Bartoš	(a) 122, 607	61.70 and 71.55; 73.40; S5.11
	(a) 122, 715	73.40 and 85; S7.12
H. Bartsch	(a) 117, 477	61.80 and 68.55; S1.61
	(a) 122, K35	68.48 and 68.55; S5.11
V.G. Baryshevskii	(b) 157, 129	61.80; 78.70; S5.12
	(a) 122, 69	61.10 and 78.70; 68.20
M.E. Bassiouni	(a) 121,K219	75.30; S1.1
C. Basu	(b) 160, 583	71.70; S9.16; S11
S.V. Batashova	(a) 120, 327	61.10 and 68.55; S7.12; S7.15
E. Bauer	(a) 117, 501	71.20; 72.20; 75.30; S1.2; S1.4
M. Bauer	(a) 122, 79	61.40 and 61.80; 64.70; S1.2; S4
S.E. Bauer	(b) 158, 503	78.65; 71.35; S9
G. Baumann	(b) 158, 573	71.10
I.J.R. Baumvol	(a) 118, 473	61.80 and 68.20; 76.80; S1.1
S. Baunack	(a) 118, 203	61.80 and 68.60; 79.20; S9.12
S.S. Bawa	(a) 118, 561	77.30 and 77.80; S12
N.G. Bebenin	(b) 158, 307	72.20; S11.2
R. Bechara Muniz	(b) 158, 743	71.25 and 77.90; S5.1; S5.11
F. Bechstedt	(b) 157, 567	68.20 and 73.20; S5; S5.11; S5.12
	(b) 159, 143	71.35 and 73.40; S7.12; S7.15
P. Becker	(a) 118, 379	61.60 and 61.70; S5.11
U. Becker	(a) 120, 653	64.75 and 78.30; S8.13
A. Beckmann	(b) 159, 765	71.35; 78.20; S7.11
M. Beer	(a) 121, 95	61.70 and 78.60; 85; S7.12
A. Beganskiene	(b) 161,K113	77.30; S12
G. Behr	(a) 118, K91	72.80; 61.50; S10.1
D.L. Beke	(b) 161, K5	66.30; S1.2; S2
A. Bel Hadj	(a) 117, 169	61.70 and 78.65; 73.60; S7.12
A. Belaidi	(b) 160, 193	71.25 and 71.60; 78.70; S7.12; S8.13
	(b) 160, 201	71.60; 78.70; S5.11; S5.12; S7.1; S8.1

A. Belaidi	(b) 160, 561	71.25; 71.60; 78.70; S5.11; S5.12
	(b) 161, 685	71.20 and 71.60; S6
	(b) 162, K43	71.60; 78.70; S4
B.D. Belan	(b) 161, 783	72.10; 64.75; 71.25; S1.4
Z. Belarbi	(b) 160, 201	71.60; 78.70; S5.11; S5.12; S7.1; S8.1
C. Bellabarba	(b) 162, 517	71.25; 78.20; S8
A. Bello	(a) 121,K235	77.30; S9.12
I.V. Beloussov	(b) 159, 91	71.36 and 78.20
Ye.V. Belozerov	(a) 121,K111	75.50; S1
A.D. Belyaev	(a) 118, 327	78.60; 62.80; 73.40; S8.11
A.E. Belyaev	(b) 160, 291	72.20; S8.15
V.K. Belyaev	(b) 158, K63	75.30; 75.60; S11.2
S. Beňačka	(a) 119,K155	74.70; 68.55; S10.15
P. Benalloul	(a) 122, 427	78.60 and 85; S8.11
S. Benci	(a) 120, 635	78.55 and 78.60; S9.12
G. Benedetto	(a) 119, 169	72.40 and 73.60; 78.65; S5.11
R. Bennaceur	(a) 117, 169	61.70 and 78.65; 73.60; S7.12
C. Benoit	(b) 162, 9	78.30; S12
J. Benoit	(a) 122, 427	78.60 and 85; S8.11
W. Benoit	(a) 122, 153	61.70 and 62.40; S1
I. Benzar	(a) 120, K73	74.70; S10.15
S. Chandra Bera	(b) 157, 723	78.50 and 78.55; 71.55; S11.1
L.I. Berezhinskii	(a) 121,K193	68.55; 73.60; 78.55; S6
H. Berger	(b) 161, 897	78.50; S8.13
H. Bergner	(b) 158, K37	72.40; S8
	(b) 162, 545	72.40; S12
T. Bergunde	(a) 121, K55	68.48 and 73.60; S8.13
S.I. Beril	(b) 158, 165	71.50 and 73.40
	(b) 161, 603	71.35 and 78.65; S8.13
G.I. Bersuker	(b) 162,K123	78.35
V.N. Berzhanskii	(b) 158, 643	72.20; 75.40; S11.2
J. Betz	(a) 122,K161	74.60; 74.70; S10.15
J. Beynon	(a) 121, 163	73.40 and 73.60; S10
	(a) 122,K169	72.80; S10.1
P.A. Bezirganyan	(a) 118, 11	61.10 and 61.70
	(a) 118, 21	61.10
S.E. Bezirganyan	(a) 118, 11	61.10 and 61.70
A. Bhalla	(a) 122, 677	77.30 and 77.90; S10.15
R.K. Bhan	(a) 117, 203	61.70 and 68.48; S5.11; S12
D. Bhandari	(b) 160, 83	63.50; S2
G.C. Bhar	(a) 119,K173	78.20; S11.1
S.N. Bhatia	(b) 159, 551	61.40; 78.70; S1.1; S1.2
R. Bhattacharya	(b) 157, 443	75.20; S4
	(b) 159, 851	75.20; 71.25; 72.20; S8
S. Bhowmik	(b) 161, 329	71.25 and 72.20; S8.15
L.B. Bhuiyan	(b) 157, 93	61.25; 61.70; S2; S4

L.B. Bhuiyan	(b) 161, 473	61.20; S2; S5.2
A.G. Bibiloni	(a) 120, 105	61.70; 61.80; S10.1
D. Bicanic	(a) 121,K231	65.90 and 77.70; S1.1; S1.2
K. Bickmann	(a) 122, 139	61.70; 61.60; S7.12
K. Biedrzycki	(a) 117, 313	77.80 and 79.40; S12
P. de Bièvre	(a) 122, 139	61.70; 61.60; S7.12
O.S. Binbrek	(b) 162, 421	63.20; S5.2
R. Bindemann	(b) 160,K183	78.65; S5.11
K.G. Binnatov	(a) 120, 305	61.55; 76.80; S1.1; S1.2
A.M. Biradar	(a) 118, 561	77.30 and 77.80; S12
J.E. Bisbee	(a) 119, 545	61.70 and 71.55; S7.12
J. Bischof	(a) 119, K67	75.50; S1.4
S. Biswas	(b) 159, 851	75.20; 71.25; 72.20; S8
S.N. Biswas	(a) 119, 63	62.20; S1.3
J. Blaesing	(b) 157,K109	74.40 and 74.50
F.H. Blak	(b) 158, K75	78.50 and 78.55; S11.2
M. Błaszak	(b) 159,K103	75.40
M.G. Blazha	(a) 119, 607	78.55; 63.20; 71.55; 78.30; S11.2
G.L. Bleris	(b) 158, K43	72.20; S5.11
K. Bluethner	(a) 121, K73	74.50; S4
K. Blum	(b) 158, 551	79.20; S4
V.I. Blynskii	(a) 121, 227	72.40; 78.40; S5.11
R.I.R. Blyth	(a) 120, K57	73.60 and 78.65; S12
M. Bobeth	(a) 117, 191	68.55
A.I. Bobrysheva	(a) 117, 587	78.55; S8
	(b) 159, 155	71.35 and 78.20; 73.40; S8.12; S8.15
G. Bochnacka	(b) 160, K29	71.38; S12.1
O.I. Bodak	(b) 161, 783	72.10; 64.75; 71.25; S1.4
I.T. Bodnar	(a) 121,K247	78.20; S8.16
I.V. Bodnar	(a) 118, 141	63.20; S8.16
	(b) 158, 469	63.20; 64.75; 78.30; S8.16
	(a) 121,K247	78.20; S8.16
U.-C. Boehnke	(a) 120, 557	74.70; 61.60; 75.30; S10.15
F.R. de Boer	(a) 121, 241	75.30; 75.60; S1.1
	(a) 121, 617	75.50; 75.30; S1.1; S1.2
H. Boerner	(a) 120, 557	74.70; 61.60; 75.30; S10.15
E. Boesman	(b) 161, K1	61.70 and 76.30; S9.11
H. Boettcher	(a) 117,K165	73.40 and 78.65; S9.15
	(a) 121, 563	72.40 and 73.60; S12
H. Boettger	(b) 158, 143	63.50 and 71.20
	(b) 162, 489	71.10; 71.20; S12.1
H. Boettner	(a) 119, 711	73.40 and 78.55; 72.40; S7.16
S.G. Boev	(a) 121, 205	61.80 and 72.20; S12
B.F. Bogacz	(a) 118, 41	61.60; 64.70; 76.80; S10
E.V. Bogdanov	(b) 160, K33	71.25 and 72.20; S4
V.N. Bogdanov	(a) 119, 621	61.40 and 78.20; 61.80; 62.65; S10.1

V.V. Bogdanov	(b) 158,K105	61.80 and 76.80; S1.3
V.A. Bogdanova	(a) 120,K121	61.14; S7.1
A.K. Bogush	(a) 118,K111	75.50; S11.1
A.A. Boguslavskii	(b) 161, K49	76.60 and 77.80; S9.16
V. Bohac	(a) 120, 371	66.30; S1.1; S1.2
I.I. Boiko	(b) 159, 805	72.10 and 73.40
E.I. Boikova	(a) 117, 259	72.40 and 79.60; S11.1
B. Bojanowski	(a) 121, K95	76.30; 75.50; S11.1
Yu.B. Bolkhovityanov	(a) 119, 139	68.20 and 68.55; S7.12
I.E. Bondarenko	(a) 122, 121	61.70; 78.70; S5.11
J.P. Bonnet	(a) 122, 371	72.20; S11.2
D. Bonnin	(a) 121, 359	61.10 and 61.55; S10.15
N.I. Borgardt	(a) 121, 369	61.14 and 61.70
G. Borghs	(b) 159, 173	71.35 and 78.20; 73.40; S7.12; S7.15
S.I. Borisenko	(a) 118, 505	72.20 and 78.20; 71.55; S7.16
S.Yu. Borisov	(a) 121,K181	68.55; S6
A. Bosacchi	(a) 118, 567	78.55 and 78.65; S7.12
F.W. Boswell	(a) 117, 351	61.55 and 68.55; S1.3
M. Boudyš	(a) 122, 341	71.55; 76.30; S10.15
N. Boukos	(b) 157, 351	72.15; S1.2
D. Boulanger	(b) 157, 323	71.70; S8.11
C. Boulesteix	(a) 122,K101	61.10; S10.15
B. Boulil	(b) 158, 629	71.50
J.-C. Boulou	(b) 158, 103	63.20 and 78.30; S9.16
K.K. Bourdelle	(a) 121, 399	61.80; 64.75; S6
R. Boyn	(a) 118, 225	71.55; 78.55; S8.13
	(a) 120, 653	64.75 and 78.30; S8.13
R. Bramley	(a) 117,K141	75.50 and 76.30; S1.1; S1.2
	(a) 122, 663	75.50 and 76.30; 75.40; S1.1; S1.2
K.P. Bransalov	(a) 121, 555	62.30 and 68.20; 78.20
K. Brańska	(a) 118, 519	75.60 and 75.70; S1.2
M. Brauer	(a) 122, 741	73.40 and 85; S5.11
R. Braunstein	(a) 119, 145	71.45 and 78.65; S8
	(a) 119, 155	71.45 and 78.65; S10
E.A. Brener	(a) 120, K7	61.70 and 66.30
J.L. Bretonnet	(b) 157, 61	61.25; 65.50; S2
N.V. Brilliantov	(a) 121,K155	61.80; 71.55; 85; S5.11
P.D. Bristowe	(b) 161, 501	61.70; 61.10; S1.3
L.H. Brixner	(a) 117, K81	78.60; S9
L.S. Brizhik	(b) 157, 417	74.20; 71.45; S12.1
	(b) 157, 649	74.20; 71.45; S12.1
E. Broese	(b) 162, K53	63.20 and 78.30; S5
M.H. Brooker	(b) 157, 167	64.60 and 78.30; 63.20; 63.50; S11.1
I. Broser	(a) 118, 343	78.55; 71.35; S8.11
	(b) 159, 205	78.20; S8.11

I. Broser	(b) 159, 363	71.55 and 78.20; 71.36; S8.11
	(b) 159, 431	78.55; 71.45; S8.11
	(b) 159, 437	71.36 and 78.20; S8.11
A.D. Brothers	(b) 158, 503	78.65; 71.35; S9
I.M. Brovchenko	(a) 119, 621	61.40 and 78.20; 61.80; 62.65; S10.1
P.D. Brown	(a) 118, 117	61.70; 61.80; S5.11
G. Bruchman	(b) 161, 575	63.20; 78.30; S10.15
V.N. Brudnyi	(a) 118, 219	71.55; 61.80; S7.12
	(a) 118, 505	72.20 and 78.20; 71.55; S7.16
V. Brueckner	(b) 158, K37	72.40; S8
R. Brun del Re	(a) 121, 483	64.75; 61.55; S1.1
B. Brunner	(a) 122,K161	74.60; 74.70; S10.15
V.V. Bryksin	(b) 161, 777	72.20; S10
H.A. Buckmaster	(b) 160, 349	76.30; 71.70; S11
A. Budkowski	(a) 117, 351	61.55 and 68.55; S1.3
A.G. Budrevich	(a) 119, K27	68.55; S4
T. Bueschel	(a) 122, 645	74.70; 74.50
M. Buivydas	(b) 161,K113	77.30; S12
A. Bukaluk	(a) 118, 99	66.30 and 68.48; 61.80; S1.2; S1.3
Z. Bukowski	(b) 158, K33	74.70; S10.15
J.C. Bureau	(a) 120, 507	68.55 and 73.60; 78.65; S10.1
J. Burianek	(a) 119, K67	75.50; S1.4
Y. Burki	(b) 161, 897	78.50; S8.13
V.M. Burlakov	(b) 161, 387	77.80; 65.40; S9.16
	(b) 161, 883	78.30; 77.80; S8.16
L.C. Burton	(a) 122, 361	72.20; 72.40; S5.11
P. Busa	(a) 121,K223	78.20; S5.11
A.V. Butashin	(a) 118, K59	78.20; 78.30; 78.45; 78.55; S10
	(a) 119, 285	78.30; S11.2
	(a) 119, 607	78.55; 63.20; 71.55; 78.30; S11.2
	(a) 120, 253	78.55; 71.55; 78.50; S10
L.V. Butov	(b) 159, 165	71.35 and 78.55; 73.40; S7.15
M.Z. Butt	(a) 120,K149	62.20; S1.2
A.P. Buturlakin	(a) 118, K9	61.80 and 78.30; S7.11
I.P. Bykov	(a) 122, 341	71.55; 76.30; S10.15
J.M. Cabrera	(a) 118, 497	71.55; 72.40; 78.50; S10
R. Cach	(a) 121, 299	77.80; 77.20; S11
M. Cadene	(a) 117,K103	73.60 and 78.65; 72.40; S8
	(a) 119, K35	72.40 and 73.40; S8
Gangang Cai	(a) 121, 57	61.80; 78.60; S9.11
M. Calamiotou	(b) 162, 243	72.20 and 74.70; 76.30; S10.15

M. Calamiotou	(b) 162,K101	76.30; 74.70; S10.15
A. Calka	(a) 117,K141	75.50 and 76.30; S1.1; S1.2
	(a) 122, 663	75.50 and 76.30; 75.40; S1.1; S1.2
F. Callens	(b) 161, K1	61.70 and 76.30; S9.11
A. Camacho	(b) 160, 127	71.25 and 73.20; S8.13
R.M. Cândea	(a) 121,K231	65.90 and 77.70; S1.1; S1.2
L.Z. Cao	(a) 119, K59	74.70; S10.15
	(a) 122, K55	74.60 and 74.70; S10.15
R. Capelletti	(a) 120, 97	66.30; 78.30; S10.15
C.X. Cardoso	(a) 119, 583	72.20 and 73.60; S12
C.J. Carlile	(a) 118, 425	63.90; 64.70; 77.80; S11.1
M. do Carmo	(a) 117,K155	75.50 and 76.50; S12
S. Casette	(a) 122, 427	78.60 and 85; S8.11
J.E. Castle	(a) 121, K61	68.55; S10.1
A. Caticha	(a) 119, 47	61.10; S1.2
	(a) 119, 643	61.10 and 78.70
S. Caticha-Ellis	(a) 119, 47	61.10; S1.2
	(a) 119, 643	61.10 and 78.70
J. Čermák	(a) 117, 387	66.30; S1.2
	(a) 119, 443	66.30; S1.1; S1.2
	(a) 120, 351	66.30; 61.70; S1.1; S1.2
	(a) 121, K31	66.30; S1.2
R. Černý	(a) 119, K67	75.50; S1.4
S. Chabik	(a) 120,K125	64.70 and 78.70; S1.2
	(a) 120,K145	61.70 ; 78.70; S4
G.S. Chaddha	(b) 161, 837	75.30
P. Chakrabarti	(a) 122, 405	73.40 and 85; S7.12; S7.16
D.K. Chakrabarty	(a) 121, 265	75.50; 76.80; S11.2
R. Chakravarthy	(b) 157, 547	64.60; 61.12; S11
	(a) 118, 425	63.90; 64.70; 77.80; S11.1
Kim Gen Chan	(a) 122, K45	74.70; S10.15
B.P. Chandra	(a) 121, 281	78.60; S8.11
S. Chandra	(a) 118, 561	77.30 and 77.80; S12
A. Chandrashekara	(a) 117, 53	61.40; 78.70; S1.1; S1.2
	(a) 120, 411	64.70; 61.40; 78.70; S11
S.K. Chang	(b) 158,K115	73.40; S8.16
	(b) 162, K9	64.75; 78.30; S8.16
	(a) 122,K135	68.55; 78.40; 78.55; S8.13
A.M. Chaplanov	(a) 120, K37	68.55; S1; S1.61
T. Chassé	(b) 160, 143	79.60; S7.1; S11
R. Chatterjee	(b) 161, 191	63.20
S. Chatterjee	(b) 160, 473	71.38 and 73.60; S1.2
S. Chaudhuri	(a) 119, K21	73.60 and 78.65; S10.1
J. Chaudhury	(b) 157, 723	78.50 and 78.55; 71.55; S11.1
N.G. Chechenin	(a) 121, 399	61.80; 64.75; S6
D.S. Chemla	(b) 159, 11	71.35 and 78.20; 73.40; S7.12
	(b) 159, 173	71.35 and 78.20; 73.40; S7.12; S7.15

Guanghua Chen	(a) 118, K17	66.30; S5.11
	(a) 118, K39	61.70 and 76.30; S5.11
	(a) 118,K115	68.60 and 76.30; S6
	(a) 120, K49	68.60; S5.11
	(a) 121, 501	61.40 and 76.30; 68.55; S5.11
	(a) 121,K177	73.60; 68.55; S5.11; S5.12
H. Chen	(a) 117, 221	68.55; S4
J. Chen	(a) 120, 617	75.60; S1.1
Kun-Ji Chen	(a) 118, K51	78.30; S5.11
Lin Chen	(b) 158, K27	74.60; 74.70; S10.15
	(a) 119,K141	74.70; S10.15
Ling Chen	(b) 157, 411	74.40; 78.70; S10.15
L.F. Chen	(a) 120, 429	62.20; 61.70; S1.2
Wuming Chen	(a) 117,K127	74.70; S10.15
Xi Chen	(a) 117, 335	61.60 and 74.70; S10.1
Yang-Fang Chen	(b) 160,K103	73.60; 72.40; S5.11
Y.F. Chen	(b) 157, 101	61.40; S5.11
Y.Y. Chen	(a) 120, K71	72.15; S4
	(a) 121, 213	72.15; 75.60; S1.2
Z. Chen	(a) 122, 361	72.20; 72.40; S5.11
Guang-Xu Cheng	(a) 118, K51	78.30; S5.11
X.A. Cheng	(a) 119, K59	74.70; S10.15
	(a) 122,K151	74.70; S10.15
Z.J. Cheng	(a) 119, K59	74.70; S10.15
Z.Y. Cheng	(a) 119, K59	74.70; S10.15
Yu.L. Chepelev	(a) 117, 259	72.40 and 79.60; S11.1
K. Cherifi	(a) 122, 311	68.55; S1.1; S1.4
N.S. Chernaya	(a) 118, 327	78.60; 62.80; 73.40; S8.11
B.P. Chernyago	(b) 160, K75	61.70 and 78.30; S9.11
A.P. Chernyakova	(b) 158, 469	63.20; 64.75; 78.30; S8.16
A. Chevy	(a) 118, 553	78.50; 78.30; S8
K.C. Chhabra	(a) 117, 203	61.70 and 68.48; S5.11; S12
	(a) 121, 533	68.55 and 78.30; S10.1
M.C. Chien	(a) 119, 35	61.40 and 64.70; 62.20; S1.1; S1.2
N. Chikov	(a) 120,K185	76.80 and 85; S5.11
M. Chirtoc	(a) 121,K231	65.90 and 77.70; S1.1; S1.2
O.D. Chistyakov	(a) 117, 299	75.50; S1.1; S1.2
L.M. Chistyakova	(b) 159, 197	71.35 and 78.20; S7.12; S7.15
D.K. Choi	(b) 157, K13	63.10; 68.20; S7.12
K.S. Choi	(a) 122,K135	68.55; 78.40; 78.55; S8.13
J. Chojcan	(b) 159, K59	71.20 and 76.80; S1.1
R.M. Chorbadjian	(b) 158,K191	75.30
T.L. Choy	(b) 161,K107	76.30; S10.1
N.E. Christensen	(b) 159, 617	61.70 and 71.55; S7.1
C. Christodoulides	(a) 118, 333	78.60
V. Christoph	(b) 159,K111	75.60
	(b) 161, K41	75.60

S. Chromik	(a) 119,K155	74.70; 68.55; S10.15
Da-ping Chu	(b) 157, 685	75.30; S11.2
T.K. Chu	(b) 160, 601	74.70; 73.40; S8.1
T.H. Chuang	(a) 117, K47	72.15; S1.2
	(a) 121, 213	72.15; 75.60; S1.2
Y.C. Chuang	(a) 120, 627	75.50; 64.80; S1.1
	(a) 121, 241	75.30; 75.60; S1.1
	(a) 121, 257	75.60; S1.1
	(a) 121, 611	75.50; 75.30; S1.2; S1.4
	(a) 121, 617	75.50; 75.30; S1.1; S1.2
	(a) 122, K69	75.50; S1.1
V.G. Chudinov	(a) 118, 415	64.70; 61.40; 61.70; S4
	(a) 119, 437	61.80 and 64.75; S1.1; S4
	(a) 122, 111	66.30; S1; S4
	(a) 122, 187	64.70; 61.70
C.H. Chung	(b) 158,K115	73.40; S8.16
I.E. Chupis	(b) 157, K65	75.30
Ya.G. Chushak	(b) 157, 557	63.20; S2
M. Chybicki	(a) 120,K159	73.25; S10.15
J. von Cieminski	(a) 120, 285	77.60; 77.20; 77.30; S10.15
P. Ciepielewski	(a) 119, 495	68.48 and 68.55; S7.12; S7.15
Z. Cimpl	(a) 117, 467	68.55; 68.60
J. Cisowski	(a) 120, 525	71.25; 72.20; S1.62
M. Ciszek	(a) 117, K43	74.90 and 79.75; S10.15
W. Ciurzyńska	(a) 118, 525	75.50; S1.1; S1.2
P. Clauws	(a) 122, 129	61.80; 78.70; S5.11
A. Climent	(a) 121,K181	68.55; S6
F. Comas	(b) 159, 731	72.10 and 73.40; S7.12
	(b) 160, 105	72.10 and 73.40; S7.12; S7.15
	(b) 160, 117	71.38 and 73.40; 63.20; S7.12
A. Cornet	(a) 120,K105	78.55; S10.1
B.V. Costa	(b) 160,K165	75.10; 75.25
M.D. de Coteau	(a) 117, 403	61.70; S5.11
R.M.J. Cotterill	(b) 162, K1	61.40
	(a) 122, 111	66.30; S1; S4
	(a) 122, 187	64.70; 61.70
J.J. Couderc	(a) 122, 545	62.20 and 78.20; S8.11
L.E. Cross	(b) 159, 861	68.30 and 75.10; 75.70
E. Cruceanu	(a) 122, 637	61.60 and 74.70; S10.15
D. de la Cruz	(b) 157, 305	71.55; 78.55; S7.12
J. Cunningham	(b) 159, 173	71.35 and 78.20; 73.40; S7.12; S7.15
F. Cussó	(b) 159, K37	61.70; S9.11
M. Cyamukungu	(b) 158,K197	76.60; S1
R. Cywiński	(b) 157, 137	61.70; 78.50; S9.11
A. Czachor	(b) 157, K1	61.40 and 61.50
W. Czaja	(b) 161, 897	78.50; S8.13
Z. Czapla	(a) 121, 299	77.80; 77.20; S11

Z.Czapla	(a) 121,K227	77.80; S11.1
J. Czaplicki	(a) 117, 555	76.60; S9.16
P. Czarnecki	(a) 117,K137	72.15; S1.2
G. D'Anna	(a) 122, 153	61.70 and 62.40; S1
D. Dădârlat	(a) 121,K231	65.90 and 77.70; S1.1; S1.2
A. Dafinei	(a) 118, 131	64.70; 65.40; S11.1
Y. Dai	(a) 118, K21	62.20; 62.40; S1.2
	(a) 118, 431	62.20 and 62.40; S1.2
P. Danesh	(a) 122, 599	68.55 and 78.65; S5.11
M. Danziger	(b) 161, 223	61.80 and 68.60; 73.60; S12
D.M. Daraselia	(b) 158,K195	76.30; S11.1
A. Dargys	(b) 162, 183	71.55 and 79.70; S5.11
G.T. Daribaeva	(b) 159, K97	76.80; 74.70; S10.15
S.A. Darmanyan	(b) 158,K201	77.80
S. Das	(a) 119,K173	78.20; S11.1
T. Das	(b) 157, 93	61.25; 61.70; S2; S4
	(b) 161, 473	61.20; S2; S5.2
B.A. Dasannacharya	(b) 157, 547	64.60; 61.12; S11
	(a) 118, 425	63.90; 64.70; 77.80; S11.1
L.I. Datsenko	(a) 121, 11	61.10 and 61.70; S5.11
P.K. Datta	(a) 119,K173	78.20; S11.1
B.K. Dauletmuratov	(a) 122, 243	73.20 and 73.40; 72.40; S8.15
D.E. Davies	(a) 120, K45	73.40; S7.12
A.S. Davydov	(b) 157, 417	74.20; 71.45; S12.1
	(b) 160, 305	71.45 and 74.20; S12.1
S.K. De	(b) 160, 473	71.38 and 73.60; S1.2
R. Debnath	(b) 157, 723	78.50 and 78.55; 71.55; S11.1
E. Dębówska	(b) 161, 281	71.25; 78.70; S1.2
S. Defregger	(b) 162, 311	78.20; 64.60; S11.1
V.T. Degtyarev	(a) 120, 397	61.70; S1.2; S9.11
S.V. Demishev	(b) 161, 705	71.55 and 78.55; 78.60; S8
P. Demkin	(b) 159, K63	71.55; S7.11
D. Demus	(a) 119, K83	77.80; S12
M.A. Demyanenko	(a) 117, 601	73.40 and 85; S5.11
S.E. Demyanov	(b) 160,K133	61.70 and 72.15; S4
Hui-Fang Deng	(b) 157, K7	66.30
A. Derouiche	(b) 157, 61	61.25; 65.50; S2
A.P. Deshpande	(b) 158, 737	78.70; S10
K. Dettmer	(b) 159, 765	71.35; 78.20; S7.11
Yu.N. Devyatko	(a) 117, 61	61.50; 75.30; S1.2
H. Dexpert	(a) 121, 359	61.10 and 61.55; S10.15
M. Dey (Bose)	(b) 157, 159	62.25 and 72.50; S8.11
	(b) 159, 817	72.10 and 72.50; 71.25; S7.13
	(b) 161, 319	72.20 and 72.50; 77.60; S8.11
L.M. Di	(a) 117, K99	64.70; S1.2

E.M. Dianov	(b) 160, 263	61.70 and 71.55; S10.1
	(b) 161, 55	61.40; 61.70; 71.50; S10.1
	(b) 162, 173	71.55; S10.1
M. Díaz	(a) 121,K235	77.30; S9.12
B.G. Dick	(a) 117,K115	71.55; S9.11
S. Dietrich	(a) 117, 417	61.70; 71.55; 72.20; S7.12
A.V. Dikarev	(b) 158, 715	78.20
U. Dillner	(a) 121, K77	72.20
	(b) 162, 147	73.60; S4
G. Dimopoulos	(a) 118, K73	73.40; S5.11
Chao Ding	(a) 117, 335	61.60 and 74.70; S10.1
Z.-J. Ding	(b) 161, 257	68.20 and 79.20; S1.3
P.H. Dionisio	(a) 118, 473	61.80 and 68.20; 76.80; S1.1
Th. Dittrich	(a) 119, K91	78.20; S7.12
	(a) 119, 479	73.60 and 78.65; S7.12
	(a) 121, K15	61.80; S7.11
	(a) 121, 547	61.80 and 78.65; S7.12
M. Diviš	(a) 119, K67	75.50; S1.4
S. Dixit	(b) 161, 583	62.20 and 64.30; 65.70; S9.11
G. Dlubek	(b) 161, K81	71.25 and 78.70; S4
A.I. Dmitriev	(b) 162, 213	72.20; 71.25; S8
N.L. Dmitruk	(b) 157, 601	68.20 and 71.36; 71.45; S7.12
V.S. Dneprovskii	(b) 159, 403	71.35 and 78.20; S8.11
I.I. Dobîndă	(a) 117, 587	78.55; S8
D. Dobrev	(a) 122, K31	68.55; S1.3
A.P. Dodokin	(b) 161,K103	71.70 and 75.50; S11.2
R. Dohle	(a) 119, 683	78.30 and 78.65; S9.12
A.E. Dolbak	(a) 119, 139	68.20 and 68.55; S7.12
K.M. Donbaev	(b) 157, K55	61.70 and 76.80; S10.1
A.O. Donghol Diallo	(b) 162, 237	61.70 and 72.10; S1.2
P. Donovan	(a) 122, 311	68.55; S1.1; S1.4
P. Dordor	(a) 118, K91	72.80; 61.50; S10.1
	(a) 122, 371	72.20; S11.2
M. Dorikens	(a) 122, 129	61.80; 78.70; S5.11
L. Dorikens-Vanpraet	(a) 122, 129	61.80; 78.70; S5.11
Yu.A. Dorofeev	(b) 158, 319	75.30; 61.55; S1.61
N.N. Dorozhkin	(b) 160, 173	71.25; 71.20; S1.2
A.P. Dostanko	(a) 117, 213	68.48 and 68.55; S1.61; S5.11
N.A. Doumanis	(b) 158,K101	61.80; 79.20
J.-P. Doumerc	(a) 118, K91	72.80; 61.50; S10.1
Ya.O. Dovgyi	(b) 157, 473	71.35 and 78.40; S9
D. Draheim	(a) 119, 209	68.55; 61.70; S9.12
V. Drchal	(b) 157, 635	71.10; 75.10
S.-L. Drechsler	(b) 159, 797	74.10 and 74.60; 74.40; S10.15
J. Dryzek	(a) 118, 61	61.70; 71.25; 78.70; S10.15
	(a) 121,K159	61.70 and 78.70; 74.50; S10.15

Jiaju Du	(a) 119, 555	74.70; 64.80; S10.15
X.Z. Du	(a) 119, K1	61.40; S1.1
P. Dubček	(a) 122, K87	78.30; S8.16
B. Dubois	(a) 121, 359	61.10 and 61.55; S10.15
S.V. Dubonos	(a) 120, 77	61.70; 71.55; S7.12
J. Dubowik	(a) 121, 173	68.55 and 78.65; 61.40; S1.2
O.I. Dubrovskii	(b) 161, 697	71.25 and 73.60; 71.10; S1.2
P.V. Dudkevich	(a) 121, 391	61.70; 77.80; S10.15
V.P. Dudkevich	(a) 121, 157	68.55; 77.80; S10
C. Dufour	(a) 122, 311	68.55; S1.1; S1.4
V.K. Dugaev	(b) 160, 415	66.30
T. Dumelow	(b) 161, 233	78.30 and 78.65; 71.36; S8.13
E.B. Dunina	(b) 157, 261	71.10 and 78.20
	(b) 157, 267	71.10 and 78.20; 71.70; S10
L.J. Dunne	(b) 157, 657	74.20; S12
	(b) 159, 781	74.30; S10.15
S. Durbin	(b) 159, 443	71.35 and 78.65; 78.55; S7.12; S8.12
K. Durczewski	(b) 159, K27	75.10 and 75.30
R. Duscher	(b) 158, 573	71.10
J.M. Dusseau	(a) 120, 151	73.60; S7
H. Dwelk	(a) 120, 185	74.70; S10.15
V.V. Dyakonov	(b) 159, K33	61.70; S9.11
M. Dybiec	(a) 120, K73	74.70; S10.15
D.L. Dzhaparidze	(b) 158,K195	76.30; S11.1
N.A. Dzhavadov	(b) 158, 119	64.60; 61.50; 65.50; S11
	(b) 158, 475	64.60; 65.50; S11
J. Echigoya	(a) 118, 371	61.60; S11.1
	(a) 122, 87	61.60; 64.75; S10.15
G. Eckold	(b) 162, 125	63.20; S8
N.V. Edneral	(a) 120, 457	68.55; 75.70; 78.20; S1.1
Sh.M. Efendiev	(b) 162, 281	78.50; 71.55; S11
K. Eftaxias	(a) 118, K13	66.30; S5.11
	(b) 160, K9	66.30; S5.12
G. Eftekhari	(a) 120,K155	73.30; S7.11
	(a) 121, 327	73.40 and 85; S7.11
	(a) 122, 751	73.30 and 85; S7.11; S7.12
A.V. Egibyan	(b) 161, 55	61.40; 61.70; 71.50; S10.1
V.D. Egorov	(b) 159, 403	71.35 and 78.20; S8.11
A.V. Egranov	(b) 160, K75	61.70 and 78.30; S9.11
O. Eibl	(a) 122, 589	68.55 and 73.40; 64.80; S10.15
G. Eichhorn	(b) 162, 125	63.20; S8
W. Eichler	(b) 159, K87	72.15 and 72.20
E.A. Eivazov	(a) 117,K147	76.30; S8.16

R.A. El Egaimi	(b) 161,K111	76.60
A.R. El Gohary	(b) 161, 233	78.30 and 78.65; 71.36; S8.13
B. El Jani	(a) 117, 169	61.70 and 78.65; 73.60; S7.12
M. El Omari	(a) 121, 415	66.30; 64.75; S9.15
E.M. Elgiar	(a) 118, 197	68.55 and 78.65; 73.60; S10.1
V.M. Elinson	(a) 122,K139	68.55 and 78.65; S5
K. Elk	(b) 159,K111	75.60
	(b) 161, K41	75.60
A.M. Elkorashy	(b) 159, 903	78.20; S8
C. Ell	(b) 159, 117	71.25 and 78.20; 73.40; S7.11; S7.15
M. Elomari	(a) 117, 409	61.60 and 66.30; 64.75; S9.15
R.M. Elshazly	(a) 122, 469	61.40 and 61.60; S10.15
E.P. Elsukov	(a) 117, 291	61.55 and 75.50; 76.80; S1.1
F. El-Akkad	(a) 118, K79	72.80; S8.13
M. El-Desouki	(b) 157, 609	68.55 and 78.65; S8
S.A. El-Hakim	(b) 158, K1	63.20 and 78.30; S8.16
S.A. El-Hamid	(b) 158, K85	78.30; 63.20; S8.16
A.M. El-Khatib	(a) 117, 119	65.40; 61.80; S11
M.K. El-Mously	(a) 121, 235	61.55 and 74.70; S10.15
N. El-Naquib	(a) 117, 437	62.20; S1.2
M.M. El-Ocker	(b) 157, 609	68.55 and 78.65; S8
V.V. Emtsev	(a) 120, K15	61.70; S5.11
	(a) 121,K163	61.70 and 78.70; S5.11
V.N. Enakii	(a) 119, 607	78.55; 63.20; 71.55; 78.30; S11.2
H. Enami	(a) 120, 199	74.70; S10.15
Y. Endo	(b) 161, 379	75.30; 75.50; 75.70
G.F. Engel	(b) 162, 311	78.20; 64.60; S11.1
C. Engler	(b) 159, 721	68.20; S5.11
E.M. Epshtein	(b) 160, 179	71.45
D.P. Erchak	(a) 121, 63	61.70; 76.30; 78.60; S5
G.N. Eritsian	(b) 160, K71	78.40; S10.1
Ş. Erkoç	(b) 157, K23	68.20; S7.12
	(b) 161, 211	63.10; 61.55; 62.20; S1.2; S1.3; S4
	(b) 162, K5	61.50; S1; S1.2
V.N. Ermakov	(b) 157, 417	74.20; 71.45; S12.1
	(b) 161, 745	74.20; 71.45
A.S. Ermolenko	(a) 121,K111	75.50; S1
M. Ermrich	(a) 118, K1	61.10; 61.50
Yu.N. Erokhin	(a) 120, K1	61.10 and 61.70; S5.11
N.V. Ershov	(b) 160, 625	76.80
	(a) 122,K143	68.20; S1.2
S.N. Ershov	(b) 161, 233	78.30 and 78.65; 71.36; S8.13
C. Esnouf	(a) 120, 419	62.40; S3
B. Esser	(b) 157, 253	71.10 and 71.35
Y. Estrin	(b) 157, 117	61.70

S. Etienne	(a) 121, 129	61.40 and 64.60; 62.40
B. Etlinger	(a) 122, K87	78.30; S8.16
R.A. Evarestov	(b) 157, 281	71.10 and 71.25; 74.10; S1.2; S10.1
	(b) 158, 201	71.20 and 71.25; 74.10; S1.2; S10
V.A. Evseev	(b) 161, 91	61.70 and 61.80; 72.20; S5.11
I.I. Evstafev	(b) 158, 643	72.20; 75.40; S11.2
B. Evtodyi	(a) 119, 523	61.70 and 68.55; 73.20; 73.40; S5.11
M.A. Zaki Ewiss	(a) 117,K133	72.15; S1
	(a) 122, K63	72.15; S1.1
K. Fabisiak	(a) 119,K127	68.55; 78.30; S5
F. Fabre	(a) 119,K181	78.20; S7.12
M. Faehnle	(a) 117, K71	75.80; S1.2
M. Fahood	(b) 158, 165	71.50 and 73.40
V.E. Fain	(b) 158, 67	61.80; S5
B.D. Fainberg	(b) 160,K169	74.90 and 78.30
	(b) 162, K67	74.90 and 78.30
G.S. Fainshtein	(a) 117, 549	75.30; 73.40; S1.2
N.N. Faleev	(a) 120, 327	61.10 and 68.55; S7.12; S7.15
M.L. Falin	(a) 121, K95	76.30; 75.50; S11.1
R. Falster	(a) 117, 403	61.70; S5.11
Zhongzhi Fan	(b) 157, 677	75.30; 75.60; S1.1; S1.4
S.S. Fanchenko	(a) 117, 341	61.10; 68.20
M.H. Fang	(a) 122,K151	74.70; S10.15
Q.F. Fang	(a) 121, 139	62.40; S4
O.V. Farberovich	(b) 159, 743	71.25; 71.50; 78.20; S1.4; S8
	(b) 160, 219	71.25; S1.62
	(b) 160, K37	71.50; 73.20; 75.25; S1.1
Z.M. Farid	(a) 117, 437	62.20; S1.2
M.P. Fateev	(b) 158, 81	64.60
A. Fattah	(a) 122, 139	61.70; 61.60; S7.12
B. Fechner	(b) 159,K103	75.40
R. Feder	(b) 157, 627	68.20 and 75.30; 61.14; S1.1
V.A. Fedorov	(b) 161, 629	66.30 and 68.45; 68.40; S8.12
A. Fedotov	(a) 119, 523	61.70 and 68.55; 73.20; 73.40; S5.11
V.K. Fedyanin	(b) 160, K29	71.38; S12.1
Yu.N. Fedyunin	(a) 117, K53	75.70; 75.60; S11.2
P. Feltham	(a) 117, 135	62.20; S1.2
D. Feng	(a) 117, 221	68.55; S4
	(a) 122, 475	61.10 and 61.70
Xiqi Feng	(b) 157,K127	76.60; S11.1
	(a) 119, 561	72.40; S11.1
	(a) 121, 429	61.70; S11.1

F. Fermi	(a) 120, 635	78.55 and 78.60; S9.12
B. Fernández	(a) 122, 235	62.20 and 62.65; S8.16
R. Ferrer	(b) 159, K93	75.10
F.J. Feuerborn	(b) 158, 503	78.65; 71.35; S9
N.A. Fidrya	(b) 157, 601	68.20 and 71.36; 71.45; S7.12
M. Fiedler	(b) 159, 143	71.35 and 73.40; S7.12; S7.15
H. Fieseler	(b) 159, 411	71.35 and 78.55; S7.15
S.R. Figarova	(b) 158,K179	72.10; 73.60
T. Figielski	(a) 121, 187	71.55
A.S. Filipchenko	(a) 121,K129	78.55; S7.13
I.M. Filippov	(a) 118, 53	61.70 and 71.55; S5.11
D. Fink	(a) 117, 67	61.10 and 61.80; S7.11; S7.12
L. Fionova	(a) 119, 523	61.70 and 68.55; 73.20; 73.40; S5.11
L.K. Fionova	(a) 120, 457	68.55; 75.70; 78.20; S1.1
H.J. Fischbeck	(b) 158, 565	71.20
F. Fischer	(a) 119, 327	78.50 and 78.55; 61.50; S11.1
G. Fischer	(b) 159, K23	75.10; S1.4
K. Fischer	(a) 119, 241	74.60 and 74.70; 61.70; S10.15
O. Fischer	(a) 122, 589	68.55 and 73.40; 64.80; S10.15
A.Ya. Fishman	(b) 160,K153	75.50; 76.20; S11.2
H.-J. Fitting	(a) 121, 305	72.20 and 79.70; 85; S8.11
	(a) 122, 297	73.40; S10.1
	(a) 122,K165	72.20; 78.60; 79.70; 85; S8.11
F. Fizzotti	(a) 119, 169	72.40 and 73.60; 78.65; S5.11
T. Flade	(a) 117, 527	71.55; 72.20; S7.12
L.I. Flaks	(a) 120, 327	61.10 and 68.55; S7.12; S7.15
N. Floros	(b) 159, 781	74.30; S10.15
E. Flouda	(b) 157, 351	72.15; S1.2
Ch. Flytzanis	(b) 159, 275	78.20; S8.15
A. Foitzik	(a) 121, 81	61.70; S8
A.G. Fokin	(b) 158, 715	78.20
J. Foltin	(b) 161, K95	74.20 and 74.70
O. Foltin	(a) 118, K43	78.30; S8
S.V. Fomichev	(b) 161, 153	62.20 and 71.25; S2; S3; S4
V.M. Fomin	(b) 159, 707	68.30 and 78.30
	(b) 161, 603	71.35 and 78.65; S8.13
	(b) 162, K63	78.30; S8.16
R.J.M. da Fonseca	(b) 158, K75	78.50 and 78.55; S11.2
	(b) 162, K59	78.55; S11.2
C. Fontaine	(a) 119,K181	78.20; S7.12
A. Forchel	(b) 159, 125	71.35 and 78.55; 73.40; S7.12; S7.15
	(b) 159, 457	71.35 and 78.20; S7.12; S7.15
J. Fousek	(a) 118, 539	77.30 and 77.80
	(a) 120, 273	77.40; 77.80; S12

C. de Francisco	(a) 118, K31	75.60; S11.2
J. Frandon	(a) 119,K181	78.20; S7.12
R. Franke	(b) 160, 143	79.60; S7.1; S11
B. Frit	(a) 117, 409	61.60 and 66.30; 64.75; S9.15
L. Fritsche	(b) 159, 617	61.70 and 71.55; S7.1
B. Fritzsch	(b) 162, K69	66.30
D. Froehlich	(b) 158, 267	71.35 and 71.70; 71.36; 78.20; S9.11
	(b) 159, 297	71.35 and 78.20; 73.40; S7.1; S10.1
G.I. Frolov	(a) 121,K145	61.40 and 75.50; S1.2
A.G. Frorip	(b) 161, 525	61.70 and 78.55; 61.80; 76.30; S9.11
L. Fruchter	(b) 157,K123	74.30; S10.15
S.Y. Fu	(a) 120, 627	75.50; 64.80; S1.1
	(a) 121, 257	75.60; S1.1
G. Fuchs	(b) 159, 797	74.10 and 74.60; 74.40; S10.15
I.Ya. Fugol	(b) 160, 245	71.70; 78.50; S5.2
W. Fuhs	(b) 160, K25	73.60; S5.11
H. Fujii	(a) 117, 563	75.30; S9.16
T. Fujino	(a) 117, 515	71.55; 73.30; S8.12
S. Fujita	(b) 158, K69	71.25 and 76.30; S5.11; S5.12
H. Fujiwara	(a) 118, 513	72.15; S1.2; S1.4
T. Fukami	(a) 117, K93	61.50 and 77.80; S11.1
	(a) 121, 383	61.60; S11
	(a) 122,K117	61.60 and 77.80; S11.1
Y. Fukuda	(a) 117,K177	78.55; 61.70; S10.1
	(b) 158, 383	61.70 and 78.60; 79.75; S9.11
M. Fukuhara	(b) 157, 357	74.20; 74.70
	(a) 122, 677	77.30 and 77.90; S10.15
J.K. Furdyna	(b) 160, 591	71.70 and 76.30; S8.12
K. Furić	(a) 122, K87	78.30; S8.16
J. Furthmueller	(a) 117, K71	75.80; S1.2
K. Futagami	(a) 121,K125	78.20; 71.50; S7.12
A.M. Gabovich	(b) 161, 293	74.50; 74.20; S10.15
S.P. Gabuda	(b) 158, 329	76.60; 71.70; S11
G.V. Gadiyak	(b) 160, 483	68.25 and 68.40; 73.20; S5.11
M. Gal	(a) 122,K157	74.70; S10.15
F. Galembeck	(a) 117,K155	75.50 and 76.50; S12
G.A. Gamal	(a) 119, K47	72.40; S8
	(a) 120,K163	72.80; S8.16
	(a) 121,K201	72.20; S8.1
H. Gamari-Seale	(b) 162, 243	72.20 and 74.70; 76.30; S10.15
	(b) 162,K101	76.30; 74.70; S10.15

M.Ya. Gamarnik	(b) 160, K1	61.55; S1.3
	(b) 161, 457	61.55; 61.10; S5; S5.11
Y.H. Gameel	(a) 119, K47	72.40; S8
	(a) 121,K201	72.20; S8.1
M. Gamoudi	(b) 160, 193	71.25 and 71.60; 78.70; S7.12; S8.13
F.C.G. Gandra	(a) 117,K155	75.50 and 76.50; S12
V.V. Ganin	(a) 118,K139	74.70; 78.30; 79.20; 79.60
V.A. Ganshin	(a) 119, 11	61.40 and 68.48; 78.20; S11.1
	(b) 161, 629	66.30 and 68.45; 68.40; S8.12
S.V. Gaponenko	(b) 158, 359	71.35 and 78.65; S8.12
	(b) 159, 267	71.35 and 78.65; S8.15; S9
	(b) 159, 449	71.35 and 78.20; S8.1
J.E. Garbarczyk	(a) 120, 387	66.30; 76.60; S11
S. García	(a) 120,K169	74.70; S10.15
A.A. Garibov	(a) 117, K57	61.80 and 76.30; S10.1
E.L. Gartstein	(a) 121, 37	61.10; 61.70; S7.12
N.M. Gasanly	(b) 158, K1	63.20 and 78.30; S8.16
	(b) 158, K85	78.30; 63.20; S8.16
A.M. Gasanov	(a) 117, K57	61.80 and 76.30; S10.1
E.M. Gasanov	(a) 122, K45	74.70; S10.15
L.G. Gasanova	(b) 158, K1	63.20 and 78.30; S8.16
F.M. Gashimzade	(b) 160,K177	78.65
	(b) 161, 861	78.30; S8.15
N.F. Gashimzade	(b) 160,K113	71.36; 78.65
V.I. Gatalskaya	(a) 119,K151	61.80 and 74.70; S10.15
	(a) 121, K85	74.70; S10.15
G.V. Gatalskii	(a) 119,K151	61.80 and 74.70; S10.15
	(a) 121, K85	74.70; S10.15
V.S. Gaviko	(a) 119,K163	75.60 and 75.80; S1.2
	(a) 121,K111	75.50; S1
V.I. Gavrilenko	(b) 162, 477	71.25 and 78.20; S6
V. Gavryushin	(b) 162, 281	78.50; 71.55; S11
H. Gdanitz	(a) 117, 395	66.30; S5.11
W. Gebhardt	(b) 158, 695	75.50; S8.15
S.Y. Gee	(b) 162, K9	64.75; 78.30; S8.16
R. Geick	(b) 162, 125	63.20; S8
V. Geist	(a) 117, 67	61.10 and 61.80; S7.11; S7.12
P.V. Geld	(a) 117, 577	75.50; S1.1; S1.2; S1.4
R.B. Gelfand	(a) 121, 63	61.70; 76.30; 78.60; S5
R. Gemperle	(a) 121,K215	75.60; S11.2
A. Genzel	(a) 117, 141	62.20; 61.70
Ch. Genzel	(a) 117, 141	62.20; 61.70
Zh. Georgiev	(a) 118, K95	72.15 and 75.70
S.S. Gerasimovich	(a) 122,K139	68.55 and 78.65; S5
W. Gerber	(b) 161, K81	71.25 and 78.70; S4
L.G. Gerchikov	(b) 160, 443	71.25 and 73.40; S7.12; S7.15

M. Gericke	(a) 119, 183	73.40 and 78.20; S7.15
E. Gerlach	(b) 157, 189	73.60; S4
	(b) 158, 531	73.60
I.N. Germanenko	(b) 159, 267	71.35 and 78.65; S8.15; S9
R. Germann	(b) 159, 457	71.35 and 78.20; S7.12; S7.15
H. Gerstenberg	(a) 118, 241	61.70 and 72.20; 71.55; S7.13
A.P. Ges	(a) 119,K151	61.80 and 74.70; S10.15
L.A. Gevorgyan	(b) 158, 213	61.80 and 71.10
V.A. Gevorkian	(b) 160, K71	78.40; S10.1
Kh.B. Gezalov	(a) 117, K57	61.80 and 76.30; S10.1
K.P. Ghatak	(b) 158,K123	73.40; S5.12
	(b) 160, 673	75.20 and 75.70; 71.25; S7.16
M.N. Ghatikar	(b) 159, 551	61.40; 78.70; S1.1; S1.2
G. Ghibaudo	(a) 117, 99	61.70; 61.80; S5.11
	(a) 119, 343	73.40 and 85
B. Ghosh	(a) 121, 657	78.60; S9.12
N.K. Ghosh	(b) 158, 223	71.38; S1.4
	(b) 161, 661	71.38
S. Ghosh	(b) 157, 159	62.25 and 72.50; S8.11
	(a) 118, 235	72.30; 62.65; S11.1
	(b) 159, 817	72.10 and 72.50; 71.25; S7.13
	(b) 161, 319	72.20 and 72.50; 77.60; S8.11
S.C. Ghosh	(b) 158, 223	71.38; S1.4
	(b) 161, 661	71.38
U.S. Ghosh	(b) 160, 583	71.70; S9.16; S11
A. Ghoshal	(b) 158,K123	73.40; S5.12
S.K. Ghoshal	(a) 120, 61	61.50; 64.60; S12
P. Gibart	(a) 117, 169	61.70 and 78.65; 73.60; S7.12
	(a) 119, K41	71.55; S7.12
A. Gilewski	(b) 159, 797	74.10 and 74.60; 74.40; S10.15
P. Gille	(b) 161, 613	73.40; S8.15
P. Gilliot	(b) 159, 259	71.35 and 78.20; S9
B. Gillot	(a) 118, K5	61.60; 78.30; S10.15
C. Giovannella	(b) 157,K123	74.30; S10.15
N.A. Gippius	(b) 159, 71	78.20; 71.35; 71.36; S8.11
W. Giriat	(b) 161, K71	78.40; S8.15
M. Giry	(b) 158, 629	71.50
D.V. Gitsu	(a) 122,K177	78.30; S8.16
B.A. Gizhevskii	(b) 158, 307	72.20; S11.2
A. Glachant	(a) 120, 507	68.55 and 73.60; 78.65; S10.1
V.V. Gladkii	(a) 121, K47	64.70; S9.15
P.S. Gladkov	(a) 118, 479	73.30; 85; S1; S7.12
S.O. Gladkov	(b) 160, K5	66.30
A. Gladun	(b) 159, 797	74.10 and 74.60; 74.40; S10.15
S.J. Gladysz	(b) 159,K119	75.30; 67.50; S2
W. Glaeser	(a) 118, 241	61.70 and 72.20; 71.55; S7.13

R. Gleichmann	(a) 121, 95	61.70 and 78.60; 85; S7.12
M.D. Glinchuk	(b) 161, 843	76.30; 61.70; S11
	(a) 122, 341	71.55; 76.30; S10.15
N.G. Gluzman	(b) 160,K141	71.30; S8.16
Yu.P. Gnatenko	(b) 161, 419	71.35 and 78.30; S8
	(b) 161, 427	78.55; 71.35; 71.55; S8
V.A. Gnatyuk	(a) 122, 243	73.20 and 73.40; 72.40; S8.15
G. Gobsch	(b) 162, 155	71.25 and 73.40
G. Godefroy	(a) 117, K85	78.20; 61.80; S11.1
	(b) 161, K55	76.30; S11.1
V.P. Gogin	(a) 118,K107	75.70; 75.60; S11.2
I.A. Gohar	(a) 122, 469	61.40 and 61.60; S10.15
R. Goldhahn	(a) 119, 183	73.40 and 78.20; S7.15
F. Golek	(a) 117, K43	74.90 and 79.75; S10.15
L.L. Golik	(b) 162, 199	72.40; 72.70; S8.11
O.A. Golikova	(a) 121, 579	71.55; 72.80; 76.80; S5
E.M. Gololobov	(a) 117,K121	61.50 and 74.70; S10.15
	(a) 122, 555	64.70 and 68.55; 74.70; S10.15
	(a) 122,K123	62.65 and 64.70; S1.2
Yu.I. Golovko	(a) 121, 157	68.55; 77.80; S10
A.V. Goltsev	(b) 161, 777	72.20; S10
A. Goltzené	(b) 160, 649	71.70 and 76.30; S7.12
G.P. Golubev	(a) 117, 587	78.55; S8
N.G. Gomaa	(a) 117, 119	65.40; 61.80; S11
V. Gomathy	(b) 160, 583	71.70; S9.16; S11
M.J.M. Gomes	(b) 158, 391	78.40; 71.35; S9
	(b) 159, 101	71.35 and 78.20; S9
K.W. Goossen	(b) 159, 11	71.35 and 78.20; 73.40; S7.12
S. Gopal	(a) 117, 53	61.40; 78.70; S1.1; S1.2
	(a) 120, 411	64.70; 61.40; 78.70; S11
V. Gopal	(a) 121, 533	68.55 and 78.30; S10.1
P.N. Gorlei	(b) 159, 837	72.30
	(b) 161, K91	72.80; S5
O.P. Gorodnichii	(b) 160, 291	72.20; S8.15
S.Ya. Goroshchenko	(b) 159, K55	72.10 and 73.40
P. Górski	(a) 121,K243	78.20; S11.1
I. Gościańska	(a) 118, 189	61.40 and 68.55; 73.60; 75.70; S1.2; S1.4
	(a) 122, 575	68.55; 61.40; 73.60; S1.1; S1.2
B.N. Goshchitskii	(a) 119, 437	61.80 and 64.75; S1.1; S4
M. Gospodinov	(b) 161, 575	63.20; 78.30; S10.15
V.I. Gostishev	(b) 160,K133	61.70 and 72.15; S4
R.D. Gould	(a) 120, K57	73.60 and 78.65; S12
P.S. Goyal	(b) 157, 547	64.60; 61.12; S11
	(a) 118, 425	63.90; 64.70; 77.80; S11.1
S.C. Goyal	(b) 157, 145	62.20; S9.12

D.B. Goyer	(a) 118, K9	61.80 and 78.30; S7.11
U. Gradmann	(b) 157, 627	68.20 and 75.30; 61.14; S1.1
V. Graefe	(a) 117, K23	61.80 and 78.65; S6
J. Graefenstein	(b) 158,K133	71.10
A. Graf	(b) 158, 695	75.50; S8.15
S. Grande	(b) 158, 707	74.70 and 76.60; S10.15
E. Grantscharova	(a) 122, K31	68.55; S1.3
E. Gratz	(a) 117, 501	71.20; 72.20; 75.30; S1.2; S1.4
M. Graurock	(b) 158, K37	72.40; S8
L.G. Grechko	(b) 157, 451	71.25 and 75.10
	(b) 162, 539	72.10 and 75.10
B.A. Greenberg	(b) 158, 441	63.20 and 71.10; 61.70; 62.20; S1
L. Grenacs	(b) 158,K197	76.60; S1
J.-C. Grenet	(a) 119, K41	71.55; S7.12
O.S. Gribkov	(a) 119, 417	61.40 and 61.80; S11.2
V.P. Gribkovskii	(b) 158, 359	71.35 and 78.65; S8.12
	(b) 159, 379	78.20; S8.11; S8.12
A.A. Grigorenko	(b) 160, 219	71.25; S1.62
D.O. Grigorev	(a) 121, 11	61.10 and 61.70; S5.11
R.V. Grigorev	(a) 119,K167	78.55; 61.80; S8.12
S.R. Grigorev	(a) 119,K167	78.55; 61.80; S8.12
N.E. Grigorian	(b) 160, K71	78.40; S10.1
R. Grill	(b) 162, 509	71.20; S8.16
B.P. Gritzenko	(a) 121, 105	61.80; 78.60; S10.1
A. Grodziński	(a) 118, 41	61.60; 64.70; 76.80; S10
R. Groessinger	(a) 120, 627	75.50; 64.80; S1.1
R. Groetzschel	(a) 117, 477	61.80 and 68.55; S1.61
V.E. Gromov	(a) 121, 437	61.70; S1.2
S. Grosswig	(a) 121, 539	68.70; 61.10; S1.1
W. Gruenewald	(a) 122,K129	68.55; 61.16; 68.60; S5
D. Gruetzmacher	(b) 159, 125	71.35 and 78.55; 73.40; S7.12; S7.15
G. Grummt	(a) 122, 765	73.40; 72.40; S7.12
J.B. Grun	(b) 159, 101	71.35 and 78.20; S9
	(b) 159, 259	71.35 and 78.20; S9
A. Grusková	(a) 122,K171	75.30 and 75.50; 76.80; S11.2
A.N. Gruzintsev	(b) 161, 705	71.55 and 78.55; 78.60; S8
Binglin Gu	(b) 161, 537	64.70; 61.60; S10
B.X. Gu	(a) 117, 283	75.50; 61.40; S1.1
	(a) 120, 159	68.55 and 75.70; 76.80; S1.1
	(a) 121, K99	75.50; 61.40; S1.1; S1.4
Shi-Wei Gu	(b) 161, K25	71.36 and 73.90; S7.11
Y.M. Gu	(b) 159, 617	61.70 and 71.55; S7.1
V.A. Gubanov	(b) 160, 459	68.45 and 73.20; 79.20; S1.61
	(b) 160, 517	71.25; S10.1
V.A. Gubanova	(a) 120, K85	76.30; S11.1

G. Guénin	(a) 117, 113	64.70; S1.2
H. Guenther	(b) 161, 179	62.20; 61.70
A.L. Guerassimov	(a) 122,K147	71.55; 61.80; S5.11
P.S. Guimarães	(b) 158, 283	61.70 and 71.55; S9.11
M. Guittard	(a) 117, 169	61.70 and 78.65; 73.60; S7.12
	(a) 121, 21	61.60; 61.10; S8
L.N. Gumen	(b) 162, 387	63.20; S12.1
R.L. Gunshor	(b) 159, 107	71.35 and 78.20; 73.40; S8.13
	(b) 159, 443	71.35 and 78.65; 78.55; S7.12; S8.12
Huifen Guo	(a) 120,K177	72.40; 61.70; S5.11
Hui-Qun Guo	(a) 119, K63	75.50; 75.60; S1.1
	(a) 121,K105	75.30; 61.40; 75.60; S1.2
Y.Y. Guo	(a) 122, 195	62.20; S4
H.C. Gupta	(b) 159, K47	62.20
	(b) 160, K19	63.20; S8.16
M.K. Gupta	(a) 121, 657	78.60; S9.12
Neelam Gupta	(b) 161, 201	63.50; S2
N. Gupta	(b) 160, 433	63.50; S2
	(b) 162, 395	63.50; S2
S.K. Gupta	(a) 121, K47	64.70; S9.15
V.B. Gupta	(b) 160, K19	63.20; S8.16
V.P. Gupta	(b) 161, 131	63.90; 62.20; 65.70; S9.11
S.A. Guretskii	(a) 119,K151	61.80 and 74.70; S10.15
L.I. Gurevich	(a) 121, 437	61.70; S1.2
Z.A. Gurskii	(b) 157, 557	63.20; S2
V.A. Gurtov	(a) 119, 113	68.55; S6
	(a) 121,K199	73.40; S5.11
A.G. Guseinov	(b) 158, K85	78.30; 63.20; S8.16
D.A. Guseinov	(a) 121, K51	65.40; 75.40; S11.2
G.I. Guseinov	(b) 158,K179	72.10; 73.60
N.G. Guseinov	(b) 159,K107	75.40; 75.50; S8
A.I. Gusev	(b) 160, 389	61.50; 61.55; S1.2; S1.3
V.E. Gusev	(b) 158, 367	62.30 and 78.20
N.A. Gushchina	(b) 158, 175	71.25 and 73.40; S7.15
N. Guskos	(b) 158, K17	72.80; S9.12
	(b) 162, 243	72.20 and 74.70; 76.30; S10.15
	(b) 162,K101	76.30; 74.70; S10.15
W. Gust	(a) 119, K9	66.30 and 68.48; S1.2
M.M. Gutman	(b) 162, 199	72.40; 72.70; S8.11
J. Gutowski	(b) 159, 205	78.20; S8.11
	(b) 159, 363	71.55 and 78.20; 71.36; S8.11
	(b) 159, 443	71.35 and 78.65; 78.55; S7.12; S8.12
	(a) 120, 11	78.65; 68.55; 71.35; 73.60; 78.55; S8.12
M. Gutowski	(a) 117,K115	71.55; S9.11

G.L. Gutsev	(b) 161, 91	61.70 and 61.80; 72.20; S5.11
I. Gutzow	(a) 117, 363	61.40; 64.75
M. Guymont	(a) 118, 29	61.60; S8.16
	(a) 121, 21	61.60; 61.10; S8
M. Guzzi	(b) 159, 577	61.70; 78.50; 78.55; S10.1
P. Haasen	(a) 121, 81	61.70; S8
F. Hache	(b) 159, 275	78.20; S8.15
E. Hacker	(a) 119, 683	78.30 and 78.65; S9.12
M. Hada	(b) 162, 407	65.40; 75.30; S1.2; S1.4
V. Hadjicontis	(a) 118, K13	66.30; S5.11
	(b) 160, K9	66.30; S5.12
H. Haefner	(a) 122, 683	73.40 and 78.55; 78.60; S7.15
J. Haertwig	(a) 121, 539	68.70; 61.10; S1.1
A.M. Hafez	(a) 122,K113	61.55; S1.4
I.M. Hafez	(a) 119, 343	73.40 and 85
H. Hagedorn	(b) 158, 753	78.55; 71.35; S5.2
Long Hai	(a) 119, K5	61.50; 74.50; S10.15
N.C. Halder	(a) 119, 545	61.70 and 71.55; S7.12
T. Halicioglu	(b) 157, K13	63.10; 68.20; S7.12
	(b) 157, K23	68.20; S7.12
	(b) 157, K77	63.10; 68.20; S7.15
C. Hamann	(a) 120, K95	78.65; S12
M. Hamerský	(a) 118, 455	62.20; S1.2
F.H. Hammad	(a) 119, 455	62.20; 61.70; 64.80; S1.2; S4
F. Hanna	(a) 117, 163	64.70; 72.20; S11
E. Haque	(b) 158, 457	63.20; S2
T.M. Haridasan	(b) 158, 427	66.30; S8.15
	(b) 162, 119	63.20; 61.70; S8.15
M. Hartmann	(b) 159, 35	71.25 and 72.10; 78.20
V.V. Harutunian	(b) 160, K71	78.40; S10.1
M. Hasebe	(b) 162, K21	65.50; 65.70; S2
M. Hasegawa	(a) 120, 199	74.70; S10.15
F. Hashimoto	(a) 120, K11	61.70; 71.55; 66.30; S5.11
H. Hashimoto	(a) 119, 511	68.55; 68.48; S5.11; S7.13; S7.15
S. Hashimoto	(a) 117, K1	61.40; S10
G. Hasnain	(b) 159, 11	71.35 and 78.20; 73.40; S7.12
A.R. Hassan	(b) 160, 185	71.25; 78.20
G.A. Hassan	(a) 119, 471	62.20; S1.3
H.H. Hassan	(a) 119, 471	62.20; S1.3
T. Hattori	(a) 120, K63	71.55 and 78.50; S5.12
W. Haubenreisser	(b) 160, 491	68.55
H. Haug	(b) 159, 117	71.25 and 78.20; 73.40; S7.11; S7.15
	(b) 159, 309	71.35 and 78.20; S7.12; S8.11; S9

R. Hayn	(b) 158, 565	71.20
Deyan He	(a) 118, K17	66.30; S5.11
	(a) 118, K39	61.70 and 76.30; S5.11
	(a) 120, K49	68.60; S5.11
Siming He	(a) 121,K167	62.40; 66.30; S1.3
Z.H. He	(a) 122, K55	74.60 and 74.70; S10.15
	(a) 122,K151	74.70; S10.15
D. Hecht	(a) 122, 297	73.40; S10.1
V. Heera	(a) 122, K35	68.48 and 68.55; S5.11
W. Heerdegen	(a) 119, 683	78.30 and 78.65; S9.12
E. Hegenbarth	(b) 159, K1	61.40; 77.70; 77.80
	(a) 120, 547	66.70 and 74.70; S10.15
	(a) 121, K17	66.70 and 74.70; S10.15
U. Heiber	(a) 118, 531	75.60; S10
R. Heidinger	(a) 121, 445	61.80; 64.80; 74.70; 76.80; S1.2
A. Heilmann	(a) 122, K83	78.90; S5.12
E. Heiner	(b) 159, 43	72.10; 72.20
	(b) 160, 225	71.10; 65.40; 71.38; 78.20; S1.3
	(b) 162, 303	78.20; 65.40; 71.10; S1.3
B. Hejda	(b) 157, 667	63.20 and 72.10; S7.1; S8.1
J. Hellberg	(b) 161, 245	61.65 and 72.20; 71.55; 75.30; 76.30; S12.1
K. Helming	(b) 157, 71	61.10; 61.12; 61.14
	(b) 157, 489	61.10; 61.12; 61.14
F. Henneberger	(b) 159, 49	71.35 and 78.20; S8.11; S8.12; S8.15
	(b) 159, 287	71.35 and 78.20; S8.11
	(b) 159, 353	71.35 and 78.20; S8.11
K. Henneberger	(b) 159, 59	72.10 and 78.20; S7.12
	(b) 159, 81	71.35 and 72.30; 78.20; S8.11
O. Henri-Rousseau	(b) 158, 629	71.50
A. Henry	(a) 120, 539	71.55; 61.70; S5.11
W. Hergert	(a) 119, 689	78.55 and 78.60; 72.10
	(b) 161, 667	71.10; 71.20; 71.25; 73.20
R. Hergt	(a) 117,K119	74.70; S10.15
	(a) 119, 241	74.60 and 74.70; 61.70; S10.15
	(a) 122, K51	74.60 and 74.70; 61.70; S10.15
	(a) 122,K161	74.60; 74.70; S10.15
H. Hermann	(a) 118, K1	61.10; 61.50
A. Hernández	(b) 157, 275	71.30 and 75.40; S11.2
A. Hernando	(a) 119, 595	72.15 and 75.80; 75.60; S1.1; S1.2
A. Herr	(b) 159, K23	75.10; S1.4
J. Herrmann	(a) 120, 557	74.70; 61.60; 75.30; S10.15

J. Herrmann	(a) 121, K91	75.30; 75.50; S1.2; S10.15
K.H. Herrmann	(a) 119, 711	73.40 and 78.55; 72.40; S7.16
	(a) 121, 635	78.20; 71.25; 78.55; S8.15
	(a) 122, 11	78.55; 78.45; S8.1; S8.15
R. Herrmann	(a) 120, 185	74.70; S10.15
	(b) 161, 613	73.40; S8.15
P. Hertel	(a) 117, K85	78.20; 61.80; S11.1
H. Heukenkamp	(a) 120, 485	73.40; 71.55; S8.15
M. Hidaka	(a) 117, 563	75.30; S9.16
H. Hikita	(a) 119, 251	76.30; S10.1
S. Hildebrandt	(a) 119, 689	78.55 and 78.60; 72.10
Th. Hingst	(a) 121, 305	72.20 and 79.70; 85; S8.11
	(a) 122, 297	73.40; S10.1
	(a) 122,K165	72.20; 78.60; 79.70; 85; S8.11
S. Hiratsuka	(a) 118, 371	61.60; S11.1
R. Hirrle	(a) 119, 317	78.55 and 78.60; 71.55; S10.15
	(a) 119, 631	78.55 and 78.60; 71.55; S10
	(a) 120, 643	78.55 and 78.60; 71.55; S11.2
	(a) 121, 289	78.55 and 78.60; 71.55; S11.2
V. Hizhnyakov	(b) 158, 725	78.40; 78.45; 78.50; S9.11
	(b) 162, K91	71.55 and 78.55; S9.11
P. Hlidek	(b) 158,K173	72.40; 71.55; S8.15
N.V. Hoang	(a) 121, 607	75.25; 75.30; S1.2; S1.4
D. Hoehl	(a) 122, K35	68.48 and 68.55; S5.11
B. Hoenerlage	(b) 158, 391	78.40; 71.35; S9
	(b) 159, 101	71.35 and 78.20; S9
	(b) 159, 259	71.35 and 78.20; S9
T. Hoenig	(b) 159, 363	71.55 and 78.20; 71.36; S8.11
W. Hoerstel	(a) 119, 589	72.20; S8.15
	(a) 120, 485	73.40; 71.55; S8.15
P. Hoeschl	(b) 158,K173	72.40; 71.55; S8.15
M. Hohenstein	(a) 122, 79	61.40 and 61.80; 64.70; S1.2; S4
J. Hollandt	(b) 159, 205	78.20; S8.11
V. Holý	(b) 162, 347	61.10 and 61.70; S7.12
T. Holz	(b) 157,K109	74.40 and 74.50
H. Homburg	(a) 120, 159	68.55 and 75.70; 76.80; S1.1
C.V. Hoof	(b) 159, 173	71.35 and 78.20; 73.40; S7.12; S7.15
G. Horváth	(a) 117, 265	74.50; 74.70
	(a) 117, 273	74.20 and 74.50
R. Horyń	(b) 158, K33	74.70; S10.15
H. Hoshizaki	(a) 120, 199	74.70; S10.15
H.Q. Hou	(a) 119, 511	68.55; 68.48; S5.11; S7.13; S7.15
R. Hrabański	(b) 162, 275	76.30; S11

M. Hrabcák	(a) 119, 601	75.50; 75.60; S1.1
A. Hu	(a) 117, 221	68.55; S4
Meisheng Hu	(a) 119, K5	61.50; 74.50; S10.15
Yifan Hu	(a) 122, 249	68.45 and 73.60; S11.1
Y.Z. Hu	(b) 159, 249	71.35 and 78.20; S8.12
Changhe Huang	(a) 121, 273	78.50; 61.70; S8.15
N.K. Huang	(a) 121, K61	68.55; S10.1
A. Hubert	(a) 118, 271	75.60; 78.20; S1.1; S1.2; S1.4; S11.2
K. Huebner	(b) 159, 917	78.30 and 78.65; S10.1
M. Huehne	(a) 117, 417	61.70; 71.55; 72.20; S7.12
J. Hugel	(b) 159, 927	71.55 and 78.50
R.E. Hummel	(a) 122, 255	61.70 and 68.55; S1.3
Vu Van Hung	(b) 161, 165	65.50; S5.2
	(b) 162, 371	65.50; 65.40; 65.70; S5.2
	(b) 162, 379	65.50; 63.90; 64.60; S5.2
Nguyen Que Huong	(b) 161, 637	71.25 and 71.35
S.A. Hussein	(a) 120,K163	72.80; S8.16
	(a) 121,K201	72.20; S8.1
J.S. Hwang	(b) 157, 701	75.60 and 76.30; S1.4
	(b) 158,K115	73.40; S8.16
M. Iannuzzi	(b) 157,K123	74.30; S10.15
E.M. Ibragimova	(a) 122, K45	74.70; S10.15
A. Ibrahim	(a) 117, 163	64.70; 72.20; S11
S. Ideshita	(b) 159, 133	71.35 and 78.20; 73.40; S7.12; S9
M. Idzikowski	(a) 117,K137	72.15; S1.2
T. Ikari	(a) 121,K125	78.20; 71.50; S7.12
Y. Ikeda	(a) 119, K71	75.50; S10
E.N. Ilicheva	(a) 117, K53	75.70; 75.60; S11.2
L. Ilkov	(a) 121, 249	64.75 and 75.50; S11.2
V. Ilković	(b) 162,K119	75.40
L. Illgen	(a) 117,K119	74.70; S10.15
Yu. Ilyashuk	(a) 119, 523	61.70 and 68.55; 73.20; 73.40; S5.11
A.S. Ilyushin	(a) 121,K219	75.30; S1.1
K. Imai	(a) 119, 177	63.20 and 68.30; 68.60; S8.12
T. Imura	(a) 120, 199	74.70; S10.15
V.L. Indenbom	(a) 118, 71	61.70; 62.20
	(a) 122, 97	61.70; 61.10
J. Iñiguez	(a) 118, K31	75.60; S11.2
H. Inokuchi	(a) 122, K97	73.40 and 85; S12
K. Inomata	(a) 117, K1	61.40; S10
M. Inoue	(b) 162, 553	72.20; 71.30; S10
T. Inoue	(b) 162, 407	65.40; 75.30; S1.2; S1.4

A. Ioanid	(a) 118, 131	64.70; 65.40; S11.1
G.V. Ionova	(b) 162, 451	64.60 and 71.20; S1.4
I. Iordanova	(a) 121, 45	61.50; 61.70; 66.30
K. Ishi	(b) 157,K137	78.30; S1.2
T. Ishihara	(b) 159, 371	71.35 and 78.20; 78.65; S9
A.K.M.A. Islam	(b) 158, 457	63.20; S2
R.K. Islamgaliev	(a) 118, K27	72.15; S1.2
T.G. Ismailov	(b) 161, 861	78.30; S8.15
M. Istrate	(b) 161, 823	72.20; 71.38
S. Istrate	(b) 161, 823	72.20; 71.38
Y. Itoh	(a) 118, 513	72.15; S1.2; S1.4
D.D. Ivanchuk	(a) 119, 215	71.55; 72.80; 75.50; S8
R.D. Ivanchuk	(a) 119, 215	71.55; 72.80; 75.50; S8
A.L. Ivanov	(b) 159, 71	78.20; 71.35; 71.36; S8.11
	(b) 159, 417	71.35 and 78.20; 71.36; S8.11
I. Ivanov	(a) 120,K185	76.80 and 85; S5.11
M.A. Ivanov	(b) 160,K153	75.50; 76.20; S11.2
S.V. Ivanov	(a) 118, 169	73.40; 78.65; S7.12; S7.15
V.I. Ivanov	(b) 158, 643	72.20; 75.40; S11.2
	(a) 119, 87	62.40; S1.2
V.V. Ivanov	(a) 122,K109	61.60; S11.2
Yu.N. Ivanov	(b) 161, K49	76.60 and 77.80; S9.16
A.L. Ivanovskii	(b) 157, K29	71.25; S9
	(b) 158,K129	71.25; S1
	(b) 160, 459	68.45 and 73.20; 79.20; S1.61
E.L. Ivchenko	(b) 161, 217	73.40 and 78.20; 71.35; S7.12; S7.15
I.N. Ivleva	(a) 118,K139	74.70; 78.30; 79.20; 79.60
J. Iwamatsu	(b) 159, 347	71.35 and 78.20; S9
S. Iwashima	(a) 122, K97	73.40 and 85; S12
T. Iwashita	(b) 158, 347	75.30
	(b) 161, 367	75.30
K. Iwauchi	(a) 119, K71	75.50; S10
	(a) 122, 503	66.30; 77.40; S10
E.N. Izakovich	(a) 118,K139	74.70; 78.30; 79.20; 79.60
V.V. Izotov	(a) 121, K95	76.30; 75.50; S11.1

I.J. Jabr	(a) 118, 467	61.80 and 73.40; S1.1; S1.2; S5.11
B. Jacobs	(a) 117, 485	73.60; 66.30; S7.15
I. Jaeger	(b) 162, 575	75.10; 68.45
K. Jahn	(a) 119, K83	77.80; S12
F. Jahnke	(b) 159, 81	71.35 and 72.30; 78.20; S8.11
K.C. Jain	(b) 160, 433	63.50; S2
	(b) 161, 201	63.50; S2
	(b) 162, 395	63.50; S2

M. Jain	(a) 119,K107	61.70
V.K. Jain	(b) 159, 495	76.30; 64.60; S11
M.A. Jakobson	(b) 161, 395	78.55; 71.35; S8.11
R. Jakubas	(a) 120, K93	77.70; S12
A.R. Jana	(b) 160, 473	71.38 and 73.60; S1.2
P. Jandl	(a) 121, 219	72.20; S8.13
P.J. Janietz	(a) 120, 687	73.40 and 85; 73.30; S7.12
V. Janiš	(b) 157, 425	75.10 and 75.40
Z. Januškevičius	(b) 159, K63	71.55; S7.11
H. Jaouen	(a) 117, 99	61.70; 61.80; S5.11
F. Jaque	(b) 159, K37	61.70; S9.11
S. Jaulmes	(a) 121, 21	61.60; 61.10; S8
G. Jeandel	(a) 117, 319	77.30; 77.70; S11.1
S.U. Jen	(a) 119, 35	61.40 and 64.70; 62.20; S1.1; S1.2
M. Jergel	(a) 119,K155	74.70; 58.55; S10.15
A. Jeżowski	(a) 119, K53	74.30; S10.15
Wei Jia	(a) 118,K115	68.60 and 76.30; S6
Sha Jian	(a) 119, K59	74.70; S10.15
Jianyi Jiang	(a) 119, 555	74.70; 64.80; S10.15
S. Jiang	(b) 161, K71	78.40; S8.15
S.S. Jiang	(a) 117, 221	68.55; S4
S.-S. Jiang	(a) 122, 475	61.10 and 61.70
Xiaoke Jiang	(a) 119, 561	72.40; S11.1
Zi-Ying Jiang	(a) 117, 105	62.40; 61.70; S4
Hanmin Jin	(b) 157, 677	75.30; 75.60; S1.1; S1.4
Jirong Jin	(a) 117,K127	74.70; S10.15
Wei-Min Jin	(b) 158, 589	71.55; 71.20; S5.11; S7.15
Xichunng Jin	(a) 119, K5	61.50; 74.50; S10.15
Xin Jin	(a) 117,K127	74.70; S10.15
R.N. Joarder	(b) 157, 93	61.25; 61.70; S2; S4
	(b) 161, 473	61.20; S2; S5.2
A. Johgo	(b) 157,K137	78.30; S1.2
U.C. Johri	(b) 161, 357	76.80; 75.50; S10.15
S. Joneliunas	(a) 118, 307	75.50; 76.50; S1.1
B.K. Jones	(a) 119, 487	73.60; 71.55; S8.15
	(a) 121, 333	73.40 and 85; S8.15
	(a) 121, 587	71.55 and 73.40; 85; S8.15
K. Jóźwikowski	(a) 122, K39	72.20; S8.15
C. Julier	(a) 118, 553	78.50; 78.30; S8
M. Julien-Pouzol	(a) 121, 21	61.60; 61.10; S8
Ch. Jung	(b) 158,K119	73.20; 73.30; 73.60; S1.2
M. Jung	(a) 122,K135	68.55; 78.40; 78.55; S8.13
W. Jung	(a) 120, 531	71.55 and 73.40; S5.11
G. Jungk	(b) 159, 213	78.20
W. Junker	(b) 161, 543	63.20 and 65.50; 64.70; 65.40; 66.30
M. Jurczyk	(a) 117, 299	75.50; S1.1; S1.2

H. Käämbre	(a) 117, K33	71.55 and 79.75; S9.11; S10.1
N.M. Kabachnik	(b) 161, 113	61.80; 78.70; 79.90; S5.11
	(b) 161, 513	61.80; S5.11
E.V. Kabanova	(a) 121, 105	61.80; 78.60; S10.1
P. Kaboš	(a) 122,K171	75.30 and 75.50; 76.80; S11.2
R. Kabulov	(a) 118, 577	72.40 and 73.40; 85; S5.11
G.A. Kachurin	(a) 117, K29	68.55; S5.11
W.A. Kaczmarek	(a) 117,K141	75.50 and 76.30; S1.1; S1.2
	(a) 119,K159	75.50 and 76.30; S1.2
	(a) 120, K79	76.80; 75.50; S11.2
	(a) 122, 663	75.50 and 76.30; 75.40; S1.1; S1.2
V.N. Kadantsev	(b) 161, 769	71.45 and 72.10; S12.1
	(b) 162, K33	71.45
G. Kaden	(a) 120, 675	73.40 and 85; S7.12
	(a) 121, 315	73.40 and 85; S7.12
A.H. Kadikova	(a) 122,K143	68.20; S1.2
H. Kadomatsu	(a) 118, 513	72.15; S1.2; S1.4
R.F. Kadrgulov	(a) 121, 111	61.60 and 64.80; 66.30; S8.16
E.I. Kafedjiiska	(a) 122,K147	71.55; 61.80; S5.11
V.D. Kagan	(b) 161, 395	78.55; 71.35; S8.11
V.M. Kaganer	(a) 118, 71	61.70; 62.20
	(a) 122, 97	61.70; 61.10
M.I. Kaganov	(b) 162, 469	71.36; 76.20
K. Kagawa	(a) 120, K63	71.55 and 78.50; S5.12
H.-Matsuo Kagaya	(b) 157, 509	61.70; S7.12
	(b) 159, K51	64.30; 64.70; S6
	(b) 162, K21	65.50; 65.70; S2
S. Kahl	(a) 120, 221	62.20 and 75.40; 61.70; S1.2
H. Kahnt	(a) 121, 73	61.40 and 66.30; S9.16
P.K. Kahol	(b) 159, 873	76.60
A. Tang Kai	(b) 157, 657	74.20; S12
J. Kaldasch	(b) 161, 613	73.40; S8.15
B.N. Kalinin	(a) 121,K129	78.55; S7.13
V.V. Kalinovskii	(b) 161, 603	71.35 and 78.65; S8.13
A. Kalninš	(b) 161, 85	61.70; 78.55; S9.11
T. Kamikawa	(b) 159, 571	61.70; 76.30; S9.11
T. Kamino	(a) 122,K105	61.16; 61.70; S1.3
A.A. Kaminskii	(b) 157, 261	71.10 and 78.20
	(b) 157, 267	71.10 and 78.20; 71.70; S10
	(a) 118, K59	78.20; 78.30; 78.45; 78.55; S10
	(a) 119, 285	78.30; S11.2
	(a) 119, 607	78.55; 63.20; 71.55; 78.30; S11.2
	(a) 120, 253	78.55; 71.55; 78.50; S10

Y. Kamiura	(a) 120, K11	61.70; 71.55; 66.30; S5.11
	(a) 120, K63	71.55 and 78.50; S5.12
T. Kanadani	(a) 117, K19	62.20; S4
	(a) 117, K97	62.20; S4
S.H. Kandil	(a) 117, 119	65.40; 61.80; S11
M.A. Kanehisa	(b) 157, 311	71.55; S7.12; S8.13
T. Kaneko	(a) 120,K117	61.60 and 62.20; S8.16
T. Kaneyoshi	(b) 160, 337	73.20 and 75.10
M.C. Kang	(a) 120, K41	68.45; S1.2; S10.1
Z.C. Kang	(a) 122,K101	61.10; S10.15
Y. Kanie	(b) 160, K97	73.20; 73.40
	(b) 162, K25	73.40
M. Kaniewska	(a) 118, 487	73.40; 71.55; 72.40; S7.12; S7.15
	(a) 120, 531	71.55 and 73.40; S5.11
J. Kaniewski	(a) 118, 487	73.40; 71.55; 72.40; S7.12; S7.15
	(a) 120, 531	71.55 and 73.40; S5.11
M.D. Kannan	(a) 121, 515	73.60; 77.55; S10
T. Kanomata	(a) 120,K117	61.60 and 62.20; S8.16
A. Kanoun	(b) 162, 523	71.55; 78.55; S8.13
B.Z. Kanter	(a) 121, 523	68.55; 61.80; S5.11
A. Kara	(a) 122, 545	62.20 and 78.20; S8.11
L.A. Karachevtseva	(a) 117, K37	72.70; 61.70; S8.15
G.R. Karagedov	(a) 119, 417	61.40 and 61.80; S11.2
M. Karai	(a) 117, 515	71.55; 73.30; S8.12
R.K. Karakhanyan	(a) 121, K1	61.14; S5.11
S. Karbanov	(a) 121, 249	64.75 and 75.50; S11.2
M. Karklinya	(b) 162, K91	71.55 and 78.55; S9.11
A.G. Karoza	(b) 158, 469	63.20; 64.75; 78.30; S8.16
S.Kh. Karpenkov	(b) 158,K169	72.15; 71.70; S1.1; S1.2
O.G. Karpinskii	(a) 117, 155	64.70; 61.10; S8
E.A. Karpova	(a) 117, 467	68.55; 68.60
S.Yu. Kasandrov	(b) 162, 435	73.20; S7.1
M. Kaschke	(b) 159, 403	71.35 and 78.20; S8.11
R. Kaschner	(b) 162, 141	68.48 and 73.40; S2
V. Kashcheev	(b) 158, 51	66.30; S11.1
V.A. Kasiyan	(b) 157, 341	72.20; 71.30; S8.12
	(b) 158, 293	71.30 and 72.60; S8.12
G. Kasner	(b) 158, 143	63.50 and 71.20
	(b) 162, 489	71.10; 71.20; S12.1
J. Kasprzak	(b) 160, K53	76.60; 72.20; S10.1
M.E. Kassem	(a) 117, 119	65.40; 61.80; S11
H.A. Kassim	(b) 162, 297	73.20 and 77.30; S10
R. Katilius	(b) 161, 395	78.55; 71.35; S8.11
D.Ya. Kats	(b) 159, 675	64.60 and 68.40
A.A. Katsnelson	(b) 160, 219	71.25; S1.62

A.A. Katsnelson	(b) 160,K117	72.15; S1.2
	(b) 161, K77	61.40; S1.2
	(a) 122, K7	61.40; S1.1
M.I. Katsnelson	(b) 158, 441	63.20 and 71.10; 61.70; 62.20; S1
	(b) 161, 153	62.20 and 71.25; S2; S3; S4
E. Katz	(a) 119, 523	61.70 and 68.55; 73.20; 73.40; S5.11
J.D. Katz	(b) 162, 429	65.50; 64.60; S6; S10.1
N. Kauser	(a) 117, 135	62.20; S1.2
K. Kawabata	(a) 117,K177	78.55; 61.70; S10.1
S. Kawabata	(a) 120, 199	74.70; S10.15
N. Kawahara	(a) 120, 199	74.70; S10.15
T. Kawashima	(a) 120,K117	61.60 and 62.20; S8.16
A.E. Kazachenko	(b) 159, 379	78.20; S8.11; S8.12
L.A. Kazakevich	(a) 118, 53	61.70 and 71.55; S5.11
M.M. Kazanin	(a) 121, 579	71.55; 72.80; 76.80; S5
A.G. Kazanskii	(b) 160, K25	73.60; S5.11
V. Kazlauskas	(b) 162, 281	78.50; 71.55; S11
T.S. Kê (Ge Tingsui)	(a) 121, 139	62.40; S4
	(a) 122, K25	62.40; S4
O.A. Keda	(b) 162, 165	71.55; 61.70; S10.1
O. Keller	(b) 157, 459	78.20
S. Kemmler-Sack	(a) 119, 317	78.55 and 78.60; 71.55; S10.15
	(a) 119, 631	78.55 and 78.60; 71.55; S10
	(a) 120, 643	78.55 and 78.60; 71.55; S11.2
	(a) 121, 289	78.55 and 78.60; 71.55; S11.2
M.A. Kenawy	(a) 120, K31	62.20 and 64.75; S4
	(a) 121, 467	62.20; S4
V.M. Kenkre	(b) 162, 429	65.50; 64.60; S6; S10.1
E.M. Kerimova	(b) 158, 73	63.20 and 78.30; 64.75; S8.16
H. Kerkow	(b) 159, 597	61.80; S5.11
E. Kessler	(b) 158, 521	72.15 and 73.60; S4
F.R. Kessler	(b) 159, 765	71.35; 78.20; S7.11
	(a) 122, 731	78.20 and 85; 78.30; S5.12; S7.13
I.B. Khaibullin	(a) 117, 477	61.80 and 68.55; S1.61
A.P. Khaimenov	(b) 162, 165	71.55; 61.70; S10.1
H.A. Khalik	(a) 117, 163	64.70; 72.20; S11
M.A. Khan	(a) 120,K149	62.20; S1.2
M.N. Khan	(a) 117, 593	78.30; 61.40; S10.15
Y. Khan	(a) 122, K1	61.55 and 75.50; S1.1
A.G. Khandozhko	(b) 157, 151	64.60; 61.70; 76.60; S8
S.D. Khanin	(b) 161, 777	72.20; S10
V.A. Kharchenko	(b) 159, 425	71.35 and 78.55; S8.11
I.Yu. Kharitonov	(a) 121,K149	61.10; S5.11
V.N. Kharkyanen	(b) 157, K99	71.55; 71.38

Z. Khás	(b) 157, 667	63.20 and 72.10; S7.1; S8.1
D.S. Khechinashvili	(b) 159, 403	71.35 and 78.20; S8.11
B. Khelifa	(b) 160, 193	71.25 and 71.60; 78.70; S7.12; S8.13
	(b) 160, 201	71.60; 78.70; S5.11; S5.12; S7.1; S8.1
	(b) 160, 561	71.25; 71.60; 78.70; S5.11; S5.12
	(b) 161, 685	71.20 and 71.60; S6
	(b) 162, K43	71.60; 78.70; S4
M.L. Khidekel	(a) 118,K139	74.70; 78.30; 79.20; 79.60
B.I. Khizhnyak	(a) 117, K37	72.70; 61.70; S8.15
S.I. Khlupin	(b) 158, 63	61.80; S5.11; S5.12
A. Khodjaoui	(a) 120,K111	78.20; 78.40; S11.1
M.Ya. Khodos	(b) 160, 517	71.25; S10.1
M.Kh. Khokonov	(b) 159, 589	61.80; S5.11
V.V. Khomyak	(b) 161, K91	72.80; S5
V.I. Khrabrov	(a) 121,K111	75.50; S1
V.A. Khramtsov	(a) 120, K15	61.70; S5.11
V.I. Khrupa	(a) 121, 11	61.10 and 61.70; S5.11
B.I. Khudik	(b) 158, K89	78.65
	(b) 158, 511	68.45
	(b) 160, 137	68.45
	(b) 162,K127	78.65
N.N. Khusnatdinov	(a) 118, 401	61.70 and 72.40; S10
A.A. Khzardzhyan	(a) 118, 11	61.10 and 61.70
	(a) 118, 21	61.10
K. Kido	(a) 117, 515	71.55; 73.30; S8.12
J. Kierul	(b) 157, 229	71.20 and 73.20; S1.2
	(a) 119,K117	72.15 and 73.60
K.A. Kikoin	(b) 159, 743	71.25; 71.50; 78.20; S1.4; S8
G.C. Kim	(a) 119,K177	77.80; S11
Hyun Kuk Kim	(a) 119,K177	77.80; S11
H.K. Kim	(b) 158,K115	73.40; S8.16
J.E. Kim	(b) 162, K9	64.75; 78.30; S8.16
J.Y. Kim	(a) 122, 255	61.70 and 68.55; S1.3
S.B. Kim	(a) 122,K135	68.55; 78.40; 78.55; S8.13
W.T. Kim	(b) 158,K115	73.40; S8.16
	(b) 162, K9	64.75; 78.30; S8.16
Y. Kimura	(a) 119, 251	76.30; S10.1
P. Kindl	(a) 118, 161	64.60; 78.70; S12
B. Kippelen	(b) 158, 391	78.40; 71.35; S9
	(b) 159, 101	71.35 and 78.20; S9
St. Kirchhoff	(b) 158, 267	71.35 and 71.70; 71.36; 78.20; S9.11
H.R. Kirchmayr	(a) 120, 627	75.50; 64.80; S1.1
I.V. Kiryushin	(a) 117, 177	68.60; S4
E. Kisdi-Koszó	(a) 118, 307	75.50; 76.50; S1.1

E. Kisdi-Koszó	(a) 119, 601	75.50; 75.60; S1.1
A.A. Kiselev	(b) 161, K35	74.20
	(b) 161, 717	74.20; 64.60; S12.1
S.A. Kiselev	(b) 161, 387	77.80; 65.40; S9.16
C. Kisielowski-Kemmerich	(b) 158, K49	71.55 and 76.30; S5.11
	(b) 161, 11	61.70; 76.30; S5.11
G. Kissinger	(a) 121,K141	64.75 and 85; 71.55; S5.11
V.G. Kistol	(b) 158, 301	72.15; S4
H. Kitagawa	(a) 120, K67	71.55; S5.11
M. Kitagawa	(a) 117, 515	71.55; 73.30; S8.12
I.V. Kityk	(b) 157, 473	71.35 and 78.40; S9
P.W. Klamut	(a) 117, K43	74.90 and 79.75; S10.15
	(a) 120, K73	74.70; S10.15
P. Kleinert	(b) 158, 539	68.15; 68.30; S5
	(b) 159, K13	71.38
E. Klier	(a) 119, K79	77.60; S7.11
A. Klimakov	(a) 119, 589	72.20; S8.15
G.M. Klimashin	(a) 121, 579	71.55; 72.80; 76.80; S5
C. Klingshirn	(b) 159, 173	71.35 and 78.20; 73.40; S7.12; S7.15
A.A. Klochikhin	(a) 119, 363	78.30 and 78.40; 64.70; 71.25; 71.35; 78.55; S9
	(b) 160, 211	71.20
T. Kloiber	(b) 158, 753	78.55; 71.35; S5.2
S.I. Klokishner	(b) 160, 641	71.70 and 75.30; 64.60; S10
G. Kluge	(b) 158, K37	72.40; S8
Th. Klupsch	(a) 120, K89	75.60; S11.2
N.I. Klyui	(b) 162, 477	71.25 and 78.20; S6
J. Knopke	(a) 121,K141	64.75 and 85; 71.55; S5.11
W.H. Knox	(b) 159, 11	71.35 and 78.20; 73.40; S7.12
M. Kobayashi	(b) 159, 443	71.35 and 78.65; 78.55; S7.12; S8.12
T. Kobayashi	(b) 160, 561	71.25; 71.60; 78.70; S5.11; S5.12
M. Kobiyama	(a) 120, 119	61.80 and 62.40; S4
S.W. Koch	(b) 159, 249	71.35 and 78.20; S8.12
V.A. Kochelap	(b) 159, 191	78.20
V.P. Kochereshko	(b) 161, 217	73.40 and 78.20; 71.35; S7.12; S7.15
E.V. Kochetkova	(a) 117, K53	75.70; 75.60; S11.2
O.E. Kochkarik	(a) 118, 141	63.20; S8.16
D.D. Koditsa	(a) 122,K177	78.30; S8.16
H.-J. Koehler	(a) 121, K73	74.50; S4
P. Koehler	(b) 158, 267	71.35 and 71.70; 71.36; 78.20; S9.11
W. Koeppen	(a) 122, 731	78.20 and 85; 78.30; S5.12; S7.13
E.M. Kogan	(b) 158, 665	72.20; 72.10

H. Koike	(a) 118, 389	61.60 and 61.70; 68.48; S1.2
S.I. Kokhanovskii	(b) 159, 197	71.35 and 78.20; S7.12; S7.15
K. Kokko	(b) 162, K29	71.10; S1.3; S2
A.A. Kolb	(a) 118, K9	61.80 and 78.30; S7.11
M. Kolbe	(a) 117, 417	61.70; 71.55; 72.20; S7.12
J. Kolenda	(b) 158, 769	78.20; S8.13
	(b) 160, K79	61.70 and 78.20; S8.12
V.M. Koleshko	(a) 117, 177	68.60; S4
M. Kolesík	(b) 160, 375	78.30; S9.15
I.I. Kolkovskii	(a) 118, 65	61.70 and 71.55; S5.11
E. Kolley	(b) 157, 389	74.20
	(b) 157, 399	74.20
	(b) 157,K113	74.20
	(b) 157,K117	74.20
W. Kolley	(b) 157, 389	74.20
	(b) 157, 399	74.20
	(b) 157,K113	74.20
	(b) 157,K117	74.20
N.P. Kolmakova	(b) 159, 845	75.30; 71.70; S10
	(b) 161,K103	71.70 and 75.50; S11.2
J. Kolník	(a) 122, 583	73.30 and 73.40; 85; S5.11; S7.12
V.N. Kolobanov	(b) 157, 745	78.20; 78.40; S11.2
A. Kolodziej	(b) 158, K5	71.25; S10.1
N.N. Kolpakova	(a) 117, 461	64.70; 76.30; S10.15
D.A. Komarkov	(b) 160, K33	71.25 and 72.20; S4
B.A. Komissarova	(b) 160, 255	71.70; 76.80; S1.1; S4
M. Konč	(a) 122, 575	68.55; 61.40; 73.60; S1.1; S1.2
L. Konczewicz	(b) 157, 593	73.40; S7.12; S7.15
E.A. Kondrashkina	(a) 121, K9	61.10; 68.20; S5.12
P.E. Kondrashov	(a) 122,K139	68.55 and 78.65; S5
V.N. Kondratev	(b) 161, 113	61.80; 78.70; 79.90; S5.11
Q.P. Kong	(a) 118, K21	62.20; 62.40; S1.2
	(a) 118, 431	62.20 and 62.40; S1.2
	(a) 122, 511	62.20; 61.70
O.I. Konkov	(a) 120, 519	68.55 and 78.55; S5.11
S.G. Konnikov	(a) 120, 327	61.10 and 68.55; S7.12; S7.15
R.F. Konopleva	(b) 161, 91	61.70 and 61.80; 72.20; S5.11
S.V. Konovalov	(a) 117, 577	75.50; S1.1; S1.2; S1.4
G. Kontrym-Sznajd	(a) 117, 227	71.20 and 78.70
B.J. Koo	(a) 122,K135	68.55; 78.40; 78.55; S8.13
P.S. Kopev	(a) 118, 169	73.40; 78.65; S7.12; S7.15
N. Koppetzki	(b) 159, 797	74.10 and 74.60; 74.40; S10.15
T. Kopte	(a) 120, 547	66.70 and 74.70; S10.15
	(a) 121, K17	66.70 and 74.70; S10.15
V.G. Koreshkov	(b) 158, 441	63.20 and 71.10; 61.70; 62.20; S1

V.G. Koreshkov	(b) 161, 153	62.20 and 71.25; S2; S3; S4
Yu.N. Korkishko	(a) 119, 11	61.40 and 68.48; 78.20; S11.1
	(b) 161, 629	66.30 and 68.45; 68.40; S8.12
J. Kornack	(b) 158, 769	78.20; S8.13
W. Korneta	(b) 160, K13	66.30
A.A. Kornienko	(b) 157, 261	71.10 and 78.20
	(b) 157, 267	71.10 and 78.20; 71.70; S10
A.V. Korolev	(a) 119,K163	75.60 and 75.80; S1.2
L.I. Koroleva	(b) 157, 431	75.30; S8.16
A.N. Korshunov	(a) 117, 509	71.55 and 73.40; S7.12
F.P. Korshunov	(a) 119,K151	61.80 and 74.70; S10.15
	(a) 121, K85	74.70; S10.15
A.V. Koryachenko	(b) 160, 641	71.70 and 75.30; 64.60; S10
I. Kosazkii	(a) 119, 363	78.30 and 78.40; 64.70; 71.25; 71.35; 78.55; S9
F. Kosek	(a) 117, 467	68.55; 68.60
V.V. Koshman	(b) 161, K91	72.80; S5
V.A. Kosobukin	(b) 161, 405	78.20; 77.30
Yu.Yu. Kostetskii	(b) 158, 595	71.70 and 76.30; S9.16
V.A. Kostylev	(b) 158, 307	72.20; S11.2
A.A. Kostyurin	(a) 119, 297	78.20; S1.2; S11.2
M. Kotrla	(b) 157, 635	71.10; 75.10
S.S. Kotsur	(b) 160, 683	78.20; S11.1
V. Koubele	(b) 158,K173	72.40; 71.55; S8.15
A. Koufoudakis	(b) 162, 243	72.20 and 74.70; 76.30; S10.15
	(b) 162,K101	76.30; 74.70; S10.15
T. Radja Koumar	(a) 122,K157	74.70; S10.15
L. Kovács	(a) 120, 97	66.30; 78.30; S10.15
V.B. Kovalchuk	(a) 117,K185	78.55; S5.11
Z.D. Kovalyuk	(b) 161, 419	71.35 and 78.30; S8
	(b) 161, 427	78.55; 71.35; 71.55; S8
	(b) 162, 213	72.20; 71.25; S8
S.V. Koveshnikov	(a) 120, 77	61.70; 71.55; S7.12
	(a) 120, 391	61.70 and 71.55; 66.30; S5.11
V.A. Kozhemyakin	(a) 119, 417	61.40 and 61.80; S11.2
P. Kozioł	(a) 120, K93	77.70; S12
S.G. Kozlova	(b) 158, 329	76.60; 71.70; S11
V.V. Kozlovskii	(a) 117, 85	61.80; 71.55; S7.11; S7.12
W. Kraak	(b) 160, K33	71.25 and 72.20; S4
	(a) 120, 185	74.70; S10.15
	(b) 161, 613	73.40; S8.15
G. Krabbes	(a) 118, K91	72.80; 61.50; S10.1
	(b) 159, 797	74.10 and 74.60; 74.40; S10.15
E. Kraetzig	(a) 117, K85	78.20; 61.80; S11.1
R. Kraienhemke	(a) 119, 327	78.50 and 78.55; 61.50; S11.1
M. Krajčí	(b) 162, K77	71.20; S4

K. Král	(b) 157, 667	63.20 and 72.10; S7.1; S8.1
	(b) 160,K109	63.20 and 72.10
R. Kramer	(a) 119, 589	72.20; S8.15
M. Kranjčec	(a) 122, K87	78.30; S8.16
A. Krapf	(a) 120, 185	74.70; S10.15
E.E. Krasovskii	(a) 118, 147	64.70 and 78.20; 71.25; S1.2; S1.3
	(a) 121, 173	68.55 and 78.65; 61.40; S1.2
L. Kraus	(b) 161, 853	75.60; 75.80
M. Kraus	(b) 160, 275	71.50 and 71.55
G. Krause	(a) 122, 731	78.20 and 85; 78.30; S5.12; S7.13
F. Kremer	(a) 118,K119	77.40; 64.70; S12
P.W. Krempl	(b) 162, 311	78.20; 64.60; S11.1
H. Kresse	(a) 118,K119	77.40; 64.70; S12
	(a) 119, K83	77.80; S12
A.I. Krieger	(b) 161, K49	76.60 and 77.80; S9.16
Yu.G. Kriger	(b) 158, 329	76.60; 71.70; S11
G.S. Krinchik	(a) 119, 297	78.20; S1.2; S11.2
J. Krištiak	(a) 117, K5	61.40 and 78.70; S1.1
K. Krištiakova	(a) 117, K5	61.40 and 78.70; S1.1
N. Kristoffel	(b) 157, K93	71.55 and 74.70; S10.15
	(b) 160, K93	64.70; 63.20; 74.70; S10.15
L.N. Kristoforov	(b) 157, K99	71.55; 71.38
H. Kronmueller	(a) 118, 253	75.60; S1.1; S1.4
	(b) 160, 661	61.70 and 75.50; S10
	(a) 120, 617	75.60; S1.1
A. Krost	(b) 158, 603	71.70; 76.30; 78.55; S8.15
S.P. Kruchinin	(b) 161, 745	74.20; 71.45
V.I. Kruglov	(b) 161, 869	71.55 and 78.50; 71.38
J. Krupski	(b) 157, 199	73.40; S10
I.B. Krynetskii	(b) 159, 845	75.30; 71.70; S10
T. Krysztoń	(b) 158, K21	74.60; S10.15
S.V. Kryuchkov	(b) 161, K85	72.40
L.N. Kryukova	(b) 160, 255	71.70; 76.80; S1.1; S4
J. Krzystek	(b) 161, 245	61.65 and 72.20; 71.55; 75.30; 76.30; S12.1
J. Kuběna	(b) 162, 347	61.10 and 61.70; S7.12
M. Kučera	(b) 157, 745	78.20; 78.40; S11.2
Z. Kučera	(b) 158,K173	72.40; 71.55; S8.15
W. Kucharczyk	(b) 161, K59	77.30; S9.11
	(a) 121,K243	78.20; S11.1
A.G. Kuchin	(a) 121,K111	75.50; S1
A.E. Kuchma	(b) 159, K77	72.10
V. Kudeyarova	(a) 122, 599	68.55 and 78.65; S5.11
G. Kudlek	(b) 159, 443	71.35 and 78.65; 78.55; S7.12; S8.12
	(a) 120, 11	78.65; 68.55; 71.35; 73.60; 78.55; S8.12

N.A. Kudryashov	(b) 157, 531	61.80
	(b) 161, 483	61.80 and 78.70; S5.11
Yu.V. Kudtyavtsev	(a) 121, 173	68.55 and 78.65; 61.40; S1.2
T.A. Kudykina	(b) 160, 365	78.20; 71.35
O. Kuehn	(b) 160, 211	71.20
G. Kuehnel	(a) 117, 527	71.55; 72.20; S7.12
	(b) 158,K111	65.70; 72.20; S7.12
L. Kuenne	(b) 157, 229	71.20 and 73.20; S1.2
T. Kuga	(b) 159, 347	71.35 and 78.20; S9
H.-A. Kuhn	(a) 119, 93	62.20; S1.2; S1.61
S.P. Kuila	(a) 118, 311	78.60; 61.70; 71.55; S9.11
P. Kuivalainen	(b) 159, 827	72.10 and 73.40; 85; S7.1
V.D. Kulakovskii	(b) 159, 125	71.35 and 78.55; 73.40; S7.12; S7.15
	(b) 159, 165	71.35 and 78.55; 73.40; S7.15
B.I. Kuliev	(b) 158,K179	72.10; 73.60
A. Kulik	(a) 119, 495	68.48 and 68.55; S7.12; S7.15
A.Yu. Kulikov	(a) 121, 227	72.40; 78.40; S5.11
N.P. Kulish	(a) 120, 315	61.55; S1.1; S1.2; S1.3; S4
S.N. Kulkov	(a) 121, 105	61.80; 78.60; S10.1
W. Kullmann	(b) 162, 125	63.20; S8
Asita Kulshreshtha	(a) 117, K77	78.30; 63.20; S12
S.K. Kulshreshtha	(b) 157, 547	64.60; 61.12; S11
	(b) 160, 313	76.80; S1.62
B.E. Kulyabin	(b) 158, 239	71.20 and 71.25; S9.11; S10.1
Anil Kumar	(b) 157,K131	75.40 and 75.60
A. Kumar	(b) 157, 209	68.30 and 73.30; S2
	(a) 119, K97	61.40; S8
K. Kumar	(b) 160, 349	76.30; 71.70; S11
M. Kumar	(b) 158, 11	64.70; 63.75; S9.11
V. Kumar	(a) 117, 251	71.55; 73.40
K. Kumazaki	(a) 119, 177	63.20 and 68.30; 68.60; S8.12
	(b) 160,K173	77.20; S8.15
P. Kundrotas	(b) 158, 51	66.30; S11.1
I.A. Kunitskii	(a) 120, 447	73.40; 73.20; S5.11; S10.1
K. Kunze	(b) 157, 71	61.10; 61.12; 61.14
	(b) 157, 489	61.10; 61.12; 61.14
M. Kupka	(b) 157, K39	74.50
K. Kurbanov	(a) 120, 253	78.55; 71.55; 78.50; S10
A.I. Kurbatov	(a) 121, K65	68.55; S5.11; S9.12
V.G. Kurdyumov	(a) 121, 455	61.70 and 62.20; 64.70; S1.2
S.I. Kurganskii	(b) 161, 697	71.25 and 73.60; 71.10; S1.2
J. Kuriata	(a) 121, K95	76.30; 75.50; S11.1
	(b) 162,K101	76.30; 74.70; S10.15
T. Kurihara	(a) 117, K9	61.70 and 78.70; S4
L.I. Kurkina	(b) 160, K37	71.50; 73.20; 75.25; S1.1
V.M. Kurliand	(a) 122, 341	71.55; 76.30; S10.15
T. Kurobori	(a) 117,K177	78.55; 61.70; S10.1

T. Kurobori	(a) 122, K91	78.55; 61.70; 71.55; S10.1
L.A. Kurochkin	(a) 119,K151	61.80 and 74.70; S10.15
K. Kurosawa	(a) 117, 515	71.55; 73.30; S8.12
M. Kurzawa	(a) 121, K95	76.30; 75.50; S11.1
P. Kus	(a) 119, 201	73.40 and 74.70; 78.65; S10.15
V.D. Kushkov	(a) 117, K13	64.70; S10.15
V.I. Kushnir	(a) 122, 391	78.70; 61.10; S5.11; S5.12
R. Kuszelewicz	(b) 159, 181	78.20; 68.55; S7.12; S7.15
V.V. Kutasov	(a) 121, 579	71.55; 72.80; 76.80; S5
V.G. Kutilin	(a) 120, 457	68.55; 75.70; 78.20; S1.1
Y. Kutomi	(b) 158, 383	61.70 and 78.60; 79.75; S9.11
T.R.N. Kutty	(a) 119, 127	68.55; 73.60; 78.65; S8.12
M. Kuwata-Gonokami	(b) 159, 347	71.35 and 78.20; S9
B.S. Kuzhel	(b) 161, 783	72.10; 64.75; 71.25; S1.4
A.A. Kuzin	(a) 122,K139	68.55 and 78.65; S5
E.A. Kuzmenkov	(a) 121, 205	61.80 and 72.20; S12
J. Kuzmik	(a) 119,K185	73.40 and 85; S7.12
A.V. Kuznetsov	(b) 159, 191	78.20
	(b) 159, 223	78.20
	(b) 159, 449	71.35 and 78.20; S8.1
P.I. Kuznetsov	(b) 158, 359	71.35 and 78.65; S8.12
	(b) 159, 449	71.35 and 78.20; S8.1
V.A. Kuznetsov	(a) 122, 563	73.40 and 75.50; S9.11; S12
N.Ya. Kuznetsova	(b) 160, K5	66.30
V.V. Kveder	(b) 158, K49	71.55 and 76.30; S5.11
Motoko Kwete	(a) 122, 129	61.80; 78.70; S5.11
W. Laasch	(b) 158, 753	78.55; 71.35; S5.2
E. Lach	(b) 159, 125	71.35 and 78.55; 73.40; S7.12; S7.15
H.K. Lachowicz	(a) 119, 595	72.15 and 75.80; 75.60; S1.1; S1.2
C. Lai	(b) 157, 379	74.60; 76.30; S10.15
P. Lais	(b) 160, 275	71.50 and 71.55
K. Lal	(a) 121, K47	64.70; S9.15
T. Lalinský	(a) 119,K185	73.40 and 85; S7.12
	(a) 122, 715	73.40 and 85; S7.12
A. Lambrecht	(a) 119, 711	73.40 and 78.55; 72.40; S7.16
S. Landgraf	(a) 118, 203	61.80 and 68.60; 79.20; S9.12
C. Landron	(a) 121, 359	61.10 and 61.55; S10.15
J. van Landuyt	(a) 120,K131	61.80; S5.11
E. Lane	(a) 117,K107	73.60; S7.12
H.Th. Langhammer	(a) 120, 285	77.60; 77.20; 77.30; S10.15
D. Lap	(b) 159, 49	71.35 and 78.20; S8.11; S8.12; S8.15

J. Lapasset	(a) 120, K25	64.70; S9
S. Lapinskas	(b) 158, 51	66.30; S11.1
E. Laredo	(a) 121,K235	77.30; S9.12
A. Largeteau	(a) 121, 627	77.20; 64.80; 77.40; S11.2
	(a) 122, 371	72.20; S11.2
G.V. Lashkarev	(b) 162, 213	72.20; 71.25; S8
A.L. Laskar	(a) 120,K133	61.70; S9.11; S10.1
J.P. Laval	(a) 117, 409	61.60 and 66.30; 64.75; S9.15
D. Lavielle	(a) 119, K41	71.55; S7.12
V. Lazauskas	(b) 159, K63	71.55; S7.11
M. Láznička	(a) 118, 567	78.55 and 78.65; S7.12
V.I. Lazorenko	(b) 162, 213	72.20; 71.25; S8
V.Yu. Lebed	(b) 159, 267	71.35 and 78.65; S8.15; S9
	(b) 159, 449	71.35 and 78.20; S8.1
A.B. Lebedev	(a) 119, 87	62.40; S1.2
J. Lech	(b) 162, 275	76.30; S11
B. Lecuyer	(b) 157,K123	74.30; S10.15
N.N. Ledentsov	(a) 118, 169	73.40; 78.65; S7.12; S7.15
J. Ledzion	(a) 119,K117	72.15 and 73.60
C.Y. Lee	(a) 119, 35	61.40 and 64.70; 62.20; S1.1; S1.2
J.H. Lee	(a) 122,K135	68.55; 78.40; 78.55; S8.13
J.Y. Lee	(b) 162, K9	64.75; 78.30; S8.16
S. Lee	(a) 120, 497	61.70 and 68.25
S.A. Lee	(b) 157, K83	68.30 and 78.35; S5
S.K. Lee	(b) 162, K9	64.75; 78.30; S8.16
A. Lehmann	(a) 119, 683	78.30 and 78.65; S9.12
	(b) 162, K53	63.20 and 78.30; S5
G. Lehmann	(b) 159, 495	76.30; 64.60; S11
J. Lehmann	(b) 158,K197	76.60; S1
L. Lehmann	(a) 122, 683	73.40 and 78.55; 78.60; S7.15
S. Lehri	(b) 159, 659	63.20; 62.20; S1
H. Leier	(b) 159, 457	71.35 and 78.20; S7.12; S7.15
V.F. Lemberg	(b) 159, 645	65.50; 63.20; 64.10; S4
H. Lemke	(a) 122, 617	71.55; S5.11
V.M. Lenchenko	(a) 118, 117	61.70; 61.80; S5.11
H. Lengfellner	(a) 122,K161	74.60; 74.70; S10.15
R. Lenk	(b) 157, 331	72.10
	(b) 158, 653	72.10
	(b) 161, 797	72.10
	(b) 162, 227	72.10
K. Lenkeit	(b) 161, 513	61.80; S5.11
	(a) 121, 523	68.55; 61.80; S5.11
M. Lenzner	(b) 158, K37	72.40; S8
	(b) 162, 545	72.40; S12
H. Leon	(b) 159, 731	72.10 and 73.40; S7.12
	(b) 160, 105	72.10 and 73.40; S7.12; S7.15
N.K. Lerinman	(b) 160,K141	71.30; S8.16

M. Leszczynski	(a) 119, 495	68.48 and 68.55; S7.12; S7.15
C. Levade	(a) 122, 545	62.20 and 78.20; S8.11
A. Levcuras	(a) 119, K41	71.55; S7.12
E.M. Levin	(b) 161, 783	72.10; 64.75; 71.25; S1.4
R.Z. Levitin	(b) 161,K103	71.70 and 75.50; S11.2
R. Lévy	(b) 158, 391	78.40; 71.35; S9
	(b) 159, 101	71.35 and 78.20; S9
	(b) 159, 259	71.35 and 78.20; S9
V. Lhotská	(a) 120, 273	77.40; 77.80; S12
Bin Li	(a) 118, K47	61.70 and 78.30; S11.1
Chang-hua Li	(a) 119, 677	78.60; 78.65; S8.11
Fu-Zhen Li	(b) 161, K29	71.70; 76.30; S11.1
F.H. Li	(a) 119, 511	68.55; 68.48; S5.11; S7.13; S7.15
G.Y. Li	(a) 122, K11	61.70 and 78.70; 62.20; S1.1
Haoliang Li	(b) 162, 545	72.40; S12
H.C. Li	(a) 122, K55	74.60 and 74.70; S10.15
Jia-liang Li	(b) 160, 329	75.10
J. Li	(a) 121, 163	73.40 and 73.60; S10
Liping Li	(a) 119, 55	64.70 and 74.70; S10.15
Mai Suan Li	(b) 160,K159	75.30 and 75.40
	(b) 161, 345	75.30 and 75.40
Shaoping Li	(b) 159, 861	68.30 and 75.10; 75.70
Xiao-Guang Li	(b) 158, K27	74.60; 74.70; S10.15
	(a) 119,K141	74.70; S10.15
X.D. Li	(a) 121,K185	68.48 and 79.60; S1.1; S10.1
X.-G. Li	(a) 122, K55	74.60 and 74.70; S10.15
Y. Li	(a) 122, 511	62.20; 61.70
Zhao-Min Li	(b) 161, K29	71.70; 76.30; S11.1
Zugang Li	(a) 119, K5	61.50; 74.50; S10.15
Kejun Liao	(a) 119, K31	68.25; S5.11
A.V. Liaptsev	(b) 161, 717	74.20; 64.60; S12.1
A.V. Liapzev	(b) 161, K35	74.20
I. Licea	(a) 118, 131	64.70; 65.40; S11.1
J. Liebler	(b) 159, 309	71.35 and 78.20; S7.12; S8.11; S9
G. Lifante	(b) 159, K37	61.70; S9.11
H. Lim	(b) 162, K9	64.75; 78.30; S8.16
M. Limonov	(b) 161, 575	63.20; 78.30; S10.15
C.L. Lin	(a) 120,K131	61.80; S5.11
K.M. Lin	(a) 120, 497	61.70 and 68.25
Xiaotong Lin	(a) 120,K177	72.40; 61.70; S5.11
K. Lindberg	(b) 159, 827	72.10 and 73.40; 85; S7.1
M. Lindberg	(b) 159, 249	71.35 and 78.20; S8.12
S.M. Lindsay	(b) 157, K83	68.30 and 78.35; S5
J.L. Lindstroem	(a) 117,K171	78.30; S5.11
	(a) 120, 539	71.55; 61.70; S5.11
J.J. Liou	(a) 118, K65	73.40 and 85; S7.12; S7.15

J.J. Liou	(a) 119, 337	73.40 and 85; S7.12; S7.15
J. Lipka	(a) 122,K171	75.30 and 75.50; 76.80; S11.2
B. Lippold	(b) 158, 707	74.70 and 76.60; S10.15
	(a) 120, 557	74.70; 61.60; 75.30; S10.15
	(a) 121, K91	75.30; 75.50; S1.2; S10.15
A.G. Lipson	(a) 122, 563	73.40 and 75.50; S9.11; S12
Yu.A. Lisovskii	(a) 120, 457	68.55; 75.70; 78.20; S1.1
V.G. Litovchenko	(b) 162, 477	71.25 and 78.20; S6
A.P. Litvinchuk	(a) 119, 363	78.30 and 78.40; 64.70; 71.25; 71.35; 78.55; S9
E. Litwin-Staszewska	(b) 157, 593	73.40; S7.12; S7.15
Hongjian Liu	(a) 119, 55	64.70 and 74.70; S10.15
H. Liu	(a) 121, K61	68.55; S10.1
Mingzhi Liu	(a) 121,K167	62.40; 66.30; S1.3
S. Liu	(b) 161, 677	71.38
S.M. Liu	(a) 118, K21	62.20; 62.40; S1.2
W. Liu	(a) 117, 221	68.55; S4
F.P. Livi	(a) 120, 609	75.60; S1.61
J. Liwo	(a) 120,K159	73.25; S10.15
J. Llopis	(a) 119, 661	78.55 and 78.60; S10.1
V.A. Lobatch	(b) 158, 239	71.20 and 71.25; S9.11; S10.1
	(b) 160, 165	71.20 and 71.25; S10
	(b) 160, 505	71.20 and 71.25; S10
	(b) 161, 271	71.20 and 71.25; S10.1
	(b) 161, 647	71.20 and 71.25; S9.11; S10.1
P.I. Loeff	(a) 117, K99	64.70; S1.2
B.M. Loginov	(a) 120, 397	61.70; S1.2; S9.11
Y.Y. Loginov	(a) 118, 117	61.70; 61.80; S5.11
M.M. Lokshin	(a) 120, 447	73.40; 73.20; S5.11; S10.1
V.M. Loktev	(b) 161, 731	74.20; 71.45; S12.1
S.K. Lomash	(a) 117, 203	61.70 and 68.48; S5.11; S12
A.T. Lonchakov	(b) 161, 303	71.30 and 72.20; S5.12
C.A. Londos	(a) 118, K13	66.30; S5.11
E. Lopez-Cruz	(b) 162,K131	78.20; S8
A.V. Losev	(a) 121,K129	78.55; S7.13
R.Sh. Lotfullin	(b) 161, K49	76.60 and 77.80; S9.16
J.E. Lowther	(b) 162, 413	65.50; 71.20; S1
Yu.E. Lozovik	(b) 161, K11	68.90
V.Z. Lozovskii	(b) 158, K89	78.65
	(b) 158, 511	68.45
	(b) 160, 137	68.45
	(b) 162,K127	78.65
H. Lu	(a) 122, 481	61.50; 79.60; S1.2
J.Z. Lu	(a) 121, 149	62.40; S4
	(a) 121, 475	62.40; S4
Mu Lu	(a) 117,K127	74.70; S10.15
W.C. Lu	(a) 120, K41	68.45; S1.2; S10.1
X.T. Lu	(b) 158, 113	63.20 and 78.50; S7.12

W. Lubczyński	(a) 120, 525	71.25; 72.20; S1.62
S. Luby	(a) 120, 371	66.30; S1.1; S1.2
C. Lucchini	(b) 157,K123	74.30; S10.15
D.G. Luchinskii	(a) 117, 587	78.55; S8
T. Luciński	(a) 118, 189	61.40 and 68.55; 73.60; 75.70; S1.2; S1.4
	(a) 122, 323	73.60; 72.15; S1.1
P.F. Lugakov	(a) 118, 53	61.70 and 71.55; S5.11
	(a) 118, 65	61.70 and 71.55; S5.11
A.M. Luginets	(a) 119,K151	61.80 and 74.70; S10.15
P. Lukáč	(a) 118, 455	62.20; S1.2
M. Łukasiak	(a) 118, K35	75.25 and 76.80; S11.2
M.M. Lukina	(b) 159, 845	75.30; 71.70; S10
B.A. Lukiyanets	(b) 159, K73	71.70
V.V. Lukyanitsa	(a) 118, 65	61.70 and 71.55; S5.11
Bao-Sheng Luo	(a) 117, 335	61.60 and 74.70; S10.1
Ningsheng Luo	(b) 158, 493	68.30; S1.2
N.H. Luong	(a) 121, 607	75.25; 75.30; S1.2; S1.4
L.N. Lupichev	(b) 161, 769	71.45 and 72.10; S12.1
	(b) 162, K33	71.45
A.Ch. Lushchik	(b) 161, 525	61.70 and 78.55; 61.80; 76.30; S9.11
P.V. Lushnikov	(b) 161, 647	71.20 and 71.25; S9.11; S10.1
F. Luty	(b) 159, 629	64.60; S9
G.I. Lyashenko	(a) 121,K193	68.55; 73.60; 78.55; S6
V.S. Lysenko	(a) 120, 447	73.40; 73.20; S5.11; S10.1
	(a) 120, 475	61.80 and 68.55; 78.30; S5.11
A.V. Lyubchenko	(a) 117, K37	72.70; 61.70; S8.15
A.S. Lyutovich	(a) 119, K27	68.55; S4
Guangquen Ma	(b) 158, 127	64.60
Teng-Ge Ma	(a) 118, 179	61.80 and 68.55; 73.60; S5.11
H. Maaref	(b) 162, 523	71.55; 78.55; S8.13
R. Mach	(a) 121, 305	72.20 and 79.70; 85; S8.11
M. Maćkowiak	(a) 119, 77	63.20 and 64.70; 76.60; S12
J.L. Macqueron	(a) 117, 113	64.70; S1.2
Y. Madhavan	(b) 158, 427	66.30; S8.15
	(b) 162, 119	63.20; 61.70; S8.15
K. Maeda	(a) 121,K125	78.20; 71.50; S7.12
M. Maeder	(a) 121, K65	68.55; S5.11; S9.12
	(a) 122,K133	68.20; 61.14; S5.11
F. Maes	(b) 161, K1	61.70 and 76.30; S9.11
P. Magdanz	(a) 122, 297	73.40; S10.1
E.L. Mager	(a) 122,K123	62.65 and 64.70; S1.2
A.Z. Magomedov	(b) 158, K1	63.20 and 78.30; S8.16
D.P. Mahapatra	(b) 158, 261	71.25 and 78.70; S7.12

K.V.S. Mahesh	(b) 159, 551	61.40; 78.70; S1.1; S1.2
I. Mahmoud	(a) 121,K219	75.30; S1.1
	(a) 122,K113	61.55; S1.4
S.A. Mahmoud	(a) 117, 437	62.20; S1.2
B.E. Maile	(b) 159, 457	71.35 and 78.20; S7.12; S7.15
E. Majkova	(a) 120, 371	66.30; S1.1; S1.2
G.M. Makarova	(a) 121,K111	75.50; S1
A.G. Makhanek	(b) 157, 641	71.70; 78.50; S11.2
V.N. Makhov	(b) 157, 745	78.20; 78.40; S11.2
K. Maki	(a) 118, 253	75.60; S1.1; S1.4
T. Maki	(a) 122, 503	66.30; 77.40; S10
L.V. Maksimov	(a) 119, 621	61.40 and 78.20; 61.80; 62.65; S10.1
G.D. Maksimović	(b) 160, 635	75.50; S1.1
Yu.E. Makushok	(a) 121,K181	68.55; S6
A.V. Malakhovskii	(b) 159, 883	78.20
I.E. Malinovskii	(b) 158, 359	71.35 and 78.65; S8.12
	(b) 159, 267	71.35 and 78.65; S8.15; S9
	(b) 159, 449	71.35 and 78.20; S8.1
V.I. Maltsev	(a) 120, 209	75.50; S11.2
P. Maly	(a) 121,K223	78.20; S5.11
G.A. Malygin	(a) 119, 423	61.70; 62.20; S4
N.T. Mamedov	(a) 117, K15	64.70 and 76.60; S8.16
	(b) 159, K83	72.60; S8
R. Manaila	(b) 161, 65	61.50; 61.55; S1.1; S1.2; S4
J. Manam	(a) 118,K129	78.60; S9.11
C. Mande	(b) 158, 737	78.70; S10
	(b) 161, 891	78.50; 71.55; S11.1
M. Manfredi	(a) 121, 407	61.70; 71.55; 78.55; S9.11
C. Manfredotti	(a) 119, 169	72.40 and 73.60; 78.65; S5.11
D. Mangalaraj	(a) 117, 495	68.55 and 73.60; 77.55; S9
	(a) 121, 515	73.60; 77.55; S10
J. Mangin	(a) 117, 319	77.30; 77.70; S11.1
	(a) 120,K111	78.20; 78.40; S11.1
Ph. Mangin	(a) 122, 311	68.55; S1.1; S1.4
I. Manna	(a) 119, K9	66.30 and 68.48; S1.2
Ll. Mañosa	(a) 117, 113	64.70; S1.2
A.M. Mansanares	(a) 117,K155	75.50 and 76.50; S12
J.R. Manson	(b) 157, K87	73.20
Y.K. Manucharova	(a) 121, K1	61.14; S5.11
M. Manzel	(a) 117,K119	74.70; S10.15
	(a) 119, 279	75.70; S1.62
I.V. Manzhos	(b) 157, K65	75.30
Yu.A. Manzhosov	(a) 117, K29	68.55; S5.11
Xianglei Mao	(b) 158, K27	74.60; 74.70; S10.15
	(a) 119,K141	74.70; S10.15
G. Marchal	(a) 122, 311	68.55; S1.1; S1.4
M. Mareyen	(b) 159, 235	71.35 and 78.20

M.L. Marín	(b) 157, 275	71.30 and 75.40; S11.2
V. Marinkovic	(a) 117, 351	61.55 and 68.55; S1.3
Yu. Markov	(b) 161, 575	63.20; 78.30; S10.15
G. Marnier	(a) 117, 319	77.30; 77.70; S11.1
	(a) 120,K111	78.20; 78.40; S11.1
S.A. Marshall	(b) 157, 379	74.60; 76.30; S10.15
	(b) 160, 591	71.70 and 76.30; S8.12
D. Martínez	(a) 118, K31	75.60; S11.2
J.L. Martínez-Cuellar	(b) 158, 481	71.25 and 73.40; S7.15
J.M. Martinez-Duart	(a) 121,K181	68.55; S6
M. Martini	(b) 159, 577	61.70; 78.50; 78.55; S10.1
K.A. Maslin	(b) 161, 233	78.30 and 78.65; 71.36; S8.13
A.V. Maslov	(a) 117, 341	61.10; 68.20
A. Masłowska	(b) 157, 593	73.40; S7.12; S7.15
S.S. Maslyanchuk	(a) 119, 215	71.55; 72.80; 75.50; S8
V.T. Maslyuk	(b) 161, 43	61.40; 65.50; 71.45; S8
V.F. Masterov	(b) 159, K97	76.80; 74.70; S10.15
Y. Masumoto	(b) 159, 133	71.35 and 78.20; 73.40; S7.12; S9
B. Matátko	(a) 122, 715	73.40 and 85; S7.12
C.D. Mateescu	(b) 160, 423	61.70; 66.30; 78.50; S9.11
A.V. Mateleshko	(a) 117, 241	71.55; 72.40; 73.60; S5
N. Mathur	(a) 122, K59	66.30 and 72.15; S1.2
E. Matsubara	(a) 117, K1	61.40; S10
M. Matsuoka	(b) 159, 347	71.35 and 78.20; S9
K. Matsuura	(a) 121, 595	71.55 and 78.55; 72.80; S8.13
S. Matthies	(b) 157, 71	61.10; 61.12; 61.14
	(b) 157, 489	61.10; 61.12; 61.14
P. Matthys	(b) 161, K1	61.70 and 76.30; S9.11
G.A. Matveev	(b) 161, 303	71.30 and 72.20; S5.12
V.A. Matveev	(a) 118,K107	75.70; 75.60; S11.2
Z.A. Matysina	(b) 159, K7	68.20 and 75.30
G. Mayer	(b) 158, 707	74.70 and 76.60; S10.15
	(b) 159, 457	71.35 and 78.20; S7.12; S7.15
J. Mayer	(a) 122, 79	61.40 and 61.80; 64.70; S1.2; S4
W. Mayr	(b) 160, 613	72.15; S1.63
Yu.I. Mazur	(b) 162, 531	72.10; 71.25; 72.40; S5
A.M. Mazzone	(a) 120, 379	61.70; 61.80; S5.11
H.A. McKinstry	(b) 157, 357	74.20; 74.70
T.R. Mchedlidze	(b) 158, K49	71.55 and 76.30; S5.11
N.I. Medvedeva	(b) 158, 239	71.20 and 71.25; S9.11; S10.1
	(b) 160, 517	71.25; S10.1
S.P. Medvetskii	(a) 119, 191	68.48; 73.60; S7.12; S10.1
I.G. Megela	(a) 118, K9	61.80 and 78.30; S7.11
W. Mehnert	(a) 122, 297	73.40; S10.1
A.K. Meikap	(b) 160, 473	71.38 and 73.60; S1.2
V.P. Meiklyar	(b) 160,K149	76.30 and 76.70; 64.60; S9.16

V.P. Meiklyar	(b) 162,K107	76.70; S9.16
A. Meisel	(b) 160, 143	79.60; S7.1; S11
	(a) 122, 331	68.48 and 73.40; 79.20; S7.12
O.G. Melikyan	(a) 122, 459	61.10
H. Mell	(b) 160, K25	73.60; S5.11
S.V. Melnichuk	(b) 157, K19	63.20; S8.13
A.V. Melnikov	(a) 117, K13	64.70; S10.15
B.Ya. Meltser	(a) 118, 169	73.40; 78.65; S7.12; S7.15
K. Melzer	(a) 118, K35	75.25 and 76.80; S11.2
Li-jian Meng	(a) 119, 677	78.60; 78.65; S8.11
D.R. Menke	(b) 159, 443	71.35 and 78.65; 78.55; S7.12; S8.12
H. Menniger	(a) 121, 95	61.70 and 78.60; 85; S7.12
A.Z. Menshikov	(b) 158, 319	75.30; 61.55; S1.61
J.-C. Merle	(b) 158, 103	63.20 and 78.30; S9.16
M. Merli	(a) 121,K135	61.80 and 85; S5.11
Z.V. Meshevich	(a) 122,K123	62.65 and 64.70; S1.2
R. Messi	(b) 157,K123	74.30; S10.15
F. Metawe	(b) 157, 609	68.55 and 78.65; S8
S. Methfessel	(a) 120, 159	68.55 and 75.70; 76.80; S1.1
	(a) 121,K211	75.50; S1.1
H. Metz	(a) 122, K73	76.30
H.-G. Meyer	(a) 121, K73	74.50; S4
F. Michard	(a) 118, 461	62.65; 78.35; S10
C. Michel	(a) 118, 225	71.55; 78.55; S8.13
J.C. Michel	(b) 159, 181	78.20; 68.55; S7.12; S7.15
J. Mielnicki	(b) 157,K103	72.10 and 75.10
	(b) 160, K61	75.10; 75.30
	(b) 160,K129	72.10 and 75.10
	(b) 161, 335	75.10
R. Mikaitis	(a) 119, 71	64.70 and 76.80; S8.16
V.V. Mikhailin	(b) 157, 745	78.20; 78.40; S11.2
G.V. Mikhailov	(b) 159, 425	71.35 and 78.55; S8.11
M.D. Mikhailov	(a) 117, 467	68.55; 68.60
A.P. Mikhailovskii	(b) 158, 229	71.25 and 75.10; S8
I.M. Mikhailovskii	(a) 119,K113	61.70; S1
M. Mikhov	(a) 117, 571	75.30; S1.2; S1.4
	(a) 118, 289	75.60
V.I. Mikityuk	(a) 119, 215	71.55; 72.80; 75.50; S8
V.I. Mikla	(a) 117, 241	71.55; 72.40; 73.60; S5
P. Mikulski	(a) 119,K127	68.55; 78.30; S5
B.V. Mill	(a) 118, K59	78.20; 78.30; 78.45; 78.55; S10
	(a) 120, 253	78.55; 71.55; 78.50; S10
D.A.B. Miller	(b) 159, 11	71.35 and 78.20; 73.40; S7.12
	(b) 159, 173	71.35 and 78.20; 73.40; S7.12; S7.15
M.I. Milyan	(b) 159, K7	68.20 and 75.30

N.K. Mina	(a) 121, 235	61.55 and 74.70; S10.15
A. Miralles	(a) 120,K105	78.55; S10.1
O.A. Mishin	(b) 161, 483	61.80 and 78.70; S5.11
Yu.M. Mishin	(a) 117, 91	61.20 and 61.70; 61.50
	(b) 159, 675	64.60 and 68.40
A. Mishra	(a) 122, 405	73.40 and 85; S7.12; S7.16
A.K. Mishra	(b) 157, 85	61.25; 72.15; S2
M.K. Mishra	(b) 162, K73	63.20; S1.2
S.K. Mishra	(b) 158, 153	63.20; 62.20; S2
J. Misiewicz	(a) 118, 185	73.20 and 73.30; 72.40; S5.11
V.R. Misko	(b) 159, 477	71.36
S.K. Misra	(b) 162, 265	76.30; S11
	(b) 162, 585	76.30; S11
R. Mitdank	(a) 117, 485	73.60; 66.30; S7.15
	(a) 122, 683	73.40 and 78.55; 78.60; S7.15
V.Ya. Mitrofanov	(b) 160,K153	75.50; 76.20; S11.2
V.I. Mitrokhin	(a) 119, 535	71.55; 62.40; S7.1
C. Mitros	(b) 162, 243	72.20 and 74.70; 76.30; S10.15
	(b) 162,K101	76.30; 74.70; S10.15
A. Mitsuishi	(a) 120, K63	71.55 and 78.50; S5.12
H. Mizubayashi	(a) 117, 125	62.40; 74.70; S10.15
	(a) 122, 221	61.40 and 62.40; S1.2
S. Mochizuki	(b) 157,K137	78.30; S1.2
K.-P. Moellmann	(a) 121, 635	78.20; 71.25; 78.55; S8.15
A.Z. Mohamed	(a) 120, 601	75.60; 64.80; S1.1
N.M. Mohamed	(a) 119, K47	72.40; S8
M.A. Mohammed	(a) 120, K45	73.40; S7.12
S. Mohan	(a) 122,K157	74.70; S10.15
S.V. Moharil	(a) 118, K55	78.60; S10
D.P. Moiseev	(b) 161, 293	74.50; 74.20; S10.15
L.P. Mokhracheva	(a) 117, 577	75.50; S1.1; S1.2; S1.4
J. Molenda	(a) 119, 221	72.15; S1.63
	(a) 119, 571	72.15; 61.70; S1.63
V.B. Molodkin	(a) 121, 11	61.10 and 61.70; S5.11
S.L. Molodtsov	(b) 159, K77	72.10
M.I. Molotskii	(a) 120, 83	61.70; 79.75; S9
R.S. Momtaz	(a) 118, 197	68.55 and 78.65; 73.60; S10.1
M. Mondal	(b) 158,K123	73.40; S5.12
	(b) 160, 673	75.20 and 75.70; 71.25; S7.16
B. Monemar	(a) 120, 539	71.55; 61.70; S5.11
O. Monnereau	(a) 122,K101	61.10; S10.15
C. Mora	(a) 120, 97	66.30; 78.30; S10.15
J.R. Morante	(a) 120,K105	78.55; S10.1
M. Mora-Ramos	(b) 160, 117	71.38 and 73.40; 63.20; S7.12
V.N. Mordkovich	(a) 120, K1	61.10 and 61.70; S5.11
R.A. Moreno	(a) 119, 583	72.20 and 73.60; S12
J. Moret	(a) 120, K25	64.70; S9

P. Moretti	(a) 117, K85	78.20; 61.80; S11.1
D.V. Morgan	(a) 120, K45	73.40; S7.12
N. Mori	(b) 159, K51	64.30; 64.70; S6
T. Mori	(a) 120,K117	61.60 and 62.20; S8.16
Yu.N. Morokov	(b) 160, 483	68.25 and 68.40; 73.20; S5.11
N.K. Moroz	(b) 159, K83	72.60; S8
N.V. Moseev	(a) 119, 437	61.80 and 64.75; S1.1; S4
L.V. Mosina	(b) 158, 337	76.30; 71.70; S9.15
S.A. Moskalenko	(b) 159, 477	71.36
A.K. Moskalev	(b) 161, K49	76.60 and 77.80; S9.16
H.A. Motaweh	(a) 121,K219	75.30; S1.1
	(a) 122,K113	61.55; S1.4
M.A. de Moura	(b) 158, K9	71.35
N.O. Mourad	(a) 120, K31	62.20 and 64.75; S4
	(a) 121, 467	62.20; S4
C.A. Moyer	(a) 120, K71	72.15; S4
P.E. Mozol	(a) 122, 243	73.20 and 73.40; 72.40; S8.15
E. Mrosan	(b) 162, 237	61.70 and 72.10; S1.2
S. Mrowec	(a) 119, 571	72.15; 61.70; S1.63
J. Mucha	(a) 118, K95	72.15 and 75.70
	(a) 119, K53	74.30; S10.15
A. Mueller	(a) 117, 447	65.90; S12
F. Mueller	(a) 121, K73	74.50; S4
G.O. Mueller	(b) 161, 395	78.55; 71.35; S8.11
	(a) 121, 305	72.20 and 79.70; 85; S8.11
H.-U. Mueller	(a) 120, 185	74.70; S10.15
M. Mueller	(b) 159, 287	71.35 and 78.20; S8.11
	(b) 160, K79	61.70 and 78.20; S8.12
R. Mueller	(b) 159, 797	74.10 and 74.60; 74.40; S10.15
S. Mueller-Pfeiffer	(b) 160, 491	68.55
E. Mugeński	(b) 157, 137	61.70; 78.50; S9.11
N.N. Mukhamadeeva	(a) 121, 111	61.60 and 64.80; 66.30; S8.16
K.K. Mukherjee	(b) 157, 85	61.25; 72.15; S2
A.A. Mukhin	(b) 159, 845	75.30; 71.70; S10
V.M. Mukhortov	(a) 121, 157	68.55; 77.80; S10
K.M. Mukimov	(a) 119, 307	78.20; 75.30; S11.2
R.R. Mulyukov	(a) 117, 549	75.30; 73.40; S1.2
	(a) 118, K27	72.15; S1.2
A. Munoz-Yague	(a) 119,K181	78.20; S7.12
F.M. Muntyanu	(b) 158, 301	72.15; S4
H. Murakami	(a) 121, 461	61.70; 78.70; S2; S6
K.R. Murali	(a) 117, 377	61.70; 78.55; 78.60; S9.11
A. Murasik	(b) 161, K99	61.60 and 75.40; S10
M.J.P. Muringer	(a) 119, 63	62.20; S1.3
J.N. Murrell	(b) 159, 781	74.30; S10.15
M.R. Murti	(a) 119, 237	72.20; 61.70; S5.11
Y.V.G.S. Murti	(a) 117, 377	61.70; 78.55; 78.60; S9.11

Y.V.G.S. Murti	(a) 120,K133	61.70; S9.11; S10.1
N.I. Musaev	(a) 117, K57	61.80 and 76.30; S10.1
N.V. Mushnikov	(a) 119,K163	75.60 and 75.80; S1.2
J. Musilová	(a) 119, 669	78.20; 71.20; S5.12
E.R. Mustafaev	(b) 162, 281	78.50; 71.55; S11
H. Mutschke	(a) 119, 683	78.30 and 78.65; S9.12
G.S. Myakenkaya	(b) 161, 91	61.70 and 61.80; 72.20; S5.11
A. Mysyrowicz	(b) 159, 387	71.35 and 78.55; S10.1
A. Mzerd	(a) 119, 487	73.60; 71.55; S8.15
	(a) 121, 333	73.40 and 85; S8.15
	(a) 121, 587	71.55 and 73.40; 85; S8.15
G. Nachtwei	(a) 120, 185	74.70; S10.15
V. Nádaždy	(a) 122, 275	73.40; 71.55; S7.12
A.F. Nadezhdina	(a) 121, K5	61.60; 65.70; S1.63
R.S. Nadirzade	(b) 161, 861	78.30; S8.15
M.Yu. Nadtochy	(a) 118, 169	73.40; 78.65; S7.12; S7.15
M.D. Nadzhafzade	(b) 159,K107	75.40; 75.50; S8
A.T. Nagat	(a) 119, K47	72.40; S8
	(a) 120,K163	72.80; S8.16
	(a) 121,K201	72.20; S8.1
G. Nagel	(a) 122, 139	61.70; 61.60; S7.12
V. Nagirnyi	(b) 162, K91	71.55 and 78.55; S9.11
L. Nagli	(b) 162, K91	71.55 and 78.55; S9.11
A.A. Nagornyi	(a) 120, 635	78.55 and 78.60; S9.12
E.P. Naiden	(a) 120, 209	75.50; S11.2
H. Naito	(a) 117, 515	71.55; 73.30; S8.12
S. Nakamura	(a) 117,K177	78.55; 61.70; S10.1
Y. Nakamura	(a) 118, 389	61.60 and 61.70; 68.48; S1.2
	(b) 162, 407	65.40; 75.30; S1.2; S1.4
K. Nakanishi	(a) 122, K97	73.40 and 85; S12
Do Nam	(b) 158, 557	71.45 and 73.20
N.S. Nam	(a) 118, 591	73.40 and 85
T. Nammalvar	(b) 157, 93	61.25; 61.70; S2; S4
M. Venkata Narayana	(a) 117, K81	78.60; S9
Sa.K. Narayandass	(a) 117, 495	68.55 and 73.60; 77.55; S9
	(a) 121, 515	73.60; 77.55; S10
K. Narita	(a) 122, 87	61.60; 64.75; S10.15
E. Laxmi Narsaiah	(a) 119, 655	78.60; S10.15
F.S. Nasredinov	(b) 159, K97	76.80; 74.70; S10.15
	(b) 161, 287	71.55 and 73.60; 78.65; S8
	(a) 121, 571	61.70 and 71.55; 76.80; S8.15
K. Navaneethakrishnan	(b) 158, 193	73.40 and 75.20; 71.55; S7.12; S7.15
P. Nayak	(b) 160,K117	72.15; S1.2
L.A. Nazarenko	(a) 119, 191	68.48; 73.60; S7.12; S10.1

A.A. Nazarov	(a) 122, 495	61.70
A.N. Nazarov	(a) 120, 447	73.40; 73.20; S5.11; S10.1
	(a) 120, 475	61.80 and 68.55; 78.30; S5.11
C. Neamţu	(a) 121,K231	65.90 and 77.70; S1.1; S1.2
E. Nebauer	(a) 117,K111	68.48; 85; S7.12
	(a) 119,K131	73.30 and 79.20; S7.12
B.D. Nechiporuk	(a) 120, 475	61.80 and 68.55; 78.30; S5.11
D.D. Nedeoglo	(b) 157, 341	72.20; 71.30; S8.12
	(b) 158, 293	71.30 and 72.60; S8.12
E.I. Neimark	(b) 158, K55	76.20
K. Nejezchleb	(a) 122, 341	71.55; 76.30; S10.15
H. Nelkowski	(b) 158, 603	71.70; 76.30; 78.55; S8.15
D.K. Nelson	(b) 159, 425	71.35 and 78.55; S8.11
R. Neuhaus	(a) 120, 221	62.20 and 75.40; 61.70; S1.2
U. Neukirch	(b) 159, 431	78.55; 71.45; S8.11
Ch. Neumann	(b) 159, 297	71.35 and 78.20; 73.40; S7.1; S10.1
G. Neumann	(b) 161, K5	66.30; S1.2; S2
H. Neumann	(b) 162, K87	71.20 and 79.60; S7.16
N. Neumann	(a) 120, 273	77.40; 77.80; S12
R.E. Newnham	(a) 122, 677	77.30 and 77.90; S10.15
Hoang Xuan Nguyen	(b) 159, 403	71.35 and 78.20; S8.11
P.H. Nguyen	(b) 162, 155	71.25 and 73.40
W.-X. Ni	(a) 117,K171	78.30; S5.11
D. Niarchos	(b) 162, 243	72.20 and 74.70; 76.30; S10.15
H. Nickel	(b) 159, 173	71.35 and 78.20; 73.40; S7.12; S7.15
W. Nieswand	(b) 158, 267	71.35 and 71.70; 71.36; 78.20; S9.11
A.N. Nigam	(a) 122, K59	66.30 and 72.15; S1.2
A.I. Nikiforov	(a) 121, 523	68.55; 61.80; S5.11
K.G. Nikiforov	(b) 158, K63	75.30; 75.60; S11.2
T.V. Nikiforova	(b) 161, 705	71.55 and 78.55; 78.60; S8
M. Nikl	(a) 117, K89	78.55; S9
	(a) 118,K133	78.40 and 78.55; S11.1
A.V. Nikolaev	(b) 162, 451	64.60 and 71.20; S1.4
V.V. Nikolaev	(a) 121, 11	61.10 and 61.70; S5.11
A. Nikolov	(b) 161, 575	63.20; 78.30; S10.15
A.Yu. Nikulin	(a) 120, K1	61.10 and 61.70; S5.11
V.K. Nikulin	(b) 158, 175	71.25 and 73.40; S7.15
Tai-Shan Ning	(a) 119, K63	75.50; 75.60; S1.1
M. Nishi	(a) 117, 563	75.30; S9.16
T. Nishikawa	(a) 122, 163	61.70; 62.20; S5.11
Y. Nishino	(a) 122, 163	61.70; 62.20; S5.11
S.V. Nistor	(b) 160, 423	61.70; 66.30; 78.50; S9.11
K. Nitsch	(a) 118,K133	78.40 and 78.55; S11.1
S. Nitsche	(a) 122,K101	61.10; S10.15

H. Nitta	(b) 159, 589	61.80; S5.11
O. Nittono	(a) 118, 389	61.60 and 61.70; 68.48; S1.2
R. Niwas	(b) 161, 583	62.20 and 64.30; 65.70; S9.11
G.P. Nizhnikova	(b) 159, 743	71.25; 71.50; 78.20; S1.4; S8
M.L. Nó	(a) 120, 419	62.40; S3
S.V. Nosenko	(a) 120, 391	61.70 and 71.55; 66.30; S5.11
B.V. Novikov	(a) 119, 363	78.30 and 78.40; 64.70; 71.25; 71.35; 78.55; S9
	(a) 119,K167	78.55; 61.80; S8.12
D.L. Novikov	(b) 160, 459	68.45 and 73.20; 79.20; S1.61
D.V. Novikov	(a) 121, K9	61.10; 68.20; S5.12
V.N. Novikov	(a) 122, 215	62.20; 61.40; S1.1; S1.2
Yu.A. Novikov	(a) 121, 495	73.30; 78.70; S4
N.I. Novokhatskaya	(a) 117, K67	75.30; S1.2
A.V. Novotelnova	(b) 161, 777	72.20; S10
P. Novotný	(a) 121,K215	75.60; S11.2
P. Nozar	(a) 117, 501	71.20; 72.20; 75.30; S1.2; S1.4
A.V. Nurmikko	(b) 159, 107	71.35 and 78.20; 73.40; S8.13
I.N. Nurutdinova	(b) 158, K55	76.20
R. Obermaier	(b) 158, 695	75.50; S8.15
A. Odajima	(a) 119, 177	63.20 and 68.30; 68.60; S8.12
P. Odier	(a) 121, 359	61.10 and 61.55; S10.15
G. Oelgart	(a) 122, 683	73.40 and 78.55; 78.60; S7.15
	(a) 122, 705	78.60; 78.55; S7.12
	(a) 122, 765	73.40; 72.40; S7.12
V.M. Ogenko	(b) 157, 451	71.25 and 75.10
	(b) 162, 539	72.10 and 75.10
S.G. Ogloblin	(b) 160, 211	71.20
T. Ohgaku	(a) 118, 153	62.20; 61.70; S9.11
M. Ohta	(a) 117, K19	62.20; S4
I. Ohtsuka	(a) 122,K105	61.16; 61.70; S1.3
B.L. Oksengendler	(a) 119, K27	68.55; S4
M. Okuda	(a) 117, 515	71.55; 73.30; S8.12
S. Okuda	(a) 122, 221	61.40 and 62.40; S1.2
	(a) 122, 535	62.40; 62.20; S1.2
A. Oleaga	(a) 120, 419	62.40; S3
J. Olejniczak	(a) 117, K43	74.90 and 79.75; S10.15
N.M. Olekhnovich	(a) 119, 27	61.10; 61.70; S9.11
A.I. Olemskoi	(b) 160, 569	71.55; 71.70
U. van Olfen	(a) 121,K121	78.35; 72.40; S11.1
J.R. de Oliveira	(b) 158, K9	71.35
L.E. Oliveira	(b) 158, 743	71.25 and 77.90; S5.1; S5.11
G. Olovsson	(b) 161, 245	61.65 and 72.20; 71.55; 75.30; 76.30; S12.1

I. Olovsson	(b) 161, 245	61.65 and 72.20; 71.55; 75.30; 76.30; S12.1
M.F. Omelyan	(b) 159, K73	71.70
M.I. Onu	(b) 158, 301	72.15; S4
J. Opfermann	(a) 121, 563	72.40 and 73.60; S12
H. Oppermann	(a) 121, K91	75.30; 75.50; S1.2; S10.15
H. Opyrchał	(a) 121, 407	61.70; 71.55; 78.55; S9.11
P.A. Orekhanov	(b) 157, 745	78.20; 78.40; S11.2
S.V. Orekhov	(a) 121,K145	61.40 and 75.50; S1.2
U. Orhun	(b) 162, 585	76.30; S11
N.S. Orlova	(a) 118, 141	63.20; S8.16
J. Ortiz-Lopez	(b) 159, 629	64.60; S9
M. Ortuño	(b) 157, 275	71.30 and 75.40; S11.2
S. Orzeszko	(a) 119,K127	68.55; 78.30; S5
Yu.N. Osetskii	(b) 158, 441	63.20 and 71.10; 61.70; 62.20; S1
V.I. Osinskii	(a) 121, 227	72.40; 78.40; S5.11
V.A. Osipov	(b) 160, K29	71.38; S12.1
Yu.A. Osipyan	(b) 158, K49	71.55 and 76.30; S5.11
P.M. Ossi	(a) 119, 463	61.80 and 64.75; 61.40; S1
S.A. Ostanin	(b) 159, 893	78.20; 71.10; S4
J. Oswald	(a) 118, 317	78.55; 68.20; S7.12
	(a) 118, 567	78.55 and 78.65; S7.12
V.V. Otroshok	(b) 160, K75	61.70 and 78.30; S9.11
J.L. Oudar	(b) 159, 181	78.20; 68.55; S7.12; S7.15
H. Over	(a) 118, 343	78.55; 71.35; S8.11
	(b) 159, 437	71.36 and 78.20; S8.11
V.N. Ovsyuk	(a) 118, 317	78.55; 68.20; S7.12
F.J. Owens	(b) 162, 565	74.30; 76.30; S10.15
M. Ožvold	(a) 122, 583	73.30 and 73.40; 85; S5.11; S7.12
A.A. O'Connel-Bronin	(b) 162, 289	71.35 and 78.40; 73.20; S9.11
G. Paasch	(b) 162, 155	71.25 and 73.40
O. Pacherová	(a) 120, 339	61.10; S5.12
H.C. Padhi	(b) 158, 261	71.25 and 78.70; S7.12
R. Paessler	(b) 158,K143	71.55; S7.11
	(b) 162, K47	71.55; S7.12
O. Paetzold	(b) 160,K183	78.65; S5.11
C.S. Pai	(a) 117,K107	73.60; S7.12
A. Paja	(b) 160,K123	72.15; S1.1
A. Pajączkowska	(a) 119, 271	75.30; 74.70; S10.15
V.I. Pakhomov	(b) 161, K49	76.60 and 77.80; S9.16
V.E. Pakseev	(b) 162, 199	72.40; 72.70; S8.11
A.K. Pal	(a) 119, K21	73.60 and 78.65; S10.1
S.K. Pal	(b) 157,K131	75.40 and 75.60

M. Palazzi	(a) 118, 29	61.60; S8.16
T. Palewski	(a) 117,K151	75.30; S1.5
M. Palpacuer	(b) 162, 9	78.30; S12
V.V. Panashchenko	(b) 159, 417	71.35 and 78.20; 71.36; S8.11
B.K. Panda	(b) 158, 261	71.25 and 78.70; S7.12
A.M. Panich	(a) 117, K15	64.70 and 76.60; S8.16
Z.V. Panova	(a) 119, 113	68.55; S6
P. Pantchev	(a) 122, 599	68.55 and 78.65; S5.11
K.-H. Pantke	(a) 118, 343	78.55; 71.35; S8.11
	(b) 159, 437	71.36 and 78.20; S8.11
N.I. Papanicolaou	(b) 157, 293	71.25; 71.10; S5.2
C. Papastaikoudis	(b) 157, 351	72.15; S1.2
A. Parashar	(b) 160, K19	63.20; S8.16
V.V. Parashchuk	(a) 122, 631	74.10 and 74.30
S.M. Paraskevas	(b) 162,K101	76.30; 74.70; S10.15
H.L. Park	(b) 158,K115	73.40; S8.16
	(a) 119,K177	77.80; S11
	(b) 162, K9	64.75; 78.30; S8.16
	(a) 122,K135	68.55; 78.40; 78.55; S8.13
H.Y. Park	(b) 162, K9	64.75; 78.30; S8.16
S. Park	(b) 162, K9	64.75; 78.30; S8.16
T.J. Parker	(b) 161, 233	78.30 and 78.65; 71.36; S8.13
V.P. Parkhutik	(a) 121,K181	68.55; S6
J.C. Parlebas	(b) 159, 927	71.55 and 78.50
	(b) 160, 11	71.25 and 71.55
P. Paroli	(a) 119, 655	78.60; S10.15
R. Parrot	(b) 157, 323	71.70; S8.11
R. Pascova	(a) 117, 363	61.40; 64.75
L.Ya. Pasenko	(b) 158, K63	75.30; 75.60; S11.2
D.V. Pashik	(b) 160,K133	61.70 and 72.15; S4
R. Passianot	(b) 162, 99	61.70 and 66.30
J. Pastrňák	(a) 118, 317	78.55; 68.20; S7.12
	(a) 118, 567	78.55 and 78.65; S7.12
S.N. Pastushonok	(a) 118,K111	75.50; S11.1
S. Patra	(b) 157, 209	68.30 and 73.30; S2
S. Patu	(a) 120, 429	62.20; 61.70; S1.2
P. Paufler	(a) 122, 469	61.40 and 61.60; S10.15
P. Păusescu	(a) 122, 637	61.60 and 74.70; S10.15
L. Pavesi	(b) 157, 615	71.25 and 73.40; S7.1
A.P. Pavlov	(a) 120, 441	68.55; 61.80; S1.2; S4
P.V. Pavlov	(a) 120, 441	68.55; 61.80; S1.2; S4
V.I. Pavlov	(a) 118,K111	75.50; S11.1
P. Pawlicki	(b) 157, K51	75.10 and 75.80
	(b) 160, K57	75.10
S.J. Pearton	(a) 117,K107	73.60; S7.12
A.P. Pechonii	(b) 161, 843	76.30; 61.70; S11
M. Pękała	(a) 118, 41	61.60; 64.70; 76.80; S10
I. Pelant	(b) 157, K69	78.55; S9

Zhifang Peng	(a) 120,K139	61.70
	(a) 121, K37	61.70
N.M. Penina	(a) 121, 63	61.70; 76.30; 78.60; S5
Yu.P. Penkov	(b) 158,K105	61.80 and 76.80; S1.3
J. Perez	(a) 121, 129	61.40 and 64.60; 62.40
R. Perez	(a) 122, 51	61.14 and 61.70; S1.3; S5.11
R. Pérez-Álvarez	(b) 158, 481	71.25 and 73.40; S7.15
Yu.E. Perlin	(a) 119, 607	78.55; 63.20; 71.55; 78.30; S11.2
D.D. Perlov	(b) 161,K103	71.70 and 75.50; S11.2
P.P. Perstnev	(a) 121, 105	61.80; 78.60; S10.1
G.V. Peschanskikh	(b) 158, 441	63.20 and 71.10; 61.70; 62.20; S1
Th. Peschel	(b) 159, 291	71.35 and 78.20
V.V. Peshev	(a) 118, 219	71.55; 61.80; S7.12
M. Pessa	(a) 118,K123	78.55; S8.16
N. Petkov	(b) 161, 575	63.20; 78.30; S10.15
A.P. Petrakov	(a) 122, 449	61.10; S5.11
M. Petrauskas	(b) 158, 769	78.20; S8.13
	(b) 160, K79	61.70 and 78.20; S8.12
P.V. Petrenko	(a) 120, 315	61.55; S1.1; S1.2; S1.3; S4
V.F. Petrenko	(a) 118, 401	61.70 and 72.40; S10
O. Petříček	(a) 118, 317	78.55; 68.20; S7.12
D.V. Petrov	(b) 158,K105	61.80 and 76.80; S1.3
M.P. Petrov	(a) 118, 357	61.40; 78.20; S12
S.V. Petrovskii	(b) 157, 531	61.80
V.A. Petryakov	(a) 117, K37	72.70; 61.70; S8.15
D.M. Petty	(b) 158,K165	72.40
	(a) 120, 593	72.40; S12
V.N. Petukhov	(b) 161, 387	77.80; 65.40; S9.16
V.Yu. Petukhov	(a) 117, 477	61.80 and 68.55; S1.61
N. Peyghambarian	(b) 159, 249	71.35 and 78.20; S8.12
H. Pfeiffer	(a) 118, 295	75.60
	(a) 119, 259	75.60; S11.2
	(a) 120, 233	75.60
	(a) 120, K89	75.60; S11.2
	(a) 121,K117	75.60
	(a) 122, 377	75.60
R. Pfranger	(a) 117, 541	75.60; S1.1
A. Philipp	(b) 160, 613	72.15; S1.63
W. Pichl	(a) 118, 109	61.70
S. Pick	(b) 160, 483	68.25 and 68.40; 73.20; S5.11
R. Pickenhain	(a) 122, 765	73.40; 72.40; S7.12
J. Piechota	(a) 119, 271	75.30; 74.70; S10.15
U. Pietsch	(b) 157, K73	61.50; 62.30; S5.11
	(a) 118, 209	65.70 and 68.60; S7.15
	(b) 158,K111	65.70; 72.20; S7.12
E. Pigorsch	(a) 117,K189	61.70 and 78.30; S10.1

S. Pilecki	(a) 119, 87	62.40; S1.2
E. Pinčík	(a) 119, 701	73.40 and 85; S7.12
	(a) 122, 275	73.40; 71.55; S7.12
	(a) 122, 715	73.40 and 85; S7.12
H. Pinkvos	(a) 120, 221	62.20 and 75.40; 61.70; S1.2
R. Pintanel	(b) 159, K93	75.10
F. Pio	(b) 159, 577	61.70; 78.50; 78.55; S10.1
T. Piotrowski	(a) 117,K181	78.55; S8.15
	(a) 120, 531	71.55 and 73.40; S5.11
R. Piotrzkowski	(b) 157, 593	73.40; S7.12; S7.15
A.S.T. Pires	(b) 160,K165	75.10; 75.25
	(b) 161, K45	75.10; 75.25
R. Pirlot	(b) 158,K197	76.60; S1
A. Pirrwitz	(a) 121, 523	68.55; 61.80; S5.11
T. Pisarkiewicz	(b) 158, K5	71.25; S10.1
P. Piskunowicz	(a) 118, 441	64.70 and 65.70; S9.16
W. Pittroff	(a) 120, K53	73.40
S.K. Plachkova	(a) 117, 155	64.70; 61.10; S8
I. Plavina	(b) 161, 85	61.70; 78.55; S9.11
K. Ploog	(b) 162, 347	61.10 and 61.70; S7.12
I.R. Plummer	(a) 122, 285	79.75; 73.20; S9.11; S11.1
D. Pľusa	(a) 117, 541	75.60; S1.1
B. Pödőr	(a) 119,K135	71.55; S7.15
E.E. Podorova	(b) 159, 267	71.35 and 78.65; S8.15; S9
A. Poeppl	(a) 121, 195	71.55; 76.30; S10.1
	(a) 121, K69	71.55; 76.30; S10.1
A. Poggi	(a) 118, 491	71.55; 72.20; S5.11
	(a) 121, 181	71.55; 72.80; 78.30; S5.11
	(a) 121,K135	61.80 and 85; S5.11
A.D. Pogrebnyak	(a) 117, 17	64.70 and 79.20; 62.20; 68.45; 68.55; S1.1; S1.2
E.P. Pokatilov	(b) 158, 165	71.50 and 73.40
	(b) 159, 707	68.30 and 78.30
	(b) 161, 603	71.35 and 78.65; S8.13
	(b) 162, K63	78.30; S8.16
J. Pokluda	(b) 160, 89	62.20; S1.1; S1.2
S.I. Pokutnyi	(b) 157, 221	73.40 and 78.55
A. Polaczek	(a) 118, 41	61.60; 64.70; 76.80; S10
K. Polak	(a) 117, K89	78.55; S9
K. Polgár	(a) 120, 97	66.30; 78.30; S10.15
A.M. Polubotko	(b) 158, 229	71.25 and 75.10; S8
L. Polyak	(a) 119, 523	61.70 and 68.55; 73.20; 73.40; S5.11
O.Yu. Polyak	(a) 122, K45	74.70; S10.15
I.O. Polyakov	(b) 157, 151	64.60; 61.70; 76.60; S8
L.A. Polyakova	(a) 120, 253	78.55; 71.55; 78.50; S10
L.M. Pomirchy	(b) 161, K11	68.90
W. Pompe	(a) 117, 191	68.55

E.A. Ponezha	(b) 161, 745	74.20; 71.45
K. Ponnammal	(b) 162, 119	63.20; 61.70; S8.15
M. Popescu	(a) 122, 637	61.60 and 74.70; S10.15
A.S. Poplavnoi	(b) 160, 153	71.25; S2
A.G. Popov	(a) 121,K111	75.50; S1
A.I. Popov	(b) 161, 85	61.70; 78.55; S9.11
A.S. Popov	(a) 122, 569	72.40 and 73.40; S7.16
I.A. Popov	(a) 118, 147	64.70 and 78.20; 71.25; S1.2; S1.3
O. Popov	(a) 118, 289	75.60
V.P. Popov	(a) 121, 523	68.55; 61.80; S5.11
Z.S. Popović	(b) 160, 635	75.50; S1.1
G.A. Popson	(a) 120,K133	61.70; S9.11; S10.1
A.Yu. Poroikov	(a) 121, 399	61.80; 64.75; S6
J.C. Portal	(a) 120, 525	71.25; 72.20; S1.62
J.-C. Portal	(a) 119, K41	71.55; S7.12
E. Possenriede	(b) 161, K55	76.30; S11.1
A.V. Postnikov	(b) 162, 477	71.25 and 78.20; S6
A.I. Potapov	(a) 118, 505	72.20 and 78.20; 71.55; S7.16
L. Potocky	(a) 118, 307	75.50; 76.50; S1.1
J. Pozo	(b) 158,K191	75.30
R.B. Prabhu	(b) 161, 309	72.15; S1.4
T.S.P.L.N. Prasad	(a) 119, 655	78.60; S10.15
A. Pratap	(b) 160, 83	63.50; S2
	(b) 160, 433	63.50; S2
B. Predel	(a) 119, K9	66.30 and 68.48; S1.2
N. Presser	(b) 159, 443	71.35 and 78.65; 78.55; S7.12; S8.12
	(a) 120, 11	78.65; 68.55; 71.35; 73.60; 78.55; S8.12
M.T. Primatarowa	(b) 159, K17	71.35; S12.1
A.M. Pritulov	(a) 119, 417	61.40 and 61.80; S11.2
V.D. Prochukhan	(b) 158, 229	71.25 and 75.10; S8
A. Prodan	(a) 117, 351	61.55 and 68.55; S1.3
V. Prosser	(b) 158,K173	72.40; 71.55; S8.15
N.A. Prytkova	(a) 117,K121	61.50 and 74.70; S10.15
S.G. Psakhie	(b) 159, 645	65.50; 63.20; 64.10; S4
W. Pueschl	(b) 162, 363	61.70
N. Puhlmann	(a) 117,K181	78.55; S8.15
	(a) 122, 705	78.60; 78.55; S7.12
E. Pulido	(a) 119, 595	72.15 and 75.80; 75.60; S1.1; S1.2
J. Puls	(b) 159, 49	71.35 and 78.20; S8.11; S8.12; S8.15
	(b) 159, 353	71.35 and 78.20; S8.11
V.I. Punegov	(a) 122, 449	61.10; S5.11
M. Puri	(a) 122, 405	73.40 and 85; S7.12; S7.16
R.P. Purlys	(b) 158, 91	63.20; 71.55; 72.20; 77.50; S8.11

A.V. Pushkarev	(a) 119, 27	61.10; 61.70; S9.11
D.I. Pushkarov	(b) 159, K17	71.35; S12.1
G. Puzonas	(b) 162, 281	78.50; 71.55; S11
Z. Pytel	(b) 160, K13	66.30
G.G. Qin	(b) 158, 113	63.20 and 78.50; S7.12
Y. Qiu	(a) 117, 221	68.55; S4
Doan Nhat Quang	(b) 157, 235	71.25 and 72.20; S7.1
	(b) 158, 669	71.25 and 72.20; S7.12
M. Quintero	(a) 121, 483	64.75; 61.55; S1.1
	(b) 162, 517	71.25; 78.20; S8
L. Quiroga	(b) 160, 127	71.25 and 73.20; S8.13
V.B. Rabukhin	(a) 119,K113	61.70; S1
G. Raciukaitis	(b) 162, 281	78.50; 71.55; S11
S.I. Radautsan	(b) 158, K63	75.30; 75.60; S11.2
	(b) 162, K63	78.30; S8.16
I.Z. Radovskii	(a) 117, 577	75.50; S1.1; S1.2; S1.4
O. Radulescu	(b) 161, 65	61.50; 61.55; S1.1; S1.2; S4
R.J. Radwański	(a) 121, 611	75.50; 75.30; S1.2; S1.4
	(a) 121, 617	75.50; 75.30; S1.1; S1.2
Z. Rafi	(a) 120,K149	62.20; S1.2
Y. Rahangdale	(a) 121, 281	78.60; S8.11
J.S.P. Rai	(a) 119, K97	61.40; S8
H. Raidt	(a) 121, 95	61.70 and 78.60; 85; S7.12
P.A. Raikerus	(a) 121,K199	73.40; S5.11
S.Yu. Rakhley	(a) 121, 227	72.40; 78.40; S5.11
H.B. Rakhmatullaev	(a) 121, 571	61.70 and 71.55; 76.80; S8.15
K. Rakkenus	(a) 118,K123	78.55; S8.16
A.V. Rakov	(a) 121, 495	73.30; 78.70; S4
K. Ramachandran	(b) 158, 427	66.30; S8.15
	(b) 160, K85	66.30; S8.13
	(b) 162, 119	63.20; 61.70; S8.15
K. Ramesh	(a) 118,K103	76.30 and 78.50; 71.55; S11.1
M. Ramrakhiani	(a) 121, 281	78.60; S8.11
A.N.C.V. Rao	(a) 118,K129	78.60; S9.11
K.S.R.Koteswara Rao	(a) 117, 251	71.55; 73.40
R.V. Gopala Rao	(b) 157, 65	61.20; 65.50; S9.11
	(a) 118, 365	61.25; S9.12
	(b) 159, 545	61.25; S4
	(b) 160, 403	61.40; 62.20; 66.30; S2
U.V. Subba Rao	(a) 119, 655	78.60; S10.15
A. Raoufi	(a) 119, K75	72.15 and 75.50; S1.1
R. Rass	(b) 159, 431	78.55; 71.45; S8.11

H. Ratajczak	(a) 118, 189	61.40 and 68.55; 73.60; 75.70; S1.2; S1.4
	(a) 122, 575	68.55; 61.40; 73.60; S1.1; S1.2
A.M. Ratner	(b) 160, 245	71.70; 78.50; S5.2
H. Rauh	(a) 117, K77	78.30; 63.20; S12
	(b) 162, 125	63.20; S8
J. Ravez	(a) 121, 627	77.20; 64.80; 77.40; S11.2
	(a) 122, 371	72.20; S11.2
V. Ravindrachary	(a) 117, 53	61.40; 78.70; S1.1; S1.2
	(a) 120, 411	64.70; 61.40; 78.70; S11
A.K. Ray	(b) 161, 463	61.50; 71.45; S2
Yu.G. Rayzanov	(a) 118,K107	75.70; 75.60; S11.2
B.S. Razbirin	(b) 159, 425	71.35 and 78.55; S8.11
I.M. Razumovskii	(a) 117, 91	61.20 and 61.70; 61.50
J.M. Réau	(a) 117, 409	61.60 and 66.30; 64.75; S9.15
	(a) 121, 73	61.40 and 66.30; S9.16
	(a) 121, 415	66.30; 64.75; S9.15
	(a) 121, 627	77.20; 64.80; 77.40; S11.2
Ch. Gopal Reddy	(a) 120, 267	78.60; S9.11
K. Narasimha Reddy	(a) 119, 655	78.60; S10.15
	(a) 120, 267	78.60; S9.11
K.V. Reddy	(a) 119, 237	72.20; 61.70; S5.11
M. Bhagavantha Reddy	(a) 120, 575	72.20; S11.2
P. Venugopal Reddy	(a) 120, 575	72.20; S11.2
Y.P. Reddy	(a) 118,K103	76.30 and 78.50; 71.55; S11.1
T.F. Refai	(b) 162, 559	74.60
D. Reichardt	(b) 157, 567	68.20 and 73.20; S5; S5.11; S5.12
J. Reichardt	(a) 119, 317	78.55 and 78.60; 71.55; S10.15
	(a) 119, 631	78.55 and 78.60; 71.55; S10
M. Reiche	(a) 118, 85	61.70; S5.11
	(a) 119,K103	61.70; S5.11
J. Reichel	(a) 118, 85	61.70; S5.11
	(a) 119,K103	61.70; S5.11
W. Reichelt	(a) 121, K91	75.30; 75.50; S1.2; S10.15
B. Reichert	(b) 158,K119	73.20; 73.30; 73.60; S1.2
A. Reif	(a) 122, 331	68.48 and 73.40; 79.20; S7.12
T.L. Reinecke	(b) 159, 125	71.35 and 78.55; 73.40; S7.12; S7.15
F.-K. Reinhart	(b) 157, 615	71.25 and 73.40; S7.1
L. Reining	(b) 162, K37	71.10
G.U. Reinsperger	(a) 121, 305	72.20 and 79.70; 85; S8.11
G. Reiter	(a) 118, 161	64.60; 78.70; S12
S.I. Rembeza	(a) 119, 535	71.55; 62.40; S7.1
A.A. Rempel	(b) 160, 389	61.50; 61.55; S1.2; S1.3
F. Ren	(a) 117,K107	73.60; S7.12

K.F. Renk	(a) 122,K161	74.60; 74.70; S10.15
P. Rennert	(b) 161, 667	71.10; 71.20; 71.25; 73.20
S. Rentsch	(b) 162, 545	72.40; S12
E.A. Repnikova	(a) 119, 113	68.55; S6
F.G. Requejo	(a) 120, 105	61.70; 61.80; S10.1
U. Retzlaff	(a) 121,K141	64.75 and 85; 71.55; S5.11
H. Reuther	(a) 118, 409	61.80 and 76.80; S1.1
M. Rezki	(b) 160, 193	71.25 and 71.60; 78.70; S7.12; S8.13
A.S. Riad	(a) 121, 507	73.60; 72.40; S12
D. Ricard	(b) 159, 275	78.20; S8.15
C.-E. Richter	(a) 119, 183	73.40 and 78.20; S7.15
E. Richter	(a) 118, 409	61.80 and 76.80; S1.1
J. Richter	(a) 122,K133	68.20; 61.14; S5.11
W. Richter	(a) 119, 683	78.30 and 78.65; S9.12
M. Rick	(a) 121, 219	72.20; S8.13
R.H. Ritchie	(b) 157, K87	73.20
M. Ritschel	(b) 159, 797	74.10 and 74.60; 74.40; S10.15
J. Rivas	(a) 118, K31	75.60; S11.2
D.R. Rizdvyanetskii	(b) 159, K7	68.20 and 75.30
J.L. Robert	(b) 157, 593	73.40; S7.12; S7.15
	(a) 120, 151	73.60; S7
P.F. Robusto	(a) 119, 145	71.45 and 78.65; S8
	(a) 119, 155	71.45 and 78.65; S10
V.E. Rodionov	(a) 118, 327	78.60; 62.80; 73.40; S8.11
	(a) 121,K193	68.55; 73.60; 78.55; S6
F. Rodríguez	(b) 160, 127	71.25 and 73.20; S8.13
H. Rodríguez-Coppola	(b) 158, 481	71.25 and 73.40; S7.15
T.V. Rodzina	(a) 121, K27	61.70; S7.13
H. Roeppischer	(a) 119, 183	73.40 and 78.20; S7.15
	(a) 119,K167	78.55; 61.80; S8.12
A.A. Rogachev	(b) 159, 471	75.20 and 78.55; 73.40; S7.12
	(a) 121,K129	78.55; S7.13
K. Rogacki	(b) 158, K33	74.70; S10.15
A. Rogalski	(a) 122, K39	72.20; S8.15
S. Rogaschewski	(a) 120, 687	73.40 and 85; 73.30; S7.12
E. Rogowska	(a) 118, 61	61.70; 71.25; 78.70; S10.15
A.B. Roitsin	(b) 158, K55	76.20
M. Rolland	(b) 162, 9	78.30; S12
A.E. Romanov	(a) 122, 495	61.70
N.G. Romanov	(b) 159, K33	61.70; S9.11
A.A. Romanyukha	(a) 118,K107	75.70; 75.60; S11.2
M. Romek	(a) 121,K159	61.70 and 78.70; 74.50; S10.15
G.I. Rosenman	(a) 117, 259	72.40 and 79.60; S11.1
J. Rosenzweig	(a) 121, 219	72.20; S8.13
O.S. Roshupkina	(a) 118,K139	74.70; 78.30; 79.20; 79.60

H. Rossmann	(b) 159, 287	71.35 and 78.20; S8.11
A.H. Rotaru	(b) 159, 477	71.36
P. Roth	(b) 159, K1	61.40; 77.70; 77.80
S.R. Rotman	(a) 119, K87	78.55; S11.2
A. Rousset	(a) 118, K5	61.60; 78.30; S10.15
P. Roussignol	(b) 159, 275	78.20; S8.15
S. Datta Roy (Paul)	(b) 162, 89	61.70; S9.11
F. Rozpłoch	(a) 119,K127	68.55; 78.30; S5
I.R. Rubin	(b) 160, 165	71.20 and 71.25; S10
	(b) 160, 505	71.20 and 71.25; S10
	(b) 161, 271	71.20 and 71.25; S10.1
	(b) 161, 647	71.20 and 71.25; S9.11; S10.1
P. Rubin	(b) 157, K93	71.55 and 74.70; S10.15
Yu.G. Rubo	(b) 162, 531	72.10; 71.25; 72.40; S5
L.A. Rubtsova	(b) 161, 697	71.25 and 73.60; 71.10; S1.2
V.Yu. Rud	(a) 121, K81	72.40 and 73.40; 85; S7.16; S10.1
	(a) 121,K171	72.40 and 73.40; S7.16
Yu.V. Rud	(b) 158, 229	71.25 and 75.10; S8
	(a) 121, K81	72.40 and 73.40; 85; S7.16; S10.1
	(a) 121,K171	72.40 and 73.40; S7.16
A.I. Rudenko	(a) 121,K155	61.80; 71.55; 85; S5.11
G.Yu. Rudko	(a) 117,K185	78.55; S5.11
P. Rudolph	(a) 118, 225	71.55; 78.55; S8.13
	(a) 120, 653	64.75 and 78.30; S8.13
W. Rudolph	(b) 159, 49	71.35 and 78.20; S8.11; S8.12; S8.15
W. Rudzińska	(a) 120,K145	64.70; 78.70; S4
J. Rudzińska-Girulska	(b) 161, 281	71.25; 78.70; S1.2
H. Ruecker	(b) 160, 117	71.38 and 73.40; 63.20; S7.12
I. Rueckmann	(b) 158, 769	78.20; S8.13
	(b) 160, K79	61.70 and 78.20; S8.12
H. Ruefer	(a) 122, 139	61.70; 61.60; S7.12
D. Ruffier	(a) 121, 359	61.10 and 61.55; S10.15
O. Ruiz	(a) 120,K105	78.55; S10.1
I. Rupsky	(a) 121, 45	61.50; 61.70; 66.30
P.C. Russell	(a) 121, 333	73.40 and 85; S8.15
	(a) 121, 587	71.55 and 73.40; 85; S8.15
S.S. Russu	(b) 159, 155	71.35 and 78.20; 73.40; S8.12; S8.15
A.E. Rustamyan	(b) 161, 55	61.40; 61.70; 71.50; S10.1
Yu.V. Ruts	(a) 122,K143	68.20; S1.2
G.I. Ryabtsev	(a) 120, 209	75.50; S11.2
G.K. Ryasnyi	(b) 160, 255	71.70; 76.80; S1.1; S4

S.C. Sabharwal	(a) 121, 657	78.60; S9.12
L.D. Sabirsyanova	(b) 160,K141	71.30; S8.16
R.F. Sabiryanov	(b) 160, 459	68.45 and 73.20; 79.20; S1.61
F. Sachslehner	(a) 119, 229	72.15; S1.2
	(a) 120, 581	72.15; S1.2
K. Sadaiyandi	(b) 160, K85	66.30; S8.13
L. Sadłowski	(a) 121, K95	76.30; 75.50; S11.1
	(b) 162,K101	76.30; 74.70; S10.15
I.Sh. Sadykhov	(b) 158, 73	63.20 and 78.30; 64.75; S8.16
E.K. Sadykov	(b) 158, 685	76.60 and 76.80; S11.1
A. Saeckl	(a) 121,K117	75.60
A.F. Safarov	(a) 117,K147	76.30; S8.16
E.Yu. Safronov	(b) 160, 357	78.50 and 78.55; 71.55; S5.12
P.D. Sahare	(a) 118, K55	78.60; S10
B.B. Sahay	(b) 157, 85	61.25; 72.15; S2
J. Sahm	(b) 158, 603	71.70; 76.30; 78.55; S8.15
J. Said	(a) 117, 99	61.70; 61.80; S5.11
M. Said	(b) 157, 311	71.55; S7.12; S8.13
A.G. Saif	(b) 162, 559	74.60
K.K. Saini	(a) 118, 561	77.30 and 77.80; S12
P. Saint-Grégoire	(a) 120, K25	64.70; S9
H. Saitovitch	(a) 120, 105	61.70; 61.80; S10.1
F. Sajovec	(a) 122, 139	61.70; 61.60; S7.12
J. Sak	(b) 161, 677	71.38
H. Saka	(a) 122,K105	61.16; 61.70; S1.3
A. Sakakibara	(a) 117, K19	62.20; S4
	(a) 117, K97	62.20; S4
D.M. Sakov	(a) 122, 563	73.40 and 75.50; S9.11; S12
E.M. Sakr	(a) 120, K31	62.20 and 64.75; S4
	(a) 121, 467	62.20; S4
M.S. Sakr	(a) 120, K31	62.20 and 64.75; S4
	(a) 121, 467	62.20; S4
M.P. Saksena	(b) 160, 83	63.50; S2
J. Sakurai	(a) 118, 513	72.15; S1.2; S1.4
R. Salam	(a) 117, 535	73.60; 71.55; S10
N.S. Saleh	(a) 118, 467	61.80 and 73.40; S1.1; S1.2; S5.11
	(a) 120, 169	61.80 and 68.55; 73.60; S1.3; S5.11
S. Saleh	(a) 117,K133	72.15; S1
	(a) 122, K63	72.15; S1.1
B. Salejda	(b) 158, 475	64.60; 65.50; S11
W. Salejda	(b) 158, 119	64.60; 61.50; 65.50; S11
E.A. Salkov	(a) 117, K37	72.70; 61.70; S8.15
E.N. Salkova	(a) 122, K83	78.90; S5.12
J.M. Sallese	(a) 119, K41	71.55; S7.12
E.G. Salman	(a) 117, 509	71.55 and 73.40; S7.12
P.T. Salo	(b) 162, K29	71.10; S1.3; S2
A. Salokatve	(a) 118, 567	78.55 and 78.65; S7.12

A.A. Samokhvalov	(b) 158, 307	72.20; S11.2
V.M. Samsonov	(b) 157, 531	61.80
	(b) 161, 483	61.80 and 78.70; S5.11
R.M. Samuel	(a) 120, 247	78.20; S8
J. San Juan	(a) 120, 419	62.40; S3
T.I. Sanadze	(b) 158,K195	76.30; S11.1
P. Šandera	(b) 160, 89	62.20; S1.1; S1.2
L.M. Sandratskii	(a) 117, 577	75.50; S1.1; S1.2; S1.4
R. Sandrik	(a) 120, 371	66.30; S1.1; S1.2
Sangeeta	(a) 121, 657	78.60; S9.12
R. Sanghvi	(a) 118, 235	72.30; 62.65; S11.1
H. Sanjeevaiah	(a) 117, 53	61.40; 78.70; S1.1; S1.2
M. Sano	(a) 121, 461	61.70; 78.70; S2; S6
K. Santra	(b) 161, 329	71.25 and 72.20; S8.15
V.B. Sapre	(b) 161, 891	78.50; 71.55; S11.1
C. Sârbu	(a) 122, 637	61.60 and 74.70; S10.15
A. Sarce	(b) 159, 557	61.70 and 66.30
R.M. Sardarly	(b) 158, 73	63.20 and 78.30; 64.75; S8.16
A. Sardo	(a) 121,K135	61.80 and 85; S5.11
A. Sarkar	(a) 119, K21	73.60 and 78.65; S10.1
C.K. Sarkar	(b) 161, 329	71.25 and 72.20; S8.15
R.L. Sarkar	(b) 158, 223	71.38; S1.4
	(b) 161, 661	71.38
S.E. Sarkisov	(a) 119, 285	78.30; S11.2
	(a) 119, 607	78.55; 63.20; 71.55; 78.30; S11.2
E.F. Sarmento	(b) 160, 337	73.20 and 75.10
M. Sasaki	(b) 162, 553	72.20; 71.30; S10
I.I. Sasovskaya	(a) 118, 147	64.70 and 78.20; 71.25; S1.2; S1.3
M.S. Sastry	(b) 157, 547	64.60; 61.12; S11
M.V. Satarić	(b) 162, K95	74.30
R. Sathyamoorthy	(a) 117, 495	68.55 and 73.60; 77.55; S9
N. Sato	(a) 122, K97	73.40 and 85; S12
A. Satpathy	(b) 159, 545	61.25; S4
A.A. Sattar	(a) 117, K61	75.60; S10
I.J. Saunders	(a) 119, 487	73.60; 71.55; S8.15
A.V. Savchuk	(a) 122, K83	78.90; S5.12
A.V. Savin	(b) 162, K33	71.45
E.J. Savino	(b) 159, 557	61.70 and 66.30
K. Saxena	(a) 118, 561	77.30 and 77.80; S12
N.S. Saxena	(b) 160, 83	63.50; S2
	(b) 160, 433	63.50; S2
	(b) 161, 201	63.50; S2
	(b) 162, 395	63.50; S2
D. Sayah	(a) 119, 487	73.60; 71.55; S8.15
	(a) 121, 333	73.40 and 85; S8.15
	(a) 121, 587	71.55 and 73.40; 85; S8.15

H. Schaeben	(b) 158, 407	61.50; 61.10; 61.12
R. Schaefer	(a) 118, 271	75.60; 78.20; S1.1; S1.2; S1.4; S11.2
W. Schaefer	(b) 159, 59	72.10 and 78.20; S7.12
D. Schalch	(a) 117, K23	61.80 and 78.65; S6
A. Scharmann	(a) 117, K23	61.80 and 78.65; S6
N. Scheludko	(a) 117, 571	75.30; S1.2; S1.4
A. Schenk	(a) 122, 413	73.40 and 85; S8.15
	(a) 122, 723	73.40 and 85; S8.15
D. Schikora	(a) 118,K123	78.55; S8.16
V. Schiller	(a) 118, 591	73.40 and 85
W. Schiller	(b) 160, 297	72.15 and 74.70; S10
	(b) 161, 755	74.30 and 74.70; S10.15
G. Schirmer	(a) 119, 683	78.30 and 78.65; S9.12
O.F. Schirmer	(b) 161, K55	76.30; S11.1
K.-H. Schlaad	(b) 159, 173	71.35 and 78.20; 73.40; S7.12; S7.15
W. Schlapp	(b) 159, 173	71.35 and 78.20; 73.40; S7.12; S7.15
D.M. Schleich	(a) 117,K107	73.60; S7.12
R. Schlesinger	(a) 120, 687	73.40 and 85; 73.30; S7.12
K. Schmalz	(a) 117, 395	66.30; S5.11
	(a) 120, K15	61.70; S5.11
	(a) 121,K163	61.70 and 78.70; S5.11
J. Schmelzer	(a) 117, 363	61.40; 64.75
	(b) 161, 173	64.75; S1.2; S4
A. Schmiede	(a) 119, 589	72.20; S8.15
H. Schmiedel	(b) 158, 707	74.70 and 76.60; S10.15
S. Schmitt-Rink	(b) 159, 11	71.35 and 78.20; 73.40; S7.12
W. Schmitz	(a) 122, 469	61.40 and 61.60; S10.15
H.A. Schneider	(a) 117, 527	71.55; 72.20; S7.12
Ch. Schnittler	(b) 161, K15	71.10 and 73.90; 71.20; 71.50
C. Schober	(b) 159, 771	71.20 and 71.25; S1.3
G. Schoeck	(a) 118, 109	61.70
A. Schoenthaler	(a) 122, 695	78.70
H. Schoetz	(a) 120, 185	74.70; S10.15
G. Schramm	(a) 120,K101	78.55; 71.35; S5.11
	(a) 120,K181	78.55; S5.11
E. Schreiber	(a) 121, 305	72.20 and 79.70; 85; S8.11
	(a) 122,K165	72.20; 78.60; 79.70; 85; S8.11
J. Schreiber	(b) 157,K109	74.40 and 74.50
K. Schreiber	(a) 117,K165	73.40 and 78.65; S9.15
B. Schroeter	(b) 162, K53	63.20 and 78.30; S5
W. Schroeter	(a) 121, 369	61.14 and 61.70
L. Schrottke	(a) 119, K91	78.20; S7.12
A. Schuelzgen	(b) 159, 287	71.35 and 78.20; S8.11
W. Schueppel	(a) 119, 259	75.60; S11.2

W. Schueppel	(a) 119, 279	75.70; S1.62
	(a) 120, K89	75.60; S11.2
	(a) 121,K117	75.60
	(a) 122, K51	74.60 and 74.70; 61.70; S10.15
	(a) 122,K161	74.60; 74.70; S10.15
F.-J. Schuette	(b) 159, 235	71.35 and 78.20
J.U. von Schuetz	(b) 161, 245	61.65 and 72.20; 71.55; 75.30; 76.30; S12.1
D. Schultze	(a) 118,K133	78.40 and 78.55; S11.1
E. Schulze	(a) 122, 683	73.40 and 78.55; 78.60; S7.15
Th. Schurig	(b) 161, 613	73.40; S8.15
C. Schwab	(b) 160, 649	71.70 and 76.30; S7.12
R. Schwabe	(b) 159, 411	71.35 and 78.55; S7.15
Ch. Schwink	(a) 120, 221	62.20 and 75.40; 61.70; S1.2
M. Seberíni	(a) 122,K171	75.30 and 75.50; 76.80; S11.2
V. Sechovsky	(a) 117, 501	71.20; 72.20; 75.30; S1.2; S1.4
D. Segers	(a) 122, 129	61.80; 78.70; S5.11
P. Seidel	(a) 122, 645	74.70; 74.50
R.P. Seisyan	(b) 159, 197	71.35 and 78.20; S7.12; S7.15
C.A. ten Seldam	(a) 119, 63	62.20; S1.3
D.G. Semak	(a) 117, 241	71.55; 72.40; 73.60; S5
A.Yu. Semchuk	(b) 157, 451	71.25 and 75.10
	(b) 162, 539	72.10 and 75.10
N.N. Semenovskaya	(b) 158, 165	71.50 and 73.40
P. Semig	(a) 119, 327	78.50 and 78.55; 61.50; S11.1
A.S. Semikolenov	(a) 120,K121	61.14; S7.1
N.A. Semikolenova	(a) 120,K121	61.14; S7.1
Ch. Semmelhack	(a) 120, 557	74.70; 61.60; 75.30; S10.15
P.K. Sen	(b) 157, 735	78.20; S7.13
S.K. Sen	(b) 157, 519	61.70; S1.2; S1.3
J. Sénégas	(a) 117, 409	61.60 and 66.30; 64.75; S9.15
	(a) 121, 415	66.30; 64.75; S9.15
S. Sengupta	(b) 162, 89	61.70; S9.11
S.P. Sentemov	(a) 122,K143	68.20; S1.2
A.V. Serebryakov	(a) 117, K67	75.30; S1.2
P.P. Seregin	(b) 159, K97	76.80; 74.70; S10.15
	(b) 161, 287	71.55 and 73.60; 78.65; S8
	(a) 121, 571	61.70 and 71.55; 76.80; S8.15
G. Serfőző	(a) 118, 307	75.50; 76.50; S1.1
T.A. Sergan	(a) 122, K83	78.90; S5.12
S.A. Sergeenkov	(b) 160,K137	74.20; 61.40; S10.15
M. Serginov	(a) 121, K81	72.40 and 73.40; 85; S7.16; S10.1
	(a) 121,K171	72.40 and 73.40; S7.16
Z.T. Serikbaeva	(a) 118, K99	61.80 and 76.80; S1.1
	(a) 120, 305	61.55; 76.80; S1.1; S1.2

J. Serna	(a) 122, 489	61.40 and 61.70; 78.70; S12
S. Seto	(a) 121, 595	71.55 and 78.55; 72.80; S8.13
B. Sfez	(b) 159, 181	78.20; 68.55; S7.12; S7.15
J. Sha	(a) 122, K55	74.60 and 74.70; S10.15
	(a) 122,K151	74.70; S10.15
G.M. Shakhlevich	(a) 117, 213	68.48 and 68.55; S1.61; S5.11
N.I. Shakhraychuk	(a) 117,K185	78.55; S5.11
V.A. Shakin	(b) 159, 379	78.20; S8.11; S8.12
T.I. Shalaeva	(b) 162, 469	71.36; 76.20
A.M. Shamah	(a) 117, 163	64.70; 72.20; S11
	(a) 121, 235	61.55 and 74.70; S10.15
Yu.F. Shamanaev	(b) 158, 441	63.20 and 71.10; 61.70; 62.20; S1
J. Shanker	(b) 158, 11	64.70; 63.75; S9.11
V.P. Shantarovich	(a) 121, 495	73.30; 78.70; S4
Huimin Shao	(a) 117,K127	74.70; S10.15
	(a) 119, K5	61.50; 74.50; S10.15
F. Sharaf	(b) 157, 609	68.55 and 78.65; S8
A.K. Sharma	(b) 157, 145	62.20; S9.12
D.K. Sharma	(b) 159, 659	63.20; 62.20; S1
S.K. Sharma	(a) 121, 29	61.60; 61.70; S9
S.S. Sharma	(a) 122, K77	76.30 and 76.80; S1.2
V.V. Shatalov	(a) 117, 213	68.48 and 68.55; S1.61; S5.11
Yu.V. Shcherbakov	(a) 121,K155	61.80; 71.55; 85; S5.11
A.D. Sheikh-Ali	(a) 117, 429	61.70; S1.2
I.R. Shein	(b) 157, K29	71.25; S9
Yu.G. Shekun	(a) 117, 587	78.55; S8
L.E. Shelimova	(a) 117, 155	64.70; 61.10; S8
Z. Shemsedini	(b) 162, K95	74.30
A.A. Shemyakov	(b) 160,K153	75.50; 76.20; S11.2
T.D. Shemyakova	(a) 119, 191	68.48; 73.60; S7.12; S10.1
Bao-Gen Shen	(a) 119, K63	75.50; 75.60; S1.1
	(a) 121,K105	75.30; 61.40; 75.60; S1.2
	(a) 121,K211	75.50; S1.1
	(a) 122, 651	75.30; 75.60; S1.1
B.G. Shen	(a) 117, 283	75.50; 61.40; S1.1
	(a) 121, K99	75.50; 61.40; S1.1; S1.4
Guoyin Shen	(b) 158,K185	71.70 and 76.30; S11
Hiumin Shen	(a) 122, 517	64.70; 62.40; 64.80; S1.2
Linjiang Shen	(a) 119, K5	61.50; 74.50; S10.15
Shuechu Shen	(b) 158, 493	68.30; S1.2
S.C. Shen	(b) 161, K71	78.40; S8.15
V.A. Shenderovskii	(b) 157, K19	63.20; S8.13
	(b) 159, 837	72.30
	(b) 161, K91	72.80; S5
V.G. Shepelevich	(a) 120,K173	61.55 and 72.15; S4
T.D. Shermergor	(b) 158, 715	78.20
C.X. Shi	(a) 120, 429	62.20; 61.70; S1.2

T.S. Shi	(a) 120,K131	61.80; S5.11
A.N. Shibko	(a) 120, K37	68.55; S1; S1.61
E. Shidlovskaya	(b) 160, 529	71.10; 71.25; S9.11; S10; S11.1
M. Shiga	(b) 162, 407	65.40; 75.30; S1.2; S1.4
C.H. Shih	(a) 122, 195	62.20; S4
N.A. Shilkova	(b) 159, 893	78.20; 71.10; S4
R. Shimano	(b) 159, 347	71.35 and 78.20; S9
N.M. Shimanskaya	(a) 117,K121	61.50 and 74.70; S10.15
R. Shimizu	(b) 161, 257	68.20 and 79.20; S1.3
H.J. Shin	(a) 122,K135	68.55; 78.40; 78.55; S8.13
S.W. Shin	(a) 119,K177	77.80; S11
T. Shinohara	(a) 120, 199	74.70; S10.15
V.P. Shirokovskii	(b) 159, 893	78.20; 71.10; S4
A.G. Shishkov	(a) 117, K53	75.70; 75.60; S11.2
A. Shluger	(b) 160, 529	71.10; 71.25; S9.11; S10; S11.1
G.M. Shmelev	(b) 160, 179	71.45
	(b) 161, K85	72.40
M.I. Shmiglyuk	(b) 159, 755	71.35 and 71.36
S.Z. Shmurak	(a) 120, 83	61.70; 79.75; S9
A.S. Shpigel	(b) 161, 293	74.50; 74.20; S10.15
L.G. Shpinkova	(b) 160, 255	71.70; 76.80; S1.1; S4
E.I. Shreder	(a) 118, 147	64.70 and 78.20; 71.25; S1.2; S1.3
T.V. Shubina	(b) 159, 197	71.35 and 78.20; S7.12; S7.15
N. Shukla	(a) 121, 29	61.60; 61.70; S9
V.I. Shulga	(b) 161, 513	61.80; S5.11
Yu.M. Shulga	(a) 118,K139	74.70; 78.30; 79.20; 79.60
V.V. Shusha	(a) 118, 65	61.70 and 71.55; S5.11
B.A. Shustrov	(a) 117, 85	61.80; 71.55; S7.11; S7.12
Yu.M. Shvera	(b) 159, 91	71.36 and 78.20
	(b) 159, 477	71.36
R. Shyam	(b) 161, 565	63.20; S1.1; S1.2
D. Siapkas	(a) 118, 553	78.50; 78.30; S8
W. Siegel	(a) 117, 527	71.55; 72.20; S7.12
	(b) 158,K111	65.70; 72.20; S7.12
H. Siemes	(b) 158, 407	61.50; 61.10; 61.12
A. Siemko	(a) 119, 595	72.15 and 75.80; 75.60; S1.1; S1.2
M. Sieskind	(b) 158, 103	63.20 and 78.30; S9.16
I. Sikakana	(b) 162, 413	65.50; 71.20; S1
O. Sild	(b) 161, 105	61.70; 78.30; 78.55; S9.11
P. Silfsten	(b) 159, K37	61.70; S9.11
	(b) 162, K15	64.70 and 78.55; S9.16
A.R. Silin	(a) 119, 621	61.40 and 78.20; 61.80; 62.65; S10.1
A.Yu. Silov	(b) 159, 471	75.20 and 78.55; 73.40; S7.12

E.C. da Silva	(a) 117,K155	75.50 and 76.50; S12
P.R.J. Silva	(a) 120, 105	61.70; 61.80; S10.1
A.V. Simashkevich	(b) 157, 341	72.20; 71.30; S8.12
	(b) 158, 293	71.30 and 72.60; S8.12
S.S. Simeonov	(a) 122,K147	71.55; 61.80; S5.11
P.D. Simova	(a) 118, 357	61.40; 78.20; S12
V.V. Singer	(a) 117, 577	75.50; S1.1; S1.2; S1.4
B.N. Singh	(b) 159, 609	61.80; 61.70
B.P. Singh	(b) 162, 329	61.50; 71.45; S7.1; S8.1; S9.11
G. Singh	(b) 161, 145	62.20; S1.3
J. Singh	(b) 161, 837	75.30
M. Singh	(b) 159, 787	74.60; S10.15
N. Singh	(b) 158, 433	63.20; S3
S.P. Singh	(b) 158, 433	63.20; S3
T.N. Singh	(b) 158, 153	63.20; 62.20; S2
A. Singhal	(a) 122, 405	73.40 and 85; S7.12; S7.16
J. Singleton	(a) 119, K41	71.55; S7.12
R.M. Singru	(b) 161, 357	76.80; 75.50; S10.15
K.P. Sinha	(b) 159, 787	74.60; S10.15
E.P. Sinyavskii	(b) 160, 357	78.50 and 78.55; 71.55; S5.12
S.K. Sipani	(b) 161, 131	63.90; 62.20; 65.70; S9.11
M.L. de Siqueira	(b) 158, 283	61.70 and 71.55; S9.11
Yu.M. Sirenko	(b) 159, 805	72.10 and 73.40
P. Sisodia	(a) 122, 207	62.20
	(a) 122, 525	62.20; S1; S8; S10
P. Sivaprasad	(a) 118,K103	76.30 and 78.50; 71.55; S11.1
L. Skala	(b) 162, 429	65.50; 64.60; S6; S10.1
A.I. Skitsko	(b) 161, K91	72.80; S5
S.Z. Skluev	(a) 121,K145	61.40 and 75.50; S1.2
W. Skorupa	(a) 122, K19	61.80; S5.11
	(a) 122, K35	68.48 and 68.55; S5.11
W. Skrotzki	(a) 121, 81	61.70; S8
V.A. Skryshevskii	(a) 119, 201	73.40 and 74.70; 78.65; S10.15
D. Skrzypek	(b) 157, 695	75.30 and 76.30; S9
A.I. Skvortsov	(b) 158, 685	76.60 and 76.80; S11.1
J. Sláma	(a) 122,K171	75.30 and 75.50; 76.80; S11.2
A. Slawska-Waniewska	(a) 118, 307	75.50; 76.50; S1.1
I. Śledzińska	(b) 161, K99	61.60 and 75.40; S10
V.V. Sleptsov	(a) 122,K139	68.55 and 78.65; S5
E.I. Slinko	(b) 157, 151	64.60; 61.70; 76.60; S8
V.Yu. Slivka	(a) 119, 71	64.70 and 76.80; S8.16
V. Slusarenko	(b) 161, 897	78.50; S8.13
Z. Smetana	(a) 119, K67	75.50; S1.4
N. Smetniansky-DeGrande	(b) 162, 99	61.70 and 66.30
G.F. Smirnova	(b) 158, 469	63.20; 64.75; 78.30; S8.16
I.M. Smolyaninov	(b) 161, 843	76.30; 61.70; S11

A.A. Snigirev	(a) 120, K1	61.10 and 61.70; S5.11
O.V. Snitko	(b) 157, 601	68.20 and 71.36; 71.45; S7.12
	(b) 162, 531	72.10; 71.25; 72.40; S5
D.W. Snoke	(b) 159, 387	71.35 and 78.55; S10.1
E. Sobeslavsky	(a) 122, K19	61.80; S5.11
V.R. Sobol	(b) 160,K133	61.70 and 72.15; S4
A.B. Sobolev	(b) 160, 165	71.20 and 71.25; S10
	(b) 161, 271	71.20 and 71.25; S10.1
	(b) 162, 165	71.55; 61.70; S10.1
B.P. Sobolev	(b) 160, 375	78.30; S9.15
H.G. Sockel	(a) 119, 93	62.20; S1.2; S1.61
S. Soederholm	(b) 161, 245	61.65 and 72.20; 71.55; 75.30; 76.30; S12.1
D. Šokčević	(b) 162, K83	71.10; 71.20; 75.10
B.Yu. Sokolov	(a) 119, 307	78.20; 75.30; S11.2
N.N. Sokolov	(a) 121, K65	68.55; S5.11; S9.12
V.O. Sokolov	(b) 160, 263	61.70 and 71.55; S10.1
	(b) 161, 55	61.40; 61.70; 71.50; S10.1
	(b) 162, 173	71.55; S10.1
M.L. Sokolova	(a) 120, 457	68.55; 75.70; 78.20; S1.1
Zh.D. Sokolovskaya	(a) 118,K139	74.70; 78.30; 79.20; 79.60
I. Sokólska	(b) 157, 137	61.70; 78.50; S9.11
A.K. Solanki	(b) 160, 549	71.25; S1.2
	(b) 162, 497	71.25; S1
M. Solanki-Moser	(a) 117, 501	71.20; 72.20; 75.30; S1.2; S1.4
R. Del Sole	(b) 162, K37	71.10
I.K. Solomin	(a) 120, 327	61.10 and 68.55; S7.12; S7.15
L.S. Solonchuk	(b) 161, K91	72.80; S5
T. Soma	(b) 157, 509	61.70; S7.12
	(b) 159, K51	64.30; 64.70; S6
	(b) 162, K21	65.50; 65.70; S2
K. Somaiah	(a) 117, K81	78.60; S9
H. Sommer	(b) 162, K87	71.20 and 79.60; S7.16
R.L. Sommer	(a) 120, 609	75.60; S1.61
R. Sommerfeldt	(a) 117, K85	78.20; 61.80; S11.1
W.J. Song	(b) 162,K113	75.10
Zhizhong Song	(a) 118, K17	66.30; S5.11
J.C. Soret	(a) 122, 427	78.60 and 85; S8.11
B. Sorkin	(a) 117, K33	71.55 and 79.75; S9.11; S10.1
A.A. Sorokin	(b) 160, 255	71.70; 76.80; S1.1; S4
	(b) 161,K103	71.70 and 75.50; S11.2
K.L. Sorokina	(b) 157, K43	71.20 and 72.20; S5.11
B. Soudini	(b) 161, 685	71.20 and 71.60; S6
P. Sovák	(a) 122, 575	68.55; 61.40; 73.60; S1.1; S1.2
R. Spagnolo	(a) 119, 169	72.40 and 73.60; 78.65; S5.11
V. Spassov	(a) 120,K185	76.80 and 85; S5.11

Ch. Spiegelberg	(b) 159, 353	71.35 and 78.20; S8.11
G. Spinolo	(b) 159, 577	61.70; 78.50; 78.55; S10.1
P. Spišák	(a) 122, 575	68.55; 61.40; 73.60; S1.1; S1.2
H.R. Sreepad	(a) 117, 53	61.40; 78.70; S1.1; S1.2
	(a) 120, 411	64.70; 61.40; 78.70; S11
V. Sreeramalu	(a) 117, 53	61.40; 78.70; S1.1; S1.2
	(a) 120, 411	64.70; 61.40; 78.70; S11
G. Srinivasan	(a) 122, 355	73.25 and 74.10; S10.15
T.T. Srinivasan	(a) 122, 355	73.25 and 74.10; S10.15
J.P. Srivastava	(a) 117, K77	78.30; 63.20; S12
K.K. Srivastava	(a) 121, 341	73.40 and 85; 61.40; S8
F. Šrobár	(b) 159, 229	78.20 and 78.90
V.A. Stadnik	(b) 159, 241	71.35 and 78.20; S8.11
J.L. Staehli	(b) 159, 411	71.35 and 78.55; S7.15
A. Stahl	(b) 159, 327	71.25 and 78.20; 71.35
S. Stamenković	(b) 162, K95	74.30
F. Stangler	(a) 120, 581	72.15; S1.2
I.M. Starchenko	(a) 119,K151	61.80 and 74.70; S10.15
M. Starke	(a) 120, K95	78.65; S12
J.W. Stasiak	(b) 157, 229	71.20 and 73.20; S1.2
I.V. Stasyuk	(b) 160, 683	78.20; S11.1
C.B. Steele	(a) 122,K169	72.80; S10.1
K. Steenbeck	(a) 122,K161	74.60; 74.70; S10.15
E. Stefanovich	(b) 160, 529	71.10; 71.25; S9.11; S10; S11.1
W.E. Steger	(a) 117,K189	61.70 and 78.30; S10.1
U. Steigenberger	(b) 162, 125	63.20; S8
N. Stein	(a) 119,K167	78.55; 61.80; S8.12
E. Steinbeiss	(a) 119, 279	75.70; S1.62
V.F. Stelmakh	(a) 121, 63	61.70; 76.30; 78.60; S5
S.I. Stenin	(a) 121, 523	68.55; 61.80; S5.11
A.G. Stepanov	(a) 121,K145	61.40 and 75.50; S1.2
A.V. Stepanov	(b) 161, 123	66.30 and 78.70
S.A. Stepanov	(a) 121, K9	61.10; 68.20; S5.12
	(a) 122, 69	61.10 and 78.70; 68.20
V.I. Stepanov	(b) 159, 471	75.20 and 78.55; 73.40; S7.12
V.S. Stepanyuk	(b) 160, 219	71.25; S1.62
	(b) 160,K117	72.15; S1.2
	(b) 161, K77	61.40; S1.2
	(a) 122, K7	61.40; S1.1
D. Stephan	(a) 117, 67	61.10 and 61.80; S7.11; S7.12
J.D. Stephenson	(a) 122, 171	61.70; S5.11
	(a) 122,K121	61.70; S5.11
I.N. Stets	(b) 161, 783	72.10; 64.75; 71.25; S1.4
M. Stiebler	(a) 119, 317	78.55 and 78.60; 71.55; S10.15
	(a) 119, 631	78.55 and 78.60; 71.55; S10

I. Stoemenos	(a) 117, 99	61.70; 61.80; S5.11
Gh. Stoicescu	(b) 160, 423	61.70; 66.30; 78.50; S9.11
A. Stoklosa	(a) 119, 221	72.15; S1.63
	(a) 119, 571	72.15; 61.70; S1.63
N.A. Stolwijk	(b) 157, 107	66.30
H. Stolz	(b) 159, 35	71.25 and 72.10; 78.20
M. Stordeur	(b) 160,K145	72.20; S8
	(b) 161, 831	72.20; 72.90; S1.63; S8.15
S.R. Stoyanov	(a) 117,K161	77.70; S12
J. Straka	(b) 159, 125	71.35 and 78.55; 73.40; S7.12; S7.15
V. Štrbik	(a) 119,K155	74.70; 68.55; S10.15
V.V. Strelchuk	(a) 120, 475	61.80 and 68.55; 78.30; S5.11
Ya.M. Strelniker	(b) 161, 597	73.20; 61.80
P. Streubel	(b) 160, 143	79.60; S7.1; S11
	(a) 122, 331	68.48 and 73.40; 79.20; S7.12
E.K. Stribuk	(a) 121, K85	74.70; S10.15
M.V. Strikha	(b) 162, 531	72.10; 71.25; 72.40; S5
M.N. Strikhanov	(b) 157, 531	61.80
	(b) 161, 483	61.80 and 78.70; S5.11
R. Strobel	(b) 158, K37	72.40; S8
S.A. Studenikin	(b) 161, K67	78.20; S7.13
A. Sturm	(a) 117, 485	73.60; 66.30; S7.15
H. Stutenbecker	(b) 158, 603	71.70; 76.30; 78.55; S8.15
R. Styrkowiec	(a) 121,K227	77.80; S11.1
Gang Su	(b) 160, K49	74.20
Wenhui Su	(a) 119, 55	64.70 and 74.70; S10.15
Z.P. Su	(a) 122, K55	74.60 and 74.70; S10.15
	(a) 122,K151	74.70; S10.15
N. Suarez	(a) 121,K235	77.30; S9.12
A.V. Subashiev	(b) 160, 443	71.25 and 73.40; S7.12; S7.15
P. Sueptitz	(b) 158, K37	72.40; S8
D. Sueva	(a) 120,K185	76.80 and 85; S5.11
M. Suhrke	(b) 159, 731	72.10 and 73.40; S7.12
B. Sujak	(a) 117, K43	74.90 and 79.75; S10.15
B. Sukumar	(b) 158, 193	73.40 and 75.20; 71.55; S7.12; S7.15
V.B. Sulimov	(b) 160, 263	61.70 and 71.55; S10.1
	(b) 161, 55	61.40; 61.70; 71.50; S10.1
	(b) 162, 173	71.55; S10.1
C. Sułkowski	(b) 158, K33	74.70; S10.15
G.D. Sultanov	(b) 159,K107	75.40; 75.50; S8
Jiansheng Sun	(b) 162, 265	76.30; S11
	(b) 162, 585	76.30; S11
X.K. Sun	(a) 120, 627	75.50; 64.80; S1.1
	(a) 121, 241	75.30; 75.60; S1.1
	(a) 121, 257	75.60; S1.1
	(a) 121, 611	75.50; 75.30; S1.2; S1.4

X.K. Sun	(a) 121, 617	75.50; 75.30; S1.1; S1.2
	(a) 122, K69	75.50; S1.1
Yuping Sun	(a) 119, 555	74.70; 64.80; S10.15
K.B. Sundaram	(b) 161, K63	78.65; S10.1
I.T. Suprun	(a) 120, 363	61.70; 62.40
M.R. Surowiec	(a) 122, K15	61.70; S7.13
A.P. Surzhikov	(a) 119, 417	61.40 and 61.80; S11.2
E. Susi	(a) 118, 491	71.55; 72.20; S5.11
	(a) 121,K135	61.80 and 85; S5.11
H. Suto	(a) 118, 371	61.60; S11.1
A.V. Suvorov	(a) 121, 399	61.80; 64.75; S6
E.V. Suvorov	(a) 122, 391	78.70; 61.10; S5.11; S5.12
J. Suwalski	(a) 118, K35	75.25 and 76.80; S11.2
B.G. Svensson	(a) 117,K171	78.30; S5.11
P. Svestarov	(b) 161, 575	63.20; 78.30; S10.15
E.V. Sviridov	(a) 121, 157	68.55; 77.80; S10
V.V. Sviridov	(a) 119, 535	71.55; 62.40; S7.1
O.G. Sviridovich	(a) 122, 555	64.70 and 68.55; 74.70; S10.15
A. Svoboda	(b) 157, K69	78.55; S9
	(a) 121,K223	78.20; S5.11
P. Svoboda	(a) 119, K67	75.50; S1.4
	(a) 120, 185	74.70; S10.15
S. Swami	(a) 122, K77	76.30 and 76.80; S1.2
K. Świderczak	(a) 118, 519	75.60 and 75.70; S1.2
J. Świerczek	(a) 118, 525	75.50; S1.1; S1.2
W.A. Sysło	(a) 121,K227	77.80; S11.1
Yu.A. Sytsko	(b) 161, 483	61.80 and 78.70; S5.11
L. Szaro	(a) 118, 185	73.20 and 73.30; 72.40; S5.11
A. Szász	(b) 160, 219	71.25; S1.62
	(b) 160,K117	72.15; S1.2
	(a) 122, K7	61.40; S1.1
S. Szasz	(b) 161, K77	61.40; S1.2
I.S. Szuecs	(a) 118, 409	61.80 and 76.80; S1.1
S. Szuszkiewicz	(b) 158, K79	78.70; S1.1
Cz. Szymański	(a) 120,K145	64.70; 78.70; S4
S. Szymura	(a) 118, 525	75.50; S1.1; S1.2

A.S. Tablin	(a) 119, 297	78.20; S1.2; S11.2
M. Tacke	(a) 119, 711	73.40 and 78.55; 72.40; S7.16
A. Tadjer	(b) 160, 193	71.25 and 71.60; 78.70; S7.12; S8.13
G.B. Taggart	(b) 157, 711	75.20
A.S. Taha	(a) 119, 455	62.20; 61.70; 64.80; S1.2; S4
S. Tahata	(b) 157, 509	61.70; S7.12
E.V. Tahirov	(b) 160,K177	78.65

Y. Tahtamoni	(b) 162, K5	61.50; S1; S1.2
G.X. Tai	(b) 162, 553	72.20; 71.30; S10
L.T. Tai	(a) 121, 607	75.25; 75.30; S1.2; S1.4
T. Takada	(a) 120, K63	71.55 and 78.50; S5.12
H. Takahashi	(a) 120,K117	61.60 and 62.20; S8.16
T. Takai	(b) 157, K13	63.10; 68.20; S7.12
	(b) 157, K77	63.10; 68.20; S7.15
S. Takamura	(a) 120, 119	61.80 and 62.40; S4
F. Takeda	(a) 121, 595	71.55 and 78.55; 72.80; S8.13
K. Takeda	(a) 119, 251	76.30; S10.1
N. Takeuchi	(a) 117,K177	78.55; 61.70; S10.1
	(b) 158, 383	61.70 and 78.60; 79.75; S9.11
	(a) 118, 153	62.20; 61.70; S9.11
	(a) 122, K91	78.55; 61.70; 71.55; S10.1
T. Takeuchi	(b) 158, 277	71.55; 61.70; S9
K. Takita	(a) 117, 125	62.40; 74.70; S10.15
V.M. Talanov	(b) 162, 61	61.50; 64.60; S11.2
	(b) 162, 339	61.60; S11.2
	(a) 122,K109	61.60; S11.2
A. Tale	(b) 161, 85	61.70; 78.55; S9.11
P. Talik	(a) 121, 489	73.60; S12
E. Tamura	(b) 157, 627	68.20 and 75.30; 61.14; S1.1
Q. Tan	(a) 122, K25	62.40; S4
S. Tanaka	(a) 120, K67	71.55; S5.11
B. Tanatar	(b) 160, 237	71.10; 71.45
C.Q. Tang	(a) 121, K21	61.70 and 78.70; S1.2
	(a) 122, K11	61.70 and 78.70; 62.20; S1.1
Dingyuan Tang	(a) 121, 273	78.50; 61.70; S8.15
D.S. Tang	(a) 122, 481	61.50; 79.60; S1.2
J.Y. Tang	(a) 122, K11	61.70 and 78.70; 62.20; S1.1
Lianan Tang	(a) 119, 561	72.40; S11.1
Linchang Tang	(a) 121,K167	62.40; 66.30; S1.3
Nguyen Tang	(b) 161, 165	65.50; S5.2
	(b) 162, 371	65.50; 65.40; 65.70; S5.2
	(b) 162, 379	65.50; 63.90; 64.60; S5.2
Yi Tang	(a) 117,K127	74.70; S10.15
Y.S. Tang	(a) 121, K61	68.55; S10.1
M. Tanibayashi	(a) 120, K19	62.20; S9.11
S. Tanigawa	(a) 117, K9	61.70 and 78.70; S4
K. Taniguchi	(b) 162, K25	73.40
R. Taouint	(a) 119,K181	78.20; S7.12
R.J. Tarento	(a) 117, 305	78.60; S5.11
A. Tari	(b) 157, 701	75.60 and 76.30; S1.4
V.G. Tashirov	(a) 118,K107	75.70; 75.60; S11.2
J. Tatibouët	(a) 121, 129	61.40 and 64.60; 62.40
E.V. Tatyanin	(a) 121, 455	61.70 and 62.20; 64.70; S1.2
E. Taute	(b) 158, 551	79.20; S4
V.L. Tavgin	(b) 161, 123	66.30 and 78.70

N.M. Tchebotaev	(a) 119, 241	74.60 and 74.70; 61.70; S10.15
G. Teichmann	(b) 161, K15	71.10 and 73.90; 71.20; 71.50
P. Tekiel	(b) 158,K155	74.60
D.E. Temkin	(a) 120, K7	61.70 and 66.30
A. Tempel	(a) 119, 209	68.55; 61.70; S9.12
	(a) 121, K65	68.55; S5.11; S9.12
Mingkang Teng	(b) 157, 411	74.40; 78.70; S10.15
E.I. Terukov	(a) 120, 519	68.55 and 78.55; S5.11
L. Tesař	(a) 122, 607	61.70 and 71.55; 73.40; S5.11
U. Teschner	(b) 159, 917	78.30 and 78.65; S10.1
	(a) 121, 641	78.30 and 78.65; S6
D.I. Tetelbaum	(a) 120, 441	68.55; 61.80; S1.2; S4
A. Teubel	(b) 157, 389	74.20
	(b) 157,K117	74.20
Do Than	(a) 119, 221	72.15; S1.63
	(a) 119, 571	72.15; 61.70; S1.63
Nguyen Toan Thang	(b) 161, 637	71.25 and 71.35
Y.T. Thathachari	(b) 157, K13	63.10; 68.20; S7.12
G. Then	(a) 121, 445	61.80; 64.80; 74.70; 76.80; S1.2
P. Thevenard	(a) 117, K85	78.20; 61.80; S11.1
W.-G. Thies	(a) 121, 445	61.80; 64.80; 74.70; 76.80; S1.2
D.B. Tran Thoai	(b) 158, 557	71.45 and 73.20
	(a) 120, 467	71.45 and 73.20; 61.80; S4
B. Thomas	(a) 119, 127	68.55; 73.60; 78.65; S8.12
	(a) 120, K45	73.40; S7.12
	(a) 121, K55	68.48 and 73.60; S8.13
I. Thurzo	(a) 119, 701	73.40 and 85; S7.12
	(a) 122, 275	73.40; 71.55; S7.12
	(a) 122, 715	73.40 and 85; S7.12
N.P. Thuy	(a) 121, 607	75.25; 75.30; S1.2; S1.4
G.Y. Tian	(a) 121, 149	62.40; S4
	(a) 121, 475	62.40; S4
R. Tiebel	(b) 159, 235	71.35 and 78.20
I.M. Tiginyanu	(b) 162, K63	78.30; S8.16
S.G. Tikhodeev	(b) 159, 71	78.20; 71.35; 71.36; S8.11
N.A. Tikhonov	(a) 122, 449	61.10; S5.11
W.A. Tiller	(b) 157, K13	63.10; 68.20; S7.12
	(b) 157, K23	68.20; S7.12
	(b) 157, K77	63.10; 68.20; S7.15
D.R. Tilley	(b) 161, 233	78.30 and 78.65; 71.36; S8.13
I.N. Timchenko	(b) 157, 341	72.20; 71.30; S8.12
	(b) 158, 293	71.30 and 72.60; S8.12
V.I. Timoshenko	(a) 121, 547	61.80 and 78.65; S7.12
S.D. Tiron	(b) 159, 755	71.35 and 71.36
A.O. Titov	(a) 120, 457	68.55; 75.70; 78.20; S1.1

K. Tittelbach-Helmrich	(a) 121,K141	64.75 and 85; 71.55; S5.11
Sie Chin Tjong	(a) 121, 119	62.20; 61.70; S1.1
A.S. Tkachenko	(a) 120, 447	73.40; 73.20; S5.11; S10.1
T.M. Tkachenko	(b) 157, K61	76.80; S1.2
V.M. Tkachuk	(b) 160, 321	75.10; 75.40
J. Tobisch	(a) 121, K65	68.55; S5.11; S9.12
E.I. Tochitskii	(a) 122, 555	64.70 and 68.55; 74.70; S10.15
O.M. Tolkachev	(b) 159,K115	75.30; S1.2
V.P. Tolstykh	(a) 121, 63	61.70; 76.30; 78.60; S5
A. Tomas	(a) 118, 29	61.60; S8.16
	(a) 121, 21	61.60; 61.10; S8
M. Tomášek	(b) 160, 483	68.25 and 68.40; 73.20; S5.11
C.N. Tomé	(b) 162, 99	61.70 and 66.30
Zh.M. Tomilo	(a) 117,K121	61.50 and 74.70; S10.15
A. Tomita	(b) 158, 383	61.70 and 78.60; 79.75; S9.11
M. Tomita	(a) 122,K105	61.16; 61.70; S1.3
J. Tomkinson	(b) 157, 547	64.60; 61.12; S11
J.W. Tomm	(a) 117,K181	78.55; S8.15
	(a) 118,K123	78.55; S8.16
	(a) 119, 711	73.40 and 78.55; 72.40; S7.16
	(a) 121, 635	78.20; 71.25; 78.55; S8.15
	(a) 122, 11	78.55; 78.45; S8.1; S8.15
K. Tompa	(b) 158,K197	76.60; S1
E. Tomzig	(a) 122, 139	61.70; 61.60; S7.12
A. Toneva	(a) 122, 599	68.55 and 78.65; S5.11
Yu.P. Toporov	(a) 122, 563	73.40 and 75.50; S9.11; S12
A.A. Toropov	(b) 159, 197	71.35 and 78.20; S7.12; S7.15
I. Tóth	(a) 122,K171	75.30 and 75.50; 76.80; S11.2
J. Toušek	(a) 122, 757	72.40 and 73.30; S8.13
R. Tovar	(a) 121, 483	64.75; 61.55; S1.1
	(b) 162, 517	71.25; 78.20; S8
K.D. Tovstyuk	(b) 157, 151	64.60; 61.70; 76.60; S8
	(b) 159, K73	71.70
	(b) 160, 415	66.30
N.K. Tovstyuk	(b) 157, 151	64.60; 61.70; 76.60; S8
S.L. Town	(a) 119, 241	74.60 and 74.70; 61.70; S10.15
I.E. Tralle	(b) 161, 869	71.55 and 78.50; 71.38
V.Yu. Trashchakov	(a) 121,K163	61.70 and 78.70; S5.11
A.V. Trefilov	(b) 158, 441	63.20 and 71.10; 61.70; 62.20; S1
	(b) 161, 153	62.20 and 71.25; S2; S3; S4
A.B. Treshchalov	(b) 162, 289	71.35 and 78.40; 73.20; S9.11
G.P. Triberis	(b) 158, K17	72.80; S9.12
	(b) 158,K149	72.10; 71.38
	(b) 162, 243	72.20 and 74.70; 76.30; S10.15

G.C. Trigunayat	(a) 121, 29	61.60; 61.70; S9
Ch. Trikalinos	(b) 161, 513	61.80; S5.11
	(b) 162, 243	72.20 and 74.70; 76.30; S10.15
Ch.G. Trikalinos	(b) 158,K101	61.80; 79.20
B.B. Tripathi	(b) 160, K19	63.20; S8.16
J.M. Triscone	(a) 122, 589	68.55 and 73.40; 64.80; S10.15
A. Tříska	(a) 118,K133	78.40 and 78.55; S11.1
Z. Trojanová	(a) 118, 455	62.20; S1.2
E. Trojnar	(b) 158, K33	74.70; S10.15
V.N. Tronin	(a) 117, 61	61.50; 75.30; S1.2
V.T. Troshchinskii	(a) 121, K85	74.70; S10.15
M. Trott	(b) 161, K15	71.10 and 73.90; 71.20; 71.50
S. Trott	(b) 161, K15	71.10 and 73.90; 71.20; 71.50
V.I. Troyan	(a) 117, 61	61.50; 75.30; S1.2
I.O. Troyanchuk	(a) 118,K111	75.50; S11.1
A.V. Trubachev	(a) 117, 291	61.55 and 75.50; 76.80; S1.1
O.S. Trushin	(b) 160,K117	72.15; S1.2
	(b) 161, K77	61.40; S1.2
	(a) 122, K7	61.40; S1.1
V.A. Tsekhomskii	(b) 159, 267	71.35 and 78.65; S8.15; S9
M.S. Tseluevskii	(a) 117,K121	61.50 and 74.70; S10.15
I.M. Tsidilkovskii	(b) 157, 17	71.30; 72.20; 75.40; S7.13; S8.15
	(b) 160,K141	71.30; S8.16
	(b) 161, 303	71.30 and 72.20; S5.12
Pei-Hsin Tsien	(a) 118, 179	61.80 and 68.55; 73.60; S5.11
E.S. Tsikhotskii	(a) 121, 391	61.70; 77.80; S10.15
T. Tsuboi	(b) 162, K15	64.70 and 78.55; S9.16
B.S. Tsukerblat	(b) 160, 641	71.70 and 75.30; 64.60; S10
A.E. Tsurkan	(a) 119, 191	68.48; 73.60; S7.12; S10.1
G.I. Tsurkan	(b) 161, K85	72.40
T. Tsvetkova	(a) 118, K83	61.80 and 72.80; S8
	(a) 119, 107	73.60; 61.80; 78.65; S8.12
R.Kh. Tukhvatulin	(a) 122, K45	74.70; S10.15
G. Tumbev	(a) 117, K5	61.40 and 78.70; S1.1
D. Tunega	(b) 160, 375	78.30; S9.15
V.P. Tupychak	(b) 160, 683	78.20; S11.1
E.Yu. Turaev	(b) 161, 287	71.55 and 73.60; 78.65; S8
	(a) 121, 571	61.70 and 71.55; 76.80; S8.15
A.N. Turanov	(a) 118, 401	61.70 and 72.40; S10
R. Turcu	(a) 119,K121	73.60; S12.1
I.K. Turdaliev	(a) 120, K85	76.30; S11.1
D.W. Turner	(a) 122, 285	79.75; 73.20; S9.11; S11.1
M. Turtenwald	(a) 122, 645	74.70; 74.50
D.M. Turtsevich	(a) 117,K121	61.50 and 74.70; S10.15
	(a) 122, 555	64.70 and 68.55; 74.70; S10.15

G.A. Turtsevich	(a) 118, 141	63.20; S8.16
Z. Tylczyński	(a) 118, 441	64.70 and 65.70; S9.16
T. Tymosz	(a) 118, 519	75.60 and 75.70; S1.2
A.P. Tyutnev	(a) 121, 205	61.80 and 72.20; S12
O.I. Tyutyunnik	(b) 162, 597	63.20 and 78.30; S9.11
V.I. Tyutyunnik	(b) 162, 597	63.20 and 78.30; S9.11
S.J. Tzeng	(a) 117, K47	72.15; S1.2
	(a) 121, 213	72.15; 75.60; S1.2
B. Uebbing	(b) 159, 297	71.35 and 78.20; 73.40; S7.1; S10.1
R. Uecker	(a) 118,K133	78.40 and 78.55; S11.1
J. Ullmann	(a) 122,K129	68.55; 61.16; 68.60; S5
J. Ulner	(b) 159, K27	75.10 and 75.30
B. Ulrici	(a) 121, 95	61.70 and 78.60; 85; S7.12
A.G. Ulyashin	(a) 121, 63	61.70; 76.30; 78.60; S5
A.M. Umpelev	(a) 120, 457	68.55; 75.70; 78.20; S1.1
H.-J. Unger	(b) 158, 609	71.50; 71.10; 71.25; S5.11; S5.12
S. Unnikrishnan	(a) 121, 265	75.50; 76.80; S11.2
J.C. Upadhyaya	(b) 159, 659	63.20; 62.20; S1
	(b) 161, 565	63.20; S1.1; S1.2
S.C. Upadhyaya	(b) 159, 659	63.20; 62.20; S1
	(b) 161, 565	63.20; S1.1; S1.2
I.N. Uraltsev	(b) 161, 217	73.40 and 78.20; 71.35; S7.12; S7.15
A. Urbaniak-Kucharczyk	(b) 157,K103	72.10 and 75.10
	(b) 158, K59	75.30; 75.70
	(b) 160,K129	72.10 and 75.10
V.V. Ursakii	(b) 162, K63	78.30; S8.16
N. Uryû	(b) 158, 347	75.30
	(b) 161, 367	75.30
A.E. Usachev	(b) 158, 337	76.30; 71.70; S9.15
	(b) 160,K149	76.30 and 76.70; 64.60; S9.16
	(b) 162,K107	76.70; S9.16
O.V. Usatenko	(b) 162, 387	63.20; S12.1
V.M. Ustinov	(a) 118, 169	73.40; 78.65; S7.12; S7.15
V.V. Ustinov	(b) 160, 97	73.60; 72.15; S4
I. Utke	(a) 120, 653	64.75 and 78.30; S8.13
A.E. Vafin	(a) 117, 577	75.50; S1.1; S1.2; S1.4
I.A. Vakarchuk	(b) 160, 321	75.10; 75.40
M.Ya. Valakh	(a) 117,K185	78.55; S5.11
	(a) 119, 363	78.30 and 78.40; 64.70; 71.25; 71.35; 78.55; S9

M. Ya. Valakh	(a) 120, 475	61.80 and 68.55; 78.30; S5.11
R.Z. Valiev	(a) 117, 429	61.70; S1.2
	(a) 117, 549	75.30; 73.40; S1.2
	(a) 118, K27	72.15; S1.2
	(a) 122, 495	61.70
S.A. Valiev	(a) 120, 447	73.40; 73.20; S5.11; S10.1
U.V. Valiev	(a) 119, 307	78.20; 75.30; S11.2
G. Vanderschaeve	(a) 122, 545	62.20 and 78.20; S8.11
A.V. Vannikov	(a) 121, 205	61.80 and 72.20; S12
E.I. Vapirev	(a) 120,K185	76.80 and 85; S5.11
A.N. Varaksin	(b) 162, 165	71.55; 61.70; S10.1
K.S. Varblianska	(a) 118, 479	73.30; 85; S1; S7.12
R.S. Varella	(a) 117,K155	75.50 and 76.50; S12
S. Varga	(b) 162, K77	71.20; S4
H. Vargas	(a) 117,K155	75.50 and 76.50; S12
V.S. Varichenko	(a) 121, 63	61.70; 76.30; 78.60; S5
J. Varma	(a) 122, K77	76.30 and 76.80; S1.2
I.A. Vartanyants	(a) 121,K149	61.10; S5.11
P. Vasek	(a) 120, 185	74.70; S10.15
A.M. Vasilev	(a) 118, 169	73.40; 78.65; S7.12; S7.15
A.N. Vasilev	(b) 161, 777	72.20; S10
E.A. Vasilev	(b) 157, K61	76.80; S1.2
F. Vasiliu	(a) 122, 637	61.60 and 74.70; S10.15
V.A. Vassilev	(a) 120, 519	68.55 and 78.55; S5.11
E. Vateva	(a) 119, 107	73.60; 61.80; 78.65; S8.12
L. Vatskichev	(a) 118, K95	72.15 and 75.70
M. Vatskicheva	(a) 118, K95	72.15 and 75.70
B. Vaupel	(a) 117,K165	73.40 and 78.65; S9.15
C. Vautier	(a) 122, 73	61.50 and 68.55; S8
G.J. Vázquez	(b) 158,K161	74.10; S5.2
	(b) 160, K89	62.20 and 65.40; S2
J. Večeř	(b) 157, K69	78.55; S9
A. de Veirman	(a) 120,K131	61.80; S5.11
A.N. Veis	(a) 122,K177	78.30; S8.16
Ts.S. Velinov	(a) 121, 555	62.30 and 68.20; 78.20
E.F. Venger	(b) 157, 601	68.20 and 71.36; 71.45; S7.12
R. Venkatesh	(b) 157, 65	61.20; 65.50; S9.11
M.P. Verma	(a) 122, 207	62.20
	(a) 122, 525	62.20; S1; S8; S10
L.I. Vershinina	(a) 121,K145	61.40 and 75.50; S1.2
V.N. Vertoprakhov	(a) 117, 509	71.55 and 73.40; S7.12
V.A. Veryazov	(b) 157, 281	71.10 and 71.25; 74.10; S1.2; S10.1
	(b) 158, 201	71.20 and 71.25; 74.10; S1.2; S10
S. Ves	(b) 159, 699	63.20 and 78.30; 64.60; S8.16
M. Vesely	(a) 120, 371	66.30; S1.1; S1.2
V.A. Vetrov	(b) 159, K33	61.70; S9.11

G. Vielsack	(b) 158, 249	71.25; 78.20; S11.1
Nguyen Ai Viet	(b) 161, 637	71.25 and 71.35
T.V. Virovets	(b) 157, 431	75.30; S8.16
A.D. Vishnoi	(b) 161, 891	78.50; 71.55; S11.1
Ya.D. Vishnyakov	(a) 117, 549	75.30; 73.40; S1.2
B. Viswanathan	(a) 117, 53	61.40; 78.70; S1.1; S1.2
D. Vladikova	(a) 121, 249	64.75 and 75.50; S11.2
Yu.I. Vladislavskii	(b) 160, 71	61.12 and 61.70; S9.11
S.I. Vlaskina	(a) 121,K193	68.55; 73.60; 78.55; S6
G. Voelkel	(a) 121, 195	71.55; 76.30; S10.1
	(a) 121, K69	71.55; 76.30; S10.1
	(a) 122, K73	76.30
F. Voelklein	(b) 158, 521	72.15 and 73.60; S4
	(a) 118, K69	66.70
	(b) 162, 147	73.60; S4
V. Voelskow	(a) 122, K35	68.48 and 68.55; S5.11
A. Vohra	(a) 121, 341	73.40 and 85; 61.40; S8
G. Voigt	(a) 121, 95	61.70 and 78.60; 85; S7.12
M. Voigt	(b) 159, K87	72.15 and 72.20
A.I. Voitenko	(b) 161, 293	74.50; 74.20; S10.15
Th. Vojta	(b) 162, 141	68.48 and 73.40; S2
A.P. Vokhmyanin	(b) 158, 319	75.30; 61.55; S1.61
A.S. Volkov	(a) 120, 519	68.55 and 78.55; S5.11
Yu.N. Vorobev	(a) 117, 291	61.55 and 75.50; 76.80; S1.1
A.P. Voronin	(a) 119, 417	61.40 and 61.80; S11.2
E.V. Voronina	(b) 160, 625	76.80
V.K. Voronkova	(b) 158, 337	76.30; 71.70; S9.15
B.E. Vugmeister	(b) 157, 183	64.60 and 77.80; S10.15
F.R. Vukajlović	(b) 160, 635	75.50; S1.1
K. Vuletić	(b) 160, 635	75.50; S1.1
S.G. Vygovskii	(b) 159, 71	78.20; 71.35; 71.36; S8.11
Yu.M. Vysochanskii	(a) 119, 71	64.70 and 76.80; S8.16
M. Wabia	(b) 162,K101	76.30; 74.70; S10.15
W. Wacławek	(a) 121, 489	73.60; S12
H. Wada	(b) 162, 407	65.40; 75.30; S1.2; S1.4
A.H. Wafik	(a) 117, K61	75.60; S10
	(a) 120, 601	75.60; 64.80; S1.1
	(a) 122, 657	75.60; S10
M. Wagner	(b) 161, 543	63.20 and 65.50; 64.70; 65.40; 66.30
P. Wahid	(b) 161, K63	78.65; S10.1
J. Walczak	(a) 121, K95	76.30; 75.50; S11.1
L. Waldoestl	(a) 120, 557	74.70; 61.60; 75.30; S10.15
G. Waller	(b) 157, 627	68.20 and 75.30; 61.14; S1.1
F. Walz	(a) 118, K31	75.60; S11.2

W. Walz	(b) 160, 661	61.70 and 75.50; S10
U. Walzer	(b) 162, 75	61.50; 71.10; 71.45; S2
T. Wamura	(b) 159, 133	71.35 and 78.20; 73.40; S7.12; S9
Dianfen Wang	(b) 162, K1	61.40
Dongsheng Wang	(b) 157,K127	76.60; S11.1
D. Wang	(b) 161, 501	61.70; 61.10; S1.3
En-Ge Wang	(b) 158, 589	71.55; 71.20; S5.11; S7.15
Guanghou Wang	(b) 157, 411	74.40; 78.70; S10.15
Hong Wang	(a) 118, K47	61.70 and 78.30; S11.1
Huafu Wang	(a) 118, K47	61.70 and 78.30; S11.1
Huai-Yu Wang	(b) 158, 589	71.55; 71.20; S5.11; S7.15
Huaqin Wang	(a) 119, K5	61.50; 74.50; S10.15
L. Wang	(b) 157, 247	71.25; 75.10
L.P. Wang	(b) 158, 113	63.20 and 78.50; S7.12
N.L. Wang	(a) 119, K59	74.70; S10.15
Qiang Wang	(b) 161, 537	64.70; 61.60; S10
R. Wang	(a) 117, 73	61.14 and 61.70; S5
R.H. Wang	(a) 120, 125	62.20; 61.70; S4
Shoushan Wang	(a) 121, 57	61.80; 78.60; S9.11
S.X. Wang	(a) 119, K59	74.70; S10.15
	(a) 122,K151	74.70; S10.15
Wanlu Wang	(a) 119, K31	68.25; S5.11
	(a) 121,K189	68.25; S5.11; S6
Xiandong Wang	(a) 119,K141	74.70; S10.15
Xiaohui Wang	(a) 117,K127	74.70; S10.15
Xiao-Guang Wang	(a) 119, K15	61.70 and 78.70; S1.1
Xu Wang	(a) 119,K141	74.70; S10.15
Yening Wang	(a) 122, 517	64.70; 62.40; 64.80; S1.2
Yifeng Wang	(a) 119, 55	64.70 and 74.70; S10.15
Yinyue Wang	(a) 121,K177	73.60; 68.55; S5.11; S5.12
Yu Wang	(a) 117, 335	61.60 and 74.70; S10.1
	(a) 119,K141	74.70; S10.15
Z.G. Wang	(a) 120, 125	62.20; 61.70; S4
	(a) 122, 195	62.20; S4
B.M. Wanklyn	(a) 117, 563	75.30; S9.16
S. Waplak	(a) 117, 461	64.70; 76.30; S10.15
Y. Waseda	(a) 117, K1	61.40; S10
S.M. Wasim	(a) 122, 235	62.20 and 62.65; S8.16
W. Wasserbaech	(b) 159, 689	62.20; S1
S. Watanabe	(b) 158, K69	71.25 and 76.30; S5.11; S5.12
M. Wattenbach	(b) 158, K49	71.55 and 76.30; S5.11
J.F. Watts	(a) 121, K61	68.55; S10.1
M. Wautelet	(b) 159, K43	64.70; 68.60; S5.11; S5.12
Ch. Weber	(b) 159, 173	71.35 and 78.20; 73.40; S7.12; S7.15
G. Weber	(a) 117, 417	61.70; 71.55; 72.20; S7.12
H.H. Weber	(a) 121,K239	78.45; 78.20; S7.15

P. Weber	(a) 121, K73	74.50; S4
R. Wedell	(b) 159, 597	61.80; S5.11
Liu Wei	(a) 119, K59	74.70; S10.15
	(a) 122, K69	75.50; S1.1
N. Weiden	(a) 117, 555	76.60; S9.16
	(a) 119, 77	63.20 and 64.70; 76.60; S12
U. Weih	(a) 121, 563	72.40 and 73.60; S12
G. Weimann	(b) 159, 125	71.35 and 78.55; 73.40; S7.12; S7.15
	(b) 159, 173	71.35 and 78.20; 73.40; S7.12; S7.15
M.W. Weiser	(b) 162, 429	65.50; 64.60; S6; S10.1
A. Weiss	(a) 117, 555	76.60; S9.16
	(a) 119, 77	63.20 and 64.70; 76.60; S12
	(b) 162, K87	71.20 and 79.60; S7.16
P. Weissbrodt	(a) 119,K131	73.30 and 79.20; S7.12
W. Weller	(b) 162, 251	74.20
Jinke Wen	(a) 118, K47	61.70 and 78.30; S11.1
J.G. Wen	(a) 117, 73	61.14 and 61.70; S5
L. Wendler	(b) 159, 143	71.35 and 73.40; S7.12; S7.15
U. Wenschuh	(b) 160, 225	71.10; 65.40; 71.38; 78.20; S1.3
	(b) 162, 303	78.20; 65.40; 71.10; S1.3
H. Wenzel	(a) 120, 661	73.40 and 85; S7.12; S7.15
H. Wenzl	(a) 122, 139	61.70; 61.60; S7.12
J. Werner	(a) 118, K91	72.80; 61.50; S10.1
J. Wernisch	(a) 122, 695	78.70
K. Weschke	(b) 160, 297	72.15 and 74.70; S10
	(b) 161, 755	74.30 and 74.70; S10.15
J.M. Wesselinowa	(b) 157, K47	75.10
	(b) 158, K95	77.80
	(b) 160, 697	77.80; S11.1
G. Wiatrowski	(b) 157,K103	72.10 and 75.10
	(b) 160, K61	75.10; 75.30
	(b) 160,K129	72.10 and 75.10
	(b) 161, 335	75.10
T. Wicht	(b) 159, 309	71.35 and 78.20; S7.12; S8.11; S9
A. Wieczorek	(a) 121,K227	77.80; S11.1
J. Wiehl	(a) 120, 643	78.55 and 78.60; 71.55; S11.2
M. Wienecke	(a) 120, 653	64.75 and 78.30; S8.13
	(a) 121, K55	68.48 and 73.60; S8.13
E. Wieser	(a) 117, 477	61.80 and 68.55; S1.61
M. Wilkens	(a) 122, 79	61.40 and 61.80; 64.70; S1.2; S4
M. Willander	(a) 117,K171	78.30; S5.11
R. Wille	(b) 159, 297	71.35 and 78.20; 73.40; S7.1; S10.1

M.R. Willis	(b) 158,K165	72.40
	(a) 120, 593	72.40; S12
P.R. Wilshaw	(a) 117, 403	61.70; S5.11
D. Windisch	(a) 118, 379	61.60 and 61.70; S5.11
W. Windsch	(a) 122, K73	76.30
U. Winter	(a) 120, 67	61.10 and 61.70; S5.11
	(a) 120, K1	61.10 and 61.70; S5.11
	(a) 121,K149	61.10; S5.11
W. Wischert	(a) 120, 643	78.55 and 78.60; 71.55; S11.2
	(a) 121, 289	78.55 and 78.60; 71.55; S11.2
A. Wiśniewski	(b) 158, K63	75.30; 75.60; S11.2
	(a) 118, 525	75.50; S1.1; S1.2
	(b) 161, K99	61.60 and 75.40; S10
H. Witte	(a) 117, 527	71.55; 72.20; S7.12
Feng Wo	(a) 119, K63	75.50; 75.60; S1.1
U. Woggon	(b) 160, K79	61.70 and 78.20; S8.12
W. Wojciechowski	(a) 118, K87	72.80; 76.30; S12
K. Wokulska	(a) 121, 539	68.70; 61.10; S1.1
Z. Wokulski	(a) 120, 175	62.20 and 68.70; S1.62
G.W. Wolf	(b) 161, 697	71.25 and 73.60; 71.10; S1.2
J. Wolf	(a) 118, 209	65.70 and 68.60; S7.15
R. Wolf	(a) 122, 139	61.70; 61.60; S7.12
J.P. Wolfe	(b) 159, 387	71.35 and 78.55; S10.1
A. Wolkenberg	(a) 120, 567	72.20; S5.11
K. Wollschlaeger	(a) 122, K35	68.48 and 68.55; S5.11
C.K. Wong	(b) 157, 101	61.40; S5.11
C.H. Woo	(b) 159, 609	61.80; 61.70
J. Woods	(a) 118, 117	61.70; 61.80; S5.11
J.C. Woolley	(a) 121, 483	64.75; 61.55; S1.1
	(b) 162, 517	71.25; 78.20; S8
A. Wrzeciono	(a) 117,K137	72.15; S1.2
C.S. Wu	(a) 117,K107	73.60; S7.12
Guo-mei Wu	(a) 122, 355	73.25 and 74.10; S10.15
Ping-Feng Wu	(b) 162, 193	71.70 and 76.30; 78.50; S11.1
P.J. Wu	(a) 122, 481	61.50; 79.60; S1.2
S.P. Wu	(a) 121, 149	62.40; S4
	(a) 121, 475	62.40; S4
X.J. Wu	(a) 119, 511	68.55; 68.48; S5.11; S7.13; S7.15
H.-J. Wuensche	(a) 120, 661	73.40 and 85; S7.12; S7.15
P. Wuensche	(a) 117, 447	65.90; S12
	(a) 120, 139	65.90; S12
M. Wurlitzer	(a) 120, 557	74.70; 61.60; 75.30; S10.15
C. Wutte	(a) 120, 581	72.15; S1.2
B. Wysłocki	(a) 117, 541	75.60; S1.1
	(a) 118, 525	75.50; S1.1; S1.2

Hua Xia	(a) 118, K51	78.30; S5.11
J.S. Xia	(a) 122,K151	74.70; S10.15
Y.B. Xia	(a) 120, 125	62.20; 61.70; S4
Jun-Kai Xie	(b) 162, 193	71.70 and 76.30; 78.50; S11.1
Y.Y. Xiong	(a) 122, 481	61.50; 79.60; S1.2
Changqing Xu	(b) 158,K185	71.70 and 76.30; S11
Jian-Jun Xu	(b) 157, 577	68.70; 61.50
Jinzhang Xu	(a) 121, 501	61.40 and 76.30; 68.55; S5.11
Liangying Xu	(a) 119, 561	72.40; S11.1
R. Xu	(a) 119, K1	61.40; S1.1
Wenlan Xu	(b) 158, 493	68.30; S1.2
You Xu	(b) 157, 685	75.30; S11.2
Yuehuang Xu	(a) 121, K37	61.70
Zhang-Long Xu	(b) 161, K25	71.36 and 73.90; S7.11
Q.K. Xue	(a) 122, 481	61.50; 79.60; S1.2

Yu.V. Yablokov	(b) 158, 337	76.30; 71.70; S9.15
	(b) 160,K149	76.30 and 76.70; 64.60; S9.16
	(b) 162,K107	76.70; S9.16
M.R. Yakheev	(b) 161, 883	78.30; 77.80; S8.16
E.B. Yakimov	(a) 120, 391	61.70 and 71.55; 66.30; S5.11
	(a) 122, 121	61.70; 78.70; S5.11
L.I. Yakovenkova	(b) 158, 441	63.20 and 71.10; 61.70; 62.20; S1
B.Y. Yakovlev	(a) 121, 105	61.80; 78.60; S10.1
D.R. Yakovlev	(b) 161, 217	73.40 and 78.20; 71.35; S7.12; S7.15
D.V. Yakovlev	(a) 121,K181	68.55; S6
R.A. Yakshibaev	(a) 121, 111	61.60 and 64.80; 66.30; S8.16
	(a) 121, K5	61.60; 65.70; S1.63
G.G. Yakushcheva	(b) 158, 359	71.35 and 78.65; S8.12
	(b) 159, 449	71.35 and 78.20; S8.1
M. Yamada	(a) 117, K19	62.20; S4
H. Yamamoto	(b) 160, K97	73.20; 73.40
	(b) 162, K25	73.40
J. Yan	(a) 122, 195	62.20; S4
J.R. Yan	(b) 162, 461	71.45
T. Yanagihara	(b) 162, 443	68.30; 68.45; S9.11
I.Y. Yanchev	(b) 158, K43	72.20; S5.11
I. Yanetka	(b) 158, 181	63.20 and 71.45
Chuan-Zhang Yang	(b) 160, 329	75.10
C.Z. Yang	(b) 162,K113	75.10
Gui-ling Yang	(b) 157, 685	75.30; S11.2
G.Q. Yang	(a) 120,K131	61.80; S5.11
Jiping Yang	(b) 162, 545	72.40; S12
Lin-Yuan Yang	(a) 119, K63	75.50; 75.60; S1.1
	(a) 121,K105	75.30; 61.40; 75.60; S1.2

P. Yang	(a) 122, 475	61.10 and 61.70
R.M. Yang	(a) 119, K1	61.40; S1.1
Xian-Jun Yang	(a) 118, 179	61.80 and 68.55; 73.60; S5.11
Zhengju Yang	(b) 158, 127	64.60
	(b) 159, 629	64.60; S9
O.V. Yanush	(a) 119, 621	61.40 and 78.20; 61.80; 62.65; S10.1
Xixian Yao	(a) 117,K127	74.70; S10.15
Y.D. Yao	(a) 117, K47	72.15; S1.2
	(a) 120, K71	72.15; S4
	(a) 121, 213	72.15; 75.60; S1.2
N.P. Yaroslavtsev	(a) 119, 535	71.55; 62.40; S7.1
Y.Y. Yeung	(b) 161,K107	76.30; S10.1
H.R. Yi	(a) 122, K55	74.60 and 74.70; S10.15
Huaqing Yin	(a) 119, 555	74.70; 64.80; S10.15
Rongchang Yin	(a) 122, 249	68.45 and 73.60; S11.1
D.R. Yoder-Short	(b) 157, 379	74.60; 76.30; S10.15
	(b) 160, 591	71.70 and 76.30; S8.12
H. Yokoyama	(a) 121,K125	78.20; 71.50; S7.12
K. Yomogita	(b) 162, 443	68.30; 68.45; S9.11
M. Yoneta	(a) 120, K11	61.70; 71.55; 66.30; S5.11
C.S. Yoon	(b) 162, K9	64.75; 78.30; S8.16
E. Yoshida	(a) 117,K177	78.55; 61.70; S10.1
M. Yoshida	(a) 119, 583	72.20 and 73.60; S12
H. Yoshikawa	(b) 161, 257	68.20 and 79.20; S1.3
S.M. Yossef	(b) 157, 609	68.55 and 78.65; S8
J.Q. You	(b) 162, 461	71.45
Zhipu You	(a) 120,K177	72.40; 61.70; S5.11
C. Yu	(b) 157, 379	74.60; 76.30; S10.15
	(b) 160, 591	71.70 and 76.30; S8.12
Wan-Lun Yu	(b) 158, K13	71.70; 76.30; S8.11; S10.1; S11.1
X.N. Yu	(a) 122, 481	61.50; 79.60; S1.2
Zhenzhong Yu	(a) 121, 273	78.50; 61.70; S8.15
Yu.M. Yuabov	(a) 118, 577	72.40 and 73.40; 85; S5.11
Fang Yuan	(a) 122, 517	64.70; 62.40; 64.80; S1.2
M. Yudasaka	(a) 122, K97	73.40 and 85; S12
A.E. Yunovich	(a) 122, 11	78.55; 78.45; S8.1; S8.15
I.N. Yuriichuk	(b) 157, K19	63.20; S8.13
E.M. Yurtaeva	(b) 160, 245	71.70; 78.50; S5.2
S.V. Yurtaeva	(b) 162,K107	76.70; S9.16
M. Ząbkowska-Wacławek	(a) 118, K87	72.80; 76.30; S12
	(a) 121, 489	73.60; S12
S.Yu. Zaginaichenko	(b) 159, K7	68.20 and 75.30
T.A. Zagorodnyaya	(b) 159, 837	72.30

A.A. Zahab	(a) 117,K103	73.60 and 78.65; 72.40; S8
	(a) 119, K35	72.40 and 73.40; S8
A.M. Zaitsev	(a) 121, 63	61.70; 76.30; 78.60; S5
A.M. Zaitseva	(b) 157, 129	61.80; 78.70; S5.12
I.N. Zakharchenko	(a) 121, 157	68.55; 77.80; S10
	(a) 121, 391	61.70; 77.80; S10.15
L.F. Zakharenkov	(a) 117, 85	61.80; 71.55; S7.11; S7.12
M. Zakharov	(b) 160, 529	71.10; 71.25; S9.11; S10; S11.1
S.P. Zakharov	(b) 160, 97	73.60; 72.15; S4
R.B. Žakula	(b) 162, K95	74.30
V.V. Zakurkin	(b) 158,K105	61.80 and 76.80; S1.3
A.J. Zaleski	(a) 117, K43	74.90 and 79.75; S10.15
	(a) 120, K73	74.70; S10.15
M. Zaluzny	(b) 160, 541	71.25; S4
	(b) 161, K19	71.25 and 73.90; S7.1
Yitong Zang	(a) 117,K127	74.70; S10.15
M.B. Zapart	(a) 121, K43	64.70 and 76.30; S11.1
W. Zapart	(a) 118, 447	64.70 and 76.30; S11.1
	(a) 121, K43	64.70 and 76.30; S11.1
A. Zarembowitch	(a) 118, 461	62.65; 78.35; S10
M.M. Zaripov	(a) 117, 477	61.80 and 68.55; S1.61
A.M. Zaslavskii	(a) 117, K13	64.70; S10.15
P. Zaumseil	(a) 120, 67	61.10 and 61.70; S5.11
	(a) 120, K1	61.10 and 61.70; S5.11
	(a) 121,K149	61.10; S5.11
D.M. Zayachuk	(a) 119, 215	71.55; 72.80; 75.50; S8
H.A. Zayed	(a) 120, K31	62.20 and 64.75; S4
	(a) 121, 467	62.20; S4
S. Zazubovich	(b) 162, K91	71.55 and 78.55; S9.11
J. Zbroszczyk	(a) 118, 525	75.50; S1.1; S1.2
A. Zehe	(b) 157, 305	71.55; 78.55; S7.12
	(a) 118, 203	61.80 and 68.60; 79.20; S9.12
	(a) 119, 209	68.55; 61.70; S9.12
	(a) 121, K65	68.55; S5.11; S9.12
	(b) 162, K69	66.30
	(a) 122,K133	68.20; 61.14; S5.11
D. Zeissig	(a) 122, 331	68.48 and 73.40; 79.20; S7.12
E. Zeitler	(b) 158, 557	71.45 and 73.20
	(a) 120, 467	71.45 and 73.20; 61.80; S4
Xian-Ting Zen	(a) 117, 335	61.60 and 74.70; S10.1
L.Z. Zeng	(a) 121,K185	68.48 and 79.60; S1.1; S10.1
W.Y. Zeng	(b) 162, 111	63.20; S5.2
J.Th. Zettler	(a) 119, K91	78.20; S7.12
H. Zeyada	(b) 158, K59	75.30; 75.70
Hong-ru Zhai	(b) 157, 685	75.30; S11.2
H.R. Zhai	(a) 117, 283	75.50; 61.40; S1.1

H.R. Zhai	(a) 120, 159	68.55 and 75.70; 76.80; S1.1
	(a) 121, K99	75.50; 61.40; S1.1; S1.4
Wen-Shan Zhan	(a) 119, K63	75.50; 75.60; S1.1
	(a) 122, 651	75.30; 75.60; S1.1
Fangqing Zhang	(a) 118, K17	66.30; S5.11
	(a) 118, K39	61.70 and 76.30; S5.11
	(a) 118,K115	68.60 and 76.30; S6
	(a) 120, K49	68.60; S5.11
	(a) 121, 501	61.40 and 76.30; 68.55; S5.11
	(a) 121,K177	73.60; 68.55; S5.11; S5.12
Hong Zhang	(a) 119, K15	61.70 and 78.70; S1.1
	(a) 121,K207	74.70 and 78.70; S10.15
H. Zhang	(a) 119,K145	74.70 and 79.60; S10.15
	(a) 122,K151	74.70; S10.15
Jian Zhang	(a) 121, 57	61.80; 78.60; S9.11
Jizhong Zhang	(a) 119, 41	61.50 and 68.45; S10.1
Jizhou Zhang	(b) 157,K127	76.60; S11.1
Ji-Sheng Zhang	(a) 118, 179	61.80 and 68.55; 73.60; S5.11
Jun-Xian Zhang	(a) 119, K63	75.50; 75.60; S1.1
	(a) 121,K105	75.30; 61.40; 75.60; S1.2
	(a) 122, 651	75.30; 75.60; S1.1
Li-De Zhang	(a) 117, 105	62.40; 61.70; S4
Li-Yuan Zhang	(b) 158, 589	71.55; 71.20; S5.11; S7.15
L.D. Zhang	(a) 121, 149	62.40; S4
	(a) 121, 475	62.40; S4
L.Z. Zhang	(b) 158, 113	63.20 and 78.50; S7.12
Qiren Zhang	(a) 121, 429	61.70; S11.1
Q.R. Zhang	(a) 119, K59	74.70; S10.15
	(a) 119,K145	74.70 and 79.60; S10.15
	(a) 122, K55	74.60 and 74.70; S10.15
	(a) 122,K151	74.70; S10.15
T. Zhang	(a) 120, K41	68.45; S1.2; S10.1
T.Y. Zhang	(a) 122, 195	62.20; S4
Wei Zhang	(a) 118, K51	78.30; S5.11
Xiaowen Zhang	(b) 161, 537	64.70; 61.60; S10
Xing-Kui Zhang	(a) 118, K51	78.30; S5.11
Yitong Zhang	(a) 119, K5	61.50; 74.50; S10.15
Yuheng Zhang	(b) 158, K27	74.60; 74.70; S10.15
	(a) 119,K141	74.70; S10.15
Y.F. Zhang	(a) 121, K21	61.70 and 78.70; S1.2
Y.N. Zhang	(b) 157, 379	74.60; 76.30; S10.15
	(b) 160, 591	71.70 and 76.30; S8.12
Y.Z. Zhang	(a) 119, K1	61.40; S1.1
Zhi-dong Zhang	(a) 121, 241	75.30; 75.60; S1.1
	(a) 121, 611	75.50; 75.30; S1.2; S1.4
	(a) 121, 617	75.50; 75.30; S1.1; S1.2
Zhuong-wu Zhang	(a) 121, 241	75.30; 75.60; S1.1
T.M. Zhantikin	(a) 118, K99	61.80 and 76.80; S1.1

T.M. Zhantikin	(a) 120, 305	61.55; 76.80; S1.1; S1.2
Jian-Gao Zhao	(a) 119, K63	75.50; 75.60; S1.1
	(a) 121,K105	75.30; 61.40; 75.60; S1.2
Sang-Bo Zhao	(b) 162, 193	71.70 and 76.30; 78.50; S11.1
Tiesong Zhao	(b) 157, 677	75.30; 75.60; S1.1; S1.4
Zhi-Gang Zhao	(a) 122, K69	75.50; S1.1
L.H. Zhe	(a) 119, K1	61.40; S1.1
N.A. Zheltova	(a) 118,K139	74.70; 78.30; 79.20; 79.60
Fanchun Zheng	(a) 119, 555	74.70; 64.80; S10.15
Wen-Chen Zheng	(b) 157, K33	71.70; 76.30; S11.1
	(b) 158,K137	71.70; 76.30; S8.11
	(b) 159, K67	71.70; 76.30; S10.1
	(b) 160, K43	71.70; 76.30; S10.1
	(b) 161, 711	71.70; 76.30; S9.16; S10
Xiang-Yang Zheng	(b) 157, K87	73.20
A.K. Zhetbaev	(b) 157, K55	61.70 and 76.80; S10.1
	(a) 118, K99	61.80 and 76.80; S1.1
	(a) 120, 305	61.55; 76.80; S1.1; S1.2
V.S. Zhigalov	(a) 121,K145	61.40 and 75.50; S1.2
D.I. Zhigunov	(a) 119,K151	61.80 and 74.70; S10.15
Yu.I. Zhirko	(b) 161, 419	71.35 and 78.30; S8
	(b) 161, 427	78.55; 71.35; 71.55; S8
Jia-lin Zhong	(b) 160, 329	75.10
J.X. Zhong	(b) 162, 461	71.45
G.F. Zhou	(a) 120, 627	75.50; 64.80; S1.1
	(a) 121, 257	75.60; S1.1
J.M. Zhou	(a) 119, 511	68.55; 68.48; S5.11; S7.13; S7.15
Kang-Wei Zhou	(b) 162, 193	71.70 and 76.30; 78.50; S11.1
X.Y. Zhou	(a) 119,K145	74.70 and 79.60; S10.15
	(a) 122,K151	74.70; S10.15
C.F. Zhu	(a) 119, K59	74.70; S10.15
	(a) 122, K55	74.60 and 74.70; S10.15
Ka-Di Zhu	(b) 161, K25	71.36 and 73.90; S7.11
W.X. Zhu	(b) 158, 113	63.20 and 78.50; S7.12
Xian-Fang Zhu	(a) 117, 105	62.40; 61.70; S4
E.A. Zhukov	(a) 117, 587	78.55; S8
V.P. Zhukov	(b) 158, 239	71.20 and 71.25; S9.11; S10.1
	(b) 160, 517	71.25; S10.1
A.F. Zhuravlev	(b) 158, K89	78.65
	(b) 162,K127	78.65
Yu.N. Zhuravlev	(b) 160, 153	71.25; S2
G. Zies	(a) 119, 503	68.55 and 73.40; S1.61; S4
P. Ziesche	(b) 159, 771	71.20 and 71.25; S1.3
	(b) 162, 237	61.70 and 72.10; S1.2
Z. Zikmund	(a) 118, 539	77.30 and 77.80
L.G. Zimin	(b) 158, 359	71.35 and 78.65; S8.12
	(b) 159, 267	71.35 and 78.65; S8.15; S9

L.G. Zimin	(b) 159, 449	71.35 and 78.20; S8.1
G. Zimmerer	(b) 158, 753	78.55; 71.35; S5.2
H. Zimmermann	(a) 118, 225	71.55; 78.55; S8.13
R. Zimmermann	(b) 159, 35	71.25 and 72.10; 78.20
	(b) 159, 317	71.35 and 78.20; S8.11
	(b) 159, 403	71.35 and 78.20; S8.11
	(b) 160,K109	63.20 and 72.10
O.S. Zinets	(b) 161, 597	73.20; 61.80
K. Zink	(b) 158, 603	71.70; 76.30; 78.55; S8.15
V. Zlatić	(b) 162, K83	71.10; 71.20; 75.10
I. Zólomy	(a) 118, 591	73.40 and 85
E. Zuber	(a) 118, 85	61.70; S5.11
V. Zubkus	(b) 158, 51	66.30; S11.1
V.E. Zubov	(a) 119, 297	78.20; S1.2; S11.2
V.V Zubritskii	(b) 158, 91	63.20; 71.55; 72.20; 77.50; S8.11
L.B. Zuev	(a) 121, 437	61.70; S1.2
S. Žurauskas	(b) 162, 183	71.55 and 79.70; S5.11
N. Žurauskienė	(b) 162, 183	71.55 and 79.70; S5.11
M. Zvára	(b) 158,K173	72.40; 71.55; S8.15
V.V. Zverev	(a) 121,K155	61.80; 71.55; 85; S5.11
M.L. Zvereva	(b) 160,K141	71.30; S8.16
A.V. Zverlin	(a) 117, K13	64.70; S10.15
V.A. Zyulkov	(b) 159, 379	78.20; S8.11; S8.12

International Classification System for Physics*)

60. Condensed matter: structure, mechanical and thermal properties

    61. Structure of liquids and solids; crystallography
    (see also 68.20. Solid surface structure, 71. Electron states)
    62. Mechanical and acoustical properties of condensed matter
    (see also 61.70. Defects in crystals, 68.30. Surfaces and interfaces)
    63. Lattice dynamics and crystal statistics
    (see also 65. Thermal properties, 66.70. Thermal conduction, 68.30. Dynamics of surface and interface vibrations, 78.30. Infrared and Raman spectra)
    64. Equations of state, phase equilibria, and phase transitions
    65. Thermal properties of condensed matter
    (see also 63. Lattice dynamics; for thermodynamic properties of quantum fluids, see 67.40; for thermal properties of solid helium, see 67.80)
    66. Transport properties of condensed matter (nonelectronic)
    67. Quantum fluids and solids; liquid and solid helium
    68. Surfaces and interfaces; thin films and whiskers
    (for impact phenomena, see 79; for crystal growth, see 61.50)

70. Condensed matter: electronic structure; electrical, magnetic, and optical properties

    71. Electronic states
    (see also 63. Lattice dynamics, 73. Electronic structure and electrical properties of surfaces, interfaces, and thin films)
    72. Electronic transport in condensed matter
    (for surfaces, interfaces, and thin films, see 73)
    73. Electronic and electrical properties of surfaces, interfaces, and thin films
    74. Superconductivity
    75. Magnetic properties and materials
    76. Magnetic resonances and relaxation in condensed matter; Mössbauer effect
    77. Dielectric properties and materials
    (for conductivity phenomena, see 72.20 and 72.80)
    78. Optical properties and condensed-matter spectroscopy and other interactions of matter with particles and radiation
    (for phonon spectra, see 63)
    79. Electron and ion emission by liquids and solids; impact phenomena

85. Devices**)

---

*) Excerpt; reproduced with permission of International Council for Scientific and Technical Information (ICSTI).

**) Outside the ICSTI Classification for Physics.

(The Substance Classification is given on cover three)

Manuscripts and letters for physica status solidi (b) — basic research and physica status solidi (a) — applied research

should be sent to

Leipziger Straße 3–4, P. B. 1233, O-1086 Berlin, FRG.

The object of "physica status solidi (b)" and "physica status solidi (a)" is to publish rapidly important and topical results concerning fundamental research in the field of solid state physics and aspects of application and experimental techniques of fundamental solid state research, respectively. The issues of "physica status solidi (b)" and "physica status solidi (a)" will be published on the 1st and the 16th day of each month, respectively. Either series will contain the following categories:

1. Review articles written by arrangement with the editors. They will be remunerated.
2. Original papers selected from those which have not been submitted for publication elsewhere, after being reviewed. (The latest date for acceptance of contributions is the 8th and the 23rd day of each month for "physica status solidi (b)" and "physica status solidi (a)", respectively, 50 days before the delivery of the journal.)
3. Short notes are selected by the editorial staff, if necessary after being reviewed. (The latest date for acceptance of manuscripts for this part is the 12th and the 27th day of each month for "physica status solidi (b)" and "physica status solidi (a)", respectively, 16 days before the delivery of the journal.)

Editorial Information: General

Manuscripts should be submitted preferentially in English. Besides, papers in German, French, and Russian are admitted. The final decision on the incorporation of the papers in the series (b) or (a) is taken by the editorial staff after consultation of the referee. Papers on liquids are only published in physica status solidi if the problems treated are also relevant for the solid state.

Authors are asked to observe the following rules:

1. Manuscripts have to be written as concisely as possible. At most, 20 pages should not be exceeded for original papers. For short notes 4 pages (28 lines per page; abstracts, tables, references, figures, and captions of figures included, where 3 figures are counted as about one page) are admitted. All manuscripts should be typewritten on one side only. Two copies are required. Badly arranged manuscripts are returned.
2. The author's name, initials, and institution should be given.
3. One abstract of about ten lines in English and one in German, French, or Russian are required for original papers.
4. Because of lack of space conclusions (if any) must not have the character of a summary (i. e. substantially a repetition of the contents of the abstract).
5. Formulae should not be written too small and not with pencil. Separate lines for formulae are desirable. Indistinct writing will lead to a return of the manuscript.
6. Drawings should be enclosed on separate sheets of paper, not exceeding the maximum and minimum sizes of 20×30 and 10×15 cm^2, respectively. Photographs should be submitted on white paper. Duplicates are required. Detailed instructions for preparing of figures are given on cover 4 of each issue No. 2.
7. Captions for the figures should be submitted on separate sheets of paper, and special instructions with regard to the size of illustrations should be given on these.
8. References should be numbered in order of appearance and should be given, on a separate sheet, in the form used in physica status solidi.
9. SI-units should be used.
10. Each paper must contain a proposal for the subject classification according to the classification scheme appearing in the January issues of each annual set.

The authors will be sent galley proofs of review articles and original papers. Proof reading of short notes will be done by the editorial staff.

For each review article, original paper, and short note a total of 50 reprints is available free of charge. Requests for additional reprints (against payment) should be given together with the submitted manuscripts. **The deadline for orders of additional reprints is the 15th and the 30th of each month** for "physica status solidi (b)" and "physica status solidi (a)", respectively, two weeks before the delivery of the journal.

Bei Fragen zur Produktsicherheit wenden Sie sich bitte an:
If you have any questions regarding product safety,
please contact:

Walter de Gruyter GmbH
Genthiner Straße 13
10785 Berlin
productsafety@degruyterbrill.com